AN INTRODUCTION

AN INTRODUCTION

To The Study of Fishes

By

Albert C. L. G.Gunther

GYAN

Published by

Gyan Publishing House
5, Ansari Road
Daryaganj, New Delhi-110002
Phone: 011-47034999, 9811692060
E-mail: books@gyanbooks.com

Distribution Network
gyanbooks.com
India, USA, Canada, UK, Australia, France

ISBN : 978-81-212-6057-2 (PB)
First Published, 1880

2nd Impression 2022

Printed at: Gyan Press, Delhi.

AN INTRODUCTION
Author: Albert C. L. G.Gunther

AN INTRODUCTION

TO THE

STUDY OF FISHES

BY

ALBERT C. L. G. GÜNTHER

M.A. M.D. Ph.D. F.R.S.

KEEPER OF THE ZOOLOGICAL DEPARTMENT IN THE BRITISH MUSEUM

Carpit aquas pinnis.

EDINBURGH

ADAM AND CHARLES BLACK

1880

PREFACE.

THE scope of the present work is to give in a concise form an account of the principal facts relating to the structure, classification, and life-history of Fishes. It is intended to meet the requirements of those who are desirous of studying the elements of Ichthyology; to serve as a book of reference to zoologists generally; and, finally, to supply those who, like travellers, have frequent opportunities of observing fishes, with a ready means of obtaining information. The article on "Ichthyology," prepared by the late Sir J. Richardson for the eighth edition of the "Encyclopædia Britannica," is the only publication which has hitherto partly satisfied such requirements; and when I undertook, some years ago, to revise, or rather rewrite that article for the new edition of that work, it occurred to me that I might at the same time prepare a Handbook of Ichthyology, whilst reserving for the article an abstract so condensed as to be adapted for the wants of the general reader.

From the general plan of the work I have only departed in those chapters which deal with the Geographical Distribution of Fishes. This is a subject which has never before been treated in a general and comprehensive manner, and seemed to demand particular attention. I have, therefore, thought it right to give nominal lists of the Faunæ, and the other details of fact on which I have based my conclusions, although all the necessary materials may be found in my "Catalogue of Fishes."

A few references only to the numerous sources which were consulted on the subjects of Chapters 1-12, are inserted in the text; more not required by the beginner; he is introduced to a merely elementary knowledge of facts well known to the advanced student.

With regard to the illustrations, about twenty have been prepared after originals published by Cuvier, J. Müller, Owen, Traquair, Duméril, Cunningham, Hasse, Poey, Siebold, and Gegenbaur. A similar number, representing extinct fishes, have been taken, with the kind permission of the author, from Owen's "Palæontology" My best thanks are due also to the Committee of Publications of the Zoological Society, and to the Editors of the "Annals and Magazine of Natural History," and of the "Journal des Museum Godeffroy," for the loan of woodcuts illustrating some of my papers on South American fishes and on larval forms. The remainder of the illustrations (about three-fourths) are either original figures, or formed part of the article on 'Ichthyology' in the former edition of the "Encyclopædia Britannica."

London, 3d *October* 1880.

CONTENTS.

INTRODUCTORY REMARKS.

CHAPTER I.

CHAPTER II.

CHAPTER III.

CHAPTER IV.

CHAPTER V.

CHAPTER VI.

CHAPTER VII.

CHAPTER VIII.

CHAPTER IX.

CHAPTER X.

CHAPTER XI.

CHAPTER XII.

CHAPTER XIII.

CHAPTER XIV.

CHAPTER XV.

CHAPTER XVI.

CHAPTER XVII.

CHAPTER XVIII.

CHAPTER XIX.

CHAPTER XX.

CHAPTER XXI.

SYSTEMATIC AND DESCRIPTIVE PART.

FIRST SUB-CLASS—PALÆICHTHYES.

SECOND SUB-CLASS—TELEOSTEI.

INTRODUCTORY REMARKS.

ACCORDING to the views generally adopted at present, all those Vertebrate animals are referred to the Class of Fishes, which living in water, breathe air dissolved in water by means of gills or branchiæ; whose heart consists of a single ventricle and single atrium; whose limbs, if present, are modified into fins, supplemented by unpaired, median fins; and whose skin is either naked, or covered with scales or osseous plates or bucklers. With few exceptions fishes are oviparous. However, there are not a few members of this Class which show a modification of one or more of these characteristics, as we shall see hereafter, and which, nevertheless, cannot be separated from it. The distinction between the Class of Fishes and that of Batrachians is very slight indeed.

The branch of Zoology which treats of the internal and external structure of fishes, their mode of life, and their distribution in space and time, is termed Ichthyology.[1]

[1] From ἰχθυς, fish, and λογος, doctrine or treatise.

CHAPTER I.

HISTORY AND LITERATURE.

Aristotle. THE commencement of the history of Ichthyology coincides with that of Zoology generally. ARISTOTLE (384-322 B.C.) had a perfect knowledge of the general structure of fishes, which he clearly discriminates from the Aquatic animals with lungs and mammæ, *i.e.* Cetaceans, and from the various groups of Aquatic Invertebrates. He says that "the special characteristics of the true fishes consist in the branchiæ and fins, the majority having four fins, but those of an elongate form, as the eels, having two only. Some, as the *Muræna*, lack the fins altogether. The Rays swim with their whole body, which is spread out. The branchiæ are sometimes furnished with an opercle, sometimes without one, as is the case in the cartilaginous fishes. . . . No fish has hairs or feathers; most are covered with scales, but some have a rough or smooth skin. The tongue is hard, often toothed; and sometimes so much adherent that it seems to be wanting. The eyes have no lids; nor are any ears or nostrils visible, for what takes the place of nostrils is a blind cavity. Nevertheless they have the senses of tasting, smelling, and hearing. All have blood. All scaly fishes are oviparous, but the cartilaginous fishes (with the exception of the Sea-devil, which Aristotle places along with them) are viviparous. All have a heart, liver, and gall-bladder; but kidneys and urinary bladder are absent. They vary much in the structure of their intestines: for whilst the mullet has a fleshy stomach like a bird, others have no stomachic dilatation. Pyloric coeca are close to the stomach, variable in number; there are even some, like the majority of the cartilaginous fishes, which have none whatever. Two bodies are situated along the spine, which have the function of testicles, and open towards the vent, and which are much enlarged in the spawning season. The scales become harder with age. Not being provided with lungs, they have no voice, but several can emit grunting sounds. They sleep like other animals. In the majority the females exceed the males in size; and in the Rays and Sharks the male is distinguished by an appendage on each side of the vent."

Aristotle's information on the habits of fishes, their migrations, mode and time of propagation, utility, is, as far as it has been tested, surprisingly correct. Unfortunately, only too often we lack the means of recognising the species of which he gives a description. His ideas of specific distinction were as vague as those of the fishermen whose nomenclature he adopted; it never occurred to him that such popular names are subject to change, or may be entirely lost with time, and the difficulty of deciphering his species is further increased by the circumstance that popular names are often applied by him to the same fish, or that different stages of growth are designated by distinct names. The number of fishes known to Aristotle seems to have been about 115, all of which are inhabitants of the Ægean Sea.

That one man should have discovered so many truths, and formed so sure a base for Zoology, is less surprising than the fact that for about eighteen centuries a science which seemed to offer particular attractions to men gifted with power of observation, was no farther advanced. Yet this is the case. Aristotle's disciples, as well as his successors, remained satisfied to be his copiers or commentators, and to collect fabulous stories or vague notions. With very few exceptions (such as *Ausonius*, who wrote a small poem, in which he describes from his own observations the fishes of the Mosel) authors entirely abandoned original research. And it was not until about the middle of the sixteenth century that Ichthyology made a new step in advance by the appearance of *Belon*, *Rondelet*, and *Salviani*, who almost simultaneously published their grand works, by which the idea of species was established definitely and for all times.

Belon. P. BELON travelled in the countries bordering on the eastern part of the Mediterranean, in the years 1547-50; he collected rich stores of positive knowledge, which he deposited in several works. The one most important for the progress of Ichthyology is that entitled "De aquatilibus libri duo" (Paris 1553; small 4to.) Belon knows about 110 fishes, of which he gives rude, but generally recognisable, figures. In his descriptions he pays regard to the classical as well as vernacular nomenclature, and states the outward characteristics, sometimes even the number of fin-rays, frequently also the most conspicuous anatomical peculiarities.

Although Belon but rarely gives definitions of the terms used by him, it is generally not very difficult to ascertain the limits which he intended to assign to each division of aquatic animals. He very properly divides them into such as are provided with blood, and into those without it: two divisions, called in modern language Vertebrate and Invertebrate aquatic animals. The former are classified by him according to sizes, the further subdivisions being based on the structure of the skeleton, mode of propagation, number of limbs, form of the body, and on the physical character of the localities inhabited by fishes. This classification is as follows:—

I. The larger fishes or Cetaceans.

A. Viviparous Cetaceans with bony skeletons (= Cetacea).
B. Viviparous Amphibians.

1. With four limbs: Seals, Hippopotamus, Beaver, Otter, and other aquatic Mammalia.
2. With two limbs: Mermaids, etc.

C. Oviparous Amphibians (= Reptiles and Frogs).

D. Viviparous Cartilaginous fishes.
1. Of an oblong form (= Sharks).
2. Of a flat form (= Rays and Lophius).

E. Oviparous Cartilaginous fishes (= Sturgeons and Silurus).

F. Oviparous Cetaceans, with spines instead of bones (= large marine fishes, like the Thunny, Sword-fish, Sciænoids, Bass, Gadoids, Trachypterus).

II. Spinous Oviparous fishes of a flat form (= Pleuronectidæ).

III. Fishes of a high form, like Zeus.

IV. Fishes of a snake-like form (= Eels, Belone, Sphyræna).

V. Small Oviparous, spinous, scaly, marine fishes.
1. Pelagic kinds.
2. Littoral kinds.
3. Kinds inhabiting rocky localities.

VI. Fluviatile and Lacustrine fishes.

The work of the Roman ichthyologist, H. SALVIANI (1514-72), is characteristic of the high social position which the author held as the physician of three popes. Its title is "Aquatilium animalium historia" (Rom. 1554-57, fol.) It treats exclusively of the fishes of Italy. Ninety-two species are figured on seventy-six plates which, as regards artistic execution, are masterpieces of that period, although those specific characteristics which now-a-days constitute the value of a zoological drawing, were entirely overlooked by the author or artist. No attempt is made at a natural classification, but the allied forms generally are placed in close proximity. The descriptions are quite equal to those given by Belon, entering much into the details of the economy and usefulness of the several species, and were evidently composed with the view of collecting in a readable form all that might Salviani.

prove of interest to the class of society in which the author moved. Salviani's work is of a high standard, most remarkable for the age in which he lived. It could not fail to convey valuable instruction, and to render Ichthyology popular in the country to the fauna of which it was devoted, but it would not have advanced Ichthyology as science generally; and in this respect Salviani is not to be compared with Rondelet or Belon.

Rondelet. G. RONDELET (1507-1557) had the great advantage over Belon in having received a medical education at Paris, and more especially in having gone through a complete course of instruction in anatomy as a pupil of Guentherus of Andernach. This is conspicuous throughout his works—"Libri de Piscibus marinis" (Lugd. 1554, fol.); and "Universæ aquatilium historiæ pars altera" (Lugd. 1555, fol.) Nevertheless they cannot be regarded as more than considerably enlarged editions of Belon's work. For although he worked independently of the latter, and differs from him in numerous details, the system adopted by him is characterised by the same absence of the true principles of classification. Rondelet had a much more extensive knowledge of details. His work is almost entirely limited to European, and chiefly Mediterranean, forms, and comprises not less than 197 marine and 47 freshwater fishes. His descriptions are more complete and his figures much more accurate than those of Belon; and the specific account is preceded by introductory chapters in which he treats in a general manner on the distinctions, the external and internal parts, and on the economy of fishes. Like Belon, he had no conception of the various categories of classification—for instance, confounding throughout his work the terms "genus" and "species;" but he had intuitively a notion of what his successors called a "species," and his principal object was to collect and give as much information as possible of such species.

For nearly a century the works of Belon and Rondelet remained the standard works of Ichthyology; but this science did not remain stationary during this period. The attention of naturalists was now directed to the products of foreign countries, especially the Spanish and Dutch possessions in the New World; and in Europe the establishment of anatomical schools and academies led to the careful investigation of the internal anatomy of the most remarkable European forms. Limited as these efforts were as to their scope, being directed either only to the fauna of some district, or to the dissection of a single species, they were sufficiently numerous to enlarge the views of naturalists, and to destroy that fatal dependency on preceding authorities which had continued to keep in bonds the minds of even such men as Rondelet and Belon.

W. Piso. G. Margrav.

The most noteworthy of those who were active in tropical countries are W. Piso and G. Margrav. They accompanied as physicians the Dutch Governor, Prince Moritz of Nassau, to Brazil (1637-44). Margrav especially studied the fauna of the country, and although he died before his return to Europe, his observations were published by his colleague, and embodied in a work "Historia naturalis Braziliæ" (Lugd. 1648, fol.), in which the fourth book treats of the fishes. He describes about 100 species, all of which had been previously unknown, in a manner far superior to that of his predecessors. The accompanying figures are not good, but nearly always recognisable, and giving a fair idea of the form of the fish. Margrav himself, with the aid of an artist, had made a most valuable collection of coloured drawings of the objects observed and described by him, but many years were allowed to pass before it was scientifically utilised by Bloch and others.

Anatomists, 1600-1700.

Of the men who left records of their anatomical researches, we may mention Borelli (1608-79), who wrote a work "De motu animalium" (Rom. 1680, 4to), in which he explained the mechanism of swimming, and the function of the air-bladder; M. MALPIGHI (1628-94), who examined the optic nerve of the sword-fish; the celebrated J. SWAMMERDAM (1637-80), who described the intestines of numerous fishes; and J. DUVERNEY (1648-1730), who entered into detailed researches of the organs of respiration.

A new era in the history of Ichthyology commences with *Ray*, *Willughby*, and *Artedi*, who were the first to recognise the true principles by which the natural affinities of animals should be determined. Their labours stand in so intimate a connection with each other that they represent only one stride in the progress of this science.

Ray and Willughby

J. RAY (born 1628 in Essex, died 1705), was the friend and guide of F. WILLUGHBY (1635-72). They had recognised that a thorough reform of the treatment of the vegetable and animal kingdoms had become necessary; that the only way of bringing order into the existing chaos was that of arranging the various forms with regard to their structure; that they must cease to be burdened with inapplicable passages and quotations of the ancient writers, and to perpetuate the erroneous or vague notions of their predecessors. They abandoned speculation, and adhered to facts only. One of the first results, and perhaps the most important, of their method was, that having recognised the "species" as such, they defined this term, and fixed it as the base, from which all sound zoological knowledge has to start.

Although they had divided their work thus that Ray attended to the plants principally, and Willughby to the animals, the "Historia piscium" (Oxford, 1686, fol.), which bears Willughby's name on the titlepage, and was edited by Ray, is clearly their joint production. A great part of the observations contained in it were collected during their common journeys in Great Britain and on the Continent, and it is no exaggeration to say that at that time these two Englishmen knew the fishes of the Continent, especially those of Germany, better than any other Continental zoologist.

By the definition of fishes as animals with blood, breathing by gills, provided with a single ventricle of the heart, covered with scales or naked; the Cetaceans are excluded. Yet, at a later period Ray appears to have been afraid of so great an innovation as the separation of whales from fishes, and, therefore, he invented a definition of fish which comprises both. The fishes proper are then arranged in the first place according to the cartilaginous or osseous nature of the skeleton; further subdivisions being formed with regard to the general form of the body, the presence or absence of ventral fins, the soft or spinous structure of the dorsal rays, the number of dorsal fins, etc. Not less than 420 species are thus arranged and described, of which about 180 were known to the authors from autopsy: a comparatively small proportion, descriptions and figures still forming at that time in a great measure a substitute for collections and museums. With the increasing accumulation of forms the want of a fixed nomenclature is now more and more felt.

PETER ARTEDI would have been a great ichthyologist if Ray or Willughby had never preceded him. But he was fully conscious of the fact that both had prepared the way for him, and therefore he derived all possible advantages from their works. Born in 1705 in Sweden, he studied with Linnæus at Upsala; from an early period he devoted himself entirely to the study of fishes, and was engaged in the arrangement and description of the ichthyological collection of *Seba*, a wealthy Dutchman who had formed the then perhaps richest museum, when he was accidentally drowned in one of the canals of Amsterdam in the year 1734, at an age of twenty-nine years. His manuscripts were fortunately rescued by an Englishman, Cliffort, and edited by his early friend Linnæus. P. Artedi.

The work is divided into the following parts :—

1. In the "Bibliotheca Ichthyologica" Artedi gives a very complete list of all preceding authors who have written on fishes, with a critical analysis of their works.

2. The "Philosophia Ichthyologica" is devoted to a description of the external and internal parts of fishes; Artedi fixes a precise terminology of all the various modifications of the organs, distinguishes between those characters which determine a genus and such as indicate a species or merely a variety; in fact he establishes the method and principles which subsequently have guided every systematic ichthyologist.

3. The "Genera Piscium" contains well-defined diagnoses of forty-five genera, for which he fixes an unchangeable nomenclature.

4. In the "Species Piscium" descriptions of seventy-two species, examined by himself, are given; descriptions which even now are models of exactitude and method.

5. Finally, in the "Synonymia Piscium" references to all previous authors are arranged for every species, very much in the same manner which is adopted in the systematic works of the present day.

Artedi has been justly called the Father of Ichthyology. Linnæus. So perfect was his treatment of the subject, that even LINNÆUS could no more improve it, only modify and add to it; and as far as Ichthyology is concerned, Linnæus has scarcely done anything beyond applying binominal terms to the species properly described and classified by Artedi.

Artedi had divided the fishes proper into four orders, viz. *Malacopterygii*, *Acanthopterygii*, *Branchiostegi*, and *Chondropterygii*, of which the third only, according to our present knowledge, appears to be singularly heterogeneous, as it comprises *Balistes*, *Ostracion*, *Cyclopterus*, and *Lophius*. Linnæus, besides separating the Cetaceans entirely from the class of fishes (at least since the 10th edition of the "Systema Naturæ") abandoned Artedi's order of Branchiostegi, but substituted a scarcely more natural combination by joining it with Artedi's Chondropterygians, under the name of "Amphibia nantes."

His classification of the genera appears in the 12th edition of the "Systema," thus—

AMPHIBIA NANTES.

Spiraculis compositis.

Petromyzon.
Raia.
Squalus.
Chimæra.

Spiraculis solitariis.

Lophius.
Acipenser.
Cyclopterus.
Balistes.
Ostracion.
Tetrodon.
Diodon.
Centriscus.
Syngnathus.
Pegasus.

PISCES APODES.

Muræna.
Gymnotus.
Trichiurus.
Anarhichas.
Ammodytes.
Ophidium.
Stromateus.
Xiphias.

PISCES JUGULARES.

Callionymus.
Uranoscopus.
Trachinus.
Gadus.
Blennius.

PISCES THORACICI.

Cepola.
Echeneis.
Coryphæna.
Gobius.
Cottus.
Scorpæna.
Zeus.
Pleuronectes.
Chætodon.
Sparus.
Labrus.
Sciæna.
Perca.
Gasterosteus.
Scomber.
Mullus.
Trigla.

PISCES ABDOMINALES.

Cobitis.
Amia.
Silurus.
Teuthis.
Loricaria.
Salmo.
Fistularia.
Esox.
Elops.
Argentina.
Atherina.
Mugil.
Mormyrus.
Exocœtus.
Polynemus.
Clupea.
Cyprinus.

Gronow and Klein.

Two contemporaries of *Linnæus* attempted a systematic arrangement of fishes; both had considerable opportunities for their study, especially in possessing extensive collections; but neither exercised any influence on the progress of Ichthyology. The one, L. T. GRONOW, a German who resided in Holland, closely followed the arrangements proposed by Artedi and Linnæus, and increased the number of genera and species from the contents of his own museum. He published two works, "Museum Ichthyologicum" (Lugd. 1754-6, fol.), and "Zoophylacium" (Lugd. 1763-81, fol.); a posthumous work, containing numerous excellent descriptions of new forms was published by J. E. Gray in 1854 under the title of "Systema Ichthyologicum." To Gronow also is due the invention of preparing flat skins of fishes in a dry state, and preserving them in the manner of a herbarium. The specimens thus prepared by him belong to the oldest which have been preserved down to our time.

Much less important are the ichthyological labours of J. T. KLEIN (1685-1759). They are embodied in five parts (*Missus*) of a work entitled "Historia naturalis piscium" (Sedæ, 1740-9, 4to.) He regarded a system merely as the means of recognising the various forms of animals, not as the expression of their natural affinities; and that method seemed to him to be the most perfect by which an animal could be most readily determined. He eschewed all reference to minute or anatomical characters. Hence his system is a series of the most unnatural combinations, and we cannot be surprised that Linnæus passed in silence over Klein's labours.

Pupils and Successors of Linnæus

The works of Artedi and Linnæus excited fresh activity, more especially in Scandinavia, Holland, Germany, and England, such as has not been equalled in the history of biological science either before or after. Whilst some of the pupils and followers of Linnæus devoted themselves to an examination and study of the fauna of their native countries, others proceeded on voyages of discovery to foreign and distant countries. Of these latter the following may be specially mentioned:—*O. Fabricius* worked out the Fauna of Greenland, *Kalm* collected in North America, *Hasselquist* in Egypt and Palestine, *Brünnich* in the Mediterranean, *Osbeck* in Java and China, *Thunberg* in Japan; *Forskål* examined and described the fishes of the Red Sea; *Steller, Pallas, S. T. Gmelin,* and *Güldenstedt* traversed nearly the whole of the Russian Empire in Europe and Asia. Others attached themselves as naturalists to the celebrated circumnavigators of the last century, like the two *Forsters* (father and son), and *Solander*, who accompanied Cook; *Commerson*, who travelled with Bougainville; and *Sonnerat*. Numerous new and startling forms were discovered by those men, and the foundation was laid of the knowledge of the geographical distribution of animals.

Of those who studied the fishes of their native country the most celebrated are *Pennant* (Great Britain), *O. F. Müller* (Denmark), *Duhamel* (France), *Meidinger* (Austria), *Cornide* (Spain), *Parra* (Cuba).

The materials brought together by those and other zoologists were so numerous that, not long after the death of Linnæus, the necessity was felt of collecting them in a compendious form. Several compilators undertook this task; they embodied the recent discoveries in new editions of Artedi's and Linné's classical works, but not possessing either a knowledge of the subject or any critical discernment, they only succeeded in covering those noble monuments under a mass of confused rubbish. For Ichthyology it was fortunate that two men at least, Bloch and Lacépède, made it a subject of long and original research.

M. E. Bloch.

MARK ELIEZER BLOCH, born in the year 1723 at Anspach in Germany, practised as a physician in Berlin; he had reached an age of fifty-six years when he commenced to write on ichthyological subjects. To commence at his age a work in which he intended not only to give full descriptions of the species known to him from specimens or drawings, but also to illustrate every species in a style truly magnificent for his time, was an undertaking of the execution of which an ordinary man would have despaired. Yet he accomplished not only this task, but even more, as we shall see hereafter.

His work consists of two divisions :—

1. "Oeconomische Naturgeschichte der Fische Deutschlands" (Berl. 1782-4, 4to. Plates in fol.)

2. "Naturgeschichte der auslændischen Fische" (Berl. 1785-95, 4to. Plates in fol.)

Bloch's work is unique, and probably will for ever remain so. Although Cuvier fifty years later undertook a similar general work on fishes, the subject had then become too extensive to allow of an attempt of giving illustrations of all the species, or illustrations of a similar size and costliness.

The first division of the work, which is devoted to a description of the fishes of Germany, is entirely original, and based upon Bloch's own observations. His descriptions as well as figures were made from nature, and are, with but few exceptions, still serviceable; many continue to be the best existing in literature.

Bloch was less fortunate and is much less reliable in his natural history of foreign fishes. For many of the species he had to rely on more or less incorrect drawings and descriptions of travellers; frequently, also, he was deceived as to the origin of specimens which he acquired by purchase. Hence his accounts contain numerous confusing errors which it would have been difficult to correct, if not nearly the whole of the materials on which his work is based had been preserved in the collections at Berlin.

After the completion of his Ichthyology Bloch occupied

himself with systematic work. He prepared a general system of fishes, in which he arranged not only those described in his great work, but also those with which he had become acquainted afterwards from the descriptions of others. The work was ably edited and published after Bloch's death by a philologist, *J. G. Schneider*, under the title "M. E. Blochii Systema ichthyologiæ iconibus ex. illustratum" (Berl. 1801, 8vo.) The number of species enumerated in it amounts to 1519. The system is based upon the number of the fins, the various orders being termed *Hendecapterygii, Decapterygii*, etc. We need not add that an artificial method like this led to the most unnatural combinations or severances.

Bloch's Ichthyology remained for many years the standard work, and, by the great number of excellent illustrations, proved a most useful guide to the student. But as regards originality of thought, Bloch was far surpassed by his contemporary, B. G. E. DE LACÉPÈDE, born at Agen, in France, in 1756, a man of great and general erudition, who died as Professor of the Museum of Natural History of Paris in 1826. Lacépède.

Lacépède had to contend with great difficulties in the preparation of his "Histoire des Poissons" (Paris, 1798-1803, 4to, in 5 vols.), which was written during the most disturbed period of the French Revolution. A great part of it was composed whilst the author was separated from collections and books, and had to rely on his notes and manuscripts only. Even the works of Bloch and other contemporaneous authors remained unknown, or at least inaccessible, to him for a long time. Therefore we cannot be surprised that his work abounds in all those errors to which a compiler is subject. The same species not only appears under two and more distinct specific names, but it sometimes happens that the author understands so little the source from which he derives his information that the description is referred to one genus and the accompanying figure to another. The names of genera are unduly multiplied; and the figures with which the work is illustrated are far inferior to those of Bloch. Thus the influence of Lacépède on the progress of Ichthyology was infinitely less than that of his fellow-labourer; and the labour caused to his successors by correcting the numerous errors into which he has fallen, probably outweighs the assistance which they derived from his work.

Anatomists. The work of the principal cultivators of Ichthyology in the period between Ray and Lacépède was chiefly systematic and descriptive, but also the internal organisation of fishes received attention from more than one great anatomist. *Haller*, *Camper*, and *Hunter*, examined the nervous system and organs of sense; and more especially *Alexander Monro* (the son) published a classical work, "The Structure and Physiology of Fishes explained and compared with those of Man and other Animals" (Edinb. 1785, fol.) The electric organs of fishes (*Torpedo* and *Gymnotus*) were examined by *Réaumur*, *Allamand*, *Bancroft*, *Walsh*, and still more exactly by *J. Hunter*. The mystery of the propagation of the Eel called forth a large number of essays, and even the artificial propagation of Salmonidæ was known and practised by *Gleditsch* (1764).

Faunists. Bloch and Lacépède's works were almost immediately succeeded by the labours of *Cuvier*, but his early publications were of necessity tentative, preliminary, and fragmentary, so that a short period elapsed before the spirit infused by this great anatomist into Ichthyology could exercise its influence on all workers in this field. Several of such antecuvierian works must be mentioned on account of their importance to our knowledge of certain Faunas: the "Descriptions and Figures of Two Hundred Fishes collected at Vizagapatam on the coast of Coromandel" (Lond. 1803; 2 vols. in fol.), by *Patrick Russel*; and "An Account of the Fishes found in the River Ganges and its branches" (Edinb. 1822; 2 vols. in

4to), by *F. Hamilton* (formerly *Buchanan*)—works distinguished by a greater accuracy of their drawings (especially in the latter), than was ever attained before. A "Natural History of British Fishes" was published by *E. Donovan* (Lond. 8vo, 1802-8); and the Mediterranean Fauna formed the study of the lifetime of *A. Risso* ("Ichthyologie de Nice." Paris, 1810, 8vo; and "Histoire naturelle de l'Europe Meridionale." Paris, 1827, 8vo). A slight beginning in the description of the fishes of the United States was made by *S. L. Mitchell*, who published, besides various papers, a "Memoir on the Ichthyology of New York," in 1815.[1]

G. CUVIER did not occupy himself with the study of fishes merely because this class formed part of the "Règne animal," but he devoted himself to it with particular predilection. The investigation of their anatomy, and especially of their skeleton, was taken up by him at an early period, and continued until he had succeeded in completing so perfect a framework of the system of the whole class that his immediate successors could content themselves with filling up those details for which their master had no leisure. Indefatigable in examining all the external and internal characters of the fishes of a rich collection, he ascertained the natural affinities of the infinite variety of fishes, and accurately defined the divisions, orders, families, and genera of the class, as they appear in the various editions of the "Règne animal." His industry equalled his genius: he opened connections with almost every accessible part of the globe; not only French travellers and naturalists, but also Germans, Englishmen, Americans, rivalled one another to assist him with collections; and for many years the Muséum of the Jardin des Plantes was the centre where all ichthyological treasures were deposited. Thus Cuvier G. Cuvier.

[1] Down to this period the history of Ichthyology is fully treated in the first volume of Cuvier and Valenciennes "Hist. nat. d. Poiss."

brought together a collection the like of which had never been seen before, and which, as it contains all the materials on which his labours were based, must still be considered to be the most important. Soon after the year 1820, Cuvier, assisted by one of his pupils, A. VALENCIENNES, commenced his great work on fishes, "Histoire naturelles des Poissons," of which the first volume appeared in 1828. The earlier volumes, in which Cuvier himself took his share, bear evidence of the freshness and love with which both authors devoted themselves to their task. After Cuvier's death in 1832 the work was left entirely in the hands of Valenciennes, whose energy and interest gradually slackened, to rise to the old standard in some parts only, as, for instance, in the treatise on the Herring. He left the work unfinished with the twenty-second volume (1848), which treats of the Salmonoids. Yet, incomplete as it is, it is indispensable to the student.

There exist several editions of the work, which, however, have the same text. One, printed in 8vo, with coloured or plain figures, is the one in common use among ichthyologists. A more luxurious edition in 4to has a different pagination, and therefore is most inconvenient to use.

As mentioned above, the various parts of the work are very unequally worked out. Many of the species are described in so masterly a manner that a greater excellency of method can hardly be conceived. The history of the literature of these species is entered into with minuteness and critical discernment; but in the later volumes, numerous species are introduced into the system without any description, or with a few words only, comparing a species with one or more of its congeners. Cuvier himself, at a late period of his life, seems to have grown indifferent as to the exact definition of his species: a failing commonly observed among Zoologists when attention to descriptive details becomes to them a tedious task. What is more surprising is, that a man of his anatomical and physiological knowledge should have overlooked the fact that secondary sexual characters are developed in fishes as in any other class of animals, and that fishes undergo great changes during growth; and, consequently, that he described almost all such sexual forms and different stages of growth under distinct specific and even generic names.

The system finally adopted by Cuvier is the following:—

A. Poissons Osseux.

I.—A branchies en peignes ou en lames.

1. A mâchoire supérieure libre.

a. *Acanthoptérygiens.*

Percoïdes.	Sparoïdes.	Branchies labyrinthiques.
Polynèmes.	Chétodonoïdes.	Lophioïdes.
Mulles.	Scomberoïdes.	Gobioïdes.
Joues cuirassées.	Muges.	Labroïdes.
Scienoïdes.		

b. *Malacoptérygiens.*

Abdominaux.	*Subbrachiens.*	*Apodes.*
Cyprinoïdes.	Sparoïdes.	Murenoïdes.
Siluroïdes.	Pleuronectes.	
Salmonoïdes.	Discoboles.	
Clupeoïdes.		
Lucioïdes.		

2. A mâchoire supérieure fixée.

Sclérodermes. Gymnodontes.

II. A branchies en forme de houppes.

Lophobranches.

B. Cartilagineux ou Chondroptérygiens.

Sturioniens. Plagiostomes. Cyclostomes.

We have to compare this system with that of Linnæus if we wish to measure the gigantic stride Ichthyology has made during the intervening period of seventy years. The various characters employed for classification have been examined throughout the whole class, and their relative importance has been duly weighed and understood. Though Linnæus had formed a category of "Amphibia nantes" for fishes with a cartilaginous skeleton, which should coincide with Cuvier's "Poissons Cartilagineux," he had failed to understand the very nature of cartilage, apparently comprising by this term any skeletal framework of less firmity than ordinary bone. Hence he considered *Lophius, Cyclopterus, Syngnathus* to be cartilaginous fishes. Adopting the position and development of the ventral fins as a highly important character, he was obliged to associate fishes with rudimentary and inconspicuous ventral fins, like *Trichiurus, Xiphias*, etc., with the true Eels. The important category of a "family" appears now in Cuvier's system fully established as that intermediate between genus and order. Important changes in Cuvier's system have been made and proposed by his successors, but in the main it is still that of the present day.

Cuvier had extended his researches beyond the living forms, into the field of palæontology; he was the first to observe the close resemblance of the scales of the fossil *Palæoniscus* to those of the living *Polypterus* and *Lepidosteus*, the prolongation and identity of structure of the upper caudal lobe in *Palæoniscus* and the Sturgeons, the presence of peculiar "fulcra" on the anterior margin of the dorsal fin in *Palæoniscus* and *Lepidosteus:* inferring from these facts that that fossil genus was allied *either* to the Sturgeons *or* to *Lepidosteus.* But it did not occur to him that there was a close relationship between those recent fishes. *Lepidosteus* and, with it, the fossil genus remained in his system a member of the order of *Malacopterygii abdominales.*

It was left to L. AGASSIZ (born 1807, died 1873) to point out the importance of the character of the structure of the scales, and to open a path towards the knowledge of a whole new sub-class of fishes, the *Ganoidei.*

Impressed with the fact that the peculiar scales of *Polypterus* and *Lepidosteus* are common to all fossil osseous fishes down to the chalk, he takes the structure of the scales generally as the base for an ichthyological system, and distinguishes four orders :—

1. *Placoids.*—Without scales proper, but with scales of enamel, sometimes large, sometimes small and reduced to mere points (Rays, Sharks, and Cyclostomi, with the fossil *Hybodontes*).

2. *Ganoids.*—With angular bony scales, covered with a thick stratum of enamel: to this order belong the fossil Lepidoides, Sauroides, Pycnodontes, and Coelacanthi; the recent Polypterus, Lepidosteus, Sclerodermi, Gymnodontes, Lophobranches, and Siluroides ; also the Sturgeons.

3. *Ctenoids.*—With rough scales, which have their free margins denticulated : Chætodontidæ, Pleuronectidæ, Percidæ, Polyacanthi, Sciænidæ, Sparidæ, Scorpænidæ, Aulostomi.

4. *Cycloids.*—With smooth scales, the hind margin of which lacks denticulation : Labridæ, Mugilidæ, Scombridæ, Gadoidei, Gobiidæ, Murænidæ, Lucioidei, Salmonidæ, Clupeidæ, Cyprinidæ.

We have no hesitation in affirming that if Agassiz had had an opportunity of acquiring a more extensive and intimate knowledge of existing fishes before his energies were absorbed in the study of their fossil remains, he himself would have recognised the artificial character of his classification. The distinctions between cycloid and ctenoid scales, between placoid and ganoid fishes are vague, and can hardly be maintained. As far as the living and post-cretacean forms are concerned, the vantage-ground gained by Cuvier was abandoned by him; and therefore his system could never supersede that of his predecessors, and finally shared the fate of every classification based on the modifications of one organ only. But Agassiz has the merit of having opened an immense new field of research by his study of the infinite variety of fossil forms. In his principal work, "Recherches sur les Poissons fossiles," (Neuchatel, 1833-43, 4to, atlas in fol.), he placed them before the world arranged in a methodical manner, with excellent descriptions and illustrations. His power of discernment and penetration in determining even the most fragmentary remains is truly astonishing; and if his order of Ganoids is an assemblage of forms very different from that as it is circumscribed now, he was at any rate the first who recognised that such an order of fishes exists.

The discoverer of the *Ganoidei* was succeeded by their explorer, JOHANNES MÜLLER (born 1801, died 1858). In his classical memoir "Ueber den Bau und die Grenzen der Ganoiden" (Berlin, 1846; 4to), he showed that the Ganoids differed from all the other osseous fishes, and agreed with the Plagiostomes, in the structure of their heart. By this primary character, all heterogeneous elements, as *Siluroids*, *Osteoglossidæ*, etc., were eliminated from the order as understood by Agassiz. On the other hand, he did not recognise the affinity of *Lepidosiren* to the Ganoids, but established for it a distinct sub-class, *Dipnoi*, which he placed at the opposite end of the system. By his researches into the anatomy of the Lampreys and Amphioxus, their typical distinctness from other cartilaginous fishes was proved; they became the types of two other sub-classes, *Cyclostomi* and *Leptocardii*.

J. Müller.

Müller proposed several other not unimportant modifications of the Cuvierian system; and although all cannot be maintained as the most natural arrangements, yet his researches have given us a much more complete knowledge of the organisation of the Teleosteous fishes, and later enquiries have shown that, on the whole, the combinations proposed by him require only some further modification and another definition to render them perfectly natural.

Under the name of *Pharyngognathi* he combined fishes with the lower pharyngeals coalesced into one bone, viz. the Labroids, Chromides, and Scombresoces. The association of the third family with the two former seemed to himself a somewhat arbitrary proceeding; and it had to be abandoned again, when a number of fishes which cannot be separated from the Acanthopterygians, were found to possess the same united pharyngeals.

A more natural combination is the union of the Cod-fishes with the Flat-fishes into the order *Anacanthini*. Flat-fishes are in fact nothing but asymmetrical Cod-fishes. Müller separates them from the remaining Malacopterygians by the absence of a connecting duct between the air-bladder and oesophagus. However, it must be admitted that the examination of those fishes, and especially of the young stages, is not complete enough to raise the question beyond every doubt, whether the presence or absence of that duct is an absolutely distinctive character between Anacanths and Malacopterygians.

Many of the families established by Cuvier were re-examined and better defined by Müller, as may be seen from the following outline of his system:—

Sub-classis I.—Dipnoi.

Ordo I.—Sirenoidei.

Fam. 1. Sirenoidei.

Sub-classis II.—Teleostei.

Ordo I.—Acanthopteri.

Fam. 1. Percoidei.
„ 2. Cataphracti.
„ 3. Sparoidei.
„ 4. Sciænoidei.
„ 5. Labyrinthiformes.
„ 6. Mugiloidei.
„ 7. Notacanthini.
„ 8. Scomberoidei.
Fam. 9. Squamipennes.
„ 10. Taenioidei.
„ 11. Gobioidei.
„ 12. Blennioidei.
„ 13. Pediculati.
„ 14. Theutyes.
„ 15. Fistulares.

Ordo II.—Anacanthini.

Sub-ordo I.—Anacanthini sub-brachii.

Fam. 1. Gadoidei.
„ 2. Pleuronectides.

Sub-ordo II.—Anacanthini apodes.

Fam. 1. Ophidini.

Ordo III.—Pharyngognathi.

Sub-ordo I.—Pharyngognathi acanthopterygii.

Fam. 1. Labroidei cycloidei.
„ 2. Labroidei ctenoidei.
„ 3. Chromides.

Sub-ordo II.—Pharyngognathi malacopterygii.

Fam. 1. Scomberesoces.

Ordo IV.—Physostomi.

Sub-ordo I.—Physostomi abdominales.

Fam. 1. Siluroidei.
„ 2. Cyprinoidei.
„ 3. Characini.
„ 4. Cyprinodontes.
„ 5. Mormyri.
„ 6. Esoces.
Fam. 7. Galaxiæ.
„ 8. Salmones.
„ 9. Scopelini.
„ 10. Clupeidæ.
„ 11. Heteropygii.

Sub-ordo II.—Physostomi apodes s. anguillares.

Fam. 12. Muraenoidei.
„ 13. Gymnotini.
„ 14. Symbranchii.

Ordo V.—Plectognathi.

Fam. 1. Balistini.
„ 2. Ostraciones.
„ 3. Gymnodontes.

Ordo VI.—Lophobranchii.

Fam. 1. Lophobranchi.

Sub-classis III.—Ganoidei.

Ordo I.—Holostei.

Fam. 1. Lepidosteini.
„ 2. Polypterini.

Ordo II.—Chondrostei.

Fam. 1. Acipenserini,
„ 2. Spatulariæ.

Sub-classis IV.—Elasmobranchi s. Selachii.

Ordo I.—Plagiostomi.

Sub-ordo I.—Squalidæ.

Fam. 1. Scyllia.
„ 2. Nyctitantes.
„ 3. Lamnoidei.
„ 4. Alopeciæ.
„ 5. Cestraciones.
Fam. 6. Rhinodontes.
„ 7. Notidani.
„ 8. Spinaces.
„ 9. Scymnoidei.
„ 10. Squatinæ.

Sub-ordo II.—Rajidæ.

Fam. 11. Squatinorajæ.
„ 12. Torpedines.
„ 13. Rajæ.
Fam. 14. Trygones.
„ 15. Myliobatides.
„ 16. Cephalopteræ.

Ordo II.—Holocephali.

Fam. 1. Chimaeræ.

Sub-classis V.—Marsipobranchii s. Cyclostomi.

Ordo I.—Hyperoartii.

Fam. 1. Petromyzonini.

Ordo II.—Hyperotreti.

Fam. 1. Myxinoidei.

Sub-classis VI.—Leptocardii.

Ordo I.—Amphioxini.

Fam. 1. Amphioxini.

Discovery of Ceratodus.

The discovery (in the year 1871) of a living representative of a genus hitherto believed to be long extinct, *Ceratodus*, threw a new light on the affinities of Fishes. The author who had the good fortune of examining this fish, was enabled to show that, on the one hand, it was a form most closely allied to *Lepidosiren;* on the other, that it could not be separated from the Ganoid fishes, and therefore that also *Lepidosiren* was a Ganoid: a relation pointed at already by Huxley in a previous paper on "Devonian Fishes." This discovery led to further considerations[1] of the relative characters of Müller's sub-classes, and to the system which is followed in the present work.

Having followed the development of the ichthyological system down to the latest time, we have to retrace our steps to enumerate the most important contributions to Ichthyology which appeared contemporaneously with or subsequently to the publication of Cuvier and Valenciennes's great work. As in other branches of Zoology, activity increased almost with every year; and for convenience's sake we may arrange these works in three rubrics.

Recent Works.

I.—Voyages, containing general Accounts of Zoological Collections.

A. *French.*

1. "Voyage autour du monde sur les Corvettes de S. M. l'Uranie et la Physicienne, sous le commandement de M. Freycinet. Zoologie: Poissons par *Quoy* et *Gaimard.*" (Paris, 1824, 4to, atlas fol.)

2. "Voyage de la Coquille. Zoologie par *Lesson.*" (Paris, 1826-30, 4to, atlas fol.)

3. "Voyage de l'Astrolabe, sous le commandement de M. J. Dumont d'Urville. Poissons par *Quoy* et *Gaimard.*" (Paris, 1834, 8vo, atlas fol.)

[1] Description of Ceratodus. "Phil. Trans.," 1871, ii.

4. "Voyage au Pôle Sud par M. J. Dumont d'Urville. Poissons par *Hombron* et *Jacquinot*." (Paris, 1853-4, 8vo, atlas fol.)

B. *English*.

1. "Voyage of H.M.S. Sulphur. Fishes by *J. Richardson*." (Lond. 1844-5, 4to.)

2. "Voyage of H.M.S.S. Erebus and Terror. Fishes by *J. Richardson*." (Lond. 1846. 4to.)

3. "Voyage of H.M.S. Beagle. Fishes by *L. Jenyns*." (Lond. 1842, 4to.)

4. "Voyage of H.M.S. Challenger. Fishes by *A. Günther*" (in course of publication).

C. *German*.

1. "Reise der österreichischen Fregatte Novara. Fische von *R. Kner*." (Wien. 1865, 4to.)

II.—Faunae.

A. *Great Britain*.

1. *R. Parnell*, "The Natural History of the Fishes of the Firth of Forth." (Edinb. 1838, 8vo.)

2. *W. Yarrell*, "A History of British Fishes." (3d edit. Lond. 1859, 8vo.)

3. *J. Couch*, "A History of the Fishes of the British Islands." (Lond. 1862-5, 8vo.)

B. *Denmark and Scandinavia*.

1. *H. Kröyer*, "Danmark's Fiske." (Kjöbnh. 1838-53, 8vo.)

2. *S. Nilsson*, "Skandinavisk Fauna." (Vol. IV. Fiskarna. Lund. 1855, 8vo.)

3. *Fries och Ekström*, "Skandinaviens Fiskar." (Stockh. 1836, 4to, with excellent plates.)

C. *Russia.*

1. *Nordmann,* "Ichthyologie Pontique," in "Voyage dans la Russie méridionale de *Demidoff.*" (Tom. iii. Paris, 1840, 8vo, atlas fol.)

D. *Germany.*

1. *Heckel* and *Kner,* "Die Süsswasser-fische der Oesterreichischen Monarchie." (Leipz. 1858, 8vo.)

2. *C. T. E. Siebold,* "Die Süsswasser-fische von Mitteleuropa." (Leipz. 1863, 8vo.)

E. *Italy and Mediterranean.*

1. *Bonaparte,* "Iconografia della Fauna Italica." Tom. iii. Pesci. (Roma, 1832-41, fol.) (Incomplete.)

2. *Costa,* "Fauna del Regno di Napoli." Pesci. (Napoli, 4to, about 1850.) (Incomplete.)

F. *France.*

1. *E. Blanchard,* "Les Poissons des eaux douces de la France." (Paris, 1866, 8vo.)

G. *Pyrenean Peninsula.*

The freshwater Fish-fauna of Spain and Portugal was almost unknown, until *F. Steindachner* paid some visits to those countries for the purpose of exploring the principal rivers. His discoveries are described in several papers in the "Sitzungsberichte der Akademie zu Wien." *B. du Bocage* and *F. Capello* contributed towards the knowledge of the marine fishes on the coast of Portugal. ("Jorn. Scienc. Acad. Lisb.")

H. *North America.*

1. *J. Richardson,* "Fauna Boreali Americana." Part III. Fishes. (Lond. 1836, 4to.) The species described in this work are nearly all from the British possessions in the North.

2. *Dekay,* "Zoology of New York." Part IV. Fishes. (New York, 1842, 4to.)

3. "Reports of the United States Commission of Fish and Fisheries." (5 vols. Washingt. 1873-79, 8vo. In progress. Contains most valuable information.)

Besides these works, numerous descriptions of North American freshwater fishes have been published in the Reports of the various U. S. Government expeditions, and in North American scientific journals, by *Storer, Baird, Girard, W. O. Ayres, Cope, Jordan, Brown Goode,* etc.; but a good general, and especially critical, account of the fishes of the United States is still a desideratum.

I.—*Japan.*

1. "Fauna Japonica." Poissons par *H. Schlegel.* (Lugd. Bat. 1850, fol.)

J.—*East Indies; Tropical parts of the Indian and Pacific Oceans.*

1. *E. Rüppell,* "Atlas zu der Reise im Nördlichen Afrika." (Frankf. 1828, fol.)

2. *E. Rüppell,* "Neue Wirbelthiere. Fische." (Frankf. 1837, fol.)

These two works form the standard works for the student of the Fishes of the Red Sea, and are distinguished by a rare conscientiousness and faithfulness of the descriptions and figures; so that there is no other part of the tropical seas, with the fishes of which we are so intimately acquainted, as with those of the Red Sea. But these works have a still wider range of usefulness, in as much as only a small proportion of the fishes is limited to that area, the majority being distributed over the Indian Ocean into Polynesia. Rüppell's works were supplemented by the two first of the following works:—

3. *R. L. Playfair* and *A. Günther,* "The Fishes of Zanzibar." (Lond. 1866, 4to); and

4. *C. B. Klunzinger*, "Synopsis der Fische des Rothen Meers." (Wien. 1870-1, 8vo.)

5. *T. Cantor*, "Catalogue of Malayan Fishes." (Calcutta, 1850, 8vo.)

6. *F. Day*, "The Fishes of India" (Lond. 1875, 4to, in progress); contains an account of the freshwater and marine species, and is not yet complete.

7. *A. Günther*, "Die Fische der Südsee." (Hamburg, 4to; from 1873, in progress.)

Unsurpassed in activity, as regards the exploration of the fish fauna of the East Indian Archipelago, is *P. Bleeker*, a surgeon in the service of the Dutch East Indian Government (born 1819, died 1878), who, from the year 1840, for nearly thirty years, amassed immense collections of the fishes of the various islands, and described them in extremely numerous papers, published chiefly in the Journals of the Batavian Society. When his descriptions and the arrangement of his materials evoked some criticism, it must be remembered that, at the time when he commenced his labours, and for many years afterwards, he stood alone, without the aid of a previously named collection on which to base his first researches, and without other works but that of Cuvier and Valenciennes. He had to create for himself a method of distinguishing species and of describing them; and afterwards it would have been difficult for him to abandon his original method and the principles by which he had been guided for so many years. His desire of giving a new name to every individual, to every small assemblage of species wherever practicable, or of changing an old name, detracts not a little from the satisfaction with which his works would be used otherwise. It is also surprising that a man with his anatomical knowledge and unusual facilities should have been satisfied with the merely external examination of the specimens. But none of his numerous articles contain anything relating to the anatomy, physiology, or habits of the fishes which came under his notice; hence his attempts at systematic arrangement are very far from indicating an advance in Ichthyology.

Soon after his return to Europe (1860) Bleeker commenced to collect the final results of his labours in a grand work, illustrated by coloured plates, "Atlas Ichthyologique des Indes Orientales Néerlandaises." (Amsterd. fol. 1862); the publication of which was interrupted by the author's death in 1878.

K.—*Africa.*

1. *A. Günther*, "The Fishes of the Nile" in Petherick's "Travels in Central Africa." (Lond. 1869, 8vo.)

2. *W. Peters*, "Naturwissenschaftliche Reise nach Mossambique. IV. Flussfische." (Berl. 1868, 4to.)

L.—*West Indies and South America.*

1. *L. Agassiz*, "Selecta genera et species Piscium, quæ in itinere per Brasiliam, collegit J. B. de Spix." (Monach. 1829, fol.)

2. *F. de Castlenau*, "Animaux nouveaux ou rares, recueillis pendant l'expedition dans les parties centrales de l'Amérique du Sud. Poissons." (Paris, 1855, 4to.)

3. *A. Günther*, "An account of the Fishes of the States of Central America." (In Trans. Zool. Soc. 1868.)

4. *L. Vaillant* and *F. Bocourt*, "Mission scientifique au Mexique et dans l'Amérique centrale. Poissons." (Paris, 1874, 4to.) (In progress.)

F. Poey, the celebrated naturalist of Havannah, devoted many years of study to the Fishes of Cuba. His papers and memoirs are published partly in two periodicals, issued by himself, under the title of "Memorias sobre la Historia natural de la Isle de Cuba" (from 1851), and "Repertorio

Fisico-natural de la Isla de Cuba" (from 1865), partly in North American scientific journals. And, finally, *F. Steindachner* has published many contributions, accompanied by excellent figures, to our knowledge of the Fishes of Central and South America.

M.—*New Zealand.*

1. *F. W. Hutton* and *J. Hector*, "Fishes of New Zealand." (Wellingt. 1872, 8vo.)

N.—*Arctic Regions.*

1. *C. Lütken*, "A revised Catalogue of the Fishes of Greenland," in "Manual of the Natural History, Geology, and Physics of Greenland." (Lond. 1875, 8vo.) Although only a nominal list, this catalogue is useful, as it contains references to all the principal works in which Arctic fishes have been described. The fishes of Spitzbergen were examined by *A. J. Malmgren* (1865).

III.—Anatomical Works.

The number of authors who worked on the anatomy of fishes is almost as great as that of faunists; and we should go beyond the limits of the present work if we mentioned more than the most prominent and successful. *M. H. Rathke, J. Müller, J. Hyrtl*, and *H. Stannius* left scarcely any organ unexamined, and their researches had a direct bearing either on the relation of the class of fishes to the other vertebrates, or on the systematic arrangement of the fishes themselves. *E. E. von Baer, F. de Filippi, C. Vogt, W. His, W. K. Parker*, and *F. M. Balfour* worked at their embryology; *A. Kölliker* and *G. Pouchet* at their histology. The osteology was specially treated by *G. Bakker, F. C. Rosenthal, L. Agassiz*, and *C. Gegenbaur;* the nervous system by *Gottsche, Philipeaux, Stannius, L. de Sanctis, L. Stieda, Baudelot* and *Miclucho-Maclay;* the organ of hearing by *E. H. Weber, C. Hasse*, and *G. Retzius.* The electric fishes were examined by *E. Geoffroy*,

C. Matteuci, P. Pacini, T. Bilharz, and *Max Schultze.* The development and metamorphosis of the Lamperns was made the subject of research by *H. Müller, M. Schultze,* and *P. Owsjannikow;* Müller's examination of *Branchiostoma* was continued by *J. Marcusen, A. Kovalevsky, L. Stieda, W. Müller, C. Hasse, T. Huxley,* and *F. M. Balfour.* The most comprehensive accounts of the anatomy of fishes are contained in the following works:—

1. *H. Stannius,* "Zootomie der Fische," 2d edit. (Berl. 1854, 8vo.)

2. *R. Owen,* "Anatomy of Vertebrates," vol. i. (Lond. 1866, 8vo.)

3. *R. Owen,* "Lectures on the Comparative Anatomy and Physiology of the Vertebrate Animals." Part I. Fishes. (Lond. 1846, 8vo.)

4. *T. Huxley,* "A Manual of the Anatomy of Vertebrated Animals." (Lond. 1871, 16mo.)

Latest Systematic Works.

It has been mentioned above that the great work of Cuvier and Valenciennes had been left incomplete. Several authors, therefore, supplied detailed accounts of the orders omitted in that work. *Müller* and *Henle* published an account of the Plagiostomes, and *Kaup* of the Murænidæ and Lophobranchii. *A. Duméril,* finally, commenced an "Histoire naturelle des Poissons ou Ichthyologie générale," of which, however, two volumes only appeared, containing a complete account of the "Plagiostomes" (Paris, 1865, 8vo), and of the "Ganoids and Lophobranchs." (Paris, 1870, 8vo.)

So great an activity had prevailed in Ichthyology since the publication of the "Histoire naturelle" by Cuvier and Valenciennes, and the results of the manifold enquiries were scattered over such a multitude of publications, that it became imperative to collect again all these materials in one comprehensive work. This was done in the "Catalogue of Fishes," published by the Trustees of the British Museum, in eight volumes (Lond. 1859-70). Beside the species previously described many new forms were added, the number total of species referred to in those volumes amounting to 8525. As regards the systematic arrangement—Müller's system was adopted in the main, but the definition of the families is much modified. This, however, need not be further entered into here, and will become sufficiently apparent in the subsequent parts of the present work.

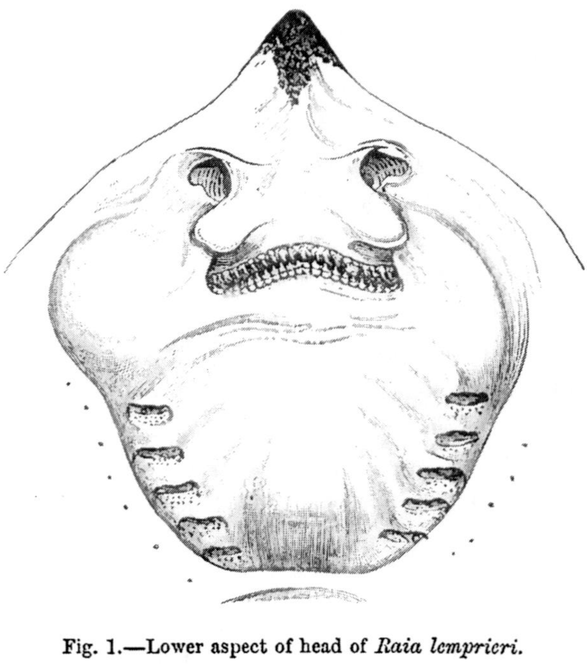

Fig. 1.—Lower aspect of head of *Raia lemprieri.*

CHAPTER II.

TOPOGRAPHICAL DESCRIPTION OF THE EXTERNAL PARTS OF FISHES.

In the body of a fish four parts are distinguished: the *head, trunk, tail,* and the *fins;* the boundary between the first and second being generally indicated by the *gill-opening,* and that between the second and third by the *vent.* The form of the body and the relative proportions of those principal parts are subject to much variation, such as is not found in any other class of Vertebrates. In fishes which are endowed with the power of steady and more or less rapid locomotion, a deviation from that form of body, which we observe in a perch, carp, or mackerel, is never excessive. The body forms a simple, equally-formed wedge, compressed or slightly rounded, well fitted for cleaving the water. In fishes which are in the habit of moving on the bottom, the whole body, or at least the head, is *vertically depressed* and flattened; the head may be so enormously enlarged that the trunk and tail appear merely as an appendage. In one family of fishes, the *Pleuronectidæ* or Flat-fishes, the body is compressed into a thin disk; they swim and move on one side only, which remains constantly directed towards the bottom, a peculiarity by which the symmetry of all parts of the body has been affected. A *lateral compression* of the body, in conjunction with a lengthening of the vertical and a shortening of the longitudinal axis, we find in fishes moving comparatively slowly through

Form of the body.

the water, and able to remain (as it were) suspended in it. This deviation from the typical form may proceed so far that the vertical axis greatly exceeds the longitudinal in length; generally all the parts of the body participate in this form, but in one kind of fish (the Sun-fish or *Orthagoriscus*) it is chiefly the tail which has been shortened, and reduced so much as to present the appearance of being cut off. An excessive lengthening of the longitudinal axis, with a shortening of the vertical, occurs in Eels and eel-like fishes, and in the so-called Band-fishes. They are bottom-fish, capable of insinuating themselves into narrow crevices and holes. The form of the body of these long fish is either cylindrical, snake-like, as in the Eels and many Codfishes, or strongly compressed as in the Band-fishes (*Trichiurus, Regalecus*, etc.) It is chiefly the tail which is lengthened, but frequently the head and trunk participate more or less in this form. Every possible variation occurs between these and other principal types of form. The old ichthyologists, even down to Linnæus, depended in great measure on them for classification; but although often the same form of body obtains in the same group of fishes, similarity of form by no means indicates natural affinity; it only indicates similitude of habits and mode of life.

Eye. *The external parts of the Head.*—The *Eye* divides the head into the *ante-orbital* and *post-orbital* portion. In most fishes, especially in those with a compressed head, it is situated on the side and in the anterior half of the length of the head; in many, chiefly those with a depressed head, it is directed upwards, and sometimes situated quite at the upper side; in very few, the eyes look obliquely downwards. In the Flat-fishes both eyes are on the same side of the head, either the right or the left, always on that which is directed towards the light, and coloured.

Fishes in general, compared with other Vertebrata, have large eyes. Sometimes these organs are enormously enlarged, their great size indicating that the fish is either nocturnal, or lives at a depth to which only a part of the sun's rays penetrate. On the other hand, small eyes occur in fishes inhabiting muddy places, or great depths to which scarcely any light descends, or in fishes in which the want of an organ of sight is compensated by the development of other organs of sense. In a few fishes, more particularly in those inhabiting caves or the greatest depths of the ocean, the eyes have become quite rudimentary and hidden under the skin.

In the *ante-orbital* portion of the head, or the *Snout*, are situated the mouth and the nostrils. **Snout.**

The *Mouth* is formed by the intermaxillary and maxillary bones, or by the intermaxillary only in the upper jaw, and by the mandibulary bone in the lower. These bones are either bare or covered by integument, to which frequently labial folds or lips are added. As regards form, the mouth offers as many variations as the body itself, in accordance with the nature of the food, and the mode of feeding. It may be narrow, or extremely wide and cleft to nearly the hind margin of the head; it may be semi-elliptical, semicircular, or straight in a transverse line; it may be quite in front of the snout (*anterior*), or at its upper surface (*superior*), or at its lower (*inferior*), or extending along each side (*lateral*); sometimes it is sub-circular, organised for sucking. The jaws of some fishes are modified into a special weapon of attack (Sword-fish, Saw-fish); in fact, throughout the whole class of fishes the jaws are the only organ specialised for the purpose of attacking; weapons on other parts of the body are purely defensive. **Mouth.**

Both jaws may be provided with skinny appendages, *barbels*, which, if developed and movable, are sensitive organs of touch.

In the majority of fishes the *Nostrils* are a double opening on each side of the upper surface of the snout; the openings **Nostrils.**

of each side being more or less close together. They lead into a shallow groove; and only in one family (the Myxinoids) perforate the palate. In this family, as well as in the Lampreys, the nasal aperture is single. In many Eels the openings are lateral, the lower perforating the upper lip. In the Sharks and Rays (Fig. 1, p. 34) they are at the lower surface of the snout, and more or less confluent; and, finally, in the Dipnoi and other Ganoids, one at least is within the labial boundary of the mouth.

The space across the forehead, between the orbits, is called the *inter-orbital* space; that below the orbit, the *infra-orbital* or *sub-orbital* region.

Gill-cover. In the *post-orbital* part of the head there are distinguished, at least in most Teleosteous Fishes and many Ganoids, (Fig. 24) the *præoperculum*, a sub-semicircular bone, generally with a free and often serrated or variously-armed margin; the *operculum*, forming the posterior margin of the gill-opening, and the *suboperculum* and *inter-operculum* along its inferior margin. All these bones, collectively called *opercles*, form the *gill-cover*, a thin bony lamella covering the cavity containing the gills. Sometimes they are covered with so thin a membrane that the single bones may be readily distinguished; sometimes they are hidden under a thick integument. In some cases the inter-operculum is rudimentary or entirely absent (Siluroids).

Gill-opening. The *Gill-opening* is a foramen, or a slit behind or below the head, by which the water which has been taken up through the mouth for the purpose of breathing is again expelled. This slit may extend from the upper end of the operculum all round the side of the head to the symphysis of the lower jaw; or it may be shortened and finally reduced to a small opening on any part of the margin of the gill-cover. Sometimes (*Symbranchus*) the two openings, thus reduced, coalesce, and form what externally appears as a single opening only. The margin of the gill-cover is provided with a cutaneous fringe, in order to more effectually close the gill-opening; and this fringe is supported by one or several or many bony rays, the *branchiostegals.* The space on the chest between the two rami of the lower jaw and between the gill-openings is called the *isthmus.*

The Sharks and Rays differ from the Teleosteous and Ganoid fishes in having five branchial slits (six or seven in *Hexanchus* and *Heptanchus*), which are lateral in the Sharks, and at the lower surface of the head in the Rays (Fig. 1, p. 34). In Myxine only the gill-opening is at a great distance from

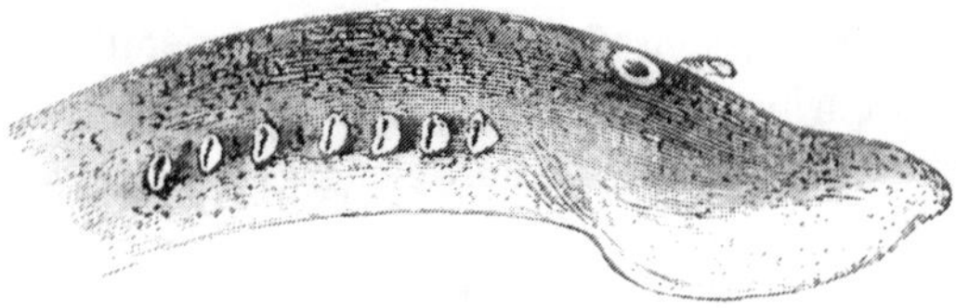

Fig. 2.—Head of *Mordacia mordax*, showing the single nostril, and seven branchial openings.

the head; it is either single in this family (Cyclostomi), or there are six and more on each side (Fig. 2).

In the *Trunk* are distinguished the *back*, the *sides*, and the *abdomen.* It gradually passes in all fishes into the *Tail;* the termination of the abdominal cavity and the commencement of the tail being generally indicated by the position of the vent. The exceptions are numerous: not only certain abdominal organs, like the sexual, may extend to between the muscles of the tail, but the intestinal tract itself may pass far backwards, or, singularly, it may be reflected forwards, so that the position of the vent may be either close to the extremity of the tail or to the foremost part of the trunk. **Tail.**

In many fishes the greater part of the tail is surrounded by the fins, leaving only a small portion (between dorsal, caudal, and anal fins) finless; this part is called the *free portion* or the *peduncle* of the tail.

Fins. The *Fins* are divided into *vertical* or *unpaired*, and into *horizontal* or *paired fins*. Any of them may be present or absent; and their position, number, and form are most important guides in determining the affinities of fishes.

The *vertical* fins are situated in the median dorsal line, from the head to the extremity of the tail, and in the ventral line of the tail. In fishes in which they are least developed or most embryonic, the vertical fin appears as a simple fold of the skin surrounding the extremity of the tail. In its further progress of development in the series of fishes, it gradually extends more forwards, and may reach even the head and vent. Even in this embryonic condition the fin is generally supported by fine rays, which are the continuations of, or articulated to, other stronger rays supported by the processes or apophyses of the vertebral column. This form of the vertical fin is very common, for instance in the Eels, many Gadoid, Blennioid and Ganoid fishes in which, besides, the rays have ceased to be simple rods, showing more or less numerous joints (simple *articulated* rays; Fig. 3). *Branched* rays are dichotomically split, the joints increasing in number towards the extremity.

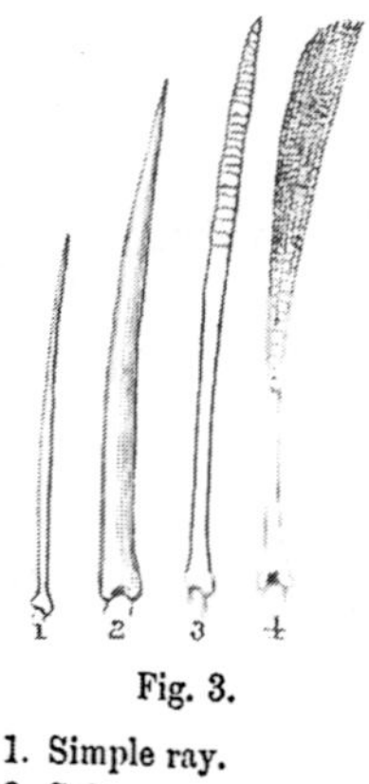

Fig. 3.

1. Simple ray.
2. Spine.
3. Simple articulated ray (soft).
4. Branched ray (soft).

The continuity of the vertical fin, however, is interrupted in the majority of fishes; and three fins then are distinguished: one in the dorsal line—the *dorsal* fin; one in the ventral line behind the anus—the *anal* fin; and one confined to the extremity of the tail—the *caudal* fin.

The *caudal* fin is rarely symmetrical, so that its upper half would be equal to its lower; the greatest degree of asymmetry obtains in fishes with heterocercal termination of the vertebral column (see subsequently, Figs. 31, 41). In fishes in which it is nearly symmetrical it is frequently prolonged into an upper and lower *lobe*, its hind margin being concave or more or less deeply excised; in others the hind margin is rounded, and when the middle rays greatly exceed in length the outer ones the fin assumes a pointed form.

Many and systematically important differences are observed in the *dorsal* fin, which is either spiny-rayed (spinous) (*Acanthopterygian*), or soft-rayed (*Malacopterygian*). In the former, a smaller or greater number of the rays are simple and without transverse joints; they may be flexible, or so much osseous matter is deposited in them that they appear hard and truly spinous (Fig. 3); these spines form always the anterior portion of the fin, which is detached from, or continuous

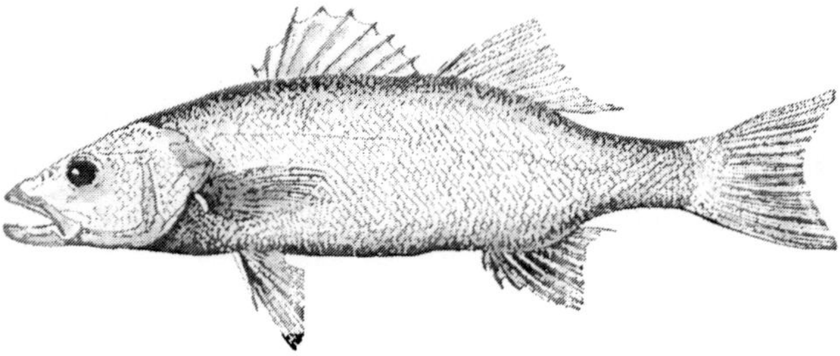

Fig. 4.—Labrax lupus (Bass), an Acanthopterygian with anterior spinous, and posterior soft dorsal fin.

with, the remaining jointed rays. The spines can be erected or depressed at the will of the fish; if in the depressed position the spines cover one another completely, their points lying in the same line, the fish is called *homacanth;* but if the spines are asymmetrical, alternately broader on one side than on the other, the fish is called *heteracanth.* The spinous division, as well as the one consisting of jointed rays, may again be subdivided. In the *Malacopterygian* type all the rays remain jointed; indeed, sometimes the foremost ray, with its preceding short supports, is likewise ossified, and a hard spine, but the articulations can nearly always be distinctly traced. Sometimes the dorsal fin of Malacopterygian fishes is very long, extending from the head to the end of the tail, sometimes it is reduced to a few rays only, and in a few cases it is

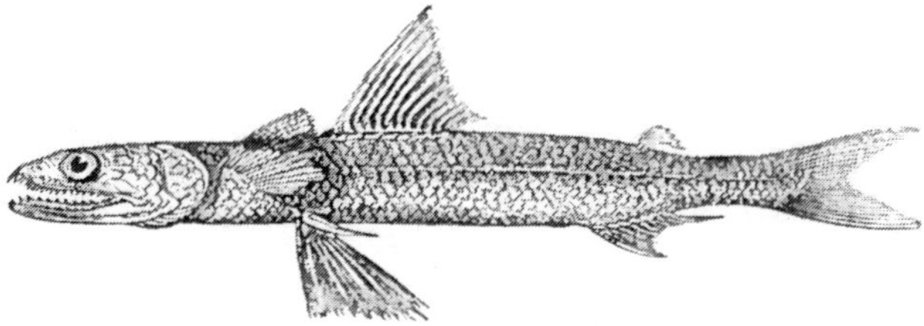

Fig. 5.—Saurus undosquamis, a Malacopterygian with anterior soft dorsal, and additional adipose fin.

entirely absent. In addition to the rayed dorsal fin, many Malacopterygian fishes (as the Salmonoids, many Siluroids, Scopeloids, etc.) have another of greater or lesser extent, without any rays; and as always fat is deposited within this fold, it is called a *fatty* fin (*pinna adiposa*).

The *anal* fin is built on the same plan as the dorsal, and may be single or plural, long or short, or entirely absent; in Acanthopterygians its foremost rays are frequently simple and spinous.

The *horizontal* or *paired* fins consist of two pairs: the pectorals and ventrals.

The *pectoral* fins (with their osseous supports) are the homologues of the anterior limbs of the higher Vertebrata. They are always inserted immediately behind the gill-opening; either symmetrical with a rounded posterior margin, or asymmetrical, with the upper rays longest and strongest; in Malacopterygians with a dorsal spine the upper pectoral ray is frequently developed into a similar defensive weapon.

The *ventral* fins are the homologues of the hind-limbs, and inserted on the abdominal surface, either behind the pectorals (*Pisces* s. *Pinnæ abdominales*), or below them (*Pisces* s. *Pinnæ thoracicæ*), or in advance of them (*Pisces* s. *Pinnæ jugulares*). They are generally narrow, composed of a small number of rays, the outer of which is frequently osseous. In some small groups of fishes, like the Gobies, the fins coalesce and form a suctorial disk.

Fig. 6.—Salmo salar (Salmon), with abdominal ventral fins.

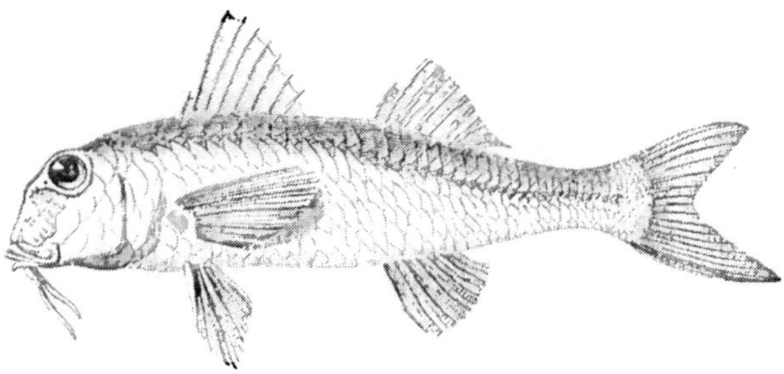

Fig. 7.—Mullus barbatus (Red Mullet), with thoracic ventral fins.

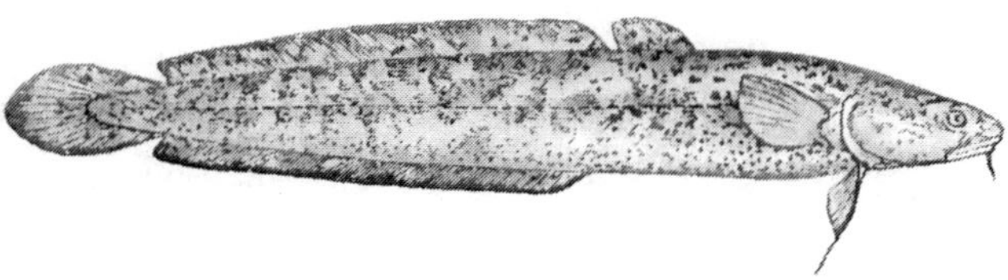

Fig. 8.—Burbot (Lota vulgaris), with jugular ventral fins.

For the definition of the smaller systematic groups, and the determination of species, the numbers of the spines and rays are generally of the greatest importance. This holds good, especially for the ventral rays, by the number of which the Acanthopterygian affinities of a fish can nearly always be determined. The numbers of the dorsal and anal rays generally correspond to the number of vertebræ in a certain portion of the spine, and are therefore constant specific, generic, or even family characters; but when their number is very great, a proportionally wide margin must be allowed for variation, and the taxinomic value of this character becomes uncertain. The numbers of the pectoral and caudal rays are rarely of any account.

Function of the Fins.

The fins are organs of motion; but it is chiefly the tail and the caudal fin by which the fish impels itself forward. To execute energetic locomotion the tail and caudal fin are strongly bent, with rapidity, alternately towards the right and left; whilst a gentle motion forwards is effected by a simply undulating action of the caudal fin, the lobes of which act like the blades of a screw. Retrograde motions can be made by fish in an imperfect manner only, by forward-strokes of the pectoral fins. When the fish wants to turn towards the left, he gives a stroke of the tail towards the right, the right pectoral acting simultaneously, whilst the left remains adpressed to the body. Thus the pectoral fins assist in the progressive motions of the fish, but rather directing its course than acting as powerful propellers. The chief function of the paired fins is to maintain the balance of the fish in the water, which is always the most unsteady where there is no weight to sink it: when the pectoral of one side, or the pectoral and ventral of the same side are removed, the fish loses its balance and falls on the side opposite; when both pectorals are removed, the fish's head sinks; on removal of the dorsal and anal fins the motion of the fish assumes a zig-zag course. A fish deprived of all fins, as well as a dead fish, floats with the belly upwards, the back being the heavier part of the body.

In numerous groups of fishes which live in mud, or are enabled to pass a longer or shorter time in soil periodically dried and hardened during the hot season, forms occur entirely devoid of, or with only rudimentary, ventral fins (Cyprinodon, Ophiocephalidæ, Galaxiidæ, Siluridæ). The chief function of these fins being to balance the body of the fish whilst swimming, it is evident that in fishes moving during a great part of their life over swampy ground, or through more or less consistent mud, this function of the ventral fins ceases, and that nature can readily dispense with these organs altogether.

In certain fishes the shape and function of the fins are considerably modified: thus, in the Rays, locomotion is almost entirely effected and regulated by the broad and expanded pectoral fins acting with an undulatory motion of their margins, similar to the undulations of the long vertical fins of the Flat-fishes; in many Blennies the ventral fins are adapted for walking on the sea-bottom; in some Gobioids (*Periophthalmus*), Trigloids, Scorpænioids, and Pediculati, the pectoral fins are perfect organs of walking; in the Gobies, Cyclopteri, and Discoboli the ventral fins are transformed into an adhesive disk, and finally in the Flying-fish, in which the pectorals act as a parachute. In the Eels and other snake-like fishes, the swimming as well as the gliding motions are effected by several curvatures of the body, alternate towards the right and left, resembling the locomotion of Snakes. In the *Syngnathi* (Pipe-fishes) and *Hippocampi*, whose body admits of but a slight degree of lateral curvature, and whose caudal fin is generally small, if present at all, locomotion is very limited, and almost wholly dependent on the action of the dorsal fin, which consists of a rapid undulating movement.

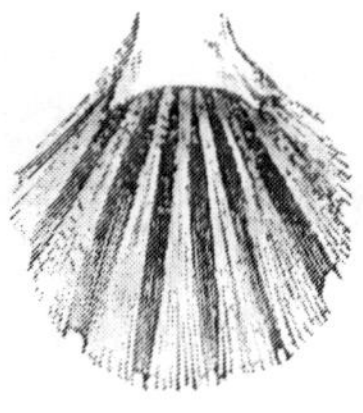

Fig. 9.—Ventrals of *Gobius*.

The *skin* of fishes is either covered with scales, or naked, or provided with more or less numerous scutes of various forms and sizes. Some parts, like the head and fins, are more Skin and Scales.

frequently naked than scaly. All fishes provided with electric organs, the majority of Eels, and the Lampreys, are naked. *Scales* of fishes are very different from those of Reptiles; the latter being merely folds of the cutis, whilst the scales of fishes are distinct horny elements, developed in grooves or pockets of the skin, like hairs, nails, or feathers. Very small or rudimentary scales are extremely thin, homogeneous in structure, and more or less imbedded in the skin, and do not cover each other. When more developed, they are imbricated (arranged in the manner of tiles), with the posterior part extruded and free, the surface of the anterior portion being usually covered by the skin to a greater or less extent. On their surface (Figs. 10 and 11) may be observed a very fine striation concentric and parallel to the margin, and coarser striæ radiating from a central point towards the hind margin. Scales without a covering of enamel, with an entire (not denticulated) posterior margin, and with a concentric striation, are called *Cycloid* scales. *Ctenoid* scales (Figs. 12-15) are generally thicker, and provided with spinous teeth on the posterior edges of the layers of which the scale consists. In some species only the layer nearest to the margin is provided with denticulations (Fig. 14). Scales, the free surface of which is spiny, and which have no denticulation on the margin, have been termed *Sparoid* scales; but their distinction from ctenoid scales is by no means sharp, and there are even intermediate

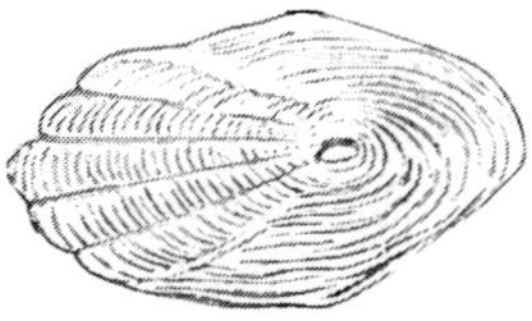

Fig. 10.—Cycloid scale of Gadopsis marmoratus (magn.)

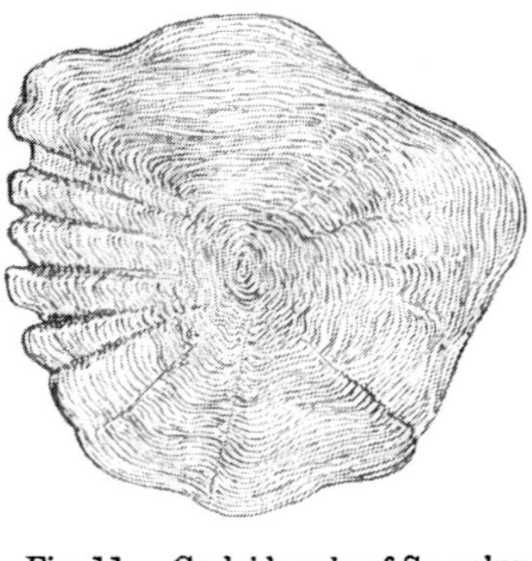

Fig. 11.—Cycloid scale of Scopelus resplendens (magn.)

forms between the cycloid and ctenoid types. Both kinds of

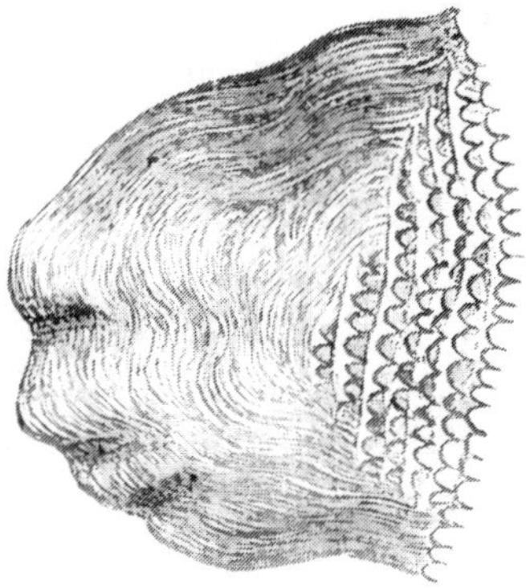

Fig. 12.—Ctenoid scale of Scatophagus multifasciatus (magn.)

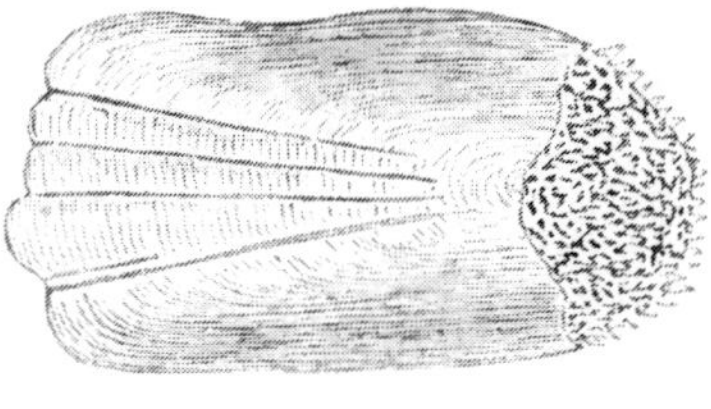

Fig. 13.—Ctenoid scale of Platycephalus cirrhonasus (magn.)

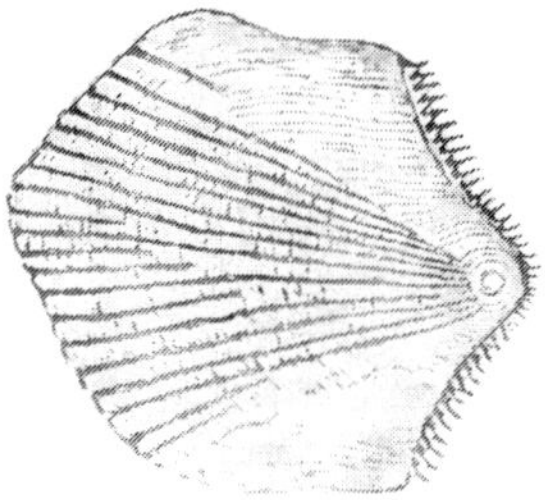

Fig. 14.—Ctenoid scale of Gobius ommaturus (magn.)

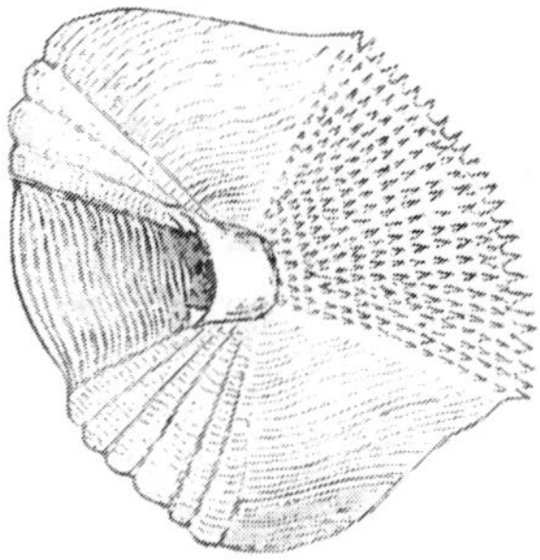

Fig. 15.—Ctenoid scale of Lethrinus (magn.)

scales may occur not only in species of the same genus of fishes, but in the same fish.

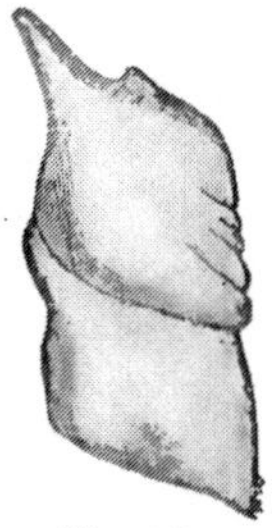

Fig. 16. Ganoid Scales.

Ganoid scales are hard and bony, covered with a layer of enamel; they are generally rhombic or quadrangular, rarely rounded and imbricate; and arranged in oblique rows, those of one row being linked together by an articulary process. This type of scales, common in fossil Ganoid fishes, occurs among recent fishes in *Lepidosteus* and *Polypterus* only.

Finally, in Sharks, the Balistidæ, and others, true scales are absent and replaced by the ossified papillæ of the cutis, which give the surface the appearance of fine-grained chagreen.

Fig. 17.—Dermal papillæ of Monacanthus trossulus.

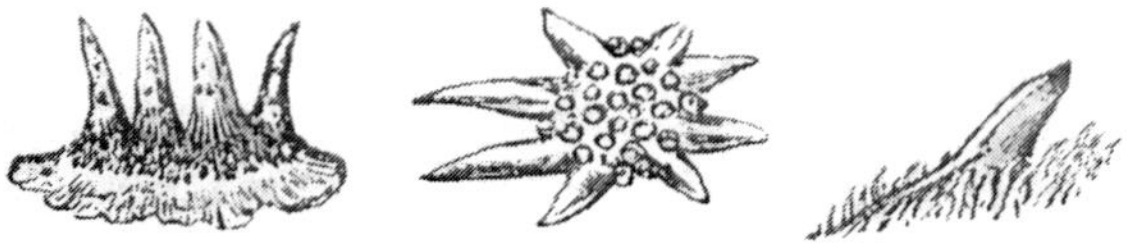

Fig. 18.—Dermal papillæ of Monacanthus hippocrepis (magn.)

These generally small bodies, as well as the large osseous scutes of the Rays, Sturgeons, etc., have been comprised under the common name *Placoid* scales; a term which deservedly is being abandoned.

Along the side of the body of osseous fishes runs a series of perforated scales, which is called the *lateral line* (Fig. 21). The perforating duct is simple at its base, and may be also simple at its outer opening (Fig. 19), or (and this is frequently the case) the portion on the free surface of the scale is ramified (Fig. 20). The lateral line runs from the head to the tail, sometimes reaching the caudal fin, sometimes stopping in front of it, sometimes advancing over its rays. It is nearer to the dorsal profile in some fishes than in others. Some species have several lateral lines, the upper one coasting the dorsal, the lower the abdominal outline, one running along the middle as usual.

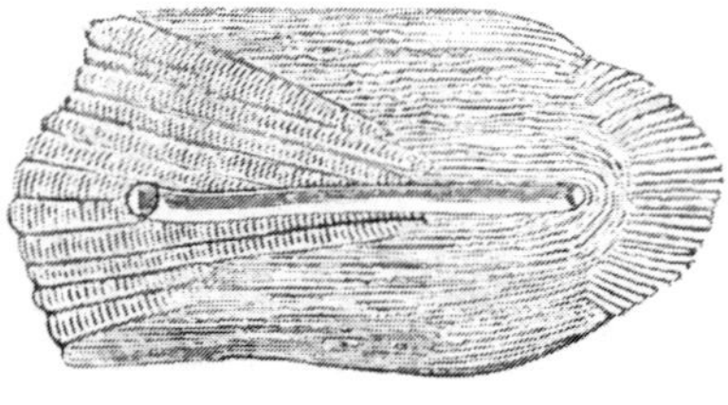

Fig. 19.—Cycloid scale from the lateral line of Odax lineatus (magn.)

The scales of the lateral line are sometimes larger than the others, sometimes smaller, sometimes modified into scutes,

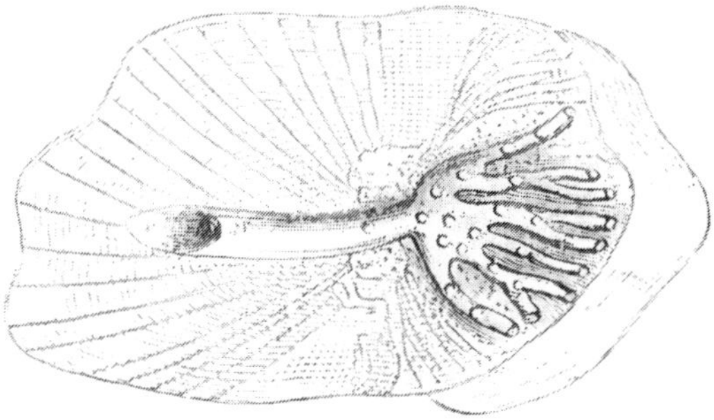

Fig. 20.—Cycloid scale from the lateral line of Labrichthys laticlavius (magn.)

sometimes there are no other scales beside them, the rest of the body being naked. The foramina of the lateral line are the outlets of a muciferous duct which is continued on to the head, running along the infraorbital bones, and sending off a branch into the præopercular margin and mandible. In many fishes, as in many Sciænoids, Gadoids, and in numerous deep-sea fishes, the ducts of this muciferous system are extraordinarily wide, and generally filled with mucus, which is congealed or contracted in specimens preserved in spirits, but swells again when the specimens are immersed in water. This system is abundantly provided with nerves, and, therefore, has been considered to be the seat of a sense peculiar to fishes, but there cannot be any doubt that its function is the excretion of mucus, although probably mucus is excreted also from the entire surface of the fish.

The scales, their structure, number and arrangement, are an important character for the determination of fishes; in most scaly fishes they are arranged in oblique transverse series; and as the number of scales in the lateral line generally corresponds to the number of transverse series, it is usual to count the scales in that line. To ascertain the number of longitudinal series of scales, the scales are counted in one of the transverse series, generally in that running from the commencement of the dorsal fin, or the middle of the back to the

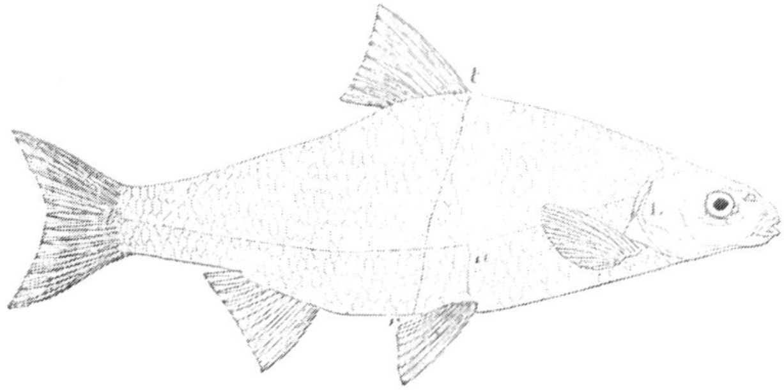

Fig. 21.—Arrangement of scales in the Roach (Leuciscus ratilus): *L l* = Lateral line; *t r* = Transverse line. *a*, Transverse line from lateral line to ventral fin.

lateral line, and from the lateral line down to the vent or ventral fin, or middle of the abdomen.[1]

The scales of many fishes are modified for special purposes, especially to form weapons of defence or a protective armour, but the details of such modifications are better mentioned under the several families in which they occur. All scales are continually growing and wasting away on the surface, and it seems that some fish, at least,—for instance, Salmonoids—"shed" them periodically; during the progress of this shedding the outlines of the scales are singularly irregular.

[1] In the formula generally preceding the description of a fish, "L. lat. 40," would express that the scales between the head and caudal fin are arranged in 40 transverse series; and probably, that the lateral line is composed of the same number of scales. "L. transv. $\frac{8}{5}$" would express that there are eight longitudinal series of scales between the median line of the back and the lateral line, and five between the lateral line and the middle of the abdomen.

CHAPTER III.

TERMINOLOGY AND TOPOGRAPHY OF THE SKELETON.

In order to readily comprehend the subsequent account of the modifications of the skeleton in the various sub-classes and groups of Fishes, the student has to acquaint himself with the terms used for the numerous bones of the fish skeleton, as well as with their relative position. The skeleton of any of the more common kinds of osseous fish may serve for this purpose; that of the Perch is chosen here.

The series of bones constituting the axis of the body, and destined to protect the spinal chord and some large longitudinal blood-vessels, is called the *vertebral* or *spinal column;* the single bones are the *vertebræ.* The *skull* consists of the bones surrounding the brain and organs of sense, and of a number of arches suspended from it, to support the commencement of the alimentary canal and the respiratory organs.

The *vertebra* (Fig. 22) consists of a body or *centrum* (*c*), with a concave anterior and posterior surface, and generally of several *processes* or *apophyses,* as—1. Two *neurapophyses* (*na*), which, on the dorsal side, rising upwards, form the *neural arch* over the canal, in which the spinal chord is lodged. 2. Two *parapophyses* (*pa*) usually projecting from the lower part of the sides of the body, or two *hæmapophyses* (*ha*) which actually coalesce to form on the ventral side the hæmal canal for a large trunk of the vascular system. 3. A *neural spine* (*ns*), which crowns the neurapophyses, or is interposed between their tips. 4. A *hæmal spine* (*hs*), having the same relation to the hæmapophyses.

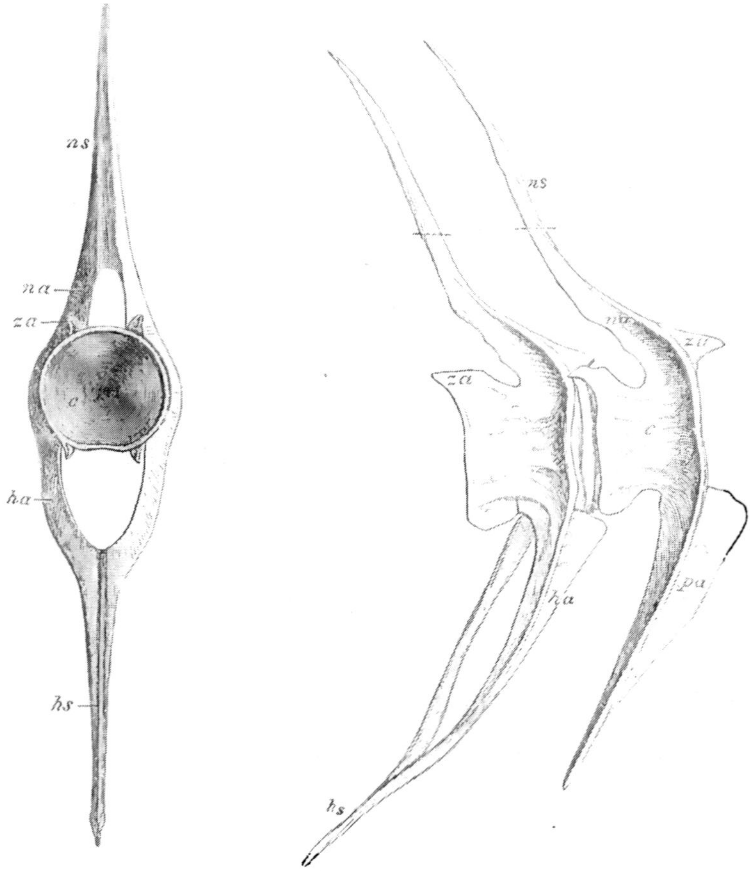

Fig. 22.—Side and Front view of Fish-vertebra.

5. Two *pleurapophyses* or floating *ribs*, suspended from, or from the base of, the parapophyses. 6. In most fishes the neural arches are connected together by articular or oblique processes, *zygapophyses* (*za*), which are developed from the base of each neurapophysis.

The vertebræ are either *abdominal* or *caudal* vertebræ, the coalescence of the parapophyses into a complete hæmal ring, and the suspension of the anal fin generally forming a sufficiently well-marked boundary between abdominal and caudal regions (Fig. 23). In the Perch there are twenty-one abdominal and as many caudal vertebræ. The centrum of the first vertebra or atlas is very short, with the apophyses scarcely indicated, and lacking ribs like the succeeding vertebra. All the other abdominal vertebræ, with the exception of the last or two last, are provided with ribs, many of which are bifid (72). A series of flat spines (74), called *interneurals*, to which the spines and rays of the dorsal fins are articulated, are supported by the neural spines, the strength of the neurals and interneurals corresponding to that of the *dermal* spines (75). The caudal vertebræ differ from the abdominal in having the hæmapophyseal elements converted into spines similar to the neurals, the anterior being likewise destined to support a series of *interhæmals* (79), to which the anal rays are articulated. The last and smallest caudal vertebra articulates with the *hypural* (70), a fan-like bone, which, together with the dilated hindmost neural and hæmal elements, supports the caudal rays.

Looking at a perch's *skull* from the side (Fig. 24), the most superficial bones will be found to be those of the jaws, a chain of thin bones round the lower half of the eye, and the opercles.

The anterior margin of the upper jaw is formed by the *intermaxillary* or *premaxillary* (17) which bears teeth, terminates in a pedicle above, to allow of a forward sliding motion of the jaw, and is dilated into a flat triangular process behind, on which leans the second bone of the upper jaw, the *maxillary* (18). This bone is toothless, articulates with the vomer and palatine bone, and is greatly dilated towards its distal extremity. Both the maxillary and intermaxillary lie and move parallel to each other, being connected by a narrow membrane; in many other fishes their relative position is very different.

The *mandible* or lower jaw consists of a right and left ramus; their union by a ligament in front is called *symphysis*. Each ramus is formed of several pieces; that which, by a sigmoid concavity articulates with the quadrate, is the *articulary* bone (35); it sends upwards a coronoid process, to which a ligament from the maxillary and the masticatory muscles are attached; and forwards a long-pointed process, to be sheathed in the deep notch of the anterior piece. A small separate piece (36) at the lower posterior angle of the mandible is termed *angular*. The largest piece (34) is tooth-bearing, and hence termed *dentary;* at its inner surface it is always deeply excavated, to receive a cylindrical cartilage, called *Meckel's cartilage*, the remains of an embryonic condition of the jaw, the articulary and angular being but ossified parts of it. In other *Teleostei* this number is still more increased by a *splenial* and other bones.

The *infraorbital* ring of bones (Fig. 23, ^{19}) consists of several (four) pieces, of which the anterior is the largest, and distinguished as *præorbital*.

The so-called *præoperculum* (30) belongs rather to the bones of the suspensorium of the mandible, presently to be described, than to the opercles proper. It is narrow, strong, angularly bent, so as to consist of a vertical and horizontal limb, with an incompletely closed canal running along both limbs. As it is quite a superficial bone, and frequently armed with various spines, its form and configuration form an important item in the descriptive details of many fishes.

The principal piece of the gill-cover is the *operculum* (28), triangular in shape, situated behind, and movably united with, the vertical limb of the præoperculum. There is an articulary cavity at its upper anterior angle for its junction with the hyomandibular. The oblong lamella below the operculum is the *suboperculum* (32), and the one in front of this latter, below the horizontal limb of the præoperculum, is the *interoperculum* (33), which is connected by ligament with the angular piece of the lower jaw, and is also attached to the outer face of the hyoid, so that the gill-covers cannot open or shut without the hyoid apparatus executing a corresponding movement.

The chain of flat bones which, after the removal of the temporal muscles, appear arranged within the inner concavity of the præoperculum (Fig. 24), are comprised with the latter under the common name of *mandibulary suspensorium*. They connect the mandible with the cranium. The uppermost, the *epitympanic* or *hyomandibular* (23), is articulated by a double articulary head with the mastoid and posterior frontal. Another articulary head is destined for the opercular joint. The *mesotympanic* or *symplectic* (31) appears as a styliform prolongation of the lower part of the hyomandibular; is entirely cartilaginous in the young, but nearly entirely ossified in the adult. The position of this bone is noteworthy, because, directly inwards of its cartilaginous junction with the hyomandibular, there is situated the uppermost piece of the hyoid arch, the stylo-hyal. The next bone of the series is the *pretympanic* or *metapterygoid* (27), a flat bone forming a bridge towards the pterygoid, and not rarely absent in the teleosteous sub-class. Finally, the large triangular *hypotympanic* or *quadrate* (26) has a large condyle for the mandibulary joint.

The palatine arch (Fig. 26) connects the suspensorium with the anterior extremity of the skull, and is formed by three bones: the *entopterygoid* (25), an oblong and thin bone attached to the inner border of the palatine and pterygoid, and increasing the surface of the bony roof of the mouth towards the median line; it constitutes also the floor of the orbit. The *pterygoid* (24) (or *os transversum*) starts from the quadrate, and is joined by suture to the *palatine*, which is toothed, and reaches to the vomer and anterior frontal.

In the occipital region there are distinguished the *basioccipital* (5), readily recognised by the conical excavation corresponding and similar to that of the atlas, with which it is articulated through the intervention of a capsule filled with a gelatinous substance (the remains of the notochord); the *exoccipitals* (10), articulated, one on each side, to the basioccipital, and expanding on the upper surface of that bone, so as to meet and support the spinal column; a superficial thin lamella (13), suturally connected with the exoccipitals, not constant in fishes, and erroneously believed by Cuvier to be the *petrosal* (*os petrosum*) of higher animals; further, the *paroccipitals* (9), which are wedged in between the exoccipitals and *supraoccipital*. This last bone (8) forms the key of the arch over the occipital foramen, and raises a strong high crest from the whole length of its mesial line; a transverse supraoccipital ridge, coming from each side of the base of this spine runs outwards laterally to the external angles of the bone. The supraoccipital separates the parietals, and forms a suture with the frontals.

In front of the basioccipital the base of the skull is formed by the *basisphenoid* (*parasphenoid* of Huxley) (6). This very long and narrow bone extends from the basioccipital beyond the brain-capsule to between the orbits, where it forms the support of the fibro-membranous interorbital septum. Anteriorly it is connate with another long hammer-shaped bone (16), the *vomer*, the head of which marks the anterior end of the palate, and is beset with teeth. The *alisphenoids* (11) are short broad bones, rising from the basisphenoid; their posterior margins are suturally connected with the anterior of the basi- and ex-occipitals.

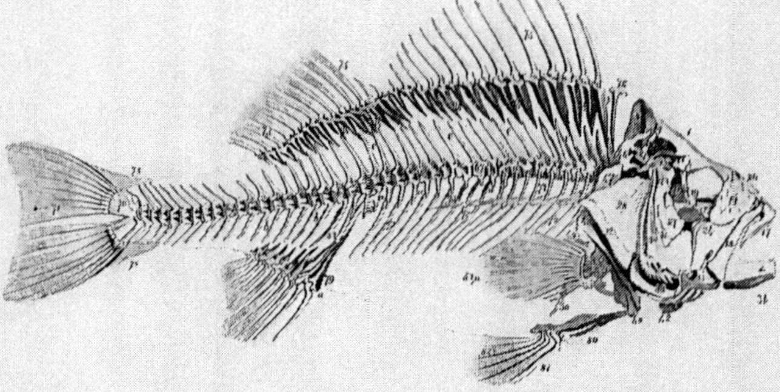

Fig. 23.—Skeleton of the Perch.

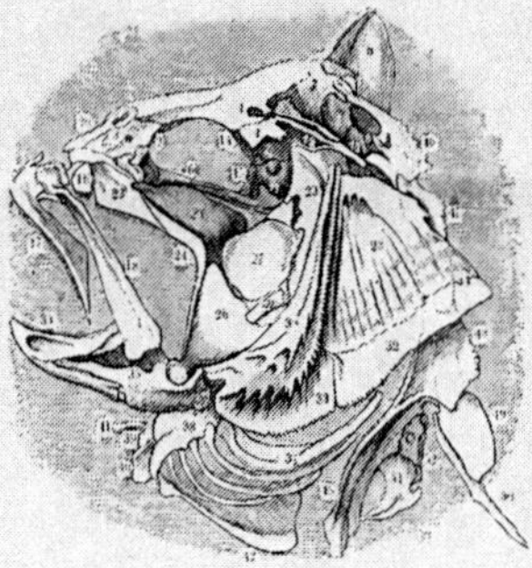

Fig. 24.—Skeleton of a Perch's Skull.

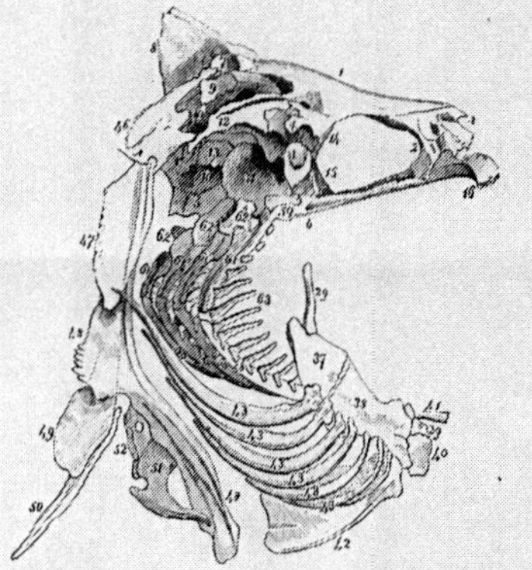

Fig. 25.—Hyoid arch, branchial apparatus, and scapulary arch of the Perch.

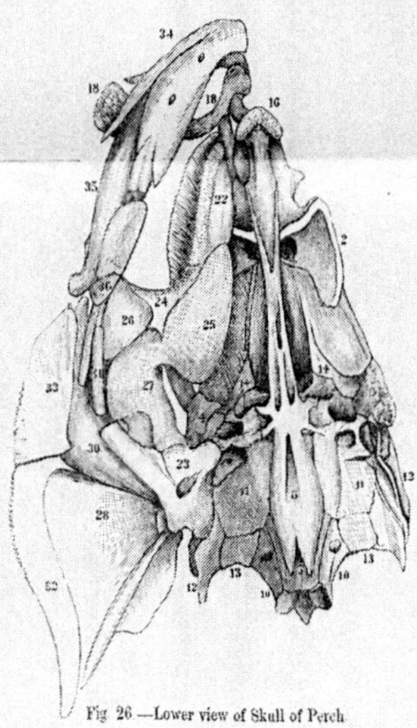

Fig. 26.—Lower view of Skull of Perch.

Fig. 27.—Hyoid bone of the Perch.

The formation of the posterior part of the side of the skull is completed by the *mastoid* and *parietal* bones. The former (12) projects outwards and backwards farther than the paroccipital, forming the outer strong process of the side of the cranium. This process lodges on its upper surface one of the main ducts of the muciferous system, and affords the base of articulation to a part of the hyomandibular. Its extremity gives attachment to the strong tendon of the dorsolateral muscles of the trunk. The *parietals* (7) are flat bones, of comparatively much smaller extent than in higher Vertebrates, and separated from each other by the anterior prolongation of the supra-occipital.

The anterior wall of the brain-capsule (or the posterior of the orbit) is formed by the *orbitosphenoids* (14), between which, superiorly, the olfactory nerves, and inferiorly, the optic, pass out of the cranium. In addition to this paired bone, the Perch and many other fishes possess another single bone (15),—the *os sphenoideum anterius* of Cuvier, *ethmoid* of Owen, and *basisphenoid* of Huxley; it is Y-shaped, each lateral branch being connected with an orbitosphenoid, whilst the lower branch rests upon the long basal bone.

A cartilage, the substance of which is thickest above the vomer, and which extends as a narrow stripe along the interorbital septum, represents the *ethmoid* of higher Vertebrata; the olfactory nerves run along, and finally perforate it.

There remain, finally, the bones distinguishable on the upper surface of the skull; the largest, extending from the nasal cavities to the occipital, are the *frontal* bones (1), which also form the upper margin of the orbit. The *postfrontals* (4) are small bones placed on the supero-posterior angle of the orbit, and serving as the point from which the infraorbital ring is suspended. The *prefrontals* (2), also small, occupy the anterior margin of the orbit. A pair of small tubiform bones (20), the *turbinals*, occupy the foremost part of the snout, in front of the frontals, and are separated from each other by intervening cartilage.

After removal of the gill-cover and mandibulary suspensorium, the hyoid arch, which encloses the branchial apparatus, and farther behind, the humeral arch are laid open to view (Fig. 25). These parts can be readily separated from the cranium proper.

The *hyoid arch* is suspended by a slender styliform bone, the *stylohyal* (29), from the hyomandibulars; it consists of three segments, the *epihyal* (37), *ceratohyal* (38), which is the longest and strongest piece, and the *basihyal*, which is formed by two juxtaposed pieces (39, 40). Between the latter there is a median styliform ossicle (41), extending forwards into the substance of the tongue, called *glossohyal* or *os linguale;* and below the junction of the two hyoid branches there is a vertical single bone (42), expanded along its lower edge, which, connected by ligament with the anterior extremity of the humeral arch, forms the *isthmus* separating the two gill-openings. This bone is called the *urohyal.* Articulated or attached by ligaments to the epihyal and ceratohyal are a number of sword-shaped bones or rays (43), the *branchiostegals,* between which the branchiostegal membrane is extended.

The *branchial arches* (Figs. 25 and 27) are enclosed within the hyoid arch, with which they are closely connected at the base. They are five in number, of which four bear gills, whilst the fifth (56) remains dwarfed, is beset with teeth, and called the *lower pharyngeal* bone. The arches adhere by their lower extremities to a chain of ossicles (53, 54, 55), *basibranchials,* and, curving as they ascend, nearly meet at the base of the cranium, to which they are attached by a layer of ligamentous and cellular tissue. Each of the first three branchial arches consists of four pieces movably connected with one another. The lowest is the *hypobranchial* (57), the next much longer one (58) the *ceratobranchial,* and, above this, a slender and a short irregularly-shaped *epibranchial* (61). In the fourth arch the hypobranchial is absent. The uppermost of these segments (62), especially of the fourth arch, are dilated, and more or less confluent; they are beset with fine teeth, and generally distinguished as the *upper pharyngeal bones.* Only the ceratobranchial is represented in the fifth arch or lower pharyngeal. On their outer convex side the branchial segments are grooved for the reception of large blood-vessels and nerves; on the inner side they support horny processes (63), called the *gill-rakers,* which do not form part of the skeleton.

The *scapular* or *humeral arch* is suspended from the skull by the (*suprascapula*) *post-temporal* (46), which, in the Perch, is attached by a triple prong to the occipital and mastoid bones. Then follows the (*scapula*) *supraclavicula* (47), and the arch is completed below by the union of the large (*coracoid*) *clavicula* (48) with its fellow. Two flat bones (51, 52), each with a vacuity, attached to the clavicle have been determined as the (*radius* and *ulna*) *coracoid* and *scapula* of higher vertebrates, and the two series of small bones (53) intervening between the forearm and the fin as *carpals* and *metacarpals.* A two-jointed appendage the (*epicoracoid*) *post-clavicula,* is attached to the clavicle: its upper piece (49) is broad and lamelliform, its lower (50) styliform and pointed.

The ventral fins are articulated to a pair of flat triangular bones, the *pubic* bones (80).

The bones of the skull of the fish have received so many different interpretations that no two accounts agree in their nomenclature, so that their study is a matter of considerable difficulty to the beginner. The following synonymic table will tend to overcome difficulties arising from this cause; it contains the terms used by Cuvier, those introduced by Owen, and finally the nomenclature of Stannius, Huxley, and Parker. Those adopted in the present work are printed in italics. The numbers refer to the figures in the accompanying woodcuts (Figs. 23-27).

	Cuvier.	*Owen.*	*Stannius.*	*Huxley, Parker, etc.*
1.	Frontal principal	*Frontal*	Os frontale	
2.	Frontal antérieur	*Prefrontal*	Os frontale anterius	Lateral ethmoid (Parker)
3.	*Ethmoid*	Nasal	Os ethmoideum	
4.	Frontal postérieur	*Postfrontal*	Os frontale posterius	Sphenotic (Parker)
5.	Basilaire	*Basioccipital*	Os basilare	
6.	Sphénoide	*Basisphenoid*	Os sphenoideum basilare	Sometimes referred to as "*Basal*"
7.	Pariétal	*Parietal*	Os parietale	
8.	Interpariétal or occipital supérieure	*Supraoccipital*	Os occipitale superius	
9.	Occipital externe	*Paroccipital*	Os occipitale externum	Epioticum (Huxley)
10.	Occipital lateral	*Exoccipital*	Os occipitale laterale	
11.	Grande aile du sphénoide	*Alisphenoid*	Ala temporalis	Prooticum (*Huxley*)
12.	Mastoidien	*Mastoid*	Os mastoideum + os extrascapulare	Opisthoticum[1] + *Squamosal* (Huxley)
13.	Rocher	Petrosal and Otosteal	Oberflächliche Knochen-lamelle	
14.	Aile orbitaire	*Orbitosphenoid*	Ala orbitalis	Alisphenoid (Huxley)
15.	Sphenoide antérieur	Ethmoid and Ethmoturbinal	Os sphenoideum anterius	*Basisphenoid* (Huxley)
16.	*Vomer*	Vomer	Vomer	
17.	Intermaxillaire	*Inter- or Pre-maxillary*	Os intermaxillare	
18.	Maxillaire supérieur	*Maxillary*	Os maxillare	
19.	Sousorbitaires	*Infraorbital ring*	Ossa infraorbitalia	
20.	Nasal	*Turbinal*	Os terminale	
22.	Palatine	*Palatin*	Os palatinum	
23.	Temporal	Epitympanic	Os temporale	*Hyomandibular* (Huxley)
24.	Transverse	*Pterygoid*	Os transversum s. pterygoideum externum	
25.	Ptérygoidien interne	*Entopterygoid*	Os pterygoideum	Mesopterygoid (Parker)
26.	Jugal	Hypotympanic	Os quadratojugale	*Quadrate* (Huxley)
27.	Tympanal	Pretympanic	Os tympanicum	*Metapterygoid* (Huxley)

[1] Pterotic of Parker.

No.	*Cuvier.*	*Owen.*	*Stannius.*	*Huxley, Parker, etc.*
28.	Operculaire	*Operculum*	Operculum	
29.	Styloide	*Stylohyal*	Os styloideum	
30.	Préopercule	*Præoperculum*	Præoperculum	
31.	Symplectique	Mesotympanic	*Os symplecticum*	
32.	Sousopercule	*Suboperculum*	Suboperculum	
33.	Interopercule	*Interoperculum*	Interoperculum	
34.	Dentaire	*Dentary*	Os dentale	
35.	Articulaire	*Articulary*	Os articulare	
36.	Angulaire	*Angular*	Os angulare	
37.	Grandes pièces latérales	*Epihyal*	Segmente der Zungenbein-Schenkel	
38.		*Ceratohyal*		
39.	Petites pièces laterales	*Basihyal*		
40.				
41.	Os lingual	*Glossohyal*	Os linguale s. entoglossum	
42.	Queue de l'os hyoide	*Urohyal*		Basibranchiostegal (Parker)
43.	Rayon branchiostège	*Branchiostegal*	Radii branchiostegi	
46.	Surscapulaire	Suprascapula	Omolita	*Post-temporal* (Parker)
47.	Scapulaire	Scapula	Scapula	*Supraclavicula* (Parker)
48.	Humeral	Coracoid	Clavicula	*Clavicula* (Parker)
49. 50.	Coracoid	Epicoracoid		*Postclavicula* (Parker)
51.	Cubital	Radius	Ossa carpi	*Coracoid* (Parker)
52.	Radial	Ulna		*Scapula* (Parker)
53.	Os du carpe	*Carpals*	Ossa metacarpi	*Basalia* (Huxley), Brachials (Parker)
53 bis. 54. 55.	Chaine intermédiaire	*Basibranchials*	Copula	
56.	Pharyngiens inférieurs	*Lower Pharyngeals*	Ossa pharyngea inferiora	
57.	Pièce interne de partie inférieure de l'arceau branchiale	*Hypobranchial*	Segmente der Kiemenbogen-Schenkel	
58.	Pièce externe ,,	*Ceratobranchial*		
59.	Stylet de première arceau branchiale	*Upper epibranchial of first branchial arch*		
61.	Partie supérieure de l'arceau branchiale	*Epibranchials*		
62.	Os pharyngian supérieur	Pharyngo-branchial	Os pharyngeum superius	*Upper pharyngeals*
63.		*Gill-rakers*		
65.	Rayons de la pectorale	*Pectoral rays*	Brustflossen-Strahlen	

	Cuvier.	Owen.	Stannius.	Huxley, Parker, etc.
67, 68.	Vertèbres abdominales	*Abdominal vertebræ*	Bauchwirbel	
69.	Vertèbres caudales	*Caudal vertebræ*	Schwanzwirbel	
70.	Plaque triangulaire et verticale	[Aggregated interhæmals]	Verticale Platte	*Hypural* (Huxley)
71.		*Caudal rays*	Schwanzflossen Strahlen	
72.	Côte	*Rib*	Rippen	
73.	Appendices or stylets	*Epipleural spines*	Muskel-Gräthen	
74.	Interépineux -	*Interneural spines*	Ossa interspinalia s. obere Flossentræger	
75.	Épines et rayons dorsales	*Dorsal rays and spines*	Rückenflossen-Strahlen u. Stacheln	
76.		*First interneural*		
78.		*Rudimentary caudal rays*		
79.	Apophyses épineuses inférieures	*Interhæmal spines*	Untere Flossentræger	
80.		*Pubic*	Becken	
81.		*Ventral spine*	Bauchflossen-Stachel	

CHAPTER IV.

MODIFICATIONS OF THE SKELETON.

THE lowermost sub-class of fishes, which comprises one form only, the Lancelet (*Branchiostoma* [s. *Amphioxus*] *lanceolatum*, possesses the skeleton of the most primitive type.

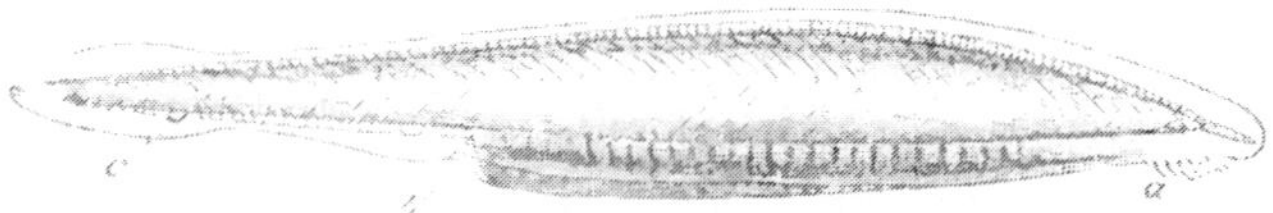

Fig. 28.—Branchiostoma lanceolatum.
a, Mouth ; *b*, Vent ; *c*, abdominal porus.

The vertebral column is represented by a simple *chorda*

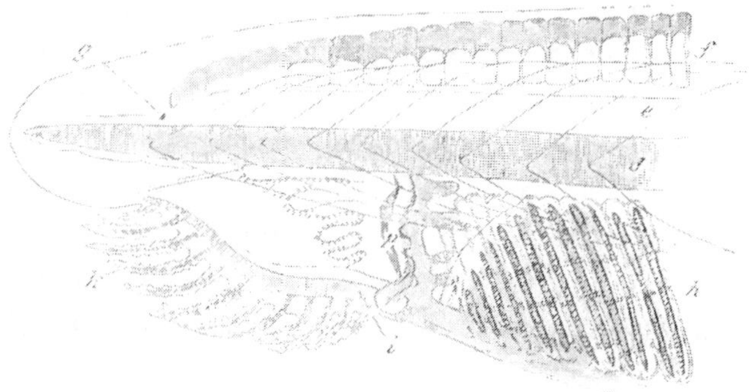

Fig. 29.—Anterior end of body of Branchiostoma (magn.)
d, Chorda dorsalis ; *e*, Spinal chord ; *f*, Cartilaginous rods ; *g*, Eye ; *h*, Branchial rods ; *i*, Labial cartilage ; *k*, Oral cirrhi.

dorsalis or *notochord* only, which extends from one extremity of the fish to the other, and, so far from being expanded into a cranial cavity, it is pointed at its anterior end as well as at its posterior. It is enveloped in a simple membrane like the spinal chord and the abdominal organs, and there is no trace of vertebral segments or ribs ; however, a series of short cartilaginous rods above the spine evidently represent apophyses. A maxillary or hyoid apparatus, or elements representing limbs, are entirely absent.

[J. Müller, Ueber den Bau und die Lebenserscheinungen des *Branchiostoma lubricum*, in Abhandl. Ak. Wiss. Berlin, 1844.]

The skeleton of the *Cyclostomata* (or Marsipobranchii) (Lampreys and Sea-hags) shows a considerable advance of development. It consists of a notochord, the anterior pointed end of which is wedged into the base of a cranial capsule, partly membranous partly cartilaginous. This skull, therefore, is not movable upon the spinal column. No vertebral segmentation can be observed in the notochord, but neural arches are represented by a series of cartilages on each side of the spinal chord. In *Petromyzon* (Fig. 30) the basis cranii emits two prolongations on each side: an inferior, extending for some distance along the lower side of the spinal column, and a lateral, which is ramified into a skeleton supporting the branchial apparatus. A stylohyal process and a subocular arch with a palato-pterygoid portion may be distinguished. The roof of the cranial capsule is membranous in *Myxine* and in the larvæ of *Petromyzon,* but more or less cartilaginous in the adult *Petromyzon* and in *Bdellostoma.* A cartilaginous capsule on each side of the hinder part of the skull contains the auditory organ, whilst the olfactory capsule occupies the anterior upper part of the roof. A broad cartilaginous lamina, starting from the cranium and overlying part of the snout, has been determined as representing the ethmo-vomerine elements, whilst the oral organs are supported by large, very peculiar cartilages (*labials*), greatly differing in general configuration and arrangement in the various Cyclostomes. There are three in the Sea-lamprey, of which the middle one is joined to the palate by an intermediate smaller one; the fore-

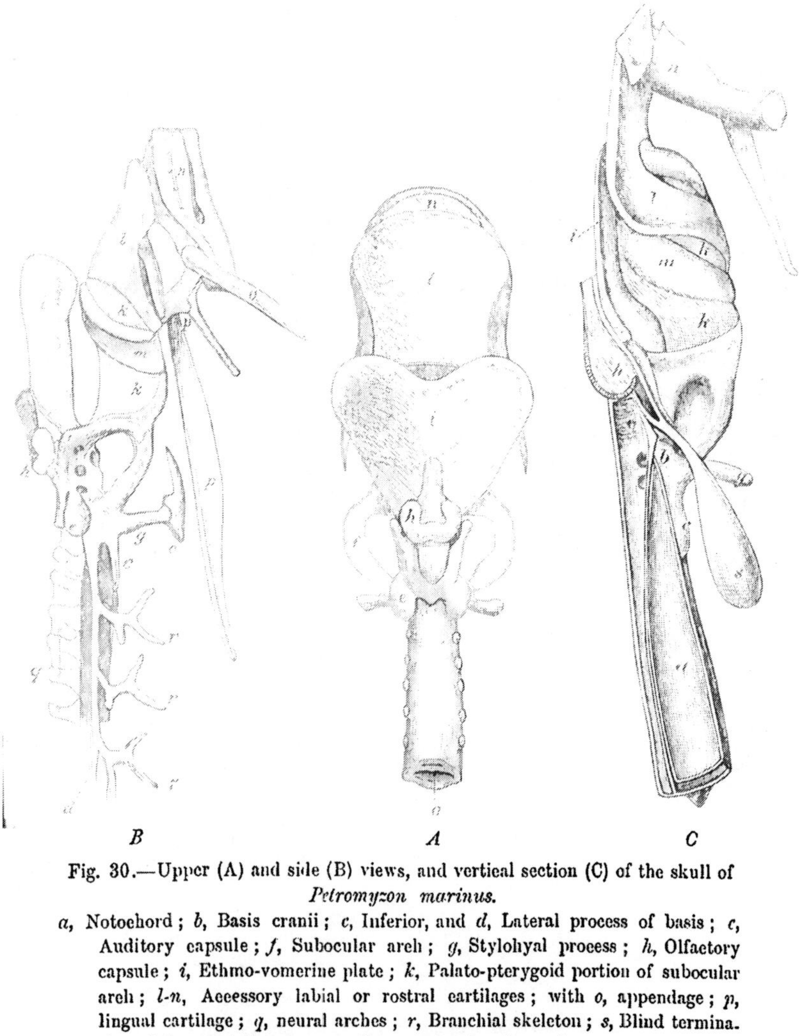

B A C

Fig. 30.—Upper (A) and side (B) views, and vertical section (C) of the skull of *Petromyzon marinus*.

a, Notochord; *b*, Basis cranii; *c*, Inferior, and *d*, Lateral process of basis; *e*, Auditory capsule; *f*, Subocular arch; *g*, Stylohyal process; *h*, Olfactory capsule; *i*, Ethmo-vomerine plate; *k*, Palato-pterygoid portion of subocular arch; *l-n*, Accessory labial or rostral cartilages; with *o*, appendage; *p*, lingual cartilage; *q*, neural arches; *r*, Branchial skeleton; *s*, Blind termination of the nasal duct between the notochord and œsophagus.

most is ring-like, tooth-bearing, emitting on each side a styliform process. The lingual cartilage is large in all Cyclostomes.

There is no trace of ribs or limbs.

[J. Müller, Vergleichende Anatomie der Myxinoiden. Erster Theil. Osteologie und Myologie, in Abhandl. Ak. Wiss. Berlin, 1835.]

The *Chondropterygians* exhibit a most extraordinary diversity in the development of their vertebral column; almost every degree of ossification, from a notochord without a trace of annular structure to a series of completely ossified vertebræ being found in this order. Sharks, in which the notochord is persistent, are the *Holocephali* (if they be reckoned to this order, and the genera *Notidanus* and *Echinorhinus*). Among the first, *Chimæra monstrosa* begins to show traces of segmentation; but they are limited to the outer sheath of the notochord, in which slender subossified rings appear. In *Notidanus* membranous septa, with a central vacuity, cross the substance of the gelatinous notochord. In the other Sharks the segmentation is complete, each vertebra having a deep conical excavation in front and behind, with a central

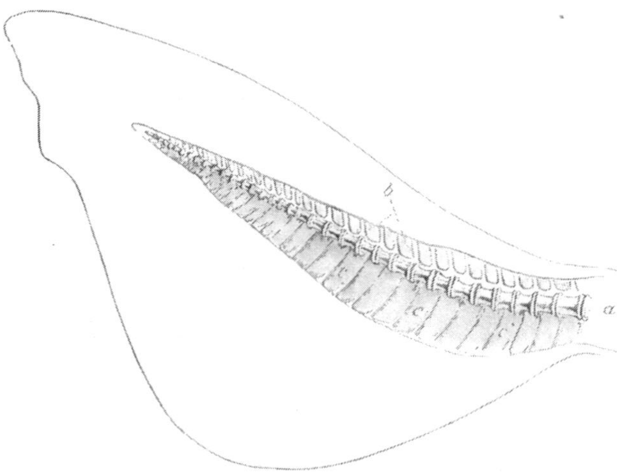

Fig. 31.—Heterocercal Tail of Centrina salviani.
a, Vertebræ; *b*, Neurapophyses; *c*, Hæmapophyses.

canal through which the notochord is continued; but the degree in which the primitive cartilage is replaced by concentric or radiating lamellæ of bone varies greatly in the various genera, and according to the age of the individuals. In the Rays all the vertebræ are completely ossified, and the anterior ones confluent into one continuous mass.

In the majority of Chondropterygians the extremity of the vertebral column shows a decidedly heterocercal condition (Fig. 31), and only a few, like *Squatina* and some Rays, possess a diphycercal tail.

The advance in the development of the skeleton of the Chondropterygians beyond the primitive condition of the previous sub-classes, manifests itself further by the presence of neural and hæmal elements, which extend to the foremost part of the axial column, but of which the hæmal form a closed arch in the caudal region only, whilst on the trunk they appear merely as a lateral longitudinal ridge.

The neural and hæmal apophyses are either merely attached to the axis, as in Chondropterygians with persistent notochord, the Rays and some Sharks; or their basal

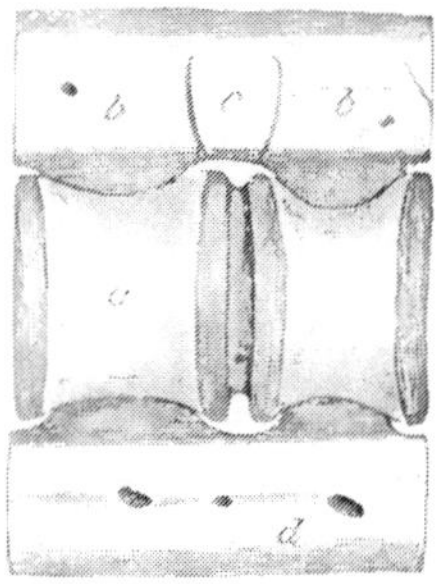

Fig. 32.

Fig. 33.

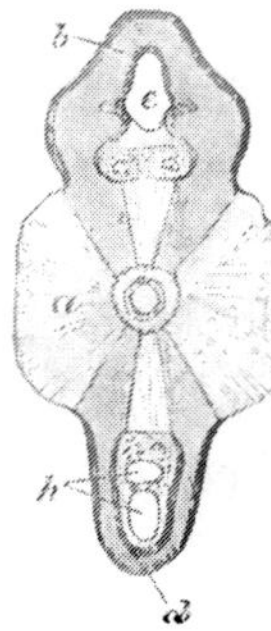

Fig. 34.

Fig. 32.—Lateral view. Fig. 33.—Longitudinal section. Fig. 34.—Transverse section of Caudal vertebra of Basking Shark (Selache maxima). (After Hasse.) *a*, Centrum; *b*, Neurapophysis; *c*, Intercrural cartilage; *d*, Hæmapophysis; *e*, Spinal canal; *f*, Intervertebral cavity; *g*, Central canal for persistent portion of notochord; *h*, Hæmal canals for blood-vessels.

portions penetrate like wedges into the substance of the centrum, so that, in a transverse section, in consequence of the difference in their texture, they appear in the form of an X.[1] The interspaces between the neurapophyses of the vertebræ are not filled by fibrous membrane, as in other fishes, but by separate cartilages, *laminæ* or *cartilagines intercrurales*, to which frequently a series of terminal pieces is superadded, which must be regarded as the first appearance of the interneural spines of the Teleostei and many Ganoids. Similar terminal pieces are sometimes observed on the hæmal arches. *Ribs* are either absent or but imperfectly represented (*Carcharias*).

The substance of the *skull* of the Chondropterygians is cartilage, interrupted especially on its upper surface by more or less extensive fibro-membranous fontanelles. Superficially it is covered by a more or less thick chagreen-like osseous deposit. The articulation with the vertebral column is effected by a pair of lateral condyles. In the Sharks, besides, a central conical excavation corresponds to that of the centrum of the foremost vertebral segment, whilst in the Rays this central excavation of the skull receives a condyle of the axis of the spinous column.

The cranium itself is a continuous undivided cartilage, in which the limits of the orbit are well marked by an anterior and posterior protuberance. The ethmoidal region sends horizontal plates over the nasal sacs, the apertures of which retain their embryonic situation upon the under surface of the skull. In the majority of Chondropterygians these plates are conically produced, forming the base of the soft projecting snout; and in some forms, especially in the long-snouted Rays and the Saw-fishes (*Pristis*) this prolongation appears in the form of three or more tubiform rods.

[1] *C. Hasse* has studied the modifications of the texture of the vertebræ and the structure of the Chondropterygian skeleton generally, and shown that they correspond in the main to the natural groups of the system, and, consequently, that they offer a valuable guide in the determination of fossil remains.

As separate cartilages there are appended to the skull a suspensorium, a palatine, mandible, hyoid, and rudimentary maxillary elements.

The suspensorium is movably attached to the side of the skull. It generally consists of one piece only, but in some Rays of two. In the Rays it is articulated with the mandible only, their hyoid possessing a distinct point of attachment to the skull. In the Sharks the hyoid is suspended from the lower end of the suspensorium together with the mandible.

What is generally called the upper jaw of a Shark is, as Cuvier has already stated, not the maxillary, but palatine. It consists of two simple lateral halves, each of which articulates with the corresponding half of the lower jaw, which is formed by the simple representative of Meckel's cartilage.

Some cartilages of various sizes are generally developed on each side of the palatine, and one on each side of the mandible. They are called *labial* cartilages, and seem to represent maxillary elements.

The *hyoid* consists generally of a pair of long and strong lateral pieces, and a single mesial piece. From the former cartilaginous filaments (representing branchiostegals) pass directly outwards. Branchial arches, varying in number, and similar to the hyoid, succeed it. They are suspended from the side of the foremost part of the spinous column, and, like the hyoid, bear a number of filaments.

The vertical fins are supported by interneural and interhæmal cartilages, each of which consists of two and more pieces, and to which the fin-rays are attached without articulation.

The *scapular arch* of the Sharks is formed by a single coracoid cartilage bent from the dorsal region downwards and forwards. In some genera (*Scyllium*, *Squatina*) a small separate scapular cartilage is attached to the dorsal extremities of the coracoid; but in none of the Elasmobranchs is the scapular arch suspended from the skull or vertebral column; it is merely sunk, and fixed in the substance of the muscles. Behind, at the point of its greatest curvature, three carpal cartilages are joined to the coracoid, which Gegenbaur has distinguished as *propterygium*, *mesopterygium*, and *metapterygium*, the former occupying the front, the latter the hind margin of the fin. Several more or less regular transverse series of styliform cartilages follow. They represent the phalanges, to which the horny filaments which are imbedded in the skin of the fin are attached.

In the Rays, with the exception of *Torpedo*, the scapular arch is intimately connected with the confluent anterior portion of the vertebral column. The anterior and posterior carpal cartilages are followed by a series of similar pieces, which extend like an arch forwards to the rostral portion of the skull, and backwards to the pubic region. Extremely numerous phalangeal elements, longest in the middle, are supported by the carpals, and form the skeleton of the lateral expansion of the so-called *disk* of the Ray's body, which thus, in fact, is nothing but the enormously enlarged pectoral fin.

The *pubic* is represented by a single median transverse cartilage, with which a tarsal cartilage articulates. The latter supports the fin-rays. To the end of this cartilage is also attached, in the male Chondropterygians, a peculiar accessory generative organ or clasper.

The *Holocephali* differ from the other Chondropterygians in several important points of the structure of their skeleton, and approach unmistakably certain Ganoids. That their spinal column is persistently notochordal has been mentioned already. Their palatal apparatus, with the suspensorium, coalesces with the skull, the mandible articulating with a short apophysis of the cranial cartilage. The mandible is simple, without anterior symphysis. The spine with which the dorsal fin is armed articulates with a neural apophysis, and is not immovably attached to it, as in the Sharks. The pubic consists of two lateral halves, with a short, rounded, tarsal cartilage.

The skeleton of the *Ganoid* Fishes offers extreme variations with regard to the degree in which ossifications replace the primordial cartilage. Whilst some exhibit scarcely any advance beyond the Plagiostomes with persistent cartilage, others approach, as regards the development and specialisation of the several parts of their osseous framework, the Teleosteans so closely that their Ganoid nature can be demonstrated by, or inferred from, other considerations only. All Ganoids possess a separate gill-cover.[1]

The diversity in the development of the Ganoid skeleton is well exemplified by the few representatives of the order in the existing Fish-fauna. Lowest in the scale (in this respect) are those with a persistent notochord, and an *autostylic* skull, that is, a skull without separate suspensorium—the fishes constituting the sub-order *Dipnoi*, of which the existing representatives are *Lepidosiren*, *Protopterus*, and *Ceratodus*, and the extinct (as far as demonstrated at present) *Dipterus*, *Chirodus* (and *Phaneropleuron ?*). In these fishes the notochord is persistent, passing uninterruptedly into the cartilaginous base of the skull. Only now and then a distinct vertical segmentation occurs in the caudal portion of the column, but it does not extend to the notochord itself, but indicates only the limits between the superadded apophyseal elements, each neural being confluent with the opposite hæmal. Some *Dipnoi* are diphy-, others hetero-cercal.

[1] The Ganoids formed at former epochs the largest and most important order of fishes, many of the fossil forms being known from very imperfect remains only. It is quite possible that not a few of the latter, in which nothing whatever of the (probably very soft) endoskeleton has been preserved, should have to be assigned to some other order lower in the scale of organisation than the Ganoids (for instance, the *Cephalaspidæ*).

Neural and hæmal elements and ribs are well developed. In *Ceratodus* each neurapophysis consists of a basal cartilaginous portion, forming an arch over the myelon, and of a superadded second portion. The latter is separated from the former by a distinct line of demarcation, and its two branches are more styliform, cartilaginous at the ends and in the centre, but with an osseous sheath, and coalesced at the top, forming a gable over an elastic fibrous band which runs along and parallel to the longitudinal axis of the column (*Ligamentum longitudinale superius*). To the top of this gable is joined a single long cylindrical neural spine. From the eleventh apophyseal segment a distinct interneural spine, of the same structure as the neural, begins to be developed, and farther on a second interneural is superadded. Towards the extremity of the column these various pieces are gradually reduced in size and number, finally only a low cartilaginous band (the rudiments of the neurapophysis) remaining. The *hæmapophyses* are in form, size, and structure, very similar to the neurapophyses; and all these long bones, including the ribs, have that in common, that they consist of a solid rod of cartilage enclosed in a bony sheath, which, after the disappearance or decomposition of the cartilage, appears as a hollow tube. Such bones are extremely common throughout the order of Ganoids, and their remains have led to the designation of a family as *Cœlacanthi* (κοιλος, hollow; and ἀκανθος, spine).

The primordial *cranium* of the *Dipnoi* is cartilaginous, but with more or less extensive ossifications in its occipital, basal, or lateral portions, and with large tegumentary bones, the arrangement of which varies in the different genera. There is no separate suspensorium for the lower jaw. A strong process descends from the cranial cartilage, and offers by means of a double condyle (Fig. 35 *s*) attachment to corresponding articulary surfaces of the lower jaw.

Maxillary and intermaxillary elements are not developed, but, perhaps, represented in *Ceratodus* by some inconstant rudimentary labial cartilages situated behind the posterior nasal opening. Facial cartilages and an infraorbital ring are developed at least in *Ceratodus.* The presence of a pair of small teeth in front indicates the vomerine portion (*v*) which remains cartilage, whilst the posterior pair of teeth are implanted in a pterygo-palatine ossification (*l*), which sometimes is paired, sometimes continuous. The base of the skull is constantly covered by a large basal ossification (*o*).

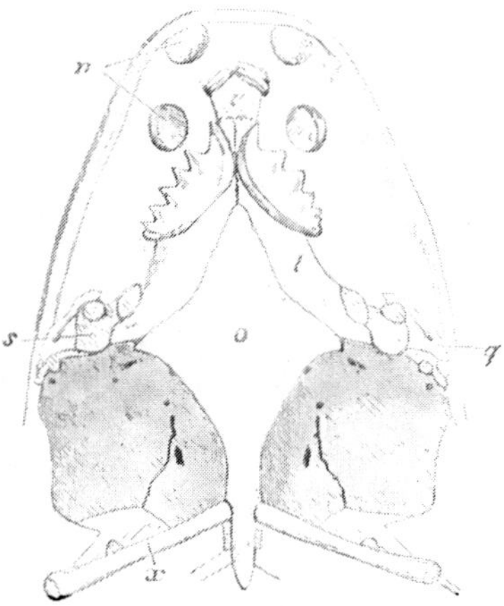

Fig. 35.—Palatal view of Skull of Ceratodus.

The *hyoid* is well developed, sometimes reduced to a pair of ceratohyals, sometimes with a basihyal and glossyhyal. The skeleton of the *branchial apparatus* approaches the Teleosteous type, less so in *Lepidosiren* than in *Ceratodus,* in which five branchial arches are developed, but with the lateral and mesial pieces reduced in number.

A large operculum, and a smaller sub- or inter-operculum are present.

The *scapular* arch consists of a single median transverse cartilage, and a pair of lateral cartilages which bear the articular condyle for the pectoral limb. The latter cartilages form the base of a large membrane-bone, and the whole arch is suspended from the skull by means of an osseous supraclavicle.

The fore-limb of the *Dipnoi* (Fig. 36) differs externally greatly from the pectoral fin of other Ganoid fishes. It is covered with small scales along the middle, from the root to its extremity, and surrounded by a rayed fringe similar to the vertical fin. A muscle split into numerous fascicles extends all the length of the fin, which is flexible in every part and in every direction. The cartilaginous framework supporting it is joined to the scapular arch by an oblong cartilage, followed by a broad basal cartilage (*a*), generally single, sometimes showing traces of a triple division. Along the middle of the fin runs a jointed axis (*b*), the joints gradually becoming smaller and thinner towards the extremity; each joint bears on each side a three, two, or one-jointed branch (*c*, *d*). This *axial* arrangement of the pectoral skeleton, which evidently represents one of its first and lowest conditions, has been termed *Archipterygium* by Gegenbaur. It is found in *Ceratodus* and other genera, but in *Lepidosiren* the jointed axis only has been preserved, with the addition of rudimentary rays in *Protopterus*.

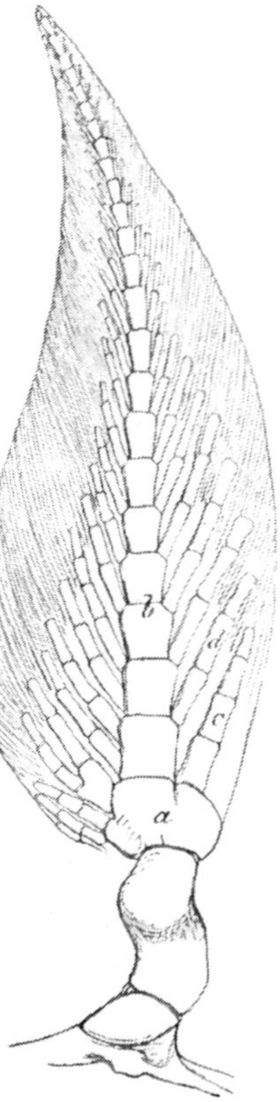

Fig. 36.—Fore-limb of Ceratodus.

The *pubic* consists of a single flattened subquadrangular cartilage, produced into a long single anterior process. Posteriorly it terminates on each side in a condyle, to which the basal cartilage of the ventral paddle is joined. The endoskeleton of the paddle is almost identical with that of the pectoral.

The Ganoid fishes with persistent notochord, but with a *hyostylic* skull (that is, a skull with a separate suspensorium) consist of the sub-order *Chondrostei*, of which the existing representatives are the Sturgeons (*Acipenser*, *Scaphirhynchus*,

Polyodon), and the extinct the *Chondrosteidæ, Palæoniscidæ*, and (according to Traquair) *Platysomidæ*.

Their spinal column does not differ essentially from that of the Dipnoi. Segmentation is represented only as far as the neural and hæmal elements are concerned. All are eminently heterocercal. Ribs are present in most, but replaced by ligaments in *Polyodon.*

The primordial cranium of the Sturgeons consists of persistent cartilage without ossifications in its substance, but superficial bones are still more developed and specialised than in the Dipnoi; so it is, at least, in the true Sturgeons, but less so in *Polyodon* (Fig. 37). The upper and lateral parts of the skull are covered by well-developed *membrane bones*, which,

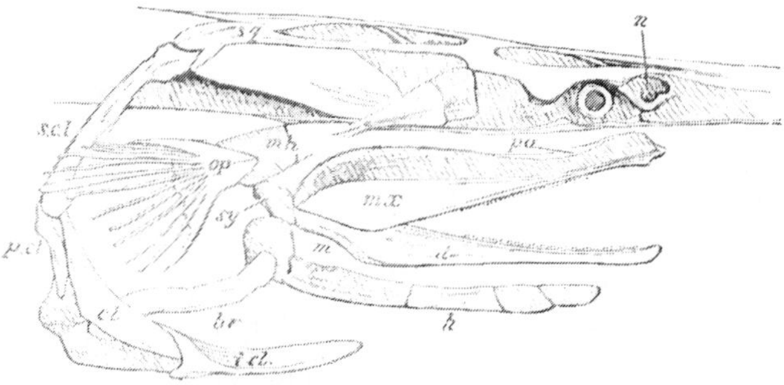

Fig. 37.—Skull of Polyodon (after Traquair).

n, Nasal cavity; *sq*, squamosal; *mh*, hyomandibular; *sy*, symplectic; *pa*, palato-pterygoid; *m*, Meckelian cartilage; *mx*, maxillary; *d*, dentary; *h*, hyoid; *op*, opercle; *br*, branchiostegal; *s.cl*, supra-clavicular; *p.cl*, post-clavicular; *cl*, clavicle; *i.cl*, infra-clavicular.

from this sub-order, upwards in the series, will be found to exist throughout the remaining forms of fishes. They are bones, the origin of which is not in cartilage but in membranous connective tissue. The lower surface of the skull is covered by an extremely large basal bone, which extends from the vomerine region on to the anterior part of the spinal column. The nasal excavation in the skull is rather lateral than inferior. The ethmoidal region is generally much produced, forming the base of the long projecting snout. The suspensorium is movably attached to the side of the skull, and consists of two pieces, a hyomandibular and a symplectic, which now appears for the first time as a separate piece, and to which the hyoid is attached. The palato-maxillary apparatus is more complex than in the Sharks and Dipnoi; a palato-pterygoid consists of two mesially-connected rami in *Polyodon*, and of a complex cartilaginous disk in *Acipenser*, being articulated in both to the Meckelian cartilage. In addition, the Sturgeons possess one or two pairs of osseous rods, which, in *Polyodon* at least, represent the maxillary, and therefore must be the representatives of the labial cartilages of the Sharks. The Meckelian cartilage is more or less covered by tegumentary bones.

In the gill-cover, besides the operculum, a sub- and inter-operculum may be distinguished in *Acipenser*.

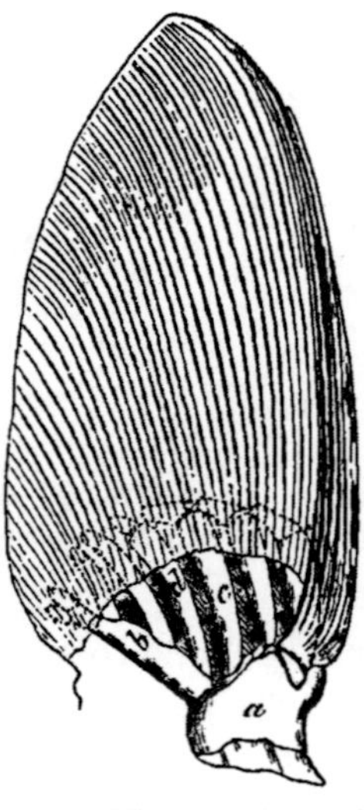

Fig. 38.—Fore-limb of Acipenser.

The hyoid consists of three pieces, of which the posterior bears a broad branchiostegal in *Polyodon*.

In the scapulary arch the primordial cartilaginous elements scarcely differ from those of the *Dipnoi*. The membrane-bones are much expanded, and offer a continuous series suspended from the skull. Their division in the median ventral line is complete.

The pectoral is supported by a cartilaginous framework (Fig. 38) similar to that of *Ceratodus*, but much more shortened and reduced in its periphery, the branches being absent altogether on one side of the axis. This modification of the fin is analogous to the heterocercal condition of the end of the spinous column. To the inner corner of a basal cartilage (*a*) a short axis (*b*) is joined, which on its outer side bears a few branches (*d*) only, the remaining branches (*e*) being fixed to the basal cartilage. The dermal fin-rays are opposed to the extremities of the branches, as in the *Dipnoi.*

The *pubic* consists of a paired cartilage, to which tarsal pieces supporting the fin-rays are attached.

The other living Ganoid fishes have the spinous column entirely or nearly entirely ossified, and have been comprised under the common name *Holostei.* However, they form three very distinct types; several attempts have been made to coordinate with them the fossil forms, but this task is beset with extreme difficulties, and its solution hitherto has not proved to be satisfactory.

The *Polypteroidei* have their spinous column formed by distinct osseous *amphicœlous* vertebræ, that is, vertebræ with concave anterior and posterior surfaces. It is nearly diphycercal; a slight degree of heterocercy obtains, inasmuch as the last vertebra is succeeded by a very thin cartilaginous filament which penetrates between the halves of one of the middle rays of the terminal fin. The rays above this cartilaginous filament are articulated to interneurals, those below lack interhæmals, and are attached either to the hæmals or vertebral centres. The neural arches, though ossified, do not coalesce with the centrum, and form one canal only, for the myelon. There are no intermediate elements between the neural spines. Interneurals developed, but simple, articulating with the dermoneurals. The abdominal vertebræ have parapophyses developed with epipleural spines. Only the caudal vertebræ have hæmal spines, which, like the interhæmals, agree in every essential respect with the opposite neurals. *Ribs* are inserted, not on the parapophyses, but on the centre, immediately below the parapophyses.

The *skull* of *Polypterus* (Fig. 39) shows a great advance

towards the Teleosteous type, the number of separable bones

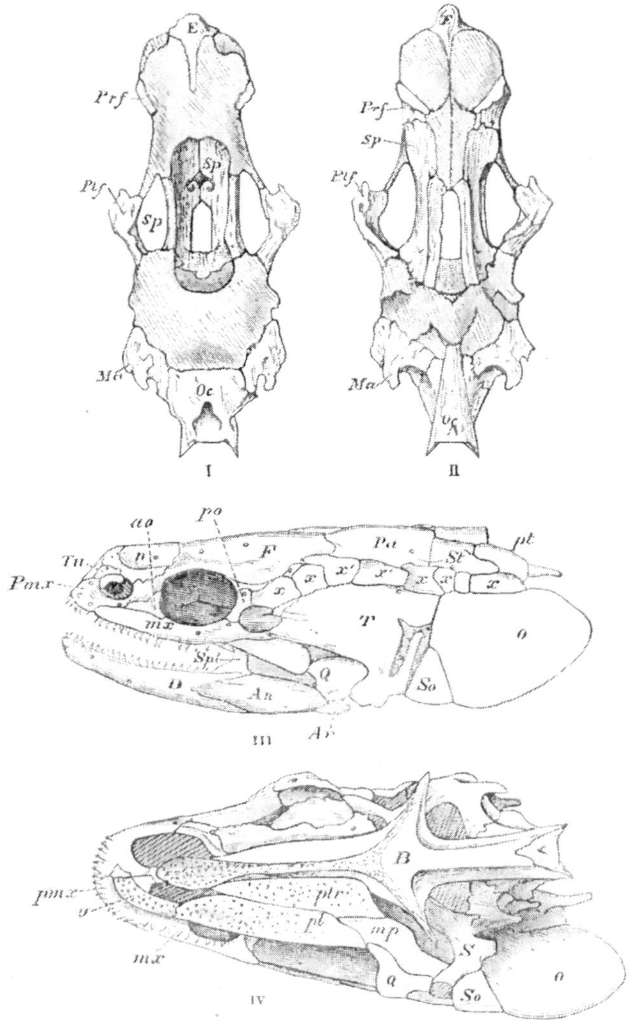

Fig. 39.—Skull of Polypterus. (After Traquair.)

Fig. I. Upper aspect of the Primordial Cranium, with the membrane-bones removed. Fig. II. Lower aspect of the same. Fig. III. Side view, with the membrane-bones. Fig. IV. Lower aspect of the Skull, part of the bones being removed on one side. The parts shaded with oblique lines are cartilage of the primordial skull.

An, Angular; *ao*, anteorbital; *Ar*, articulary; *B*, basal; *D*, dentary; *E*, ethmoid; *F*, frontal; *Ma*, mastoid; *Mp*, metapterygoid; *Mx*, maxillary; *N*, nasal; *O*, operculum; *Oc*, occipital; *Pa*, parietal; *Pl*, palatine; *Pmx*, præmaxillary; *po*, postorbital; *Prf*, prefrontal; *Pt*, post-temporals; *Ptf*, postfrontal; *Ptr*, pterygoid; *Q*, quadrate; *S*, suspensorium; *So*, suboperculum; *Sp*, sphenoid; *Spl*, splenial; *St*, supratemporals; *T*, tympanic lamina; *Tu*, turbinal; *v*, vomer; *x x*, small ossicles; *x′ x′*, spiraculars.

being greatly increased. They are arranged much in the same fashion as in Teleostei. But a great portion of the primordial cranium remains cartilaginous. The membrane-bones which cover the upper and lower surfaces of the brain-case are so much developed as to cause the underlying cartilage to disappear, so that a large vacuity or fontanelle exists in the substance of the upper as well as lower cartilaginous wall. Of ossifications belonging to the primordial skull must be noticed the single occipital with a mastoid on each side. They are separated by persistent cartilage from the sphenoids and postfrontals; the former, which are the largest ossification of the primordial cranium, enclose the anterior half of the brain cavity. Finally, the nasal portion contains a median ethmoid and a pair of præfrontal bones.

Only a very small portion of the bones described are visible externally, nearly the whole of the primordial cranium being covered by the membrane-bones. Of these are seen on the upper surface a pair of parietals, frontals, "nasals," and turbinals; on the lower surface a large cross-shaped basal, anteriorly bordered on each side by a pterygoid, parallel to a palatine which forms a suture with the double vomer. The suspensorium has in front a metapterygoid and quadrate bone, and an operculum and suboperculum are attached to it behind.

Præmaxillaries and maxillaries are now fully developed, but immovably attached to the skull. The lower jaw is ossified, and consists of an articulary, angular, dentary, and splenial. Of labial cartilages a rudiment at the angle of the mouth has remained persistent.

The side of the skull, in front of the operculum, is covered by a large irregularly-shaped bone (*T*) (corresponding to the "tympanic lamina" of *Ceratodus*, Fig. 35, *q*), held by some to be the præoperculum; along its upper circumference lies a series of small ossicles, of which two may be distinguished as spiraculars, as they form a valve for the protection of the spiracular orifice of these fishes. An infraorbital ring is represented by a præ- and post-orbital only.

Each *hyoid* consists of three pieces, none of which bear branchiostegals, the single median piece being osseous in front and cartilaginous behind. Four branchial arches are developed, the foremost consisting of three, the second and third of two, and the last of a single piece. There is no lower pharyngeal. Between the rami of the lower jaw the throat is protected by a pair of large osseous laminæ (*gular plates*), which have been considered to represent the urohyal of osseous fishes.

The scapulary arch is almost entirely formed by the well-developed membrane-bones, which in the ventral line are suturally united. The pectoral fin is supported by three bones, pro-, meso-, and metapterygium, of which the dilated middle one alone bears rays, and is excluded from the articulation with the shoulder-girdle.

The pubic consists of paired bone, to which tarsal bones supporting the fin-rays are attached.

In the *Lepidosteoidei* the vertebræ are completely ossified, and *opisthocœlous*, having a convexity in front and a concavity behind, as in some Amphibians. Though the end of the body externally appears nearly diphycercal, the termination of the vertebral column is, in fact, distinctly heterocercal (Fig. 40). Its extremity remains cartilaginous, is turned upwards, and lies immediately below the scutes which cover the upper margin of the caudal fin. It is preceded by a few rudimentary vertebræ which gradually pass into the fully developed normal vertebræ. The caudal fin is suspended from hæmapophyses only, and does not extend to the neural side of the vertebral

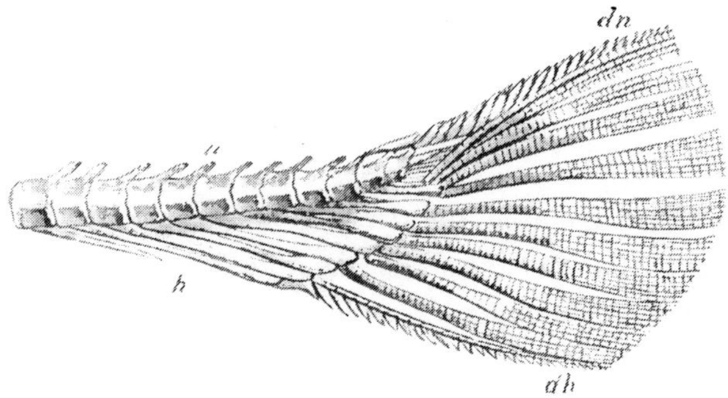

Fig. 40.—Heterocercal Tail of Lepidosteus.
n, Vertebral column ; *h*, hæmal spines ; *dn*, fulcra ; *dh*, lower fulcra.

column. The neural arches coalesce with the centrum ; inter-neurals simple. The abdominal vertebræ have parapophyses, to which the ribs are attached. Only the caudal vertebræ have hæmal spines.

In the skull of *Lepidosteus* the cartilage of the endo-cranium is still more replaced by ossifications than in *Polypterus;* those ossifications, moreover, being represented by a greater number of discrete bones; especially the membrane-bones are greatly multiplied: the occipital, for instance, consists of three pieces; the vomer is double as in *Polypterus;* the maxillary consists of a series of pieces firmly united by suture. The symplectic reaches the lower jaw, so that the articulary is provided with a double joint, viz. for the symplectic and quadrate; the component parts of the lower jaw are as numerous as in reptiles, a dentary, splenial, articulary, angular, supra-angular, and coronary being distinct. The sides of the head are covered with numerous bones, and a præoperculum is developed in front of the gill-cover which, again, consists of an operculum and suboperculum.

Each hyoid consists of three pieces, of which the middle is the longest, the upper bearing the largest of the three branchiostegals which *Lepidosteus* possesses; a long and large glossohyal is intercalated between the lower ends of the hyoids. There are five branchial arches, the hindmost of which is modified into a lower pharyngeal; upper pharyngeals are likewise present as in the majority of Teleosteous fishes. No gular plate.

Of the scapulary arch the two halves are separated by a suture in the median line; the membrane-bones are well developed, only a remnant of the primordial cartilage remaining; the supraclavicle is very similar to that of Teleosteous fishes, less so the post-temporal. The base to which the limb is attached is a single osseous plate, supporting on its posterior margin semi-ossified rods in small number, which bear the pectoral rays.

The pubic consists of paired bone, the anterior ends of which overlap each other, the extremity of the right pubis being dorsad to that of the left. The elements representing a tarsus are quite rudimentary and reduced in number (two or three).

The vertebral column of the *Amioidei* shows unmistakable characters of the Palæichthyic type. The arrangement of its component parts is extremely simple. The centra of the amphicœlous vertebræ are well ossified, but the neural and hæmal arches do not coalesce with the centra, from which they are separated by a thin layer of cartilage. Singularly, not every vertebra has apophyses: in the caudal portion of *Amia* the vertebræ are alternately provided with them and lack them. The heterocercal condition of the spinous column is well marked: as in the other Holostei the hindmost vertebræ are turned upwards, become smaller and smaller in size, and lose their neural arches, the hæmals remaining developed to the end. Finally, the column terminates in a thin cartilaginous band, which is received between the lateral halves of the fifth or sixth upper caudal ray. Interneurals and interhæmals simple. Only the abdominal vertebræ have parapophyses, with which the ribs are articulated.

The configuration of the skull, and the development and arrangement of its component parts, approaches so much the Teleosteous type that, perhaps, there are greater differences in skulls of truly Teleosteous fishes than between the skulls of *Amia* and many *Physostomi*. Externally the cranium is entirely ossified; and the remains of the cartilaginous primordial cranium (which, however, has no vacuity in its roof) can only be seen in a section, and are of much less extent than in many Physostomous fishes. The immovable intermaxillary, the double vomer, the plurality of ossifications representing the articulary, the double articulary cavity of the mandible for junction with the quadrate and symplectic bones, remind us still of similar conditions in the skull of *Lepidosteus*, but the mobility and formation of the maxillary, the arrangement of the gill-covers, the development of the opercles, the suspensorium, the palate, the insertion of a number of branchiostegals on the long middle hyoid piece, the composition of the branchial framework (with upper and lower pharyngeals), are as in the Teleosteous type. A gular plate replaces the urohyal.

The scapular arch is composed entirely of the membrane-bones found in the *Teleostei*, and the two sides are loosely united by ligament. The base to which the limb is attached is cartilaginous; short semi-ossified rods are arranged along its hinder margin and bear the pectoral rays.

The skeleton of the hind-limb agrees entirely with that of *Lepidosteus*.

[T. W. Bridge, The Cranial Osteology of *Amia calva*; in Journ. Anat. and Physiol. vol. xi.]

In the *Teleosteous* fishes the spinous column consists of completely ossified amphicœlous vertebræ; its termination is *homocercal*—that is, the caudal fin appears to be more or less symmetrical, the last vertebra occupying a central position in the base of the fin, and being coalesced with a flat osseous lamella, the *hypural* (Fig. 23, 70), on the hind margin of which the fin-rays are fixed. The hypural is but a union of modified hæmapophyses which are directed backwards, and the actual termination of the notochord is bent upwards, and lies along the upper edge of the hypural, hidden below the last rudimentary neural elements. In some Teleosteans, as the *Salmonidæ*, the last vertebræ are conspicuously bent upwards: in fact, strictly speaking, this homocercal condition is but one of the various degrees of heterocercy, different from that of many Ganoids in this respect only, that the caudal fin itself has assumed a higher degree of symmetry.

The neural and hæmal arches generally coalesce with the centrum, but there are many exceptions, inasmuch as some portion of the arches of a species, or all of them, show the original division.

The vertebræ are generally united with one another by zygapophyses, and frequently similar additional articulations exist at the lower parts of the centra. Parapophyses and ribs are very general, but the latter are inserted on the centra and the base of the processes, and never on their extremities. The point of insertion of the rib, more especially on the anterior vertebræ, may be still higher—viz. at the base of the neural arch, as in *Cotylis* and allied genera, and even on the top of the neurapophysis, as in *Batrachus*.

There is a great amount of variation as regards the degree in which the primordial cranium persists; it is always more or less replaced by bone; frequently it disappears entirely, but in some fishes, like the *Salmonidæ* or *Esocidæ*, the cartilage persists to the same or even to a greater extent than in the Ganoidei holostei. Added to the bones preformed in cartilage are a great number of *membrane-bones*. The different kinds of these membrane-bones occur with greater or less constancy throughout this sub-class; they often coalesce with, and are no more separable from, the neighbouring or underlying cartilage-bones. All these bones have been topographically enumerated in Chapter IV.

Many attempts have been made to classify the bones of the Teleosteous skull, according to their supposed relation to each other, or with the view to demonstrate the unity of plan on which the skull has been built; but in all either the one or the other of the following two principles has been followed:—

A. The "vertebral doctrine" starts from the undeniable fact that the skull is originally composed of several segments, each of which is merely the modification of a vertebra. The component parts of such a cranial segment are considered to be homologous to those of a vertebra. Three, four, or five cranial vertebræ have been distinguished, all the various bones of the fully-developed and ossified skull being referred, without distinction as to their origin, to one or the other of those vertebral segments. The idea of the typical unity of the osseous framework of Vertebrates has been worked out with the greatest originality and knowledge of detail, by *Owen*, who demonstrates that the fish-skull is composed of *four* vertebræ.

The bones of the fish-skull are, according to him, primarily divisible into those of the *neuroskeleton, splanchnoskeleton,* and *dermoskeleton.*

The bones of the *neuro-* or proper *endo-skeleton* are arranged in a series of four horizontally succeeding segments: the occipital, parietal, frontal, and nasal vertebræ; each segment consisting of an upper (neural) and a lower (hæmal) arch, with a common centre, and with diverging appendages.

The *neural* arches of the four vertebræ, in their succession from the occiput towards the snout, are:—

1. *Epencephalic arch,* composed of the occipitals.

2. *Mesencephalic arch*, composed of basisphenoid, alisphenoid, parietal, and mastoid.

3. *Prosencephalic arch*, composed of presphenoid, orbitosphenoid, frontal, and postfrontal.

4. *Rhinencephalic arch*, composed of vomer, prefrontal, and nasal.

The *hæmal* arches in the same order of succession are :—

1. *Scapular* or *scapulo-coracoid* arch, composed of suprascapula, scapula, and coracoid ; its appendage consists of the ulna, radius and carpal.

2. *Hyoid* or *stylo-hyoid* arch, composed of stylo-hyal, epihyal, ceratohyal, basihyal, glossohyal, and urohyal ; its appendage is the branchiostegals.

3. *Mandibular* or *tympano-mandibular* arch, composed of epi-, meso-, pre-, and hypo-tympanic, and the bones of the lower jaw ; its appendage consists of the præoperculum and the other opercles.

4. *Maxillary* or *palato-maxillary* arch, composed of palatine, maxillary, and premaxillary ; its appendage consists of the pterygoid and entopterygoid.

Parts of the *splanchnoskeleton* are held to be the ear-capsule or petrosal and the otolite, the eye-capsule or sclerotic, the nose-capsule or "ethmoid" and turbinal ; the branchial arches.

The bones of the *dermoskeleton* are the supratemporals, supraorbitals, suborbitals, and labials.

B. In the second method of classifying the bones of the skull prominence is given to the facts of their different origin as ascertained by a study of their development. The parts developed from the primordial skull, or the cartilaginous case protecting the nervous centre are distinguished from those which enclose and support the commencement of the alimentary canal and the respiratory apparatus, and which, consisting of several arches, are comprised under the common name of *visceral skeleton of the skull.* Further, a distinction is made between the bones preformed in cartilage and those originating in tegumentary or membranous tissue. It is admitted that the primordial cranium is a coalition of several segments, the number of which is determined by that of the visceral arches, these representing the hæmal arches of the vertebral column; but the membrane-bones are excluded from a consideration of the vertebral division of the primordial skull, as elements originally independent of it, although these additions have entered into special relations to the cartilage-bones.

With these views the bones of the Teleosteous skull are classified thus:—

1. *Cartilage-bones of the primordial skull.*—The *basi-occipital* (5 in Figs. 23-26) has retained the form of a vertebral centrum; it is generally concave behind, the concavity containing remains of the notochord; rarely a rounded articulary head of the first vertebra fits into it, as in *Symbranchus,* and still more rarely it is provided with such an articulary head (*Fistularia*); frequently it shows two excavations on its inner surface for the reception of the *saccus vestibuli.* The *exoccipitals* (10) are situated on the side of the basi-occipital, and contribute the greater portion of the periphery of the foramen magnum; frequently they articulate with the first vertebra, or meet in the upper median line, so as to exclude the supraoccipital from the foramen magnum. The *supraoccipital* (8) is intercalated between the exoccipitals, and forms a most prominent part by the median crest, which sometimes extends far forwards on the upper side of the skull, and offers attachment to the dorsal portion of the large lateral muscle of the trunk. When the interior portions of this bone remain cartilaginous, some part of the semicircular canals may be lodged in it.

The region of the skull which succeeds the bones described encloses at least the greater portion of the labyrinth, and its component parts have been named with reference to it by some anatomists.[1] The *alisphenoids* (11) (*Prooticum*) form sutures posteriorly with the basi- and ex-occipitals, and meet each other in the median line at the bottom of the cerebral cavity; they contribute to the formation of a hollow in which the hypophysis cerebri and the saccus vasculosus are received; in conjunction with the exoccipital it forms another hollow for the reception of the vestibulum; generally it is perforated by the Trigeminal and Facial nerves. The *paroccipitals* (9) (*Epioticum*) lodge a portion of the posterior vertical semicircular canal, and form a projection of the skull on each side of the occipital crest, to which a terminal branch of the scapular arch is attached. The *Mastoid* (12 + 13) (*Opisthoticum*) occupies the postero-external projection of the head; it encloses a part of the external semicircular canal; is generally coalesced with a membrane-bone, the superficial *squamosal*, which emits a process for the suspension of the scapular arch, and is frequently, as in the Perch, divided into two separate bones.

The anterior portion of the skull varies greatly as regards form, which is chiefly dependent on the extent of the cerebral cavity; if the latter is advanced far forwards, the lateral walls of the primordial cranium are protected by more developed ossifications than if the cerebral cavity is shortened by the presence of a wide and deep orbit. In the latter case parts which normally form the side of the skull are situated in front of the brain-case, between it and the orbit, and generally reduced in extent, often replaced by membranes; especially the interorbital septum may be reduced to membrane. The most constant ossifications of this part of the skull are the *orbitosphenoids* (14), which join the upper anterior margin of the alisphenoids. They vary much with regard to their development—they are small in Gadoids; larger in the Perch, Pike, Salmonoids, Macrodon, and the Clupeoids; and

[1] As first proposed by Huxley.

very large in Cyprinoids and Siluroids, in which they contribute to the formation of the side of the brain-case. The single Y-shaped *Sphenoideum anterius* (15) is as frequently absent as present; it forms the anterior margin of the fossa for the hypophysis. Finally, the *post-frontal* (4) belongs also to this group of cartilage-bones.

The centre of the foremost part of the skull is occupied by the *ethmoid* (3), which shows great variations as regards its extent and the degree of ossification; it may extend backwards into the interorbital septum, and reach the orbitosphenoids, or may be confined to the extremity of the skull; it may remain entirely cartilaginous, or ossify into a lamina which separates the two orbits and encloses an anterior prolongation of the brain-case, along which the olfactory nerves pass: modifications occurring again in higher vertebrates. A paired ossification attached to the fore-part of the ethmoid is the *prefrontals* (2), which form the base of the nasal fossa.

2. *Membrane-bones attached to the primordial skull.*—To this group belong the *parietals* (7) and *frontals* (1). The *squamosal* (12) has been mentioned above in connection with the mastoid. The *supraorbital* is always small, and frequently absent. The lower surface of the skull is protected by the *basisphenoid* (parasphenoid) (6) and the *vomer* (16), both of which, especially the latter, may be armed with teeth.[1]

3. *Cartilage bones of the alimentary portion of the visceral skeleton of the skull.*—The suspensorium consists of three cartilage-bones, and affords a base for the opercular apparatus as well as a point of attachment to the hyoid, whilst in front it is connected with the palato-pterygo-palatine arch. They are the *hyomandibular* (23), *symplectie* (31), and *quadrate*

[1] Stannius (pp. 60, 65) doubts the pure origin of these two bones from membranous tissue, and is inclined to consider them as "the extreme end of the abortive axial system."

(26), connected by means of the *metapterygoid* (27) with the *ecto-* (24) and *ento-pterygoid* (25), the foremost bone of the arch being the *palatine* (22). All these bones have been sufficiently described above (p. 55), and it remains only to be mentioned that the bones of the palatine arch are but rarely absent, as for instance in *Muraenophis;* and that the symplectic does not extend to the articulary of the mandible, as in *Amia* and *Lepidosteus,* though its suspensory relation to the Meckelian cartilage is still indicated by a ligament which connects the two pieces. Of the mandibulary bones the *articulary* (35) is distinctly part of Meckel's cartilage. Frequently another portion of cartilage below the articulary remains persistent, or is replaced by a separate membrane-bone, the angular.

4. *Membrane-bones of the alimentary portion of the visceral skeleton of the skull.*—The suspensorium has one tegumentary bone attached to it, viz. the *praeoperculum* (30); it is but rarely absent, for instance in *Muraenophis.* The *premaxillary* (17) and *maxillary* (18) of the Teleostei appear to be also membrane-bones, although they are clearly analogous to the upper labial cartilages of the Sharks. The premaxillaries sometimes coalesce into a single piece (as in *Diodon, Mormyrus*), or they are firmly united with the maxillaries (as in all *Gymnodonts, Serrasalmo,* etc.) The relative position and connection of these two bones differs much, and is a valuable character in the discrimination of the various families. In some, the front margin of the jaw is formed by the premaxillary only, the two bones having a parallel position, as it has been described in the Perch (p. 53); in others, the premaxillary is shortened, allowing the maxillary to enter, and to complete, the margin of the upper jaw; and finally, in many no part of the maxillary is situated behind the premaxillary, but the entire bone is attached to the end of the premaxillary, forming its continuation. In

the last case the maxillary may be quite abortive. The mobility of the upper jaw is greatest in those fishes in which the premaxillary alone forms its margin. The form of the premaxillary is subject to great variation : the beak of *Belone*, *Xiphias* is formed by the prolonged and coalesced premaxillaries. The maxillary consists sometimes of one piece, sometimes of two or three. The principal membrane-bone of the mandible is the *dentary* (34), to which is added the *angular* (36) and rarely a smaller one, the *splenial* or *os opercularе*, which is situated at the inside of the articulary.

5. *Cartilage-bones of the respiratory portion of the visceral skeleton of the skull.*—With few exceptions all the ossifications of the hyoid and branchial arches, as described above (p. 58), belong to this group.

6. *Membrane-bones of the respiratory portion of the visceral skeleton of the skull.*—They are the following: the opercular pieces, viz. *operculum* (28), *suboperculum* (32), and *interoperculum* (33). The last of these is the least constant; it may be entirely absent, and represented by a ligament extending from the mandible to the hyoid. The *urohyal* (42) which separates the musculi sternohyoidei, and serves for an increased surface of their insertion ; and finally the *branchiostegals* (43), which vary greatly in number, but are always fixed to the cerato- and epi-hyals.

7. *Dermal bones of the skull.*—To this category are referred some bones which are ossifications of, and belong to, the cutis. They are the *turbinals* (20), the *suborbitals* (19), and the *supratemporals*. They vary much with regard to the degree in which they are developed, and are rarely entirely absent. Nearly always they are wholly or partly transformed into tubes or hollows, in which the muciferous canals with their numerous nerves are lodged. Those in the temporal and scapulary regions are not always developed; on the other hand, the series of those ossicles may be continued on

to the trunk, accompanying the lateral line. In many fishes those of the infraorbital ring are much dilated, protecting the entire space between the orbit and the rim of the præoperculum; in others, especially those which have the angle of the præoperculum armed with a powerful spine, the infraorbital ring emits a process towards the spine, which thus serves as a stay or support of this weapon (*Scorpænidæ, Cottidæ*).

The *pectoral* arch of the Teleosteous fishes exhibits but a remnant of a primordial cartilage, which is replaced by two ossifications,[1] the *coracoid* (51) and *scapula* (52); they offer posteriorly attachment to two series of short rods, of which the proximal are nearly always ossified, whilst the distal frequently remain small cartilaginous nodules hidden in the base of the pectoral rays. The bones, by which this portion is connected with the skull, are membrane-bones, viz. the *clavicle* (49), with the *postclavicle* (49 + 50), the *supraclavicle* (47), and *post-temporal* (46). The order of their arrangement in the Perch has been described above (p. 59). However, many Teleosteous fish lack pectoral fins, and in them the pectoral arch is frequently more or less reduced or rudimentary, as in many species of *Murænidæ*. In others the membrane-bones are exceedingly strong, contributing to the outer protective armour of the fish, and then the clavicles are generally suturally connected in the median line. The postclavicula and the supraclavicula may be absent. Only exceptionally the shoulder-girdle is not suspended from the skull, but from the anterior portion of the spinous column (*Symbranchidæ, Murænidæ, Notacanthidæ*). The number of basal elements of each of the two series never exceeds five, but may be less; and the distal series is absent in Siluroids.

The *pubic* bones of the Teleosteous fishes undergo many modifications of form in the various families, but they are essentially of the same simple type as in the Perch.

1 Parker's nomenclature is adopted here.

CHAPTER V.

MYOLOGY.

In the lowest vertebrate, *Branchiostoma*, the whole of the *muscular* mass is arranged in a longitudinal band running along each side of the body; it is vertically divided into a number of flakes or segments (*myocommas*) by aponeurotic septa, which serve as the surfaces of insertion to the muscular fibres. But this muscular band has no connection with the notochord except in its foremost portion, where some relation has been formed to the visceral skeleton. A very thin muscular layer covers the abdomen.

Also in the *Cyclostomes* the greatest portion of the muscular system is without direct relation to the skeleton, and, again, it is only on the skull and visceral skeleton where distinct muscles have been differentiated for special functions.

To the development of the skeleton in the more highly organised fishes corresponds a similar development of the muscles; and the maxillary and branchial apparatus, the pectoral and ventral fins, the vertical fins, and especially the caudal, possess a separate system of muscles. But the most noteworthy is the muscle covering the sides of the trunk and tail (already noticed in *Branchiostoma*), which Cuvier described as the "great lateral muscle," and which, in the higher fishes, is a compound of many smaller segments, corresponding in number with the vertebræ. Each lateral muscle is divided by a median longitudinal groove into a dorsal and ventral half; the depression in its middle is filled by an embryonal muscular substance which contains a large quantity of fat and blood-vessels, and therefore differs from ordinary muscle by its softer consistency, and by its colour which is reddish or grayish. Superficially the lateral muscle appears crossed by a number of white parallel tendinous zigzag stripes, forming generally three angles, of which the upper and lower point backwards, the middle one forwards. These are the outer edges of the aponeurotic septa between the myocommas. Each septum is attached to the middle and the apophyses of a vertebra, and, in the abdominal region, to its rib; frequently the septa receive additional support by the existence of epipleural spines. The fibres of each myocomma run straight and nearly horizontally from one septum to the next; they are grouped so as to form semiconical masses, of which the upper and lower have their apices turned backwards, whilst the middle cone, formed by the contiguous parts of the preceding, has its apex directed forward; this fits into the interspace between the antecedent upper and lower cones, the apices of which reciprocally enter the depressions in the succeeding segment, whereby all the segments are firmly locked together (*Owen*).

In connection with the muscles reference has to be made to the *Electric organs* with which certain fishes are provided, as it is more than probable, not only from the examination of peculiar muscular organs occurring in the Rays, *Mormyrus*, and *Gymnarchus* (the function of which is still conjectural), but especially from the researches into the development of the electric organ of *Torpedo*, that the electric organs have been developed out of muscular substance. The fishes possessing fully developed electric organs, with the power of accumulating electric force and communicating it in the form of shocks to other animals, are the electric Rays (*Torpedinidæ*), the electric Sheath-fish of tropical Africa (*Malapterurus*), and the electric Eel of tropical America

(*Gymnotus*). The structure and arrangement of the electric organ is very different in these fishes, and will be subsequently described in the special account of the several species.

The phenomena attending the exercise of this extraordinary faculty also closely resemble muscular action. The time and strength of the discharge are entirely under the control of the fish. The power is exhausted after some time, and it needs repose and nourishment to restore it. If the electric nerves are cut and divided from the brain the cerebral action is interrupted, and no irritant to the body has any effect to excite electric discharge; but if their ends be irritated the discharge takes place, just as a muscle is excited to contraction under similar circumstances. And, singularly enough, the application of strychnine causes simultaneously a tetanic state of the muscles and a rapid succession of involuntary electric discharges. The strength of the discharges depends entirely on the size, health, and energy of the fish: an observation entirely agreeing with that made on the efficacy of snake-poison. Like this latter, the property of the electric force serves two ends in the economy of the animals which are endowed with it; it is essential and necessary to them for overpowering, stunning, or killing the creatures on which they feed, whilst incidentally they use it as the means of defending themselves from their enemies.

CHAPTER VI.

NEUROLOGY.

THE most simple condition of the nervous central organ known in Vertebrates is found in *Branchiostoma*. In this fish the spinal chord tapers at both ends, an anterior cerebral swelling, or anything approaching a brain, being absent. It is band-like along its middle third, and groups of darker cells mark the origins of the fifty or sixty pairs of nerves which accompany the intermuscular septa, and divide into a dorsal and ventral branch, as in other fishes. The two anterior pairs pass to the membranous parts above the mouth, and supply with nerve filaments a ciliated depression near the extremity of the fish, which is considered to be an olfactory organ, and two pigment spots, the rudiments of eyes. An auditory organ is absent.

The *spinal chord* of the *Cyclostomes* is flattened in its whole extent, band-like, and elastic; also in *Chimæra* it is elastic, but flattened in its posterior portion only. In all other fishes it is cylindrical, non-ductile, and generally extending along the whole length of the spinal canal. The Plectognaths offer a singular exception in this respect that the spinal chord is much shortened, the posterior portion of the canal being occupied by a long cauda equina; this shortening of the spinal chord has become extreme in the Sun-fish (*Orthagoriscus*), in which it has shrunk into a short and conical appendage of the brain. Also in the Devil-fish (*Lophius*) a long cauda equina partly conceals the chord which terminates on the level of about the twelfth vertebra.

The *brain* of fishes is relatively small; in the Burbot (*Lota*) it has been estimated to be $\frac{1}{720}$th part of the weight of the entire fish, in the Pike the $\frac{1}{1305}$th part, and in the large Sharks it is relatively still smaller. It never fills the entire cavity of the cranium; between the dura mater which adheres to the inner surface of the cranial cavity, and the arachnoidea which envelops the brain, a more or less considerable space remains, which is filled with a soft gelatinous mass generally containing a large quantity of fat. It has been observed that this space is much less in young specimens than in adult, which proves that the brain of fishes does not grow in the same proportion as the rest of the body; and, indeed, its size is nearly the same in individuals of which one is double the bulk of the other.

The brain of *Osseous fishes* (Fig. 41) viewed from above shows three protuberances, respectively termed *prosencephalon, mesencephalon*, and *metencephalon*, the two anterior of which are paired, the hindmost being single. The foremost pair are the *hemispheres*, which are solid in their interior, and provided with two swellings in front, the *olfactory lobes*. The second pair are the *optic lobes*, which generally are larger than the hemispheres, and succeeded by the third single portion, *the*

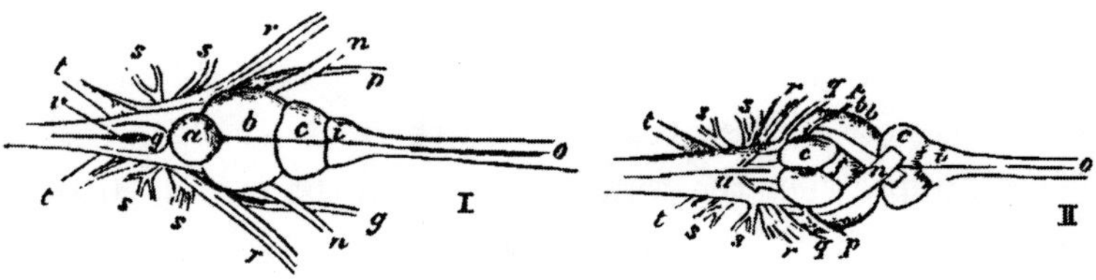

Fig. 41.—Brain of Perch.

I. Upper aspect. II. Lower aspect.

a, cerebellum; *b*, optic lobes; *c*, hemispheres; *e*, lobi inferiores; *f*, hypophysis; *g*, lobi posteriores; *i*, Olfactory lobes; *n*, *N.* opticus; *o*, *N.* olfactorius; *p*, *N.* oculo-motorius; *q*, *N.* trochlearis; *r*, *N.* trigeminus; *s*, *N.* acusticus; *t*, *N.* vagus; *u*, *N.* abducens; *v*, Fourth ventricle.

cerebellum. In the fresh state the hemispheres are of a grayish colour, and often show some shallow depressions on their surface; a narrow commissure of white colour connects them with each other. The optic lobes possess a cavity (*ventriculus lobi optici*), at the bottom of which some protuberances of variable development represent the corpora quadrigemina of higher animals. On the lower surface of the base of the optic lobes, behind the *crura cerebri*, two swellings are observed, the *lobi inferiores*, which slightly diverge in front for the passage of the *infundibulum*, from which a generally large *hypophysis* or *pituitary gland* is suspended. The relative size of the cerebellum varies greatly in the different osseous fishes: in the Tunny and Silurus it is so large as nearly to cover the optic lobes; sometimes distinct transverse grooves and a median longitudinal groove are visible. The cerebellum possesses in its interior a cavity which communicates with the anterior part of the fourth ventricle. The *medulla oblongata* is broader than the spinal chord, and contains the *fourth ventricle*, which forms the continuation of the central canal of the spinal chord. In most fishes a perfect roof is formed over the fourth ventricle by two longitudinal pads, which meet each other in the median line (*lobi posteriores*), and but rarely it remains open along its upper surface.

The brain of *Ganoid fishes* shows great similarity to that of the Teleostei; however, there is considerable diversity of the arrangement of its various portions in the different types. In the Sturgeons and *Polypterus* (Fig. 42) the hemispheres are more or less remote from the mesencephalon, so that in an upper view the crura cerebri, with the intermediate entrance into the third ventricle (*fissura cerebri magna*), may be seen. A vascular membranous sac, containing lymphatic fluid (*epiphysis*), takes its origin from the third ventricle, its base being expanded over the anterior interspace of the optic lobes, and the apex being fixed to the cartilaginous roof of the cranium. This structure is not peculiar to the Ganoids, but found in various stages of development in Teleosteans, marking, when present, the boundary between prosencephalon and mesencephalon. The lobi optici are essentially as in Teleosteans. The cerebellum penetrates into the ventriculus lobi optici, and extends thence into the open sinus rhomboidalis. At its upper surface it is crossed by a commissure formed by the *corpora restiformia* of the medulla.

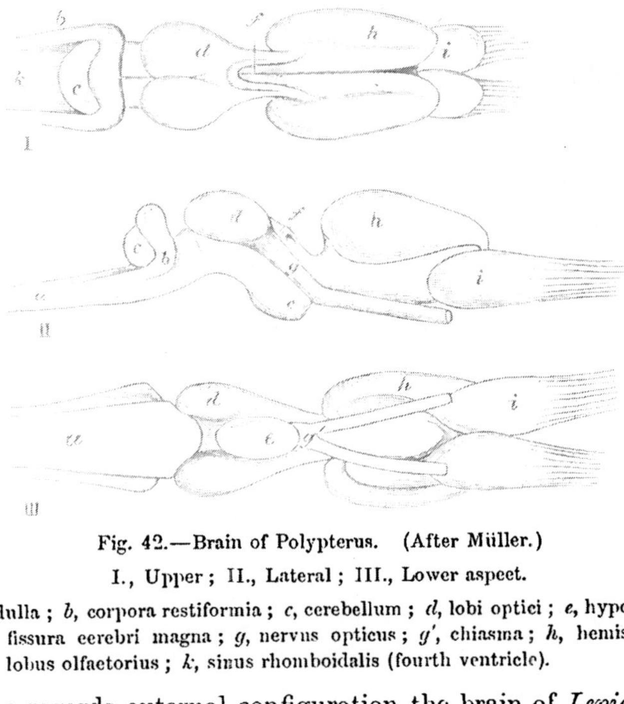

I

II

III

Fig. 42.—Brain of Polypterus. (After Müller.)

I., Upper; II., Lateral; III., Lower aspect.

a, Medulla; *b*, corpora restiformia; *c*, cerebellum; *d*, lobi optici; *e*, hypophysis; *f*, fissura cerebri magna; *g*, nervus opticus; *g'*, chiasma; *h*, hemispheres; *i*, lobus olfactorius; *k*, sinus rhomboidalis (fourth ventricle).

As regards external configuration, the brain of *Lepidosteus* and *Amia* approach still more the Teleosteous type. The prosencephalon, mesencephalon, and metencephalon are contiguous, and the cerebellum lacks the prominent transverse commissure at its upper surface. The sinus rhomboidalis is open.

The brain of the *Dipnoi* shows characters reminding us of that of the Ganoids as well as the Chondropterygians, *Ceratodus* agreeing with *Protopterus* in this respect, as in most other points of its organisation. The hemispheres form the largest part of the brain; they are coalescent, as in Sharks, but possess two lateral ventricles, the separation being externally indicated by a shallow median groove on the upper surface. The olfactory lobes take their origin from the upper anterior end of the hemispheres. Epiphysis and hypophysis well developed. The lobi optici are very small, and remote from the prosencephalon, their division into the lateral halves being indicated by a median groove only. The cerebellum is very small, overlying the front part of the sinus rhomboidalis.

The brain of *Chondropterygians* (Fig. 43) is more developed than that of all other fishes, and distinguished by well-marked characters. These are, first, the prolongation of the olfactory lobes into more or less long pedicles, which dilate into great ganglionic masses, where they come into contact with the olfactory sacs; secondly, the space which generally intervenes between prosencephalon and mesencephalon, as in some Ganoids; thirdly, the large development of the metencephalon.

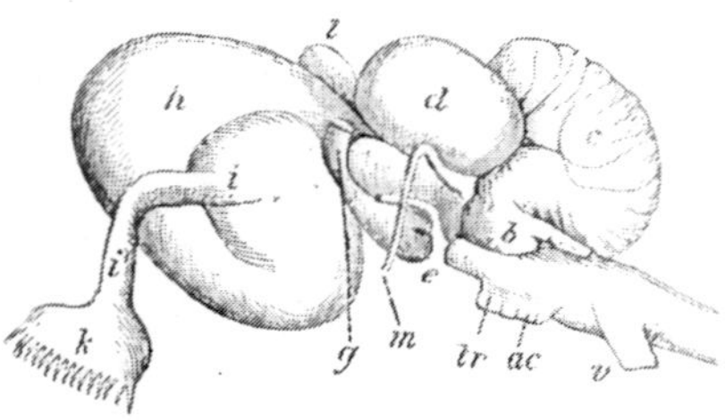

Fig. 43.—Brain of Carcharias. (After Owen.)

ac, Nerv. acusticus; *b*, corpus restiforme; *c*, cerebellum; *d*, lobus opticus; *e*, hypophysis; *g*, nervus opticus; *h*, hemisphere; *i*, lobus olfactorius; *i'*, olfactory pedicle; *k*, nerv. olfactorius; *l*, epiphysis; *m*, nerv. oculo-motorius; *tr*, nerv. trigeminus; *v*, nerv. vagus.

The hemispheres are generally large, coalescent, but with a median, longitudinal, dividing groove. Frequently their surface shows traces of gyrations, and when they are provided with lateral ventricles, tubercles representing *corpora striata* may be observed. The olfactory pedicles take their origin from the side of the hemispheres, and are frequently hollow, and if so, their cavity communicates with the ventricle of the hemisphere. The optic lobes are generally smaller than the hemispheres, coalescent, and provided with an upper median groove like the prosencephalon. At their base a pair of lobi inferiores are constant, with the hypophysis and *saccus vasculosus* (a conglomeration of vascular loops without medullary substance) between them.

The cerebellum is very large, overlying a portion of the optic lobes and of the sinus rhomboidalis, and is frequently transversely grooved. The side-walls of the fourth ventricle, which are formed by the corpora restiformia, are singularly folded, and appear as two pads, one on each side of the cerebellum (*lobi posteriores* s. *lobi nervi trigemini*).

The brain of the *Cyclostomes* (Figs. 44, 45) represents a type different from that of other fishes, showing at its upper surface

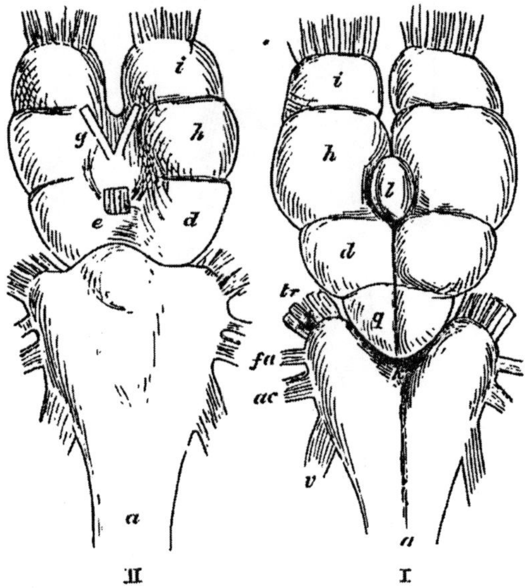

Fig. 44.—Brain of Bdellostoma. (Enlarged, after Müller.)
I., Upper; II., Lower aspect. Letters as in Fig. 45.

three pairs of protuberances in front of the cerebellum; they are all solid. Their homologies are not yet satisfactorily determined, parts of the Myxinoid brain having received by the same observers determinations very different from those given to the corresponding parts of the brain of the Lampreys. The

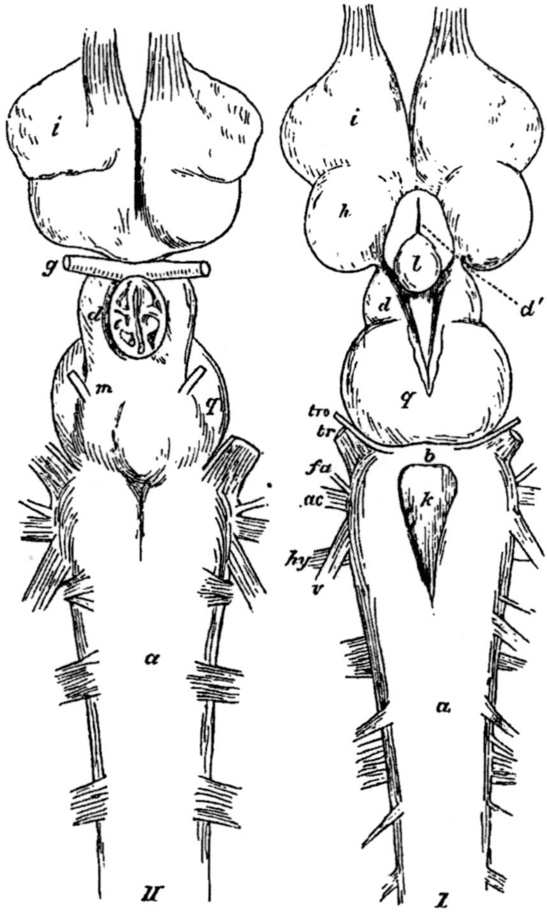

Fig. 45.—Brain of Petromyzon. (Enlarged, after Müller.)
I., Upper; II., Lower aspect.

a, Medulla oblongata; *ac*, nerv. acusticus; *b*, corpus restiforme or rudimentary cerebellum; *d*, lobus ventriculi tertii; *d'*, entrance into the third ventricle; *c*, hypophysis; *fa*, nerv. facialis; *g*, nerv. opticus; *h*, hemisphere; *hy*, nerv. hypoglossus (so named by Müller); *i*, lobus olfactorius; *k*, sinus rhomboidalis; *l*, epiphysis; *m*, nerv. oculo-motorius; *q*, corpora quadrigemina; *tr*, nerv. trigeminus; *tro*, nerv. trochlearis; *v*, nerv. vagus.

foremost pair are the large olfactory tubercles, which are exceedingly large in *Petromyzon*. They are followed by the hemispheres, with a single body wedged in between their posterior half; in *Petromyzon*, at least, the vascular tissue leading to an epiphysis seems to be connected with this body. Then follows the lobus ventriculi tertii, distinctly paired in Myxinoids, less so in *Petromyzon*. The last pair are the *corpora quadrigemina*. According to this interpretation the cerebellum would be absent in Myxinoids, and represented in *Petromyzon* by a narrow commissure only (Fig. 45, *b*), stretching over the foremost part of the sinus rhomboidalis. In the Myxinoids the *medulla oblongata* ends in two divergent swellings, free and obtuse at their extremity, from which most of the cerebral nerves take their origin.

The *Nerves* which supply the organs of the head are either merely continuations or diverticula of the brain-substance, or proper nerves taking their origin from the brain, or receiving their constituent parts from the foremost part of the spinal chord. The number of these spino-cerebral nerves is always less than in the higher vertebrates, and their arrangement varies considerably.

A. *Nerves which are diverticula of the brain* (Figs. 41-45).

The *olfactory* nerves (*first pair*) always retain their intimate relation to the hemispheres, the ventricles of which are not rarely continued into the tubercle or even pedicle of the nerves. The different position of the olfactory tubercle has been already described as characteristic of some of the orders of fishes. In those fishes in which the tubercle is remote from the brain, the nerve which has entered the tubercle as a single stem leaves it split up into several or numerous branches, which are distributed in the nasal organ. In the other fishes it breaks up into branchlets spread into a fan-like expansion at the point, where it enters the nasal cavity. The nerve always passes out of the skull through the ethmoid.

The *optic* nerves (*second pair*) vary in size, their strength corresponding to the size of the eye; they take their origin from the *lobi optici*, the development of which again is proportionate to that of the nerves. The mutual relation of the two nerves immediately after their origin is very characteristic of the sub-classes of fishes. In the *Cyclostomes* they have no further connection with each other, each going to the eye of its own side.[1] In the *Teleostei* they simply cross each other (*decussate*), so that the one starting from the right half of the brain goes to the left eye and *vice versa*. Finally, in *Palæichthyes* the two nerves are fused together, immediately after their origin, into a *chiasma*. The nerve is cylindrical for some portion of its course, but in most fishes gradually changes this form into that of a plaited band, which is capable of separation and expansion. It enters the bulbus generally behind and above its axis. The foramen through which it leaves the skull of Teleostei is generally in a membranous portion of its anterior wall, or, where ossification has taken place, in the orbito-sphenoid.

B. *Nerves proper taking their origin from the brain* (Figs. 41-45).

The *Nervus oculorum motorius* (*third pair*) takes its origin from the Pedunculus cerebri, close behind the lobi inferiores; it escapes through the orbito-sphenoid, or the membrane replacing it, and is distributed to the musculi rectus superior, rectus internus, obliquus inferior, and rectus inferior. Its size corresponds to the development of the muscles of the eye. Consequently it is absent in the blind *Amblyopsis*, and the Myxinoids. In *Lepidosiren* the nerves supplying the muscles of the eye have no independent origin, but are part

[1] According to *Langerhans* "Untersuchungen über Petromyzon planeri" (Freiburg, 1873) an optic chiasma exists in that species.

of the ophthalmic division of the Trigeminus. In *Petromyzon* these muscles are supplied partly from the Trigeminus, partly by a nerve representing the Oculo-motor and Trochlearis, which are fused into a common trunk.

The *Nervus trochlearis* (*fourth pair*), if present with an independent origin, is always thin, taking its origin on the upper surface of the brain from the groove between lobus opticus and cerebellum; it goes to the Musculus obliquus superior of the eye.

C. *Nerves taking their origin from the Medulla oblongata* (Figs. 41-45).

The *Nervus abducens* (*sixth pair*) issues on the lower surface of the brain, taking its origin from the anterior pyramids of the Medulla oblongata, and supplies the Musculus rectus externus of the eye, and the muscle of the nictitating membrane of Sharks.

The *Nervus trigeminus* (*fifth pair*) and the *Nervus facialis* (*seventh pair*) have their origins close together, and enter into intimate connection with each other. In the Chondropterygians and most Teleostei the number of their roots is four, in the Sturgeons five, and in a few Teleostei three. When there are four, the first issues immediately below the cerebellum from the side of the Medulla oblongata; it contains motory and sensory elements for the maxillary and suspensorial muscles, and belongs exclusively to the trigeminal nerve. The second root, which generally becomes free a little above the first, supplies especially the elements for the Ramus palatinus, which sometimes unites with parts of the Trigeminal, sometimes with the Facial nerve. The third root, if present, is very small, and issues immediately in front of the acustic nerve, and supplies part of the motor elements of the facial nerve. The fourth root is much stronger, sometimes double, and its elements pass again partly into the Trigeminal, partly into the Facial nerve. On the passage of these stems through the skull (through a foramen or foramina in the alisphenoid) they form a ganglionic plexus, in which the palatine ramus and the first stem of the Trigeminus generally possess discrete ganglia. The branches which issue from the plexus and belong exclusively to the Trigeminus, supply the organs and integuments of the frontal, ophthalmic, and nasal regions, and the upper and lower jaws with their soft parts. The Facial nerve supplies the muscles of the gill-cover and suspensorium, and emits a strong branch accompanying the Meckelian cartilage to the symphysis, and another for the hyoid apparatus.

The *Nervus acusticus* (*eighth pair*) is strong, and takes its origin immediately behind, and in contact with, the last root of the seventh pair.

The *Nervus glosso-pharyngeus* (*ninth pair*) [1] takes its origin between the roots of the eighth and tenth nerves, and issues in Teleostei from the cranial cavity by a foramen of the exoccipital. In the Cyclostomes and Lepidosiren it is part of the Nervus vagus. It is distributed in the pharyngeal and lingual regions, one branch supplying the first branchial arch. After having left the cranial cavity it swells into a ganglion, which in Teleostei is always in communication with the sympathic nerve.

The *Nervus vagus* or *pneumogastricus* (*tenth pair*) rises in all Teleostei and Palæichthyes with two discrete strong roots: the first constantly from the swellings of the corpora restiformia, be they thinner or thicker and overlying the sinus rhomboidalis, or be they developed into lateral plaited pads, as in Acipenser and Chondropterygians. The second much thicker root rises from the lower tracts of the medulla oblongata. Both stems leave the cranial cavity by a common fora-

[1] This nerve is not shown in the figure of the brain of the Perch (Fig. 41), as reproduced above from Cuvier.

men, situated in Teleosteous fishes in the exoccipital; and form ganglionic swellings, of which those of the lower stem are the more conspicuous. The lower stem has mixed elements, motory as well as sensory, and is distributed to the muscles of the branchial arches and pharynx, the œsophagus and stomach; it sends filaments to the heart and to the air-bladder where it exists. The first (upper) stem forms the *Nervus lateralis.* This nerve, which accompanies the lateral mucous system of the trunk and tail, is either a single longitudinal stem, gradually becoming thinner behind, running superficially below the skin (Salmonidæ, Cyclopterus), or deeply between the muscles (Sharks, Chimæra), or divided into two parallel branches (most Teleostei): thus in the Perch there are two branches on each side, the superficial of which supplies the lateral line, whilst the deep-seated branch communicates with the spinal nerves and supplies the septa between the myocommas and the skin. In fishes which lack the lateral muciferous system and possess hard integuments, as the Ostracions, the lateral nerve is more or less rudimentary. It is entirely absent in Myxinoids, but the gastric branches of the Vagus are continued, united as a single nerve, along the intestine to the anus.

No fish possesses a *Nervus accessorius.* Also a separate *Nervus hypoglossus* (*twelfth pair*)[1] is absent, but elements from the first spinal nerve are distributed in the area normally supplied by this nerve in higher vertebrates.

The number of *Spinal* nerves corresponds to that of the vertebræ, through or between which they pass out. Each nerve has two roots, an anterior and posterior, the former of which has no ganglion, and exclusively contains motor elements. The posterior or dorsal has a ganglionic enlarge-

[1] Müller considers a nerve rising jointly with the Vagus in Petromyzon to be this nerve (Fig. 45, *hy*).

ment, and contains sensory elements only. After leaving the vertebral canal each spinal nerve usually divides into a dorsal and ventral branch. The Gadoids show that peculiarity that each of the posterior roots of some or many of the spinal nerves possesses two separate threads, each of which has a ganglion of its own; the one of these threads joins the dorsal and the other the ventral branch. In fishes in which the spinal chord is very short, as in Plectognaths, Lophius, the roots of the nerves are extremely long, forming a thick *Cauda equina.* The additional function which the (five) anterior spinal nerves of *Trigla* have to perform in supplying the sensitive pectoral appendages and their muscles has caused the development of a paired series of globular swellings of the corresponding portion of the spinal chord. A similar structure is found in *Polynemus.*

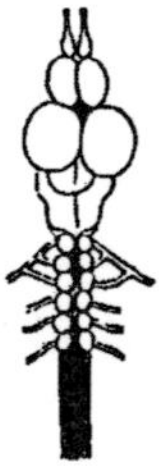

Fig. 46. Brain and anterior portion of the spinal chord of Trigla (Gurnard), showing the globular swellings at the base of the anterior spinal nerves.

A *Sympathic nervous* system appears to be absent in *Branchiostoma,* and has not yet been clearly made out in *Cyclostomes.* It is well developed in the *Palæichthyes,* but without cephalic portion. This latter is present in all Osseous fishes, in which communication of the Sympathic has been found to exist with all cerebral nerves, except the olfactory, optic, and acustic. The sympathic trunks run along each side of the aorta and the back of the abdomen into the hæmal canal; communicate in their course with the ventral branches of each of the spinal nerves; and, finally, often blend together into a common trunk beneath the tail. At the points of communication with the cerebral and spinal nerves frequently ganglia are developed, from which nerves emerge which are distributed to the various viscera.

CHAPTER VII.

THE ORGANS OF SENSE.

CHARACTERISTIC of the *Organ of Smell* in Fishes is that it has no relation whatever to the respiratory function, with the exception of the *Dipnoi*, in which possibly part of the water received for respiration passes through the nasal sac.

The olfactory organ is single in *Branchiostoma* and the *Cyclostomes*. In the former a small depression on the front end of the body, clothed with a ciliated epithelium, is regarded as a rudimentary organ of smell. In the adult *Petromyzon* a membranous tube leads from the single opening on the top of the head into the cartilaginous olfactory capsule, the inside of which is clothed by membranes prolonged into a posterior blind tube (Fig. 30, *s*), which penetrates the cartilaginous roof of the palate, but not the mucous membrane of the buccal cavity. In the Myxinoids the outer tube is strengthened by cartilaginous rings like a trachea; the capsule is lined by a longitudinally folded pituitary membrane, and the posterior tube opens backwards on the roof of the mouth; the opening is provided with a valve.

In all other Fishes the organ of smell is double, one being on each side; it consists of a sac lined with a pituitary membrane, and without, or with one or two, openings. The position of these openings is very different in the various orders or suborders of Fishes.

In the *Dipnoi* the nasal sac opens downwards by two wide openings which are within the boundaries of the cavity of the mouth. The pituitary membrane is transversely folded, the transverse folds being divided by one longitudinal fold. The walls of the sac are strengthened by sundry small cartilages.

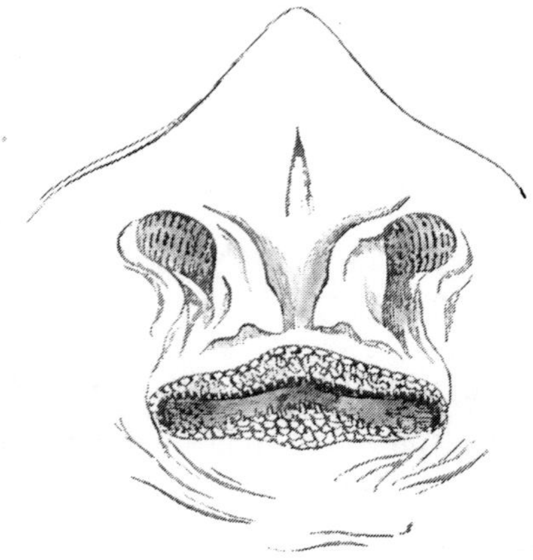

Fig. 47.—Nostrils of *Raia lemprieri*, with nasal flaps reverted.

Also in *Chondropterygians* the openings, of which there is one to each sac, are on the lower part of the snout, and in the Rays, Holocephali, and some Sharks, each extends into the cleft of the mouth. The openings are protected by valvular flaps, supported by small cartilages, and moved by muscles, whence it may be concluded that these fishes are able to scent (actively) as well as to smell (passively).

In the majority of *Teleostei* the olfactory capsules are lateral or superior on the snout, covered externally by the skin, each usually pierced by two openings, which are either close together, or more or less remote from each other; the posterior is generally open, the anterior provided with a valve or tube. In the *Chromides* and *Labroidei ctenoidei* a single opening only exists for each sac. In the *Murænidæ* the two openings of each side are either superior, or lateral, or labial, that is, they are continued downwards and pierce the margin of the upper lip. In many Tetrodonts nasal openings are absent, and replaced by a conical papilla, in which the olfactory nerve terminates.

It is certain that fishes possess the faculty of perceiving odours, and that various scents attract or repel them. A mangled carcase or fresh blood attracts Sharks as well as the voracious Serrasal monoids of the South American rivers. There is no reason to doubt that the seat of that perception is in the olfactory sac; and it may be reasonably conjectured that its strength depends mainly on the degree of development indicated by the number and extent of the interior folds of the pituitary membrane.

Organ of Sight.—The position, direction, and dimensions of the eyes of fishes vary greatly. In some they have an upward aspect, and are often very close together; in others they are lateral, and in a few they are even directed downwards. The Flat-fishes represent the extraordinary anomaly that both eyes are on the same side of the head, and rarely on the same level, one being generally placed more forward than the other. In certain species of marine fishes the eyes are of an extraordinary size, a peculiarity indicating that the fish either lives at a great depth, to which only a small proportion of the rays of light penetrate, or that it is of nocturnal habits. In fishes which have descended to such great depths that no rays whatever can reach them, or in fresh-water fishes living in caves, or in species which grovel and live constantly in mud, the eyes are more or less aborted, sometimes quite rudimentary, and covered by the skin. In very few this organ appears to be entirely absent. In some Gobioids and Trachinoids (*Periophthalmus, Boleophthalmus, Uronoscopus*, etc.) the eyes, which are on the upper side of the head, can be elevated and depressed at the will of the fish. In the range of their vision and acuteness of sight, Fishes are very inferior to the higher classes of Vertebrates, yet at the same time it is evident that they perceive their prey or approaching danger from a considerable distance; and it would appear that the visual powers of a *Periophthalmus*, when hunting insects on mud-flats of the tropical coasts, are quite equal to that of a frog. Again, the discrimination with which fishes sometimes prefer one colour or kind of artificial fly to another affords sufficient evidence that the vision, at least of certain species is by no means devoid of clearness and precision.

The eye of *Branchiostoma* is of the most rudimentary condition. It is simply a minute speck coated by dark pigment, and receiving the end of a short nerve. In *Myxinoids* the minute rudiment of the eye is covered by the skin and muscles. This is also the case in many of the blind Teleosteous fishes; however, whilst in the former fishes the organ of sight has not attained to any degree of development, the rudimentary eye of blind Teleostei is a retrogressive formation, in which often a lense and other portions of the eye can be recognised. In fishes with a well-developed eye it is imbedded in a layer of gelatinous and adipose substance, which covers the cavity of the orbit. A lacrymal gland is absent. In the orbit of one fish only, *Chorismodentex*, an organ has been found which can be compared to a *saccus lacrymalis.* It is a round, blind, wide sac, of the size of a pea, situated below the anterior corner of the orbit, between the maxillary bone and the muscles of the cheek, communicating by a rather wide foramen with the orbital cavity. The membrane by which it is formed is continuous with that coating the orbita. In the Chondropterygians the eyeball is supported by and moves on a cartilaginous peduncle of the orbital wall. In the majority of Teleosteans, and in Acipenser, a fibrous ligament attaches the sclerotic to the wall of the orbit. The proper muscles of the eyeball exist in all fishes, and consist of the four *Musculi recti* and the two *M. obliqui.* In many Teleostei the former rise from a subcranial canal, the origin of the *M. rectus externus* being prolonged farthest backwards. The *Recti* muscles are extraordinarily long in the Hammerheaded Sharks, in which they extend from the basis cranii along the lateral prolongations of the head to the eyes, which are situated at the extremities of the hammer.

In all fishes the general integument of the head passes

over the eye, and becomes transparent where it enters the orbit; sometimes it simply passes over the orbit, sometimes it forms a circular fold. The anterior and posterior portions may be especially broad and the seat of an adipose deposit (*adipose eyelids*), as in *Scomber*, *Caranx*, *Mugil*, etc. In many of these fishes the extent of these eyelids varies with the seasons; during the spawning season they are so much loaded with fat as nearly to hide the whole eye. Many Sharks possess a *nictitating* membrane, developed from the lower part of the palpebral fold, and moved by a proper set of muscles.

The form of the *bulbus* (Fig. 48) is subhemispherical, the cornea (*co*) being flat. If it were convex, as in higher Vertebrates, it would be more liable to injury; but being level with the side of the head the chances of injury by friction are diminished. The *sclerotica* (*sc*) is cartilaginous in Chondropterygians and Acipensers, fibrous and of varying thickness in Teleosteans, in the majority of which it is supported by a pair of cartilaginous or ossified hemispheroid cups (*c*). In a few fishes, as in *Ceratodus*, *Xiphias*, the cups are confluent into one cup, which possesses a foramen behind to allow the passage of the optic nerve (*o*). The *cornea* of *Anableps* shows an unique peculiarity. It is crossed by a dark horizontal stripe of the conjunctiva, dividing it into an upper and lower portion; also the iris is perforated by two pupils. This fish is observed to swim frequently with half of its head out of the water, and it is a fact that it can see out of the water as well as in it.

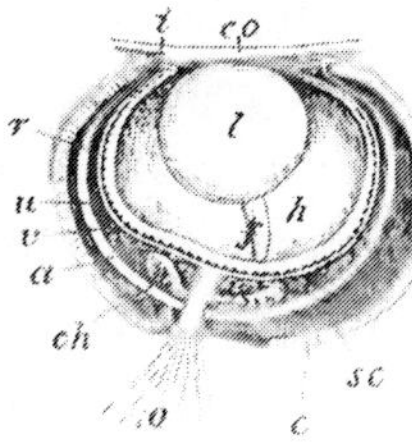

Fig. 48.
Vertical section through eye of Xiphias. (After Owen.)
co, Cornea; *sc*, sclerotica; *o*, nervus opticus; *c*, sclerotic capsule; *a*, membrana argentea; *v*, membrana vasculosa; *u*, membrana uvea; *ch*, choroid gland; *r*, retina; *f*, processus falciformis; *h*, humor vitreus; *l*, lens; *i*, iris.

The membranes situated between the sclerotica and retina are collectively called *choroidea*, and three in number. The one in immediate contact with the sclerotic, and continued upon the iris, is by no means constantly present; it is the *membrana argentea* (*a*), and composed of microscopical crystals reflecting a silvery or sometimes golden lustre. The middle layer is the *membrana vasculosa* s. *halleri* (*v*), the chief seat of the ramifications of the choroid vessels; the innermost layer is the *membrana ruyscheana* or *uvea* (*u*), which is composed of hexagonal pigment-cells, usually of a deep brown or black colour.

In many *Teleostei* a *rete mirabile* surrounds the entry of the optic nerve; it is situated between the membrana argentea and vasculosa, and called the *choroid gland* (*ch*). It receives its arterial blood from the artery issuing from the pseudobranchia; the presence of a choroid gland always being combined with that of a pseudobranchia. Teleosteans without pseudobranchia lack a choroid gland. In the Palæichthyes, on the other hand, the pseudobranchia is present and a choroid gland absent.

The *iris* (*i*) is merely the continuation of the choroid membrane; its capability of contracting and expanding is much more limited than in higher Vertebrates. The *pupil* is generally round, sometimes horizontally or vertically elliptical, sometimes fringed. In the Rays and Pleuronectidæ a lobe descends from the upper margin of the pupil, and the outer integument overlying this lobe is coloured and non-transparent; a structure evidently preventing light from entering the eye from above.

In most *Teleostei* a fold of the *Choroidea*, called the *Processus falciformis* (*f*), extends from the vicinity of the entrance of the optic nerve to the lens. It seems to be constantly absent in Ganoids.

The *retina* (*r*) is the membrane into which the optic nerve penetrates, and in which its terminal filaments are distributed. It consists of several layers (Fig. 49). The outermost is an extremely delicate membrane (*a*), followed by a layer of nerve-cells (*b*), from which the terminal filaments issue, passing through several granular strata (*c, d, e*), on which the innermost stratum rests. This stratum is composed of cylindrical rods (*f*) vertically arranged, between which twin fusiform corpuscles (*g*) are intercalated. This last layer is thickly covered with a dark pigment. The retina extends over a portion of the iris, and a well-defined raised rim runs along its anterior margin.

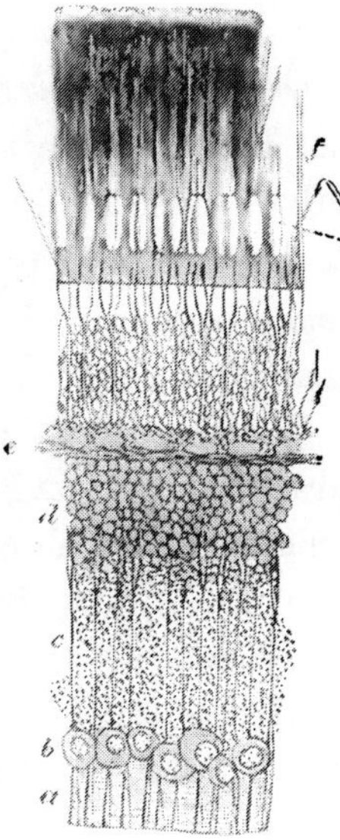

Fig. 49.—Vertical section of the Retina of the Perch, magn. X 350.

The *vitreous humour* (Fig. 48, *h*) which fills the posterior cavity of the eyeball, is of a firmer consistency than in the higher Vertebrates. The *lens* is spherical, or nearly so; firm, denser towards the centre, and lies in a hollow of the vitreous humour. When a falciform process is present, it is with one end attached to the lens, which is thus steadied in its position. It consists of concentric layers consisting of fibres, which in the nucleus of the body have marginal teeth, by which they are interlocked together. In *Petromyzon* this serrature is absent, or but faintly indicated.

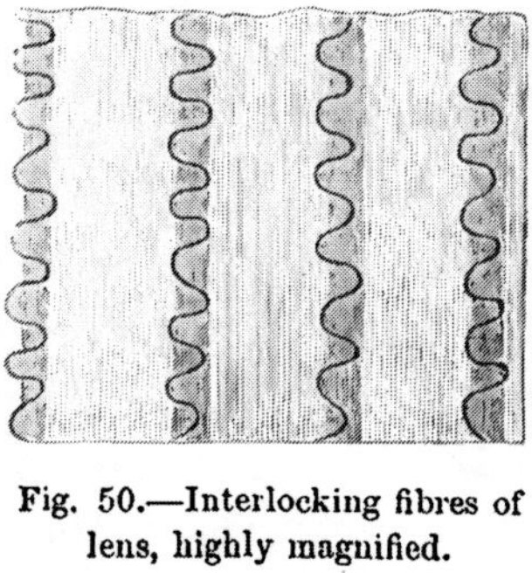

Fig. 50.—Interlocking fibres of lens, highly magnified.

The anterior cavity of the eye is very small in Fishes, in consequence of the small degree of convexity of the cornea; the quantity of the aqueous humour, therefore, is very small, just sufficient to float the free border of the iris; and the lessened refractive power of the aqueous humour is compensated by the greater convexity of the lens.

Organ of Hearing.—No trace of an organ of hearing has been found in *Branchiostoma.* In the *Cyclostomes* the labyrinth is enclosed in externally visible cartilaginous capsules laterally attached to the skull; it consists of a single *semicircular canal* in the Myxinoids, whilst the Petromyzontes possess two semicircular canals with a *vestibulum.*

In all other fishes the labyrinth consists of a vestibule and three semicircular canals, the vestibule dilating into one or more sacs which contain the otoliths. A tympanum, tympanic cavity, and external parts, are entirely absent in the class of fishes.

In the *Chondropterygians* and *Dipnoi,* the labyrinth is enclosed in the cartilaginous substance of the skull. In the former the excavation in the cartilage is larger than the membranous labyrinth, but nearly corresponds to it in form; the part which receives the membranous vestibulum is called *Vestibulum cartilagineum,* from which a canal issues and penetrates to the surface of the skull, where it is closed by the skin in Sharks, but opens by a minute foramen in Rays. The otolithic contents are soft and chalklike.

In the *Holocephali* part of the labyrinth is enclosed in the cartilage of the skull, another part being in the cranial cavity, as in Ganoids and Teleosteans. The membranous vestibulum is continued by a canal to a single opening in the roof of the skull, from which two smaller canals are continued to two small foramina in the skin covering the occipital region.

In the *Teleosteans* the sac which contains the otoliths lies on each side of the base of the cranial cavity and is often divided by a septum into two compartments of unequal size, each containing a firm and solid *otolith;* these bodies (Fig. 51), possess indented margins, frequently other impressions and grooves, in which nerves from the N. acusticus are lodged; they vary much in size and form, but in both respects show a remarkable constancy in the same kind of fishes. The vestibule is outwards in contact with the osseous side wall of the skull, inwards with the metencephalon and medulla oblongata; it contains another firm concretion, and opens by five foramina into the three semicircular canals. The terminations of the acustic nerve are distributed over the vestibular concretion and the ampulliform ends (Fig. 52 *p*) of the semicircular canals, without being continued into the latter, which are filled with fluid. The semicircular canals (Fig. 52 *g*), are sometimes lodged in the cranial bones, sometimes partly free in the cranial cavity. Many Teleostei have fontanelles in the roof of the skull, closed by skin or very thin bone only at the place where the auditory organ approaches the surface, by which means sonorous undulations must be conducted with greater ease to the ear.

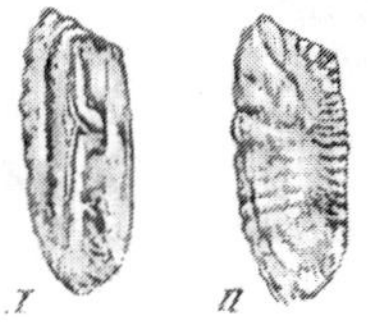

Fig. 51.—Otolith of Haddock (Gadus æglefinus). I. Outer, II. Inner aspect.

In many Teleostei a most remarkable relation obtains between the organ of hearing and the air-bladder. In the most simple form this connection is established in Percoids and the allied families, in which the two anterior horns of the air-bladder are attached to fontanelles of the occipital region of the skull, the vestibulum occupying the opposite side of the membrane by which the fontanelle is closed. The condition is similar, but more complicated in many Clupeoids. The anterior narrow end of the air-bladder is produced into a canal at the base of the skull, and divided into two very narrow branches, which again bifurcate and terminate in a globular swelling. An appendage of the vestibulum meets the anterior of these swellings, and comes into close contact with it. Besides, the two vestibules communicate with each other by a transverse canal, crossing the cranial cavity below the brain.

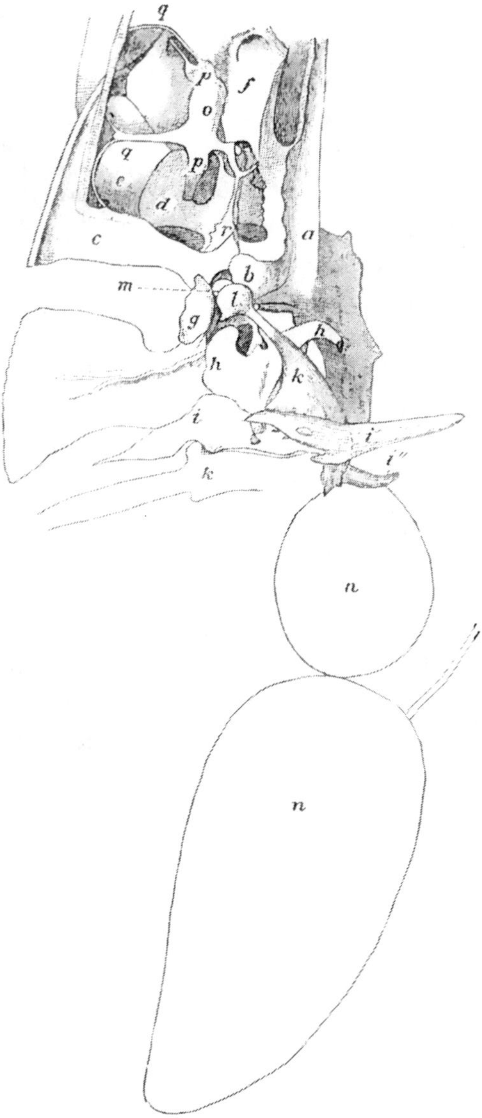

Fig. 52.—Communication between auditory organ and air-bladder in the Carp. (After E. H. Weber.)

a, Basisphenoid; *b*, Occipital; *c*, Supraoccipital; *d*, Exoccipital; *e*, Paroccipital; *f*, Alisphenoid; *g*, Neural arch of first vertebra; *h*, *i*, *k*, second, third, and fourth vertebra; *h′*, *i′*, Parapophyses of second and third vertebra; *i″*, process of the third vertebra for the attachment of the air-bladder; *k*, *l*, *m*, Chain of ossicles; *n*, Air-bladder; *o*, vestibulum; *p*, *p*, Ampullæ; *q*, *q*, Canales semicirculares; *r*, Sinus impar.

The connection is effected by means of a chain of ossicles

in *Siluridæ*, *Characinidæ*, *Cyprinidæ* and *Gymnotidæ*. A canal issues from the communication between vestibule and its sac, and meeting that from the other side forms with it a common *sinus impar* (Fig. 52, *r*), lodged in the substance of the basi-occipital; this communicates on each side by a small orifice with two subspherical atria, on the body of the atlas, close to the foramen magnum. Each atrium is supported externally by a small bone (*m*); a third larger bone (*k*) completes the communication with the anterior part of the air-bladder. From the sinus impar a bifid canal penetrates into the alisphenoids, in which it terminates. In *Cobitis* and several Loach-like Siluroids the small air-bladder consists of two globular portions placed side by side, and wholly included within two bullæ, formed by the modified parapophyses of the second and third vertebræ. The three ossicles on each side are present, but concealed by the fore part of the osseous bulla.

Organ of Taste.—Some fishes, especially vegetable feeders, or those provided with broad molar-like teeth, masticate their food; and it may be observed in Carps and other Cyprinoid fish, that this process of mastication frequently takes some time. But the majority of fish swallow their food rapidly, and without mastication, and therefore we may conclude that the sense of taste cannot be acute. The tongue is often entirely absent, and even when it exists in its most distinct state, it consists merely of ligamentous or cellular substance, and is never furnished with muscles capable of producing the movements of extension or retraction as in most higher Vertebrates. A peculiar organ on the roof of the palate of Cyprinoids, is perhaps an organ adapted for perception of this sense; in these fishes the palate between and below the upper pharyngeal bones is cushioned with a thick, soft contractile substance, richly supplied with nerves from the Nervi vagus and glossopharyngeus.

Organs of Touch.—The faculty of touch is more developed than that of taste, and there are numerous fishes which possess special organs of touch. Most fishes are very sensitive to external touch, although their body may be protected by hard horny scales. They perceive impressions even on those parts which are covered by osseous scutes, in the same manner as a tortoise perceives the slightest touch of its carapace. The seat of the greatest sensitiveness, however, appears to be the snout and the labial folds surrounding the mouth. Many species possess soft and delicate appendages, called barbels, which are almost constantly in action, and clearly used as organs of touch. Among the *Triglidæ* and allied families, there are many species which have one or more rays of the pectoral fin detached from the membrane, and supplied with strong nerves. Such detached rays (also found in the *Polynemidæ, Bathypterois*) are used partly for locomotion, partly for the purpose of exploring the ground over which the fish moves.

Some fish appear to be much less sensitive than others, or at least lose their sensitiveness under peculiar circumstances. It is well known that a Pike, whose mouth has been lacerated and torn by the hook, continues to yield to the temptation of a bait immediately afterwards. The Greenland Shark when feeding on the carcase of a whale allows itself to be repeatedly stabbed in the head without abandoning its prey. A pair of Congers are so dead to external impression at the time of copulation, and so automatically, as it were, engaged, that they have been taken by the hand together out of the water.

CHAPTER VIII.

THE ORGANS OF NUTRITION AND DIGESTION.

FISHES are either exclusively carnivorous or herbivorous, but not a few feed on vegetable substances as well as animal, or on mud containing alimentary substance in a living or decomposing state. Generally they are very voracious, especially the carnivorous kinds, and the rule of "eat or be eaten" applies to them with unusual force. They are almost constantly engaged in the pursuit and capture of their prey, the degree of their power in these respects depending on the dimensions of the mouth and gullet and the strength of the teeth and jaws. If the teeth are sharp and hooked, they are capable of securing the most slender and agile animals; if this kind of teeth is combined with a wide gullet and distensible stomach, they are able to overpower and swallow other fish larger than themselves; if the teeth are broad, strong molars, they are able to crush the hardest aliments; if they are feeble, they are only serviceable in procuring some small or inert and unresisting prey. Teeth may be wanting altogether. Whatever the prey, in the majority of cases it is swallowed whole; but some of the most voracious fishes, like some Sharks and *Characinidæ*, are provided with cutting teeth, which enable them to tear their prey to pieces if too large to be swallowed whole. Auxiliary organs for the purpose of overpowering their prey, which afterwards is seized or torn by the teeth, like the claws of some carnivorous mammals and birds, are not found in this class; but in a few fishes the jaws themselves are modified for that purpose. In the Sword-fishes the bones of the upper jaw form a long dagger-shaped weapon, with which they not only attack large animals, but also frequently kill fishes on which they feed. The Saw-fishes are armed with a similar but still more complicated weapon, the saw, which is armed on each side with large teeth implanted in deep sockets, specially adapted for killing and tearing the prey before it is seized and masticated by the small teeth within the mouth. Fishes show but little choice in the selection of their food, and some devour their own offspring indiscriminately with other fishes. Their digestive powers are strong and rapid, but subject in some degree to the temperature, which, when sinking below a certain point, lowers the vital powers of these cold-blooded animals. On the whole, marine fishes are more voracious than those inhabiting fresh waters; and whilst the latter may survive total abstinence from food for weeks or months, the marine species succumb to hunger within a few days. The growth of fishes depends greatly on the nature and supply of food, and different individuals of the same species may exhibit a great disparity in their respective dimensions. They grow less rapidly and to smaller dimensions in small ponds or shallow streams than in large lakes and deep rivers. The young of coast fishes, when driven out to sea, where they find a much smaller supply of food, remain in an undeveloped condition, assuming an hydropic appearance. The growth itself seems to continue in most fishes for a great length of time, and we can scarcely set bounds to—certainly we know not with precision—the utmost range of the specific size of fishes. Even among species in no way remarkable for their dimensions we sometimes meet with old individuals, favourably situated, which more or less exceed the ordinary weight and measurement of their kind. However, there are certain evidently short-lived species of fishes which attain a remarkably uniform size within a very short time; for instance, the Stickleback, many species of *Gobius* and *Clupea*.

The organs of nutrition, manducation, and deglutition, are lodged in two large cavities—an anterior (the mouth or *buccal cavity*), and a posterior (the *abdominal cavity*). In the former the alimentary organs are associated with those fulfilling the respiratory functions, the transmission of food to the stomach and of water to the gills being performed by similar acts of deglutition. The abdominal cavity commences immediately behind the head, so, however, that an extremely short thoracic cavity for the heart is partitioned off in front. Beside the alimentary organs it contains also those of the urogenital system and the air-bladder. The abdominal cavity is generally situated in the trunk only, but in numerous fishes it extends into the tail, being continued for some distance along each side of the hæmal apophyses.

In numerous fishes the abdominal cavity opens outwards by one or two openings. A single *porus abdominalis* in front of the vent is found in *Lepidosiren* and some Sturgeons; a paired one, one on each side of the vent, in *Ceratodus*, some species of Sturgeon, *Lepidosteus*, *Polypterus*, *Amia*, and all Chondropterygians. As in these fishes semen and ova are discharged by proper ducts, the abdominal openings may serve for the expulsion of semen, and those ova only which, having lost their way to the abdominal aperture of the oviduct, would be retained in the abdominal cavity. In those *Teleosteans* which lack an oviduct a single *porus genitalis* opens *behind* the vent.

The *mouth* of fishes shows extreme variation with regard to form, extent, and position. Generally opening in front, it may be turned upwards, or may lie at the lower side of the snout, as in most Chondropterygians, Sturgeons, and some Teleosteans. Vogt regards this position as a persistent fœtal condition. In most fishes the jaws are covered by the skin, which, before passing over the jaws, is often folded, forming more or less fleshy lips. In the Sharks the skin retains its external character even within the teeth, but in other fishes it changes into a mucous membrane. A *tongue* may exist as a more or less free and short projection, formed by the glosso-hyal and a soft covering, or may be entirely absent. *Salivary* glands and a *velum palati* are absent in fishes.

With regard to the *dentition*, the class of Fishes offers an amount of variation such as is not found in any of the other classes of Vertebrates. As the teeth form one of the most important elements in the classification of fishes, their special arrangement and form will be referred to in the account of the various families and genera. Whilst not a few fishes are entirely edentulous, in others most of the bones of the buccal cavity, or some of them, may be toothed, as the bones of the jaws, the palatines, pterygoids, vomers, basi-sphenoid, glosso-hyal, branchial arches, upper and lower pharyngeals. In others teeth may be found fixed in some portion of the buccal membrane without being supported by underlying bone or cartilage; or the teeth have been developed in membrane overlying one of the dentigerous bones mentioned, without having become anchylosed to the bone. When the tooth is fixed to the bone the attachment has generally been effected by the ossification of the bone of the tooth, but in some fishes a process of the bone projects into the cavity of the tooth; in others the teeth are implanted in alveoli. In these, again, frequently a process of bone rises from the bottom, on which the tooth rests.

Many fishes, especially predatory fishes with long, lancet-shaped teeth, have all or some of the teeth capable of being bent inwards towards the mouth. Such "hinged" teeth resume at once the upright position when pressure is removed from them. They are, however, depressible in one direction only, thus offering no obstacle to the ingress, but opposing the egress of prey. Mr. C. S. Tomes has shown that the means by which this mechanism is worked are different in different fishes; for whilst, in the *Pediculati* and *Gadoids* (Hake) the elasticity resides solely in the tissue of the hinge (the tooth being as resilient as ever after everything else is severed), in the Pike the hinge is not in the least endowed with elasticity, but the bundles of fibres proceeding from the interior of the dentine cap are exceedingly elastic.

As regards texture the teeth of fishes show considerable variation. The conical teeth of the Cyclostomes and the setiform teeth of many Teleosteans consist of a horny albuminous substance. The principal substance of the teeth of other fishes consists of dentine, with numerous dividing and anastomosing tubercles, sometimes covered by a stratum of unvascular dentine. An enamel-like substance has been observed on the crown of the teeth of *Sargus* and *Balistes*, and an ossification of the capsule of their matrix covers the enamel with a thin coating of cement. The teeth either possess a cavity in which the matrix is received, or, more frequently, they are solid, in which case vascular canals of the underlying bone are continued into the substance of the tooth. In the teeth of some fishes numerous sets of canals and tubes are so arranged that they do not anastomose with one another, each set being surrounded by a layer of dentine and cement. These apparently simple teeth are evidently composed of numerous small teeth, and called *compound* teeth.

The teeth may be, and generally are, very different as regards size or form in the different parts of the mouth; they may be also different according to the age or sex of the fish (*Raja*). The teeth may be few in number and *isolated*, or placed in a single, double, or triple *series*, *distant* from one another or *closely set*; they may form narrow or broad *bands*, or *patches* of various forms. As regards form, they may be cylindrical or conical, pointed, straight, or curved, with or without an angular bent near their base; some are compressed laterally or from the front backwards; the latter may be triangular in shape, or truncated at the top like incisors of mammals; they may have one apex (cusp) only, or be bi- or tri-lobate (bi- or tri-cuspid); or have the margins denticulated or serrated. Compressed teeth may be confluent, and form a cutting edge in both jaws, which assume the shape of a parrot's beak (Fig. 53). In some the apex is hooked or provided with barbs. Again, some teeth are broad, with flat or convex surface, like molar teeth. With regard to size, the finest teeth are like fine flexible bristles, *ciliiform* or *setiform;* or, if very short and anchylosed to the bone, they appear only as inconspicuous asperities of the bone. Very fine conical teeth arranged in a band are termed *villiform* teeth; when they are coarser, or mixed with coarser teeth, they are *card-like* (dents en rape or en cardes) (Fig. 54); molar-like teeth of very small size are termed *granular*.

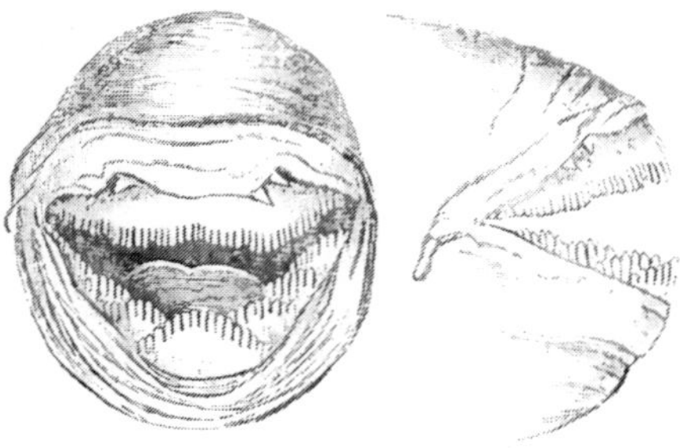

Fig. 53.—Jaws of Calliodon.

In all fishes the teeth are constantly shed or renewed during the whole course of their life. In fishes which have compound teeth, as the Dipnoï, Chimæroids, Scari,[1] Gymnodonts, as well as in those which have apparently permanent teeth, as in the saw of *Pristis*, the detrition of the surface is made up by a constant growth of the tooth from its base. When the teeth are implanted in alveoli, they are generally succeeded by others in the vertical direction, but in others they succeed one another, side by side. In the majority of

[1] On the development and structure of the dentition of *Scarina*, see *Boas*, "Die Zähne der Scaroiden," in Zeitschr. f. Wiss. Zoolog. xxxii. (1878).

fishes the new tooth is not developed (as in reptiles and mammals) in a diverticulum of the sack of its predecessor, but like this from the free surface of the buccal membrane. Generally there are more than one tooth growing, which are in various stages of development, and destined to replace the one in function. This is very conspicuous in Sharks, in which the whole phalanx of their numerous teeth is ever marching slowly forwards (or in some backwards), in rotatory progress, over the alveolar border of the jaw, the teeth being successively cast off after having reached the outer margin, and fulfilled for a longer and shorter period their special function.

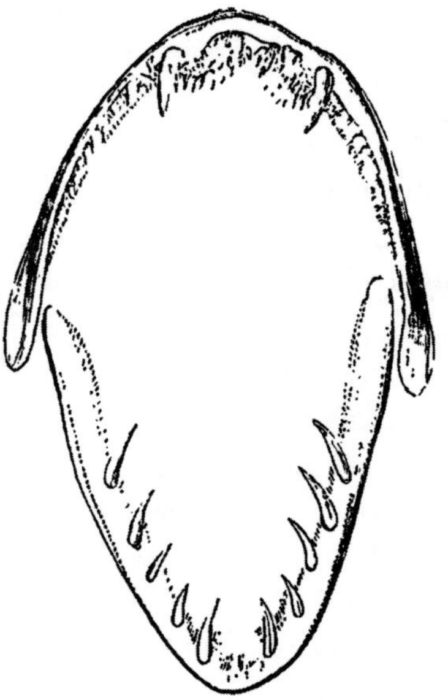

Fig. 54.—Cardlike teeth of Plectropoma dentex, with canines.

[The richest materials for our knowledge of the teeth of fishes are contained in *Owen's* "Odontography." Lond. 1840. 8vo.]

The *intestinal tract* is divided into four portions: œsophagus, stomach, small and large intestine; two or more of these divisions may coalesce in fishes and become indistinguishable. But it is characteristic of the class that the urinary apertures are constantly situated behind the termination of the intestinal tract.

In *Branchiostoma* the whole intestinal tract is straight, and coated with a ciliated mucous membrane. The wide pharynx passes into a narrow œsophagus, this into a gastric cavity, the remainder being again narrower and terminating in the anal aperture, which lies somewhat to the left of the median line. The liver is represented by a green coloured cœcal diverticulum of the stomachic dilatation. A mesenterium is absent.

In the *Cyclostomi* the intestinal tract is likewise straight, and without clearly defined divisions; however, in *Petromyzon* the œsophagus shows numerous longitudinal folds, and the intestine proper is provided with a single longitudinal fold. A mesentery, which is present in the Myxinoids, is represented by a short median fold only, by means of which the hindmost part of the intestine is fixed.

The *Palæichthyes* show differences in the structure of their intestinal tract as considerable as are found among the *Teleostei*, but they have that in common that the absorbent surface of their intestine is enlarged by the development of a spiral valve, evidence of the presence of which in extinct Palæichthyes is still preserved in the fossilised fæces or *coproliths*, so abundant in some of the older strata.

In *Chondropterygians* (Fig. 55) the stomach is divided into a cardiac and pyloric portion, the former frequently terminating in a blind sac, and the latter varying in length. The pyloric portion is bent at its origin and end, and separated from the short duodenum (called *Bursa entiana* in these fishes) by a valve; the ductus hepaticus and pancreaticus enter the duodenum. This is succeeded by the straight intestine provided with the spiral valve, the coils of which may be either longitudinal and wound vertically about the axis of the intestine, as in *Carcharias, Galeocerdo, Thalassorhinus*, and *Zygœna*, or they may be transverse to that axis, as in the other genera. The number of gyrations in the latter case varies: there may be as many as forty. The short rectum passes into a cloaca, which contains also the orifices of the urogenital ducts. Only the commencement and end of the intestinal tract are fixed by mesenterial folds.

In the *Holocephali* and *Dipnoi*, the intestinal tract is short, straight, and wide, without stomachic dilatation, a pyloric valve, close to which the ductus choledochus enters, indicating

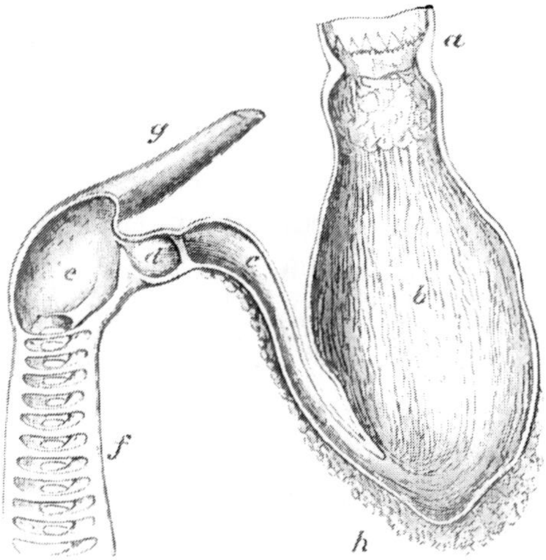

Fig. 55.—Siphonal stomach and spiral valve of Basking-Shark (Selache). (After Home and Owen.)

a, Œsophagus ; *b*, Cardiac portion of stomach ; *c*, pyloric portion ; *d*, pouch intermediate between stomach and duodenum, with circular valves at both ends ; *e*, Duodenum ; *f*, Valve of intestine ; *g*, Ductus hepaticus ; *h*, Spleen.

the boundary of the intestine proper (Fig. 57, *p*). The spiral valve is perfect, and makes from three (*Chimæra*) to nine (*Ceratodus*) gyrations. A cloaca is present, as in Chondropterygians. A mesentery fixing the dorsal side of the intestine is absent.

The other *Ganoids* resemble again more the Chondropterygians in the structure of their intestinal tract. The stomach has always a distinct pyloric portion, and has a still more complicated structure in *Acipenser*. The duodenal portion receives the contents of *Appendices pyloricæ*, which are confluent into a gland-like mass in *Acipenser*, but separate in *Polyodon*, and numerous and short in *Lepidosteus*, whilst *Polypterus* possesses one such appendage only. A spiral valve is developed in the Sturgeons and Polypterus, but in *Amia*, in which the intestine performs several convolutions, the four gyrations of the valve are situated far back towards the end of the intestine. In *Lepidosteus* the valve is rudimentary, and indicated only by three raised lines crossing the terminal portion of the intestine. In all these Ganoids the rectum has a separate opening, without cloaca.

The structure of the intestinal tract of *Teleosteous* fishes is subject to so numerous modifications that we should go beyond the limits of the present work if we would attempt to enter into details. Great differences in this respect may be found even in groups of the same natural families. Frequently the intestinal tract remains of nearly the same width throughout its course, and only the entrance of the various ducts serves as a guide for the distinction of its divisions. An intestine of such uniform width may be straight and short, as in *Scombresocidæ*, *Symbranchidæ*, or it may be more or less convoluted and long, as in many *Cyprinidæ*, *Doradina*, etc. On the whole, carnivorous fishes have a much shorter and simpler intestinal tract than herbivorous.

In the majority of Teleosteans, however, œsophagus, stomach, duodenum, small intestine and rectum, can be more or less distinctly, even externally distinguished.

There are two predominant forms of the stomach, intermediate forms being, however, numerous. In the first, the *siphonal*, it presents the form of a bent tube or canal, one-half of the horse-shoe being the cardiac, the other the pyloric portion. In the second, the *cæcal*, the cardiac division is prolonged into a long descending blind sac, the cardiac and pyloric openings of the stomach lying close together (*Clupea*, *Scomber*, *Thynnus*, etc.)

The duodenum receives always the hepatic and pancreatic secretions, and, besides, those of the appendices pyloricæ, which, in varying numbers (from 1 to 200), are of very common occurrence in Teleosteans (Fig. 56). They vary also in length

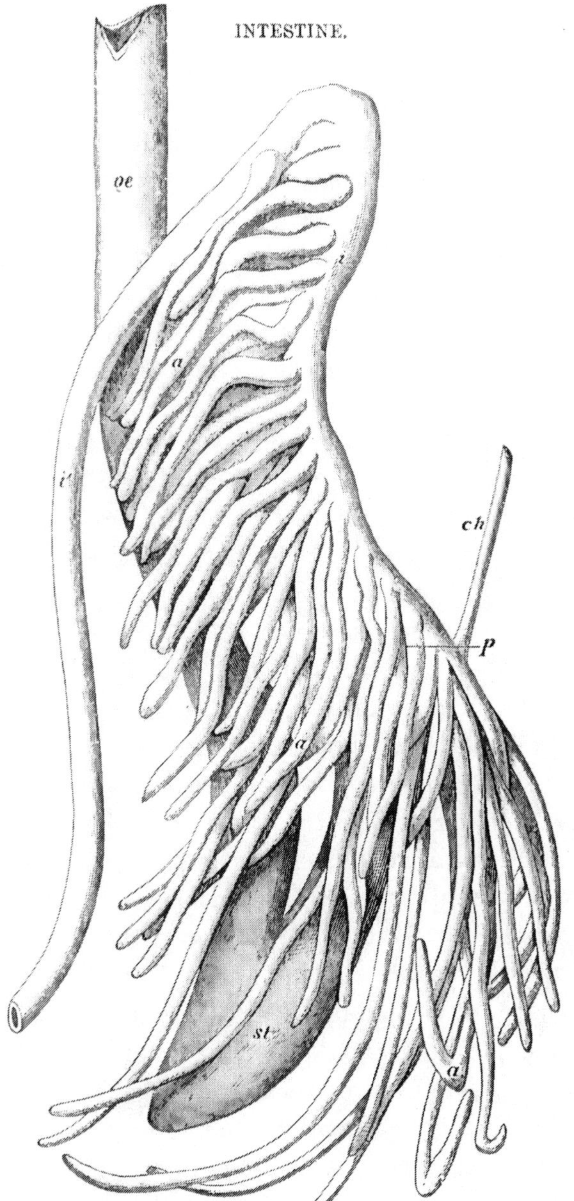

Fig. 56.—Siphonal Stomach and Pyloric Appendages of a Female Salmon, 3⅓ feet long. *a a a*, Pyloric appendages; *ch*, ductus choledochus; *oe*, œsophagus; *st*, lower end of stomach; *p*, pyloric region; *i*, ascending; and *i'*, descending portion of intestine.

and width, and whilst the narrowest serve only as secretory organs, the widest are frequently found filled with the same contents as the intestine. When few in number, each opens by a separate duct into the duodenum; when their number is greater two or more coalesce into a common duct; in the latter case the appendages cease to be free, and are connected with one another by a more or less firm tissue.

Cœcal appendages at the end of the intestinal tract are of exceedingly rare occurrence in fishes (*Box*). There is no *cloaca* in Teleosteans.

In the majority of Teleosteous fishes the *vent* is situated on the boundary between trunk and tail, behind the ventral fins. In a few it lies farther backwards, not far from the caudal fin; more frequently it is advanced forwards, under the middle of the abdomen or to the scapular arch. In two fishes, *Aphredoderus* and *Amblyopsis*, it lies before the pectoral fins.

A peritoneum envelops all the divisions of the intestinal tract within the abdominal cavity. A broad, well-developed *omentum* has hitherto been found in *Gobiesox cephalus* only.

Liver.—The existence of a liver in *Branchiostoma* as a long diverticulum of the intestine has been mentioned above. In the Myxinoids the liver is divided into two glandular bodies, an anterior rounded smaller one, and a posterior larger one of an elongate shape. The gall-bladder lies between both, and receives a cystic duct from each of them. In the other fishes the proportionally large liver is a single large gland, from which only now and then small portions are found to be detached. It is either simple, or with a right and left lobe, or with a third lobe in the middle; each lobe may have incisions or subdivisions, which, however, are very inconstant. The liver of fishes is distinguished by the great quantity of fluid fat (oil) which it contains. The gall-bladder is but rarely absent, and attached to the right lobe, or towards the centre; however, in some fishes it is detached from the liver and connected with it by the cystic duct only. The bile may be conveyed by one or more hepatic ducts into a common duct which is continued towards the gall-bladder as *ductus cysticus*, and towards the duodenum as *ductus choledochus;* or some of the hepatic ducts enter directly the gall-bladder, or directly the duodenum, without communicating with the common duct. Individual variations in this respect are of common occurrence.

A *pancreas* has been found hitherto in all Chondropterygians, Acipenser, and many Teleosteans. In the first it is a glandular mass of considerable size behind the stomach, close to the spleen; its duct leads into the duodenum. In the Sturgeons the pancreas is attached to the duodenum, and opens close to the ductus choledochus. In *Silurus glanis* it is very large, and the ductus choledochus passes through its substance; it is smaller in *Belone* and *Pleuronectes*, and situated in the mesentery; its duct accompanies the terminal portion of the ductus choledochus. In the Salmon, which possesses a large lobed pancreas, the duct is so intimately connected with the ductus choledochus that both appear externally as a single duct only.

The *spleen*, which is substantially a lymphatic gland, may be mentioned here, as it is constantly situated in the immediate vicinity of the stomach, generally near its cardiac portion. With the exception of *Branchiostoma*, it is found in all fishes, and appears as a rounded or oblong organ of dark-red colour. In the Sharks frequently one or more smaller pieces are detached from the principal body. In the Dipnoi a thin layer of a very soft substance of brownish-black colour below the mucous membrane of the stomach and upper part of the intestine has been regarded as the homologue of the spleen (Fig. 57, *m*). In most *Teleostei* the spleen is undivided, and appended by its vessels and a fold of the peritoneum to the pyloric bend of the stomach or the beginning of the intestine.

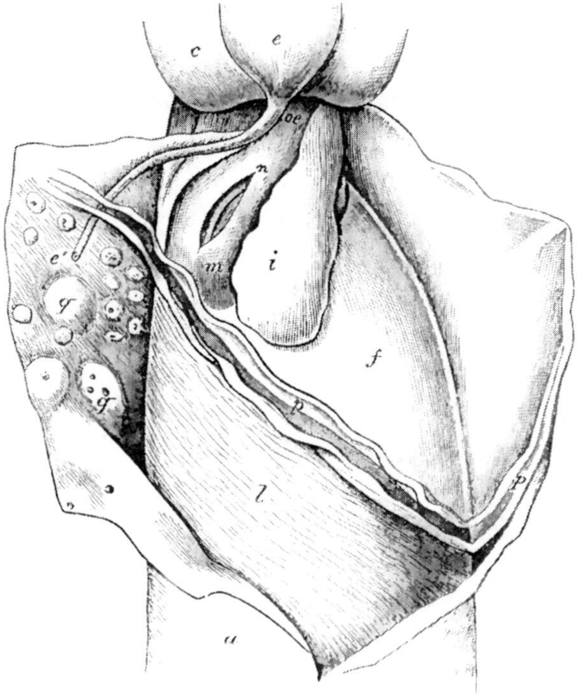

Fig. 57.—Upper part of Intestine of Ceratodus. The anterior wall of the intestine is opened, the liver (*c*) and gall-bladder (*e*) being drawn forward. A slit is made at *n*, through which part of the next compartment of the spirally wound intestine may be seen.

e', Mouth of ductus choledochus; *f*, stomach; *i*, adipose agglomeration; *l*, first compartment of intestinal spire; *m*, spleen; *oc*, lower part of œsophagus, opened; *p*, double pyloric fold; *q q*, glandular patches.

CHAPTER IX.

ORGANS OF RESPIRATION.

FISHES breathe the air dissolved in water by means of gills or branchiæ. The oxygen consumed by them is not that which forms the chemical constituent of the water, but that contained in the air which is dissolved in water. Hence fishes transferred into water from which the air has been driven out by a high temperature, or in which the air absorbed by them is not replaced, are speedily suffocated. The absorption of oxygen by fishes is comparatively small, and it has been calculated that a man consumes 50,000 times more than is required by a Tench. However, some fishes evidently require a much larger supply of oxygen than others: Eels and Carps, and other fishes of similar low vitality, can survive the removal out of their elements for days, the small quantity of moisture retained in their gill-cavity being sufficient to sustain life, whilst other fishes, especially such as have very wide gill-openings, are immediately suffocated after being taken out of the water. In some fishes noted for their muscular activity, like the *Scombridæ*, the respiratory process is so energetic as to raise the temperature of their blood far beyond that of the medium in which they live. A few fishes, especially such as are periodically compelled to live in water thickened into mud by desiccation and vitiated by decomposing substances, breathe atmospheric air, and have generally special contrivances for this purpose. These are so much habituated to breathing air that many of them, even when brought into pure water of normal condition, are obliged to rise to the surface at frequent intervals to take in a quantity of air, and if they be kept beneath the surface by means of a gauze net, they perish from suffocation. The special contrivances consist of additional respiratory organs, lodged in cavities either adjoining the gill-cavity or communicating with the ventral side of the œsophagus, or of the air-bladder which enters upon respiratory functions (Dipnoi, Lepidosteus, Amia).

The water used by fishes for respiration is received by the mouth, and by an action similar to that of swallowing driven to the gills, and expelled by the gill-openings, of which there may be one or several on each side behind the head; rarely one only in the median line of the ventral surface.

The *gills* or *branchiæ* consist essentially of folds of the

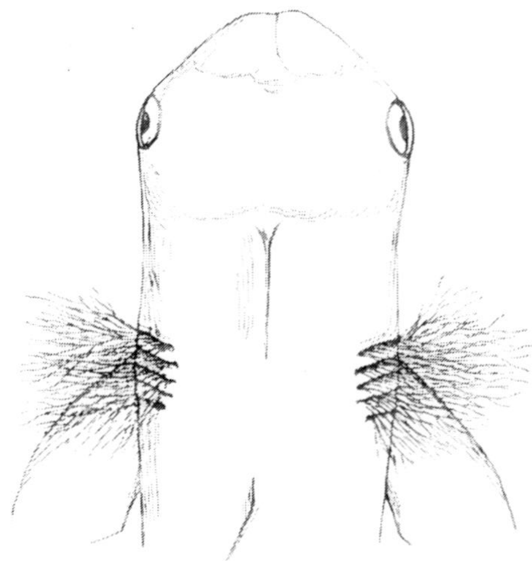

Fig. 58.—Fore-part of the body of an embryon of Carcharias, showing the branchial filaments (natural size).

mucous membrane of the gill-cavity (*laminæ branchiales*), in which the capillary vessels are distributed. In all fishes the gills are lodged in a cavity, but during the embryonic stage the Chondropterygians have the gill-laminæ prolonged into long filaments projecting beyond the gill-cavity (Fig. 58), and in a few young Ganoids external gills are superadded to the internal.

In *Branchiostoma* the dilated pharynx is perforated by numerous clefts, supported by cartilaginous rods (Fig. 29, *h*). The water passes between these clefts into the peritoneal cavity, and makes its exit by the porus abdominalis situated considerably in advance of the vent. The water is propelled by cilia.

In the *Cyclostomes* the gills of each side are lodged in a series of six or more antero-posteriorly compressed sacs, separated from each other by intervening septa. Each sac communicates by an inner duct with the œsophagus, the water being expelled by an outer duct. In *Bdellostoma* each outer duct has a separate opening, but in *Myxine* all the outer ducts pass outwards by one common gill-opening on each side. In the Lampreys the ducts are short, the outer ones having separate openings (Fig. 2, p. 39). The inner ducts lead into a single diverticulum or bronchus, blind behind, situated below the œsophagus, and communicating in front with the pharynx, where it is provided with two valves by which the regurgitation of the water into the buccal cavity is prevented.

The same type of branchial organs persists in *Chondropterygians*, which possess five, rarely six or seven, flattened pouches with transversely plaited walls. The septa between them are supported by cartilaginous filaments rising from the hyoidean and branchial arches. Each pouch opens by a cleft outwards, and by an aperture into the pharynx, without intervening ducts. The anterior wall of the first pouch is supported by the hyoidean arch. Between the posterior wall of the first and the anterior of the second sac, and between the adjacent walls of the succeeding, a branchial arch with its two series of radiating cartilaginous filaments is interposed. Consequently the first and last pouch have one set of gill-laminæ only, viz. the first on its posterior and the last on its anterior wall. The so-called *spiracles* on the upper surface of the head of Chondropterygians are to be referred to in connection with the respiratory organs. They are the external openings of a canal leading on each side into the pharynx, and situated generally close to and behind the orbit. They frequently possess valves or an irregularly indented margin, and are found in all species during the embryonic stage, but remaining persistent in a part only. The spiracles are the remains of the first visceral cleft of the embryo, and in the fœtal state long branchial filaments have been observed to protrude, as from the other branchial clefts.

The *Holocephali* and *Ganoidei* show numerous deviations from the Chondropterygian type, all leading in the direction towards the Teleosteans. As a whole they take an intermediate position between the preceding types and the Teleosteans, but they show a great variation among themselves, and have in common only the imperfect separation of the branchial sacs and the presence of a single outer branchial aperture.

In *Chimæra* the septum separating the branchial sacs is confluent with the wall of the gill-cavity in a part of its extent only, and still more imperfect is the separation of those branchial divisions in *Ceratodus* (Fig. 60). The other Ganoids show no such division whatever. In *Chimæra* the first gill is incomplete (uniserial), and belongs to the hyoid; then follow three complete gills; the last, belonging to the fourth branchial arch, being again incomplete. *Acipenser*, *Scaphirhynchus*, *Lepidosiren*, *Protopterus*, and *Lepidosteus*, possess likewise an anterior incomplete gill (*opercular gill*), followed by four complete gills in the Sturgeons and *Lepidosteus*, whilst in

Lepidosiren and *Protopterus* a part of the branchial arches is gill-less. In *Polyodon, Ceratodus,* and *Polypterus,* an opercular gill is absent, the two former having four complete gills, the latter three and a half only. *Spiracles* are still in some Ganoids present, viz. in the Sturgeons and *Polypterus.* In all the Ganoids an osseous gill-cover is now developed.

In the *Teleostei* the gills with their supporting branchial arches lie in one undivided cavity; more or less wide clefts between the arches lead from the pharynx to the gills, and a more or less wide opening gives exit to the water after it has washed the gills. The interbranchial clefts have sometimes nearly the same extent as the branchial arches; sometimes they are reduced to small openings, the integuments stretching from one arch to the other. Sometimes there is no cleft behind the fourth arch, in which case this arch has only an uniserial gill developed. The gill-opening likewise varies much in its extent, and when reduced to a foramen may be situated at any place of the posterior boundary of the head. In the *Symbranchidæ* the gill-openings coalesce into a single narrow slit in the median line of the isthmus. In the majority of Teleosteans the integument of the concave side of the branchial arches develops a series of horny protuberances of various form, the so-called *gill-rakers.* They are destined to catch any solid corpuscles or substances which would be carried into the gill-cavity with the water. In some fishes they are setiform, and form a complete sieve, whilst in others they are merely rough tubercles, the action of which must be very incomplete if they have any function at all.

Most Teleosteans possess four complete gills, but frequently the fourth arch is provided with an uniserial gill only, as mentioned above, or even entirely gill-less. The most imperfect gills are found in *Malthe,* which has two and a half gills only, and in *Amphipnous cuchia,* in which one small gill is fixed to the second arch.

The gills of the Teleosteans as well as of the Ganoids are supported by a series of solid cartilaginous or horny pointed rods, arranged along the convex edges of the branchial arches. Arches bearing a complete gill have two series of those rods, one along each edge; those with uniserial gills bear one row of rods only. The rods are not part of the arch, but fixed in its integument, the several rods of one row corresponding to those of the other, forming pairs (*feuillet*, Cuvier) (Fig. 59). Each rod is covered by a loose mucous membrane passing from one rod to its fellow opposite, which again is finely transversely plaited, the general surface being greatly increased by these plaits. In most Teleostei the branchial lamellæ are compressed, and taper towards their free end, but in the Lophobranchs their base is attenuated and the end enlarged. The mucous membrane contains the finest terminations of the vessels, which, being very superficial, impart the blood-red colour to living gills. The *Arteria branchialis*, the course of which lies in the open canal in the convexity of the branchial arch, emits a branch (*a*) for every pair of lamellæ which ascends (*b*) along the inner edge of the lamella, and supplies every one of the transverse plaits with a branchlet. The latter break up into a fine net of capillaries, from which the oxygenised blood is collected into venous branchlets, returning by the venous branch (*d*), which occupies the outer edge of the lamella.

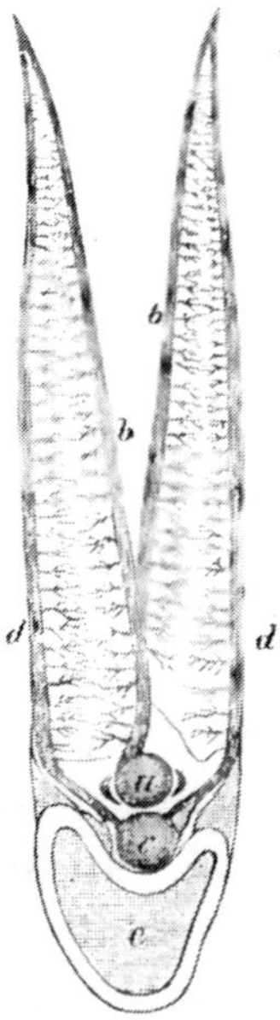

Fig. 59.—A pair of branchial lamellæ (magnified) of the Perch.
a, Branch of Arteria branchialis; *b*, Ascending branch of the same; *c*. Branch of Vena branchialis; *d*, Descending branch of the same; *e*, Transverse section through the branchial arch.

The so-called *Pseudobranchiæ* (Fig. 60) are the remains of an anterior gill which had respiratory functions during the embryonic life of the individuals. By a change in the circulatory system these organs have lost those functions, and appear in the adult fish as retia mirabilia, as they receive oxygenised blood, which, after having passed through their capillary system, is carried to other parts of the head. In Palæichthyes the pseudobranchia is a rete mirabile caroticum for the brain and eye; in Teleosteans a rete mirabile ophthalmicum only. Pseudobranchiæ are as frequently absent as present in Chondropterygians as well as Teleosteans. As to the Ganoids, they occur in *Ceratodus*, *Acipenser*, *Polyodon*, and *Lepidosteus*, and are absent in *Lepidosiren*, *Protopterus*, *Scaphirhynchus*, *Polypterus*, and *Amia*.

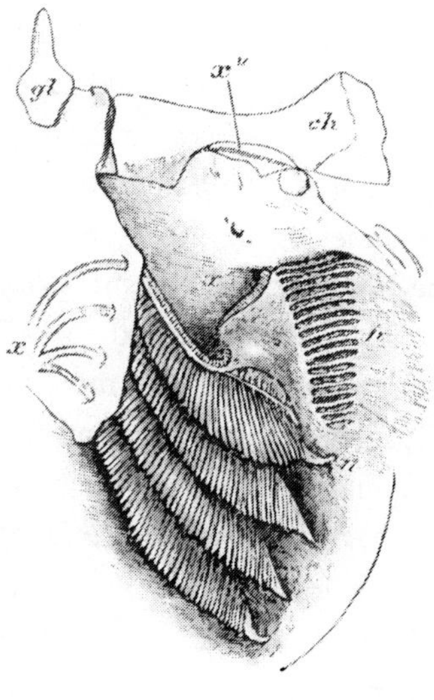

Fig. 60.—Gills of Ceratodus. *x*, Arcus aortæ; *gl*, Glossohyal; *ch*, Ceratohyal; *u*, Attachment of the first gill to the walls of the gill-cavity; *h*, Pseudobranchia; *x′*, *x″*, two series of gill-rakers belonging to the Pseudobranchia.

In Chondropterygians and Sturgeons the pseudobranchiæ are situated within the spiracles; in those, in which spiracles have become obliterated, the pseudobranchiæ lie on the suspensorium, hidden below cellular tissue; but pseudobranchiæ are not necessarily co-existent with spiracles. In the other Ganoids and Teleosteans the pseudobranchiæ (Fig. 60, *h*) are within the gill-cavity, near the base of the gill-cover; in *Ceratodus* even rudiments of the gill-rakers (x', x'') belonging to this embryonic gill are preserved, part of them (x'') being attached to the hyoid arch. Pseudobranchiæ are frequently hidden below the integuments of the gill-cavity, and have the appearance of a glandular body rather than of a gill.

[See Müller, "Vergleichende Anatomie des Gefäss-systems der Myxinoiden;" and "Ueber den Bau und die Grenzen der Ganoiden."]

Accessory respiratory organs for retaining water or breathing air, such as are found in the *Labyrinthici*, *Ophiocephalidæ*, certain *Siluridæ*, and *Lutodira*, are structures so specialised that they are better described in the accounts of the Fishes in which they have been observed.

Air-Bladder.—The air-bladder, one of the most characteristic organs of fishes, is a hollow sac, formed of several tunics, containing gas, situated in the abdominal cavity, but without the peritoneal sac, entirely closed or communicating by a duct with the intestinal tract. Being compressible, its special functions consist in altering the specific gravity of the fish or in changing the centre of gravity. In a few fishes it assumes the function of the organ of higher Vertebrates, of which it is the homologue—viz. of a *lung*.

The gas contained in the air-bladder is secreted from its inner surface. In most fresh-water fishes it consists of nitrogen, with a very small quantity of oxygen and a trace of carbonic acid; in sea-fishes, especially those living at some depth, oxygen predominates, as much as 87 per cent having been found. Davy found in the air-bladder of a fresh-run Salmon a trace of carbonic acid and 10 per cent of oxygen, the remainder of the gas being nitrogen.

An air-bladder is absent in *Leptocardii*, *Cyclostomi*, *Chondropterygii*, and *Holocephali;* but occurs in all Ganoids, in which, besides, its respiratory functions more or less clearly manifest themselves. Its occurrence in Teleosteans is most irregular, closely allied species sometimes differing from each other in this respect; it shows in this sub-class the most extraordinary modifications, but has no respiratory function whatever.

Constantly situated within the abdominal cavity, below the vertebral column, but without the sac of the peritoneum which covers only its ventral portion, the air-bladder is frequently prolonged into the tail, the prolongation being either simple and lodged between the non-united parapophyses, or double and penetrating between the muscles and hæmapophyses of each side. In the opposite direction processes of the air-bladder may penetrate into the skull, as has been mentioned above (p. 117). In some fishes the air-bladder is almost loose in the abdominal cavity, whilst in others it adheres most intimately by firm and short tissue to the vertebral column, the walls of the abdomen, and the intestines. In the Cobitina and many Siluroids it is more or less completely enclosed in osseous capsules formed by the vertebræ.

The tunics of the majority of air-bladders are an extremely fine internal one, frequently shining silvery, containing crystalline corpuscles, sometimes covered with a pavement-epithelium; and a thicker outer one of a fibrous texture, which sometimes attains to considerable thickness and yields isinglass. This wall is strengthened in many fishes by muscular layers for the compression of the whole organ or of some portion of it.

A distinction has been made between air-bladders which communicate by a duct with the intestinal tract and those which are entirely closed. However, it is to be remembered that at an early stage of development all air-bladders are provided with such a duct, which in a part of the fishes more or less completely obliterates, being then represented by a fine ligament only. In young *Lucioperca* of six to eight inches in length the duct may be found still open for a considerable distance; and, on the other hand, in adult *Physostomi*, that is Teleosteous fishes with a ductus pneumaticus, not rarely the whole duct is found very narrow, or, for some part of its length, even entirely closed.

Air-bladders without duct are found in Acanthopterygians, Pharyngognaths, Anacanths, and Lophobranchs. They may consist of a single cavity or divided by constrictions into two or three partitions situated behind one another; they may consist of two lateral partitions, assuming a horseshoe-like form, or they may be a single sac with a pair of simple or bifid

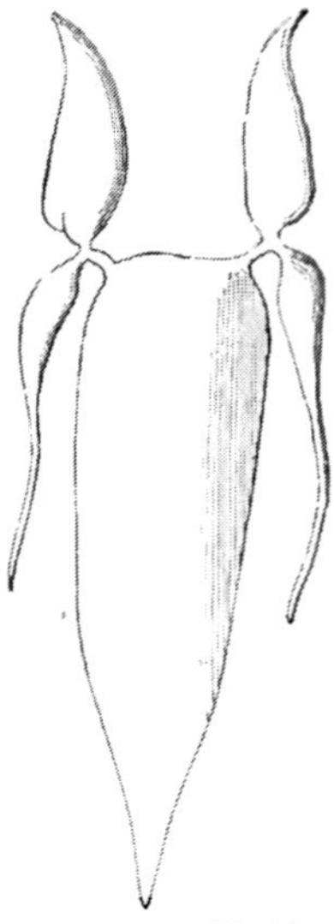

Fig. 61.—Air-bladder of Otolithus sp.

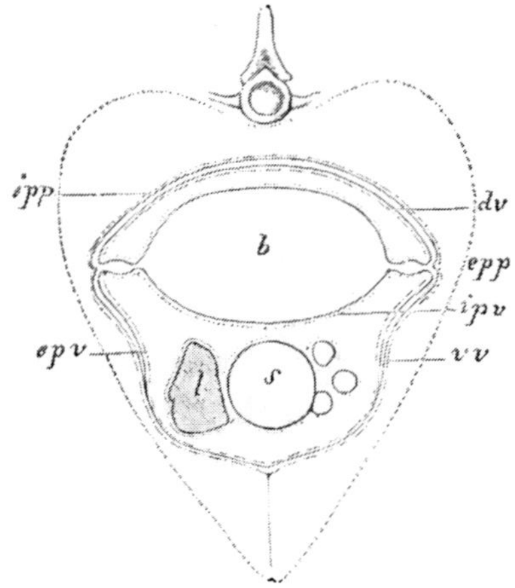

Fig. 62.—Vertical section through abdominal cavity of Collichthys lucida. *b*, air-bladder; *l*, liver; *s*, stomach; *epp* and *ipp*, external and internal laminæ of peritoneum parietale; *epv* and *ipv*, external and internal laminæ of peritoneum viscerale; *dv*, dorsal air-vessels; *vv*, ventral air-vessels.

processes in front or behind (Fig. 61). The families of *Sciænidæ* and *Polynemidæ* possess air-bladders with a most extraordinary development of appendages rising from each side of the air-bladder. In the Sciænoid (Fig. 63) fifty-two branches issue from each side, each branch being bifurcate and bearing smaller appendages. In *Pogonias chromis* (Fig. 64) the sides of the anterior half is provided with irregular broad-fringed appendages, the hindmost of which communicates by a narrow duct with the posterior extremity of the air-bladder. In *Collichthys lucida* (Fig. 62) twenty-five appendages issue from each side; the anterior ones are directed towards the front, but the lateral assume a more posterior direction, the nearer they are to the posterior extremity of the air-bladder, where they form an assemblage giving the appearance of a cauda equina. All these appendages soon bifurcate in a dorsal and ventral stem; these stems bifurcate again and again, and either terminate after the first or second bifurcation or are so far prolonged as to reach the median line of the ventral and dorsal sides, anastomosing with the branches of the other side. The branches being enveloped in laminæ of the peritonæum, form a dorsal and ventral sac of beautiful appearance, caused by the regular arrangement of the air-vessels. The dorsal sac is situated between the air-bladder and the roof of the abdominal cavity without being attached to the latter. The ventral sac receives within its cavity the intestine, liver, and ovaries.—A peculiar mechanism has been observed in the air-bladder of the *Ophidiidæ*, the anterior portion of which can be prolonged by the contraction of two muscles attached to its anterior extremity, with or without the addition of a small bone.

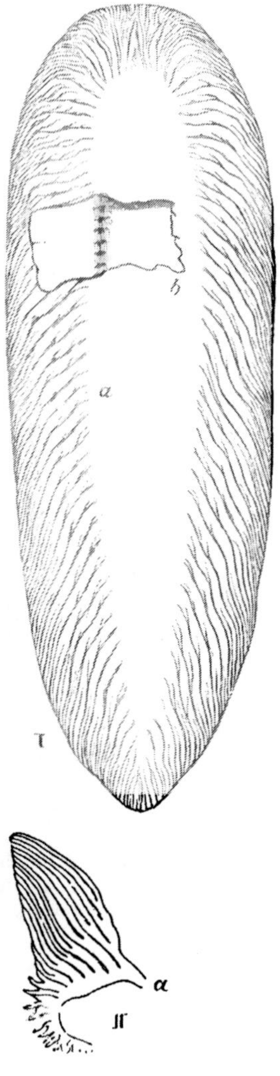

Fig. 63.—Air-bladder of a Sciænoid.
I. Visceral surface opened at *b*, to show openings of the lateral branches.
II. Isolated lateral branch; *a*, its opening into the cavity of the air-bladder.

Air-bladders with a pneumatic duct are found in Ganoids and Physostomes, the duct entering the dorsal side of the intestinal tract, with the exception of *Polyp-*

terus and the *Dipnoi*, in which it enters on the ventral side of the œsophagus. In the majority the orifice is in the œsophagus, but in some, as in *Acipenser*, in the cardiac portion of the stomach, or in its blind sac, as in many Clupeoids. The air-bladder may be single, or consist of two divisions situated one behind the other (Fig. 52); its inner surface may be perfectly smooth, or form manifold pouches and cells. If

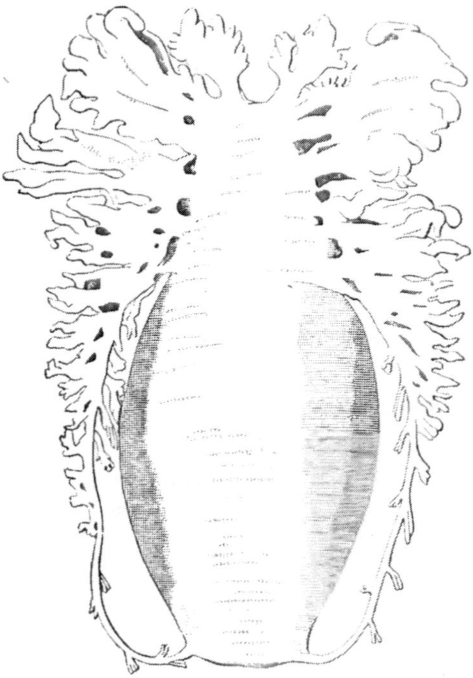

Fig. 64.—Air-bladder of Pogonias chromis.

two divisions are present the anterior possesses a middle elastic membrane which is absent in the posterior; each division has a muscular layer, by which it can be separately compressed, so that part of the contents of the posterior may be driven into the elastic anterior division, and *vice versa*. The posterior division being provided with the ductus pneumaticus does not require the elasticity of the anterior.

Some Siluroids possess a peculiar apparatus for voluntarily exercising a pressure upon the air-bladder. From the first vertebra a process takes its origin on each side, expanding at its end into a large round plate; this is applied to the side of the air-bladder, and by pressing upon it expels the air through the duct; the small muscle moving the plate rises from the skull.

The connection of the air-bladder with the organ of hearing in some Physostomes has been described above, p. 117.

In the modifications of the air-bladder, hitherto mentioned, the chief and most general function is a mechanical one; this organ serves to regulate the specific gravity of the fish, to aid it in maintaining a particular level in the water, in rising or sinking, in raising or depressing the front part of its body as occasion may serve. Yet a secretion of gas from the blood into its cavity must take place; and if this be so, it is not at all impossible that also an exchange of gases between the two kinds of blood is effected by means of the extraordinary development of *retia mirabilia* in many air-bladders.

In all fishes the arteries of the air-bladder take their origin from the aorta or the system of the aorta, and its veins return either to the portal, or vertebral, or hepatic veins; like the other organs of the abdominal cavity it receives arterial blood and returns venous blood. However, in many fishes the arteries as well as veins break up below the inner membrane into *retia mirabilia* in various ways. The terminal ramifications of the arteries may dissolve into fan-like tufts of capillaries over almost every part of the inner surface, as in Cyprinoids. Or these tufts of radiating capillaries are more localised at various places, as in *Esocidæ;* or the tufts are so aggregated as to form gland-like, *red bodies,* the capillaries reuniting into larger vessels, which again ramify freely round the border of the red body; the red bodies are formed not only by minute arteries but also by minute veins, both freely anastomosing with its kind, and being inextricably inter-

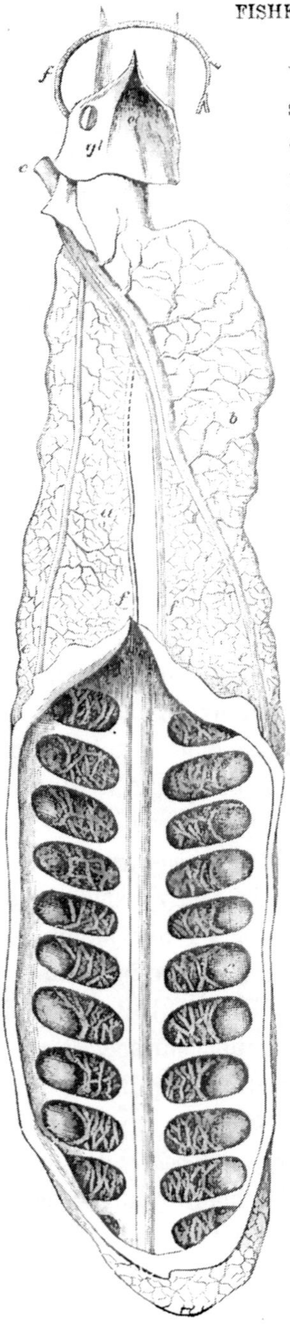

Fig. 65.—Lung of Ceratodus opened in its lower half to show its cellular pouches. *a*, Right half; *b*, Left half; *c*, Cellular pouches; *e*, Vena pulmonalis; *f*, Arterial blood-vessel; *oe*, Œsophagus opened, to show glottis (*gl.*)

woven. The rest of the inner surface of the air-bladder receives its blood, not from the red bodies, but from normally ramifying vessels. This kind of rete mirabile or "vaso-ganglion" is found in the Perch and Gadoids; it is generally distributed in closed air-bladders, but also sometimes observed in air-bladders with pneumatic duct. In *Anguilla* and *Conger* two similar vaso-ganglia are situated at the sides of the opening of the pneumatic duct.

Whilst the air-bladders of some *Ganoids*, anatomically as well as functionally, closely adhere to the Teleosteous type, that of *Amia* is more cellular and lung-like in its interior than the Teleosteous air-bladder, and *Polypterus* approaches the Dipnoi not only in having a laterally divided air-bladder but also in its pneumatic duct entering the *ventral* side of the œsophagus. The air-bladder of the *Dipnoi* possesses still more the anatomical characteristics of a lung and assumes its functions, though, as it coexists with gills, only periodically or in an auxiliary manner. The ductus pneumaticus is a membranous bronchus, entering the ventral side of the œsophagus, and provided at its entrance with a *glottis*. In *Ceratodus* (Fig. 65) the lung is still a single cavity, but with a symmetrical arrangement of its internal pouches; it has no pulmonal artery, but receives branches from the *arteria cœliaca*. Finally, in *Lepidosiren* and *Protopterus* the lung is completely divided into lateral halves, and by its cellular structure approaches most nearly that of a reptile; it is supplied with venous blood by a true pulmonary artery.

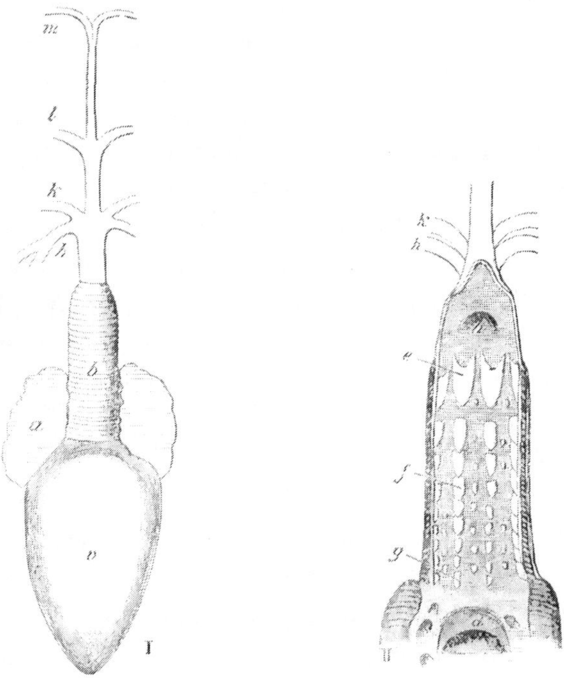

Fig. 66.—Heart of Lepidosteus osseus.

I. External aspect. II. Conus arteriosus opened.

a, Atrium ; *b*, Conus arteriosus ; *v*, Ventricle ; *h*, Branchial artery for 3d and 4th gill ; *k*, for the second ; *l*, for the first ; *m*, branch for the opercular gill ; *d*, Single valve at the base of the conus ; *e-g*, Transverse rows of Ganoid valves.

CHAPTER X.

ORGANS OF CIRCULATION.

The *Blood-corpuscles* of fishes are, with one exception, of an elliptic shape; this exception is *Petromyzon*, which possesses circular, flat, or slightly biconvex blood-corpuscles. They vary much in size; they are smallest in Teleosteans and Cyclostomes, those of *Acerina cernua* measuring $\frac{1}{2481}$ of an inch in their longitudinal, and $\frac{1}{3000}$ in their transverse diameter. As far as it is known at present the *Salmonidæ* have the largest blood-corpuscles among Teleosteans, those of the salmon measuring $\frac{1}{1524}$ by $\frac{1}{2400}$ in., approaching those of the Sturgeon. Those of the Chondropterygians are still larger; and finally, *Lepidosiren* has blood-corpuscles not much smaller than those of Perennibranchiates, viz.—$\frac{1}{570}$ by $\frac{1}{941}$ in. Branchiostoma is the only fish which does not possess red blood-corpuscles.

[See G. Gulliver, "Proc. Zool. Soc.," 1862, p. 91; and 1870, p. 844; and 1872, p. 333.]

Fishes, in common with the other Vertebrates, are provided with a complete circulation for the body, with another equally complete for the organs of respiration, and with a particular abdominal circulation, terminating at the liver by means of the *vena portæ;* but their peculiar character consists in this, that the branchial circulation alone is provided at its base with a muscular apparatus or *heart*, corresponding to the right half of the heart of Mammalia and Birds.

The *Heart* is situated between the branchial and abdominal cavities, between the two halves of the scapulary arch, rarely farther behind, as in *Symbranchidæ*. It is enclosed in a *pericardium*, generally entirely separated from the abdominal cavity by a diaphragma, which is, in fact, the anterior portion of the peritoneum, strengthened by aponeurotic fibres. However, in some fishes there is a communication between the pericardial and peritoneal sacs, viz. in the Chondropterygians and Acipenser, whilst in the Myxinoids the pericardial sac is merely a continuation of the peritoneum.

The heart is, relatively to the size of the body, very small, and consists of three divisions: the *atrium*, with a large *sinus venosus* into which the veins enter; the *ventricle;* and a conical hollow swelling at the beginning of the arterial system, the structure of which forms one of the most important characters used in the classification of fishes. In all *Palæichthyes* (Figs. 66 and 67) this swelling is still a division of the pulsating heart, being provided with a thick muscular stratum; it is not separated from the ventricle by two valves opposite to each other, but its interior is fitted with a plurality of valves, arranged in transverse series more or less numerous in the various groups of *Palæichthyes*. *Lepidosiren* and *Protopterus* offer an example of a modification of this valvular arrangement, their valves being longitudinal, each valve in fact being formed by the confluence of several smaller ones situated behind one another. This Palæichthyan type is called *conus arteriosus*.

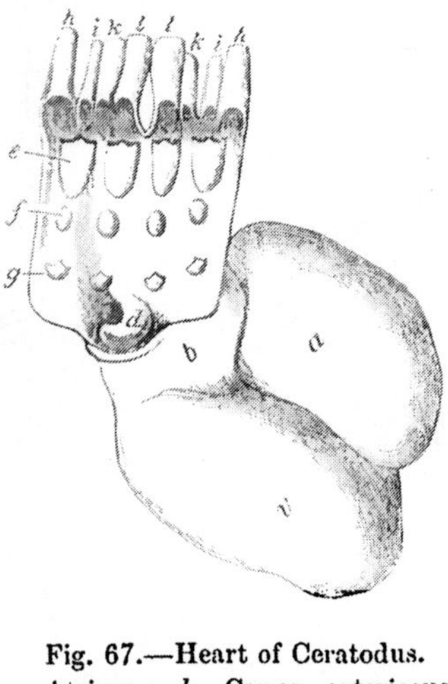

Fig. 67.—Heart of Ceratodus. *a*, Atrium; *b*, Conus arteriosus; *d*, Papillary valve within the conus; *c-g*, Transverse rows of Ganoid valves; *h*, *i*, Anterior arcus aortæ; *k*, *l*, Posterior arcus aortæ; *v*, Ventricle.

In Cyclostomes and Teleosteans (Fig. 68) the enlargement

is a swelling of the artery, without muscular stratum and without contractility; with the exception of the Myxinoids its walls are thick, fibrous, with many trabeculæ and pouches, but it has no valves in its interior, and is separated from the ventricle by two valves opposite to each other. This Teleostean type is called *bulbus aortæ*.

The sinus venosus sends the whole of the venous blood by a single orifice of its anterior convexity into the atrium;

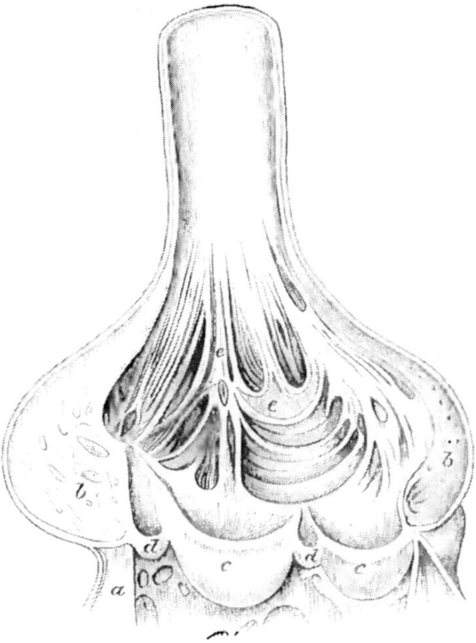

Fig. 68.—Bulbus aortæ of Xiphias gladius, opened.

a, Section through part of the wall of ventricle; *b*, Section through the bulbus; *c*, Teleosteous valves of the ostium arteriosum; *d*, Accessory valves, of rudimentary nature and inconstant; *e*, Trabeculæ carneæ of the bulbus.

two thin membranous valvules turned towards the atrium, prevent the blood from re-entering the sinus. A pair of other valves between atrium and ventricle have the same function. The walls of the ventricle are robust, and, internally, it is furnished with powerful fleshy trabeculæ.

The bulbus or conus arteriosus is prolonged into the branchial artery which soon divides, sending off a branch to each branchial arch. On returning from the respiratory organ the branchial veins assume the structure and functions of arteries. Several branches are sent off to different portions of the head and to the heart, but the main trunks unite to form the great artery which carries the blood to the viscera and all the parts of the trunk and tail, and which, therefore, represents the *aorta* of higher animals.

In the majority of Teleosteans the *aorta* has proper walls formed by its own membranes, but in the Sturgeons it is independent at its commencement only, and replaced by a canal formed by hæmal elements of the vertebral column, and clothed inside with a perichondrium. In many Chondropterygians and some Teleosteans (*Esox, Clupea, Silurus*), the aorta possesses its own firm membranes along its ventral side, dorsally being protected by a very thin membrane only, attached to the concavity of the centra of the vertebræ.

The circulatory system of *Branchiostoma* and of the *Dipnoi* shows essential differences from that of other fishes.

Branchiostoma is the only fish which does not possess a muscular heart, several cardinal portions of its vascular system being contractile. A great vein extends forwards along the caudal region below the notochord, and exhibits contractility in a forward direction; it is bent anteriorly, passing into another tube-like pulsatile trunk, the branchial heart, which runs along the middle of the base of the pharynx, sending off branches on each side to the branchiæ; each of these branches has a small contractile dilatation (*bulbillus*) at its base. The two anterior branches pass directly into the aorta, the others are branchial arteries, the blood of which returns by branchial veins emptying into the aorta. The blood of the intestinal veins is collected in a contractile tube, the portal vein, situated below the intestine, and distributed over the rudimentary liver. Of all other fishes, only in *Myxinoids* the portal vein is contractile. All the blood-corpuscles of *Branchiostoma* are colourless and without nucleus.

In *Dipnoi* a rudimentary division of the heart into a right and left partition has been observed; this is limited to the ventricle in *Ceratodus*, but in *Lepidosiren* and *Protopterus* an incomplete septum has been observed in the atrium also. All Dipnoi have a pulmonal vein, which enters the atrium by a separate opening, provided with a valve. The pulmonal artery rises in *Lepidosiren* and *Protopterus* from an arch of the aorta, but in *Ceratodus* it is merely a subordinate branch, rising from the *Arteria cœliaca.*

CHAPTER XI.

URINARY ORGANS.

In *Branchiostoma* no urinary organs have been found.

In Myxinoids these organs are of a very primitive structure: they consist of a pair of ducts, extending from the urogenital porus through the abdominal cavity. Each duct sends off at regular intervals from its outer side a short wide branch (the uriniferous tube), which communicates by a narrow opening with a blind sac. At the bottom of this sac there is a small vaso-ganglion (*Malpighian corpuscle*), by which the urine is secreted.

In the Lampreys the kidneys form a continuous gland-like body, with irregular detached small portions. The ureters coalesce before they terminate in the urogenital papilla.

In Chondropterygians the kidneys occupy the posterior half or two-thirds of the back of the abdominal cavity, without the sac of the peritoneum (as in all fishes) which forms a firm tendinous horizontal septum. The kidneys of the two sides are never confluent, and generally show a convoluted or lobulated surface. The ureters are short; each is dilated into a pouch, and communicating with its fellow terminates by a single urethra (which also receives the vasa deferentia) behind the end of the rectum in the large common cloaca.

In Ganoids the kidneys occupy a similar position as in Chondropterygians, but these fishes differ considerably with regard to the termination and the arrangement of the ends of the urogenital ducts. The Dipnoi possess a cloaca. In *Ceratodus* the ureters open into it by a common opening, separate from the genital opening; and no closed urinary bladder has been developed. *Lepidosiren* has a small urinary bladder; the ureters do not communicate directly with it, but terminate separately on small papillæ in the dorsal compartment of the cloaca. The other Ganoids lack a cloaca, and the urogenital opening is behind the vent as in Teleosteans. In all the genital and urinary ducts coalesce towards their end. The Sturgeons have no urinary bladder, whilst it is present in *Amia*, the ureters opening separately into it.

The *kidneys* of Teleosteans are situated likewise without the peritoneal cavity, immediately below some part of the vertebral column, and vary exceedingly with regard to form and extent. Sometimes they reach from the skull to between the muscles of the tail, sometimes they are limited to the foremost part of the abdominal cavity (in advance of the diaphragm), but generally their extent corresponds to that of the abdominal portion of the vertebral column. Frequently they are irregular on their dorsal surface, filling every available recess, flat, attenuated on the sides, more or less coalescent towards the middle; in other fishes they are more compact bodies. The ureters terminate, either separate or united, in a urinary bladder, varying in shape, which opens by a short urethra behind the vent. The urinary opening may be separate or confluent with that of the genital ducts, and is frequently placed on a more or less prominent papilla (*papilla urogenitalis*). If separate, the urinary opening is behind the genital; and if a papilla is developed, its extremity is perforated by the urethra, the genital opening being situated nearer the base. A few Teleosteans show an arrangement similar to that of Chondropterygians and Dipnoi, the urogenital openings being in the posterior wall of the rectum (*Symbranchidæ*, *Pediculati*, and some *Plectognathi*).

CHAPTER XII.

ORGANS OF REPRODUCTION.

ALL fishes are *dioecious*, or of distinct sex. Instances of so-called *hermaphroditism* are, with the exception of *Serranus*, abnormal individual peculiarities, and have been observed in the Cod-fish, some Pleuronectidæ, and in the Herring. Either the generative organ of one side was found to be male, that of the other female; or the organ of one or both sides was observed to have been developed partly into an ovary partly into a testicle. In the European species of Serranus a testicle-like body is attached to the lower part of the ovary; but many specimens of this genus are undoubtedly males, having normally developed testicles only.

The majority of fishes are oviparous, comparatively few viviparous; the embryos being developed either in the ovarium or in some dilated portion of the oviduct. In viviparous fishes actual copulation takes place, and the males of most of them are provided with copulatory or intromittent organs. In oviparous fishes the generative products are, during sexual excitement, discharged into the water, a very small quantity of semen being sufficient for effectual impregnation of a number of ova dispersed in a considerable quantity of water; circumstances which render *artificial impregnation* more practicable than in any other class of animals.

In *Branchiostoma* the generative organs occupy the ventral side of the abdominal cavity, into which they discharge their contents. No ducts are developed in either sex.

In the *Cyclostomes* the generative organ is single, and fixed to or suspended from the median line of the back of the visceral cavity by a duplicature of the peritoneum (*mesoarium*); the testicle and ovary being distinguishable by their contents only. These escape by dehiscence of the cells or capsules and rupture of the peritoneal covering into the abdominal cavity, and are expelled by reciprocal pressure of the intertwined sexes through the *porus genitalis*, which is sunk between two labia of the skin in *Myxine*, and produced into a long papilla in Petromyzon.

The ova of the Lampreys are small, globular, like those of Teleosteans. Those of Myxine have a very peculiar shape when mature; they are of an oval form, about 15 millimetres long and 8 millimetres broad, enveloped in a horny case, which at each end is provided with a bundle of short threads, each thread ending in a triple hook. Whilst in the mesoarial fold the eggs are attached to one another by means of these hooks, and after being expelled they probably fix themselves by the same means to other objects. As in all fishes producing ova of large size, the number of ova matured in one season is but small.

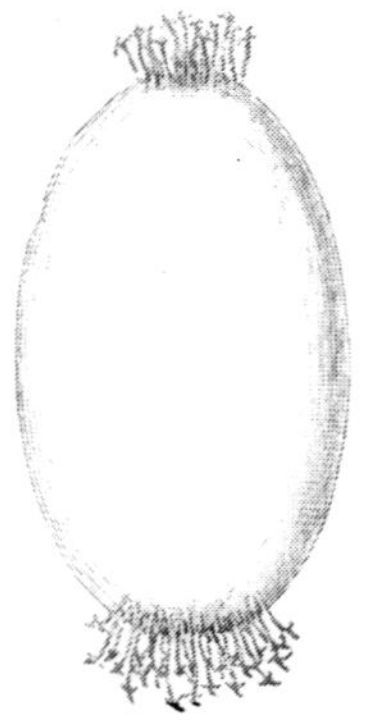

Fig. 69.—Ovum of Myxine glutinosa, enlarged.

In *Teleosteans* the generative organs are comparatively large. In some families the ovaries are without closed covering and without oviducts, as in *Salmonidæ*, *Galaxiidæ*, *Notopteridæ*, *Murænidæ*, and others. The surface of such an open ovary—as, for instance, that of the Salmon—is transversely plaited, the ova being developed in capsules in the stroma of the laminæ; after rupture of the capsules the mature ova drop into the abdominal cavity, and are expelled by the porus genitalis. The ovaries of the other Teleosteans are closed sacs, continued into oviducts. Frequently such ovaries coalesce into a single body, or one in which the division is effected internally only by a more or less complete septum. Fixed by a mesoarium, the ovaries occupy generally a position outwards of the intestine or air-bladder; their form varies as well as the thickness and firmness of their covering, which frequently is an extremely thin transparent membrane. The inner surface of the ovarian sac is transversely or longitudinally plaited or covered with fringes, on which the ova are developed, as in the open ovaries. In the viviparous Teleosteans the embryons are likewise developed within the ovary, notably in the *Embiotocidæ*, many *Blenniidæ*, and *Cyprinodontidæ*, *Sebastes vivi-*

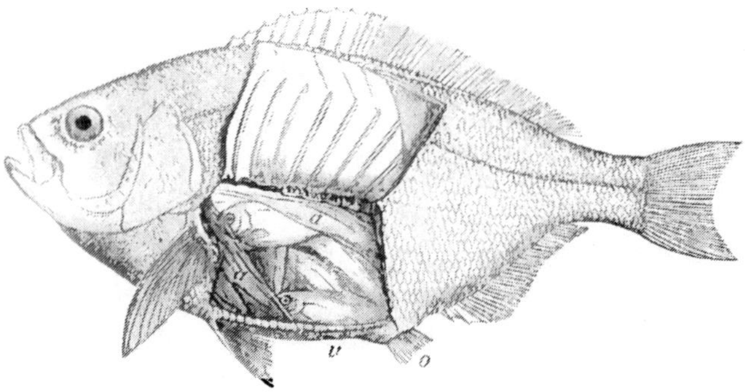

Fig. 70.—Ditrema argenteum, with fully developed young, ready for expulsion by the genital orifice, *o*; *a*, folds of the ovarian sac ; *v*, vent.

parus, etc. Among the Cyprinodonts the end of the oviduct is attached to the anterior anal rays, which are modified into supports of its termination. In *Rhodeus* the oviduct is periodically prolonged into a long oviferous tube, by means of which the female deposits her ova into the shells of living Bivalves.

The *ova* of Teleosteous Fishes are extremely variable in size, quite independently of the size of the parent species. The ova of large and small individuals of the same species, of course, do not differ in size; but, on the whole larger individuals produce a greater number of ova than smaller ones of the same species. The larger the size of the ova is in a species, the smaller is the number produced during one season. The ova of the Eel are almost microscopic. The small sized roe in the Herring, Lump-fish, Halibut, and Cod-fish, have been estimated at respectively 25,000, 155,000, 3,500,000, and 9,344,000. Larger in size and fewer in number are those of *Antennarius, Salmo, Aspredo, Lophobranchs,* etc. Comparatively largest are those of *Gastrosteus;* and the Siluroid genus *Arius,* the males of which take care of their progeny, produces ova from 5 to 10 millimeters in diameter. The ova of all Teleosteans are perfectly globular and soft-shelled. Teleosteans without oviduct, deposit them separated from one another; whilst in many Teleosteans with an oviduct the ova are enveloped in a glutinous substance, secreted by its glands, swelling in the water and forming lumps or cords, in which the ova are aggregated.

Fig. 71. Ovum of Arius boakii (Ceylon), showing embryo. Nat. size.

Instances of the female taking care of her progeny are extremely scarce in fishes. At present only two examples are known, that of the Siluroid genus *Aspredo,* and of *Solenostoma.* In the former, during the time of propagation, the integuments of the lower side of the flat trunk of the female assume a soft and spongy texture. After having deposited the eggs, the female attaches them to, and presses them into, the spongy integument, by merely lying over them. She carries them on her belly, as the Surinam Toad (*Pipa*) carries her ova on the back. When the eggs are hatched the excrescence on the skin disappears, and the abdomen becomes as smooth as before. In *Solenostoma* the inner side of the long and broad ventral fins coalesces with the integuments of the body, a large pouch being formed for the reception of the eggs. There is a peculiar provision for the retention of the eggs in the sac, and probably for the attachment of the embryo.

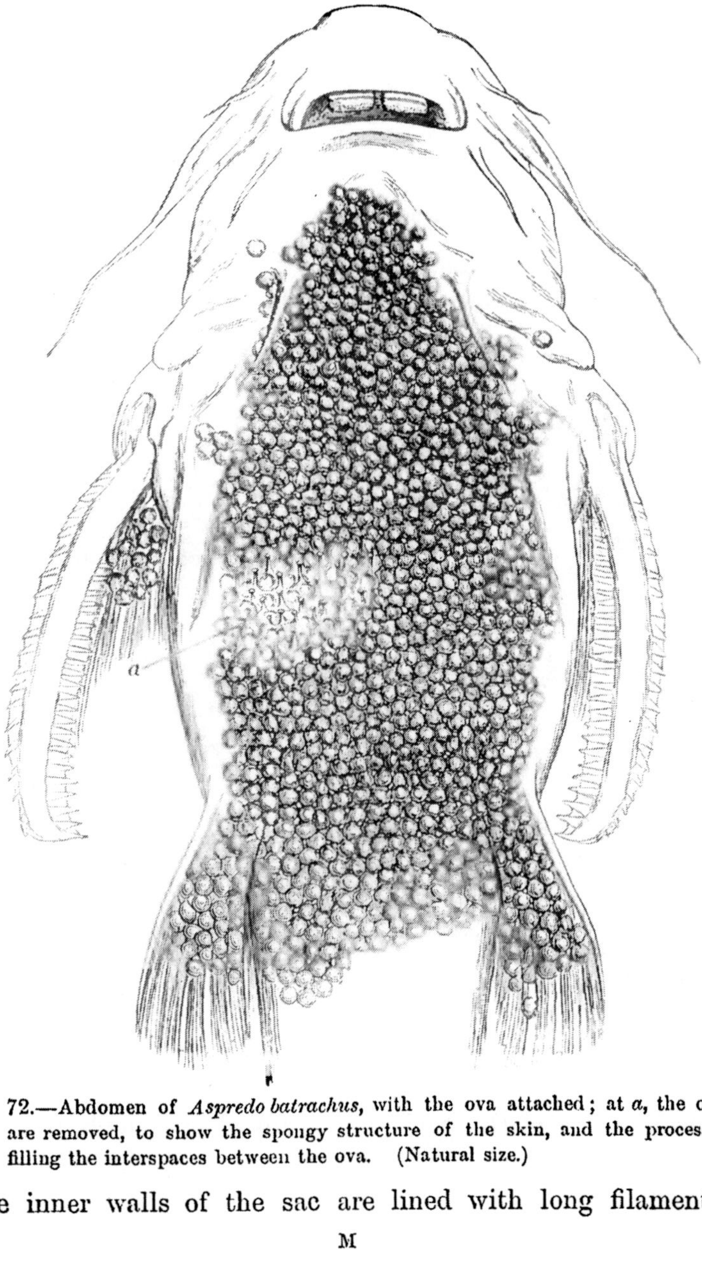

Fig. 72.—Abdomen of *Aspredo batrachus*, with the ova attached; at *a*, the ova are removed, to show the spongy structure of the skin, and the processes filling the interspaces between the ova. (Natural size.)

The inner walls of the sac are lined with long filaments,

arranged in series along the ventral rays, and more nume-

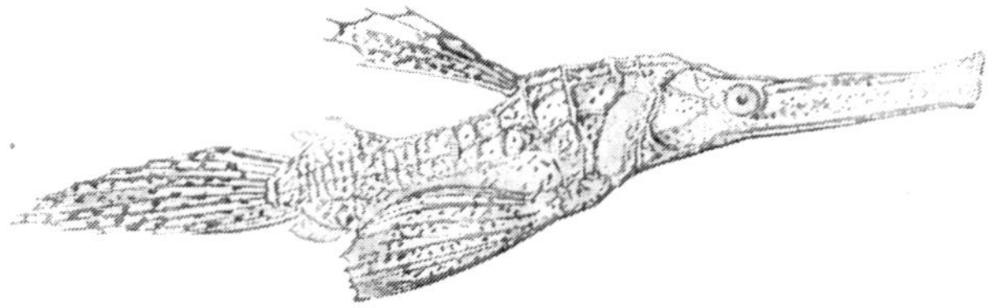

Fig. 73.—Solenostoma cyanopterum ♀ (Indian Ocean).

rous and longer at the base of the rays than in the middle of their length, behind which they disappear entirely. They are also more developed in examples in which eggs are deposited in the sac than in those which have the sac empty. The filaments most developed have a length of half an inch, and are beset with mamilliform appendages. A slightly undulated canal runs along the interior of the filament.

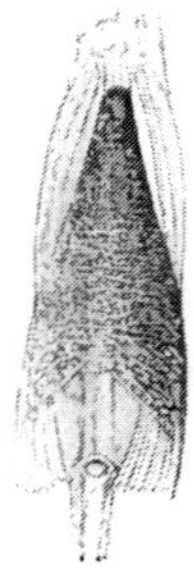

Fig. 74.—Pouch with ova, formed by the ventral fins of *Solenostoma*. Lower aspect; the edges of the fins have been pushed aside to allow of a view of the inside of the pouch. (Natural size.)

The *Testicles* of the *Teleosteans* are always paired, and occupy the same position as the ovaries. Their size varies extraordinarily at the different seasons of the year. Vasa deferentia are constant. In the males of viviparous Teleosteans the urogenital papilla is frequently enlarged, and clearly serves as an intromittent organ. In *Clinus despicillatus* the vas deferens widens within the abdomen into a cavity occupied by a complex network of loose fasciculi, rising from the mucous membrane. The cavity can be compressed by a special powerful muscle, the accumulated semen being thus expelled with considerable force through the narrow aperture of the penis. In many Cyprinodonts the vas deferens runs along the anterior anal rays, which may be thickened, and prolonged into a long slender organ.

Many Teleostei take care of their progeny, but with the

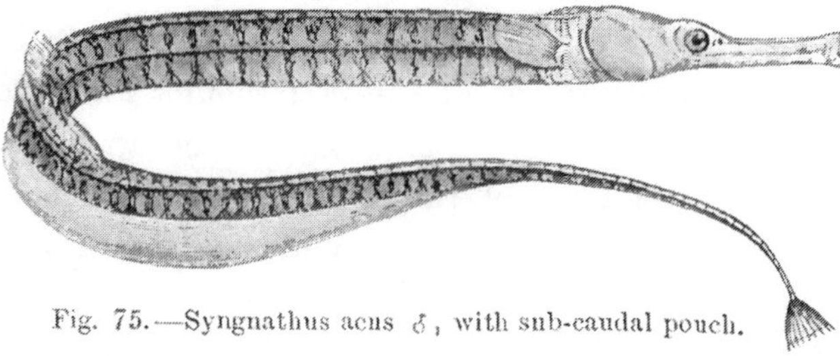

Fig. 75.—Syngnathus acus ♂, with sub-caudal pouch.

exception of *Aspredo* and *Solenostoma*, mentioned above (p. 160), it is the male on which this duty devolves. In some, as in *Cottus, Gastrosteus, Cyclopterus, Antennarius, Ophiocephalus, Callichthys*, the male constructs with more or less skill a nest, and jealously guards the ova deposited in it by the female. The male of some species of *Arius* carries the ova (Fig. 71) about with him in his capacious pharynx. The species of *Chromis*, inhabiting the sea of Galilee, are said to take care of their ova in the same manner. And, finally, in the Lophobranchs, nature has aided this instinct by the development of a pouch on the abdomen or lower side of the tail. In the Syngnathidæ this pouch is formed by a fold of the skin developed from each side of the trunk and tail, the free margins of the fold being firmly united in the median line, whilst the eggs are being hatched in the inside of the pouch. In *Hippocampus* the pouch is completely closed, with a narrow anterior opening.

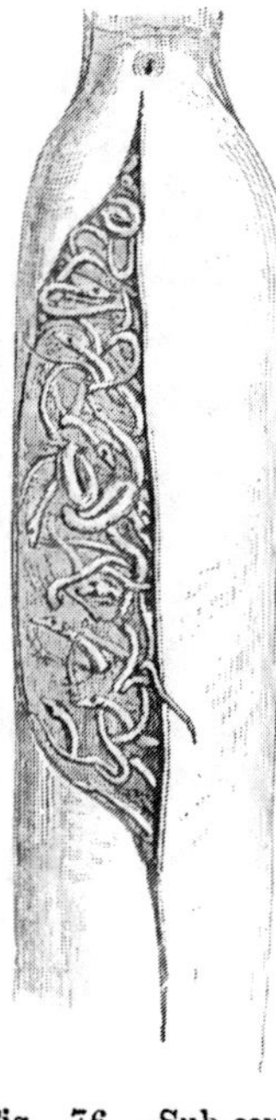

Fig. 76.—Sub-caudal pouch of Syngnathus acus, with the young, ready to leave the pouch. One side of the membrane of the pouch is pushed aside to admit of a view of its interior. (Natural size.)

The genital organs of *Ganoids* show similar diversity of structure as those of Teleosteans, but on the whole they approach the Batrachian type. The ovaries are not closed, except in *Lepidosiren;* all Ganoids possess oviducts. In the Sturgeons the oviduct as well as the vas deferens is represented by a funnel-shaped prolongation of the peritoneum, which communicates with the wide ureter. The inner aperture of the funnel is on a level of the middle of the testicle or ovary, the outer within the ureter; and it is a noteworthy fact that only at certain periods of the life of the fish this outer aperture is found to be open,—at other times the peritoneal funnel appears as a closed blind sac within the ureter. The mode of passage of the semen into the funnel is not known.

In *Polypterus* and *Amia*, proper oviducts, with abdominal apertures in about the middle of the abdominal cavity, are developed; they coalesce with the ureters close to the common urogenital aperture.

In *Ceratodus* (Fig. 77), a long convoluted oviduct extends to the foremost limit of the abdominal cavity, where it opens by a slit at a considerable distance from the front end of the long ovary; this aperture is closed in sexually immature specimens. The oviducts unite close to their common opening in the cloaca. During their passage through the oviduct the ova receive a gelatinous covering secreted by its mucous membrane. This is probably also the case in *Lepidosiren*, which possesses a convoluted oviduct with secretory glands in the middle of its length. The oviduct begins with a funnel-shaped dilatation, and terminates in a wide pouch, which posteriorly communicates with that of the other side, both opening by a common aperture behind the urinary bladder.

The ova of Ganoids, as far as they are known at present, are small, but enveloped in a gelatinous substance. In the Sturgeon have been counted as many as 7,635,200. Those of *Lepidosteus* seem to be the largest, measuring 5 millimetres in diameter with their envelope, and 3 millimetres without it. They are deposited singly, like those of Newts.

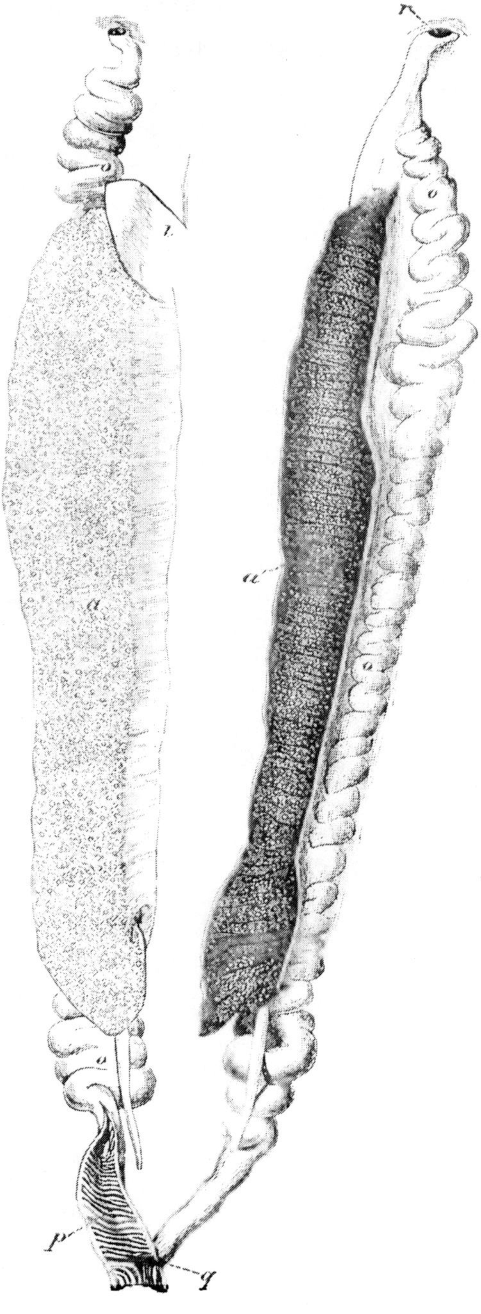

Fig. 77.—Ovaries of Ceratodus.

a, Right ovary shown from the inner surface, which is covered by the peritoneum; *a'*, Left ovary, showing its outer surface; *l*, Portion of liver; *o*, Oviduct; *p*, the lower part of the oviduct is opened to show the folds of its inner membrane; *q*, Opening of the left oviduct into the right; *r*, Abdominal orifice of the oviduct.

In *Chondropterygians* (and *Holocephali*) the organs of reproduction assume a more compact form, and are more free from a lengthened attachment to the back of the abdominal cavity. The ovaries of the majority are paired, single in the *Carchariidæ* and *Scylliidæ*, one remaining undeveloped. But the oviducts are always paired, beginning immediately behind the diaphragma with a common aperture. They consist of two divisions, separated by a circular valve; the upper is narrow, and provided within its coats with a gland which secretes the leathery envelope in which most of the Chondropterygian ova are enclosed; the lower forms the uterine dilatation, in which the embryoes of the viviparous species are developed. Generally the vitelline sac of the embryoes is free, and without connection with the uterus, which in these cases has merely the function of a protecting pouch; but in Carcharias and Mustelus lævis a *placenta uterina* is formed, the vascular walls of the vitelline sac forming plaits fitting into those of the membrane of the

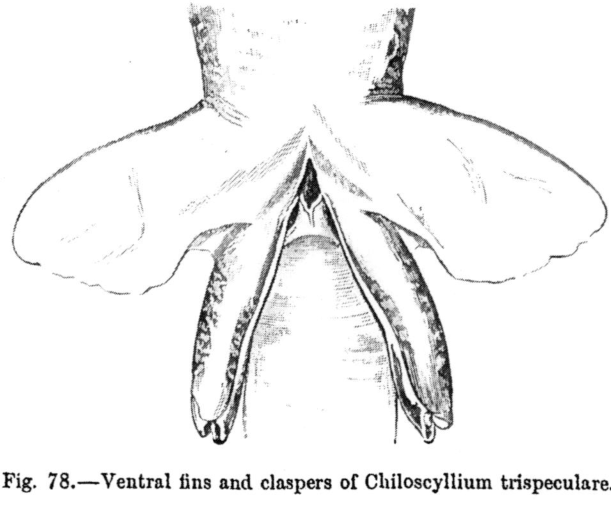

Fig. 78.—Ventral fins and claspers of Chiloscyllium trispeculare.

uterus. The ends of the uteri open by a common aperture behind the ureter into the cloaca.

The testicles are always paired, rounded, and situated in the anterior part of the abdominal cavity, covered by the liver. *Vasa efferentia* pass the semen into a much-convoluted *epididymis,* which is continued into the *vas deferens;* this, at the commencement of its course, is spirally wound, but becomes straight behind, and has its end dilated into a seminal reservoir. It opens with the urethra in a papilla within the cloaca.

The so-called *claspers* of Chondropterygians (Fig. 78) are characteristic of all male individuals. They are semi-ossified appendages of the pubic, with which they are movably joined, and special muscles serve to regulate their movements. Sometimes they are armed with hook-like osseous excrescences (*Sclache*). They are irregularly longitudinally convoluted, and, when closely adpressed to each other, form a canal open at their extremity. A gland, abundantly discharging a secretion during the season of propagation, is situated at, and opens into, the base of the canal. It is still doubtful whether the generally-adopted opinion that their function consists in holding the female during copulation is correct, or whether they are not rather an intromittent organ, the canal of which not only conducts the secretion of their proper gland but also the impregnating fluid.

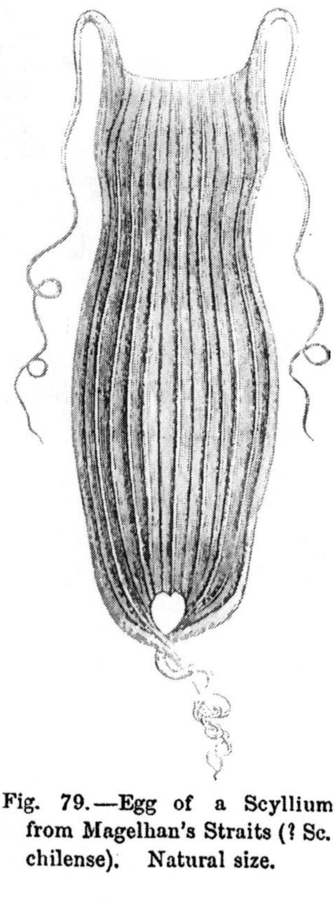

Fig. 79.—Egg of a Scyllium from Magelhan's Straits (? Sc. chilense). Natural size.

The ova of the oviparous Chondropterygians are large and few in number; they are successively impregnated, and the impregnation must take place before they are invested with a tough leathery envelope which would be

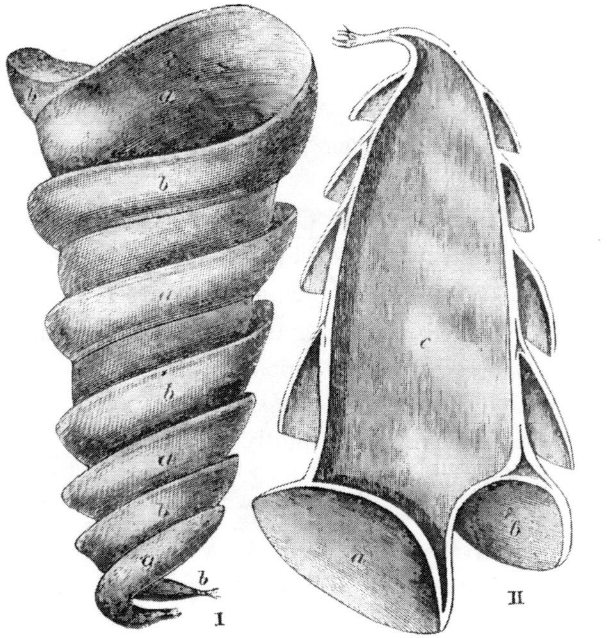

Fig. 80.—Egg-shell of Cestracion philippi, half natural size, linear.
I. External view. II. Vertical section.
a, One spiral ridge ; *b*, The other spiral ridge ; *c*, Cavity for the ovum.

impenetrable to the semen, that is, before they enter the uterus; therefore, copulation must take place in all these fishes. The form of the egg-shell differs in the various genera; generally (Fig. 79) they are flattened, quadrangular, with each of the four corners produced, and frequently prolonged into long filaments which serve for the attachment of the ova to other fixed objects. In *Notidanus* the surfaces are crossed by numerous ridges. In *Cestracion* (Fig. 80) the egg is pyriform, with two broad ridges or plates, wound edgewise round it, the two ridges forming five spires. The eggs of *Callorhynchus*

(Fig. 81) have received a protective resemblance to a broad-leaved fucus, forming a long depressed ellipse, with a plicated and fringed margin.

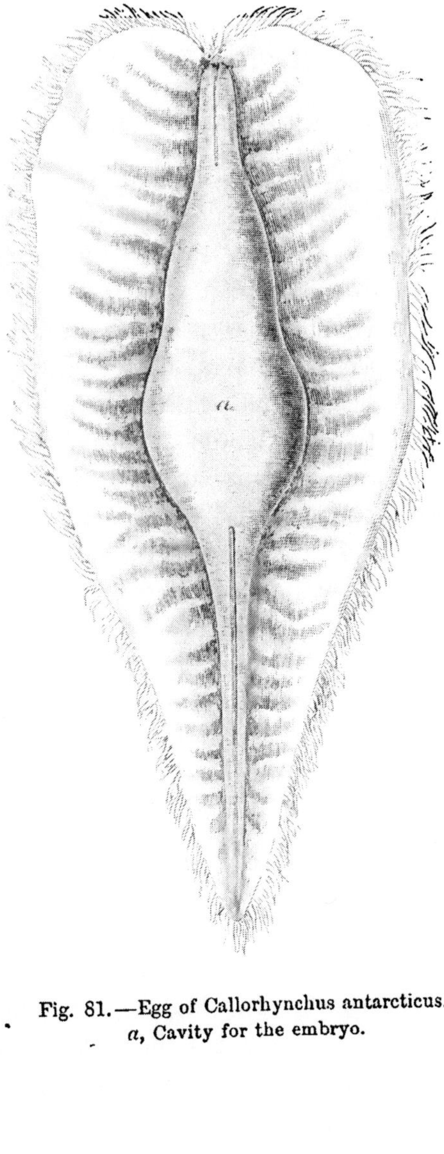

Fig. 81.—Egg of Callorhynchus antarcticus.
a, Cavity for the embryo.

CHAPTER XIII.

GROWTH AND VARIATION OF FISHES.

CHANGES of form normally accompanying growth (after absorption of the vitelline sac) are observed in all fishes; but in the majority they affect only the proportional size of the various parts of the body. In young fishes the eyes are constantly larger than in adult relatively to the size of the head; and again, the head is larger relatively to that of the body. Changes amounting to metamorphosis have been hitherto observed in *Petromyzon* only. In the larval condition (*Ammocœtes*) the head is very small, and the toothless buccal cavity is surrounded by a semicircular upper lip. The eyes are extremely small, hidden in a shallow groove; and the vertical fins form a continuous fringe. In the course of three or four years the teeth are developed, and the mouth changes into a perfect suctorial organ; the eyes grow; and the dorsal fin is divided into two divisions. In Malacopterygians and Anacanths the embryonal fringe from which the vertical fins are developed, is much longer per-

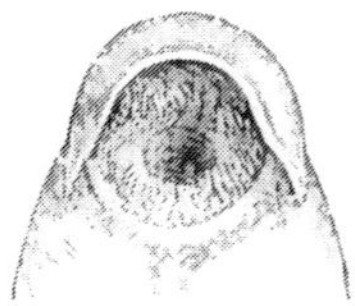

Fig. 82.—Mouth of Larva of Petromyzon branchialis.

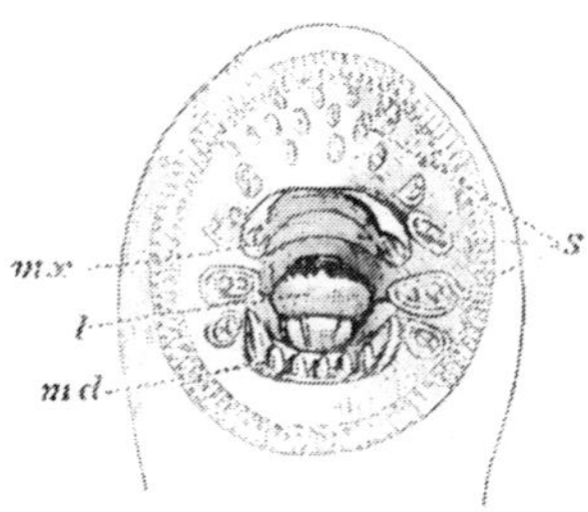

Fig. 83.—Mouth of Petromyzon fluviatilis.
mx, Maxillary tooth; *md*, Mandibulary tooth; *l*, Lingual tooth; *s*, Suctorial teeth.

sistent than in Acanthopterygians. A metamorphosis relating to the respiratory organs, as in Batrachians, is indicated in the class of Fishes by the external gills with which fœtal Plagiostomes (Fig. 58, p. 136) and the young of some Ganoids, viz. the *Protopterus* and *Polypterus*, are provided.

One of the most extraordinary changes by which, during growth, the form and position of several important organs are affected, occurs in Flat-fishes (*Pleuronectidæ*); their young are symmetrically formed, with a symmetrical mouth, and with one eye on each side, and, therefore, keep their body in a vertical position when swimming. As they grow they live more on the bottom, and their body, during rest, assumes a horizontal position; in consequence, the eye of the lower side moves towards the upper, which alone is coloured; and in many genera the mouth is twisted in the opposite direction, so that the bones, muscles, and teeth are much more developed on the blind side than on the coloured. In a great number of other *Teleostei* certain bones of the head show a very different form in the young state. Ossification proceeds in those bones in the direction of lines or radii which project in the form of spines or processes; as the interspaces between

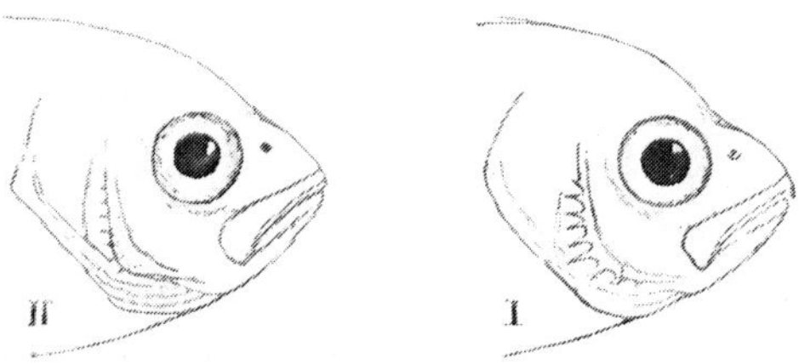

Fig. 84.—Armature of præoperculum of young Caranx ferdau. (Magnified.) I. Of an individual, 1¼ inch long. II. Of an individual, 2 inches long.

these processes are filled with bone, the processes disappear entirely, or at least project much less in the older than in the younger individuals (Fig. 84). The young of some fishes may be armed with a long powerful præopercular or scapular spine, or show a serrature of which nothing remains in the adult fish except some ridges or radiating lines. These processes seem to serve as weapons of defence during a period in the life of the fish in which it needs them most. In not a few instances a portion of this armature is so much

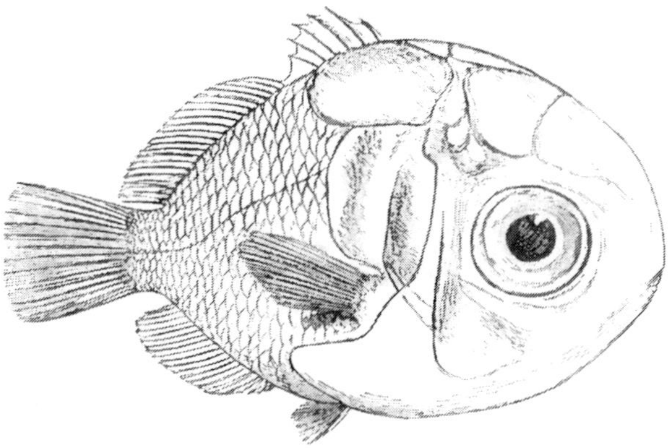

Fig. 85.—Tholichthys osseus. Six times the natural size.

developed that the disappearance of its most projecting parts with the growth of the fish is not only due to its being surrounded by other bone, but, partially at least, caused by absorption. The *Carangidæ*, *Cyttidæ*, *Squamipinnes*, *Xiphiidæ*, offer instances of such remarkable changes. A fish, described as *Tholichthys osseus* (Fig. 85), is probably the young of a Cyttoid, the supra-scapula, humerus, and præoperculum forming enormously enlarged plates. In the fish

Fig. 86.—Tholichthys-stage of Heniochus (?).

Fig. 86 those bones appear still enlarged, and the frontals develop a remarkably long and curved horn above the orbit. In the *Tholichthys*-stage of *Pomacanthus* (specimens 10 millimetres long, Fig. 87), the frontal bone is prolonged into a straight lancet-shaped process, nearly half as long as the body; the suprascapular and præopercular processes cover and hide the dorsal and ventral fins. The plates attached to the shoulder-girdle remain persistent until the young fish

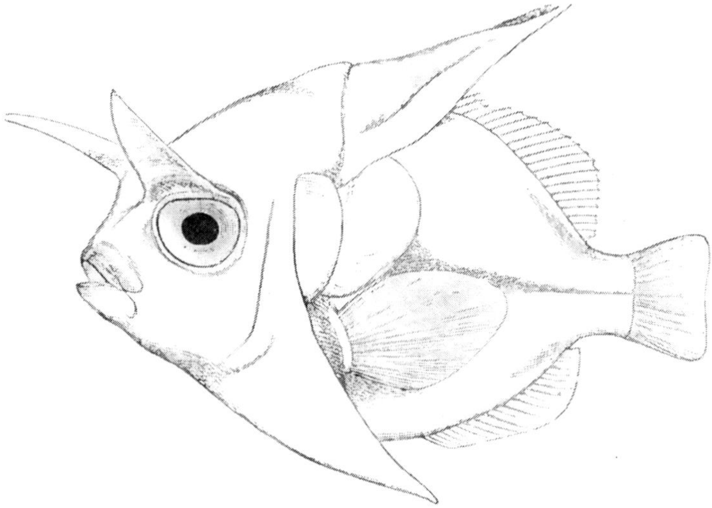

Fig. 87.—Tholichthys-stage of Pomacanthus (magn.) Atlantic.

has assumed the form of the adult; thus they are still visible in young *Chætodon citrinellus*, 30 millimetres long, in which the specific characters are already fully developed.—The Sword-fishes with ventral fins (*Histiophorus*) belong to the Teleosteans of the largest size; in young individuals, 9 millimetres long (Fig. 89), both jaws are produced, and armed with pointed teeth; the supra-orbital margin is ciliated; the parietal and præoperculum are prolonged into long spines; the dorsal and anal fins are a low fringe, and the ventrals make their appearance as a pair of short buds. When 14 millimetres long (Fig. 90) the young fish has still the same armature of the head, but the dorsal fin has become much higher, and the ventral

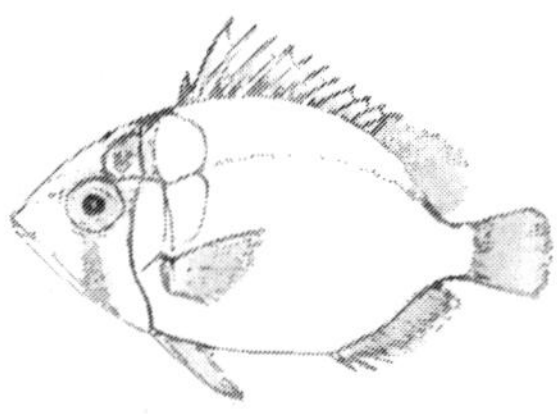

Fig. 88.—Young Chætodon citrinellus (30 mill. long).

filaments have grown to a great length. At a third stage,

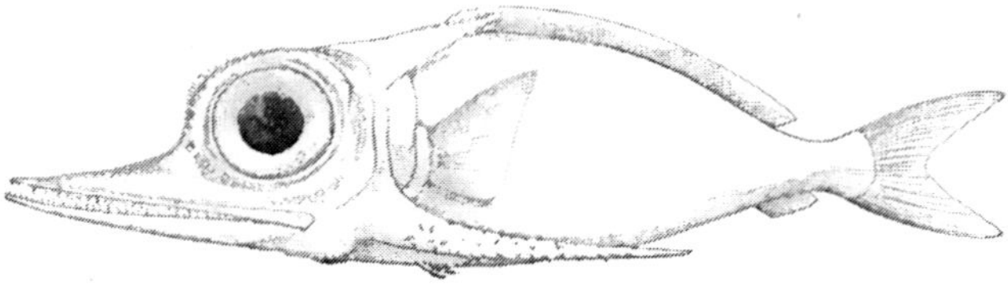

Fig. 89.—Young Sword-fish (Histiophorus), 9 mill. long. Atlantic. (Magn.)

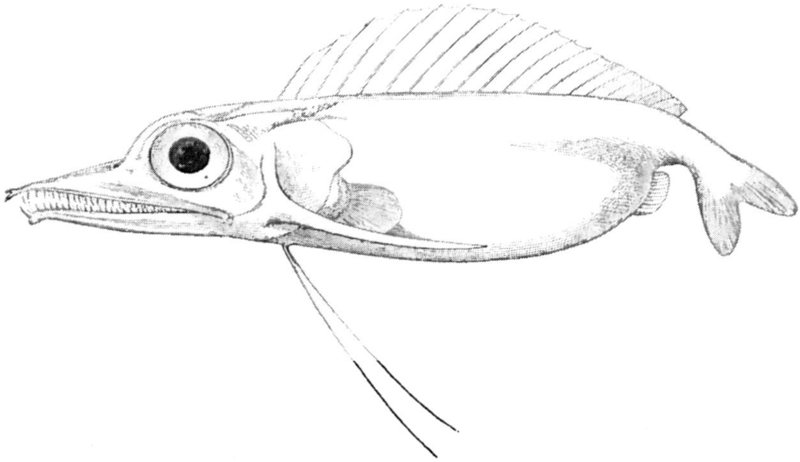

Fig. 90.—Young Sword-fish (Histiophorus), 14 mill. long. South Atlantic. (Magn.)

when the fish has attained to a length of 60 millimetres,

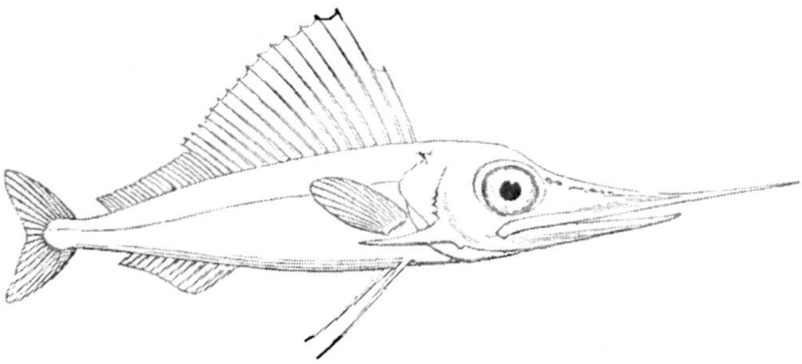

Fig. 91.—Young Sword-fish (Histiophorus), 60 mill. long. Mid-Atlantic.

the upper jaw is considerably prolonged beyond the lower,

losing its teeth; the spines of the head are shortened, and the fins assume nearly the shape which they retain in mature individuals. Young Sword-fishes without ventral fins (*Xiphias*) undergo similar changes; and, besides, their skin is covered with small rough excrescences longitudinally arranged, which continue to be visible after the young fish has assumed the form of the mature in other respects (Fig. 92).

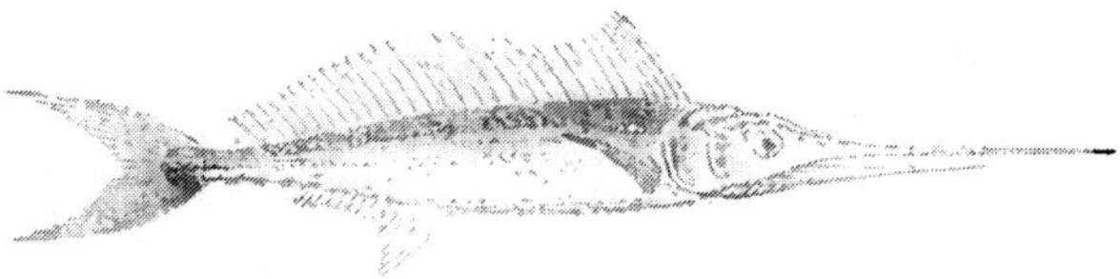

Fig. 92.—Xiphias gladius, young, about 8 inches long.

The Plectognaths show no less extraordinary changes: an extraordinary form taken in the South Atlantic, and named *Ostracion boops*, is considered by Lütken to be the young of a Sunfish (*Orthagoriscus*). In very young more advanced Sun-fishes (18 to 32 millimetres) the vertical diameter of the body exceeds, or is not much less than, the longitudinal; and small conical spines are scattered over its various parts. The caudal

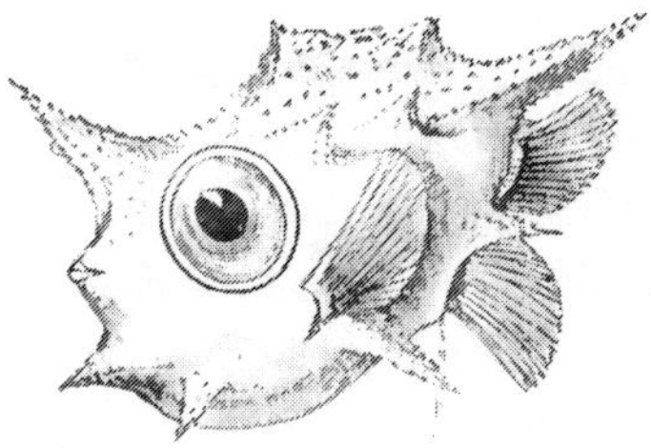

Fig. 93.—"Ostracion boops" (much magnified).

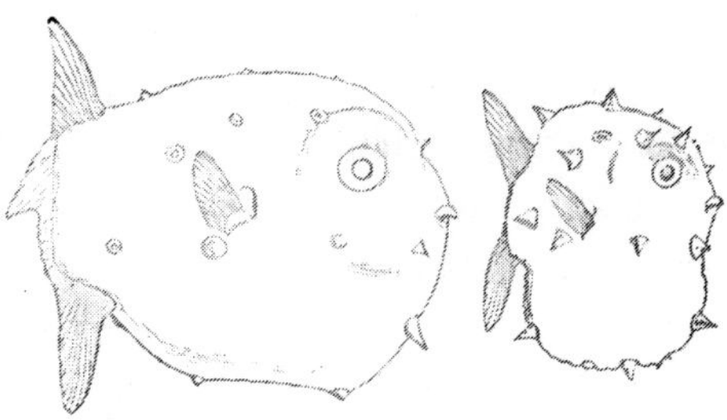

Fig. 94.—Young of Orthagoriscus, 18 and 32 mill. long. (Natural size.)

fin is developed long after the other vertical fins.

Similar changes take place in a number of other fishes, and in many cases the young are so different that they were described as distinct genera: thus *Priacantichthys* has proved to be the young of *Serranus*, *Rhynchichthys* that of *Holocentrum*, *Cephalacanthus* of *Dactylopterus*, *Dicrotus* of *Thyrsites*, *Nauclerus* of *Naucrates*, *Porthmeus* of *Chorinemus*, *Lampugus* of *Coryphæna*, *Acronurus* of *Acanthurus*, *Keris* of *Nascus*, *Porobronchus* of *Fierasfer*, *Couchia* of *Motella*, *Stomiasunculus* of *Stomias*, etc.

The fins are most frequently subject to changes; but, whilst in some fishes parts of them are prolonged into filaments with age, in others the filaments exist during the early life-periods only; whilst in some a part of the dorsal or the ventral fins is normally developed in the young only, in others those very parts are peculiar to the mature age. The integuments are similarly altered: in some species the young only has asperities on the skin, in others the young are smooth and the old have a tubercular skin; in some the young only have a hard bony head; in others (some Siluroids) the osseous carapace of the head and neck, as it appears in the adult, is more or less covered with soft skin whilst the fish is young.

In not a few fishes the external changes are in relation to the sexual development (*Callionymus*, many *Labyrinthici*, Cyprinodonts). These *secondary sexual differences* show themselves in the male individual, only when it commences to enter upon his sexual functions, and it may require two or more seasons before its external characteristics are fully developed. Immature males do not differ externally from the old female. The male secondary sexual characters consist principally in the prolongation of some of the fin-rays, or of entire fins; and in *Salmonidæ* in the greater development of the jaw-bones. The coloration of the male is in many fishes much brighter and more variegated than that of the female, but in comparatively few permanent (as in some *Callionymus*, *Labrus mixtus*); generally it is acquired immediately before and during the season of propagation only, and lost afterwards. Another periodical change in the integuments, also due to sexual influence and peculiar to the male, is the excrescence of wart-like tubercles on the skin of many *Cyprinoids*; they are developed chiefly on the head, but sometimes extend over the whole body and all the fins.

With regard to size, it appears that in all Teleosteous fishes the female is larger than the male; in many Cyprinodonts the male may be only one-sixth or even less of the bulk of the female. The observations on the relative size of the sexes are few in Palæichthyes, but such as have been made tend to show that, if a difference exists at all, the male is generally the larger (*Lepidosteus*). In the Rays (*Raja*) the sexes, after they have attained maturity, differ in the development of dermal spines and the form of the teeth, the female being frequently much rougher than the male. There is much variation in this respect in the different species; but the males are constantly distinguished by an oblong patch of erectile clawlike spines on each pectoral fin, and by having the teeth (all, or only a portion) pointed, and not obtuse, like those of the females. In Sharks no secondary sexual differences have been observed; the male *Chimæridæ* (see Fig. 96, p. 184), possess a singular comblike cartilaginous appendage on the top of the head, which can be erected or depressed into a groove, both the appendage and the anterior part of the groove being armed with hooklets. The use of this singular organ is not known.

The majority of Teleostei are *mixogamous*—that is, the males and females congregate on the spawning-beds, and the number of the former being in excess, several males attend to the same female, frequently changing from one female to another. The same habit has been observed in

Lepidosteus. *Gastrosteus* is truly polygamous, several females depositing their ova into the same nest, guarded by one male only. Some Teleostei (*Ophiocephalus*), and probably all Chondropterygians, are monogamous; and it is asserted that the connection between the pair is not merely temporary, but lasts until they are separated by accident. Monogamous are probably also all those Teleosteans which bring forth living young, and those, the males of which, for the attraction of the female, are provided with appendages, or ornamented with a bright coloration.

Hybridism is another source of changes and variations within the limits of a species, and is by no means so scarce as has been believed hitherto; it is only apparently of exceptional occurrence, because the life of fishes is more withdrawn from our direct observation than that of terrestrial animals. It has been observed among species of *Serranus*, *Pleuronectidæ*, *Cyprinidæ*, *Clupeidæ*, and especially *Salmonidæ*. As in other animals, the more certain kinds of fishes are brought under domestication, the more readily do they interbreed with other allied species. It is characteristic of hybrids that their characters are very variable, the degrees of affinity to one or the other of the parents being inconstant; and as these hybrids are known readily to breed with either of the parent race, the variations of form, structure, and colour are infinite. Of internal organs the dentition, gill-rakers, pyloric appendages, are those particularly affected by such mixture of species.

Some fishes are known to grow rapidly (in the course of from one to three years) and regularly to a certain size, growth being definitely arrested after the standard has been attained. Such fishes may be called "full-grown," in the sense in which the term is applied to warm-blooded Vertebrates—the Sticklebacks, most Cyprinodonts, and many Clupeoids (Herring,

Sprat, Pilchard) are examples of this regular kind of growth.[1] But in the majority of fishes the rate of growth is extremely irregular, and it is hardly possible to know when growth is actually and definitely arrested. All seems to depend on the amount of food and the more or less favourable circumstances under which the individual grows up. Fishes which rapidly grow to a definite size are short-lived, whilst those which steadily and slowly increase in size attain to a great age, Teleosteans as well as Chondropterygians. Carp and Pike have been ascertained to live beyond a hundred years.

It is evident that such diversity and irregularity of growth in the same species is accompanied by considerable differences in the appearance and general development of the fish. No instance is more remarkable than that of the so-called *Leptocephali*, which for a long time have been regarded either as a distinct group of Fishes, or as the larval stages of various genera of fishes.

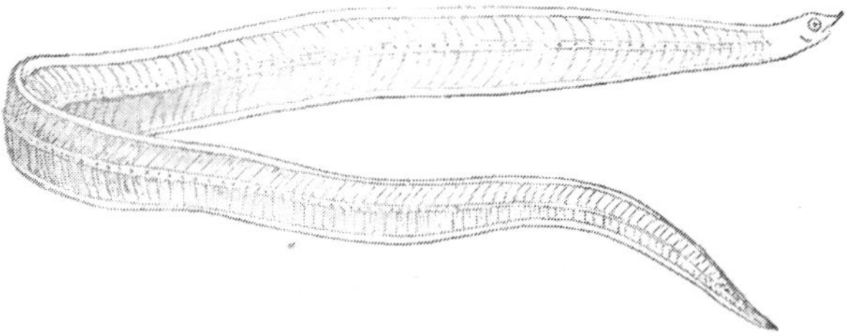

Fig. 95.—Leptocephalus.

The *Leptocephali* proper are small, narrow, elongate, more or less band-shaped fishes, pellucid in a fresh state, but assum-

[1] This applies to individuals only growing up under normal conditions. Dr. H. A. Meyer has made observations on young Herrings. Individuals living in the sea had attained at the end of the third month a length of 45 to 50 millimetres, whilst those reared from artificially-impregnated ova were only from 30 to 35 millimetres long. When the latter had been supplied with more abundant food, they grew proportionally more rapidly in the following months, so that at the end of the fifth month they had reached the same length as their brethren in the sea, viz. a length of 65 to 70 millimetres.

ing a white colour when preserved in spirits, resembling a tapeworm, being quite as soft and flexible. The skeleton is entirely cartilaginous, or slight ossifications are only now and then visible, especially towards the end of the vertebral column. The latter is replaced by a chorda dorsalis which, in many specimens, is found to be divided into numerous segments. Neural arches are sometimes present in their rudimentary condition. The anterior end of the chorda passes into the cartilaginous base of the skull, the connection not being by means of joint and ligaments. Hæmal arches are found on the caudal portion. Ribs none. The skull, like the vertebral column, is nearly entirely cartilaginous. The basi-sphenoid, frontal, and jaw-bones are the first which may be distinguished, and the mandible has generally ossifications.

The muscles are generally not attached to the chorda, which is surrounded by a thick gelatinous mass, separating the lateral sets of muscles from each other. These muscles are attached to the external integument, each forming a thin flat angular band, the angle being directed forwards. However, specimens are frequently found in which the muscles are more developed, evidently at the expense of the gelatinous matter, which is diminished in quantity. They are attached to the chorda, and the entire fish has a more cylindrical form of the body (*Helmichthys*).

The nervous, circulatory, and respiratory organs are well developed. In those with a sub-cylindrical body the blood is red, in those with a flat body the blood-corpuscles show but rarely a faint coloration. There are four branchial arches, and in some (*Tilurus*) pseudobranchiæ have been found. The gill-openings are more or less narrow. The nostrils are double on each side, and the posterior is close to the eye.

The stomach has a large blind sac, and in *Leptocephalus*

two lateral cæca. The intestine is straight, running close to the abdominal profile, with a small appendix directed forward and a larger one directed backwards. The vent is nearly always very small, and, in preserved examples at least, cannot always be discovered. Its position is variable, even in examples entirely similar in other points. Air-bladder none. No trace of generative organs.

The vertical fins, when present, are confluent, with more or less conspicuous traces of rays; sometimes they are merely a fold of the skin, without any rays. Pectoral fins sometimes present, sometimes rudimentary, sometimes entirely absent. Ventrals none.

Most examples have series of round black dots along each side of the abdominal profile, along the lateral line, and sometimes along the dorsal fin. They remind us of the luminous organs of many *Scopelidæ*, *Stomiatidæ*, and other pelagic fishes, but are composed entirely of pigmentary cells.

These fishes are found floating in the sea, frequently at a great distance from land. Their movements are slow and languid. The largest specimen of Leptocephalus observed was 10 inches, but specimens of that size are very rare.

[See Kölliker, Zeitschr. wiss. Zool. iv. 1852, p. 360; and Carus, Ueber die Leptocephaliden. Leipz. 1861. 4to.]

Taking into account all the various facts mentioned, we must come to the conclusion that the Leptocephalids are the offspring of various kinds of marine fishes, representing, not a normal stage of development (larvæ), but an arrest of development at a very early period of their life; they continue to grow to a certain size without corresponding development of their internal organs, and perish without having attained the characters of the perfect animal. The cause by which this abnormal condition is brought about is not known; but it is quite within the limits of probability that fishes usually spawning in the vicinity of land sometimes spawn in the open ocean, or that floating spawn is carried by currents to a great distance from land; and that such embryoes, which for their normal growth require the conditions afforded by the vicinity of the shore, if hatched in mid-ocean, grow into undeveloped hydropic creatures, such as the Leptocephales seem to be.

Abundance or scarcity of food, and other circumstances connected with the localities inhabited by fishes, affect considerably the colour of their muscles and integuments; the periodical changes of colour in connection with their sexual functions have been referred to above (p. 176). The flesh of many Teleostei is colourless, or but slightly tinged by the blood; that of Scombridæ, most Ganoids and Chondropterygians, is more or less red; but in badly-fed fishes, as well as in very young ones, the flesh is invariably white (anæmic). Many fishes, like the *Salmonidæ,* feed at times exclusively on Crustaceans, and the colouring substance of these Invertebrates, which by boiling and by the stomachic secretion turns red, seems to pass into the flesh of the fishes, imparting to it the well-known "salmon" colour. Further, the coloration of the integuments of many marine fish is dependent on the nature of their surroundings. In those which habitually hide themselves on the bottom, in sand, between stones or seaweeds, the colours of the body readily assimilate to those of the vicinity, and are thus an important element in the economy of their life. The changes from one set or tinge of colours to another may be rapid and temporary, or more or less permanent; in some fishes—as in the Pediculati, of which the Sea-Devil, or *Lophius,* and *Antennarius* are members—scarcely two individuals are found exactly alike in coloration, and only too frequently such differences in coloration are mistaken for specific characters. The changes of colours are produced in two ways: either by an increase or decrease of the black, red, yellow, etc., pigment-cells, or *chromatophors*, in the skin of the fish; or by the rapid contraction or expansion of the chromatophors which happen to be developed. The former change is gradual, like every kind of growth or development; the latter rapid, owing to the great sensitiveness of the cells, but certainly involuntary. In many bright-shining fishes—as Mackerels, Mullets—the colours appear to be brightest in the time intervening between the capture of the fish and its death: a phenomenon clearly due to the pressure of the convulsively-contracted muscles on the chromatophors. External irritation readily excites the chromatophors to expand—a fact unconsciously utilised by fishermen, who, by scaling the Red Mullet immediately before its death, produce the desired intensity of the red colour of the skin, without which the fish would not be saleable. However, it does not require such strong measures to prove the sensitiveness of the chromatophors to external irritation, the mere change of darkness into light is sufficient to induce them to contract, the fish appearing paler, and *vice versa*. In Trout which are kept or live in dark places, the black chromatophors are expanded, and, consequently, such specimens are very dark-coloured; when removed to the light they become paler almost instantaneously.

Total absence of chromatophors in the skin, or *Albinism*, is very rare among fishes; much more common is *incipient Albinism*, in which the dark chromatophors are changed into cells with a more or less intense yellow pigment. Fishes in a state of domestication, like the Crucian Carp of China, the Carp, Tench, and the Ide, are particularly subject to this abnormal coloration, and are known as the common Goldfish, the Gold-Tench, and the Gold-Orfe. But it occurs also not rarely in fishes living in a wild state, and has been observed in the Haddock, Flounder, Plaice, Carp, Roach, and Eel.

It will be evident, from the foregoing remarks, that the amount of variation within the limits of the same species—either due to the natural growth and development, or to external physical conditions, or to abnormal accidental circumstances—is greater in fishes than in any of the higher classes of Vertebrates. The amount of variation is greater in certain genera or families than in others, and it is much greater in Teleosteans and Ganoids than in Chondropterygians. Naturally, it is greatest in the few species which have been domesticated, and which we shall mention in the succeeding chapter.

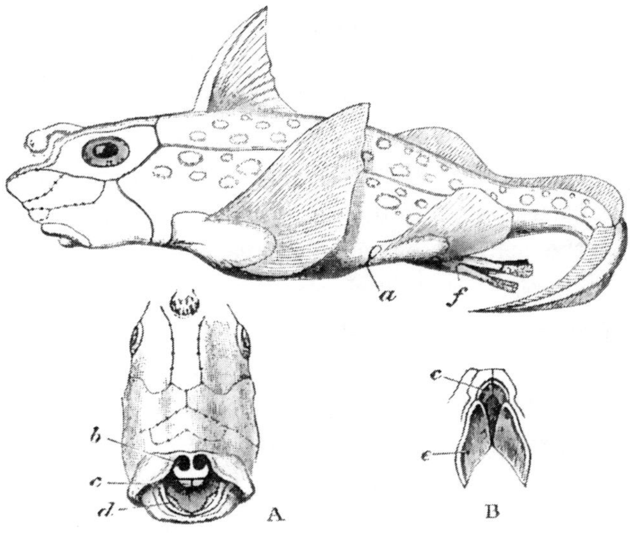

Fig. 96.—Chimæra colliei ♂, west coast of North America. A. Front view of head. B. Palate. *a*, Peritoneal aperture ; *b*, Nostrils ; *c*, Vomerine teeth ; *d*, Mandibular teeth ; *e*, Palatine teeth ; *f*, Claspers.

CHAPTER XIV.

DOMESTICATED AND ACCLIMATISED FISHES; ARTIFICIAL IMPREGNATION OF OVA—TENACITY OF LIFE AND REPRODUCTION OF LOST PARTS—HYBERNATION—USEFUL AND POISONOUS FISHES.

A FEW fishes only are thoroughly domesticated — that is, bred in captivity, and capable of transportation within certain climatic limits—viz. the Carp, Crucian Carp (European and Chinese varieties), Tench, Orfe or Ide, and the Goramy. The two former have accompanied civilised man almost to every place of the globe where he has effected a permanent settlement.

Attempts to acclimatise particularly useful species in countries in which they were not indigenous have been made from time to time, but were permanently successful in a few instances only; the failures being due partly to the choice of a species which did not yield the profitable return expected, partly to the utter disregard of the difference of the climatic and other physical conditions between the original and new homes of the fish. The first successful attempts of acclimatisation were made with domestic species, viz. the Carp and Goldfish, which were transferred from Eastern Asia to Europe. Then, in the first third of the present century, the Javanese Goramy was acclimatised in Mauritius and Guiana, but no care seems to have been taken to insure permanent advantages from the successful execution of the experiment. In these cases fully developed individuals were transported to the country in which they were to be acclimatised. The most successful attempt of recent years is the acclimatisation of the Trout and Sea-Trout, and probably also of the Salmon, in Tasmania and New Zealand, and of the Californian Salmon (*Salmo quinnat ?*), in Victoria, by means of artificially-impregnated ova. The ova were transported on ice, in order to retard their development generally, and thus to preserve them from destruction during the passage of the tropical zone.

Artificial impregnation of fish-ova was first practised by J. L. JACOBI, a native of Westphalia, in the years 1757-63, who employed exactly the same method which is followed now; and there is no doubt that this able observer of nature conceived and carried out his idea with the distinct object of advantageously restocking water-courses which had become unproductive, and increasing production by fecundating and preserving all ova, of which a great proportion, in the ordinary course of propagation, would be left unfecundated or accidentally perish. Physiology soon turned to account Jacobi's discovery, and artificial impregnation has proved to be one of the greatest helps to the student of embryology.

Fishes differ in an extraordinary degree with regard to tenacity of life. Some will bear suspension of respiration—caused by removal from water, or by exposure to cold or heat—for a long time, whilst others succumb at once. Nearly all marine fishes are very sensitive to changes in the temperature of the water, and will not bear transportation from one climate to another. This seems to be much less the case with some freshwater fishes of the temperate zones : the Carp may survive after being frozen in a solid block of ice, and will thrive in the southern parts of the temperate zone. On the other hand, some freshwater fishes are so sensitive to a change in the water that they perish when transplanted from their native river into another apparently offering the same physical conditions (Grayling, *Salmo hucho*). Some marine fishes may be abruptly transferred from salt into fresh water, like Sticklebacks, some Blennies, and *Cottus*, etc.; others survive the change when gradually effected, as many migratory fishes; whilst again, others cannot bear the least alteration in the composition of the salt water (all pelagic fishes). On the whole, instances of marine fishes voluntarily entering brackish or fresh water are very numerous, whilst freshwater fishes proper but rarely descend into salt water.

Abstinence from food affects different fishes in a similarly different degree. Marine fishes can endure hunger less than freshwater fishes, at least in the temperate zones, no observations having been made in this respect on tropical fishes. Goldfishes, Carps, Eels, are known to be able to subsist without food for months, without showing a visible decrease of bulk; whilst the Trigloids, Sparoids, and other marine fishes, survive abstinence from food for a few days only. In freshwater fishes the temperature of the water is of great influence on their vital functions generally, and consequently on their appetite,—many cease to feed altogether in the course of the winter; a few, like the Pike, are less inclined to feed during the heat of the summer than when the temperature is lowered.

Captivity is easily borne by most fishes, and the appliances introduced in our modern aquaria have rendered it possible to keep in confinement, and even to induce to propagate, fishes which formerly were considered to be intolerant of captivity.

Wounds affect fishes generally much less than higher Vertebrates. A Greenland Shark continues to feed whilst his head is pierced by a harpoon or by the knife, as long as the nervous centre is not touched; a Sea-perch or a Pike (Fig. 97) will survive the loss of a portion of its tail; a Carp that of half of its snout. However, some fishes are much more sensitive, and perish even from the superficial abrasion caused by the meshes of the net during capture (*Mullsn.*)

The power of *reproduction* of *lost parts* in Teleosteous fishes is limited to the delicate terminations of their fin-rays and the various tegumentary filaments with which some are provided. These filaments are sometimes developed in an extraordinary degree, mimicking the waving fronds of the sea-weed in which the fish hides. Both the ends of the fin-rays

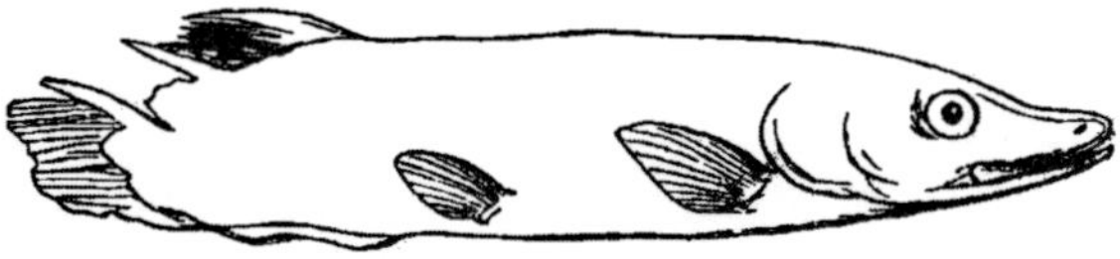

Fig. 97.—Pike caught in the Thames, which, when young, had lost part of the tail with the caudal fin.

and the filaments are frequently lost, not only by accident, but merely by wear and tear; and as these organs are essential for the preservation of the fish, their reproduction is necessary.

In Dipnoi, *Ceratodus*, and *Protopterus*, the terminal portion of the tail has been found to have been reproduced, but without the notochord.

Hybernation has been observed in many Cyprinoids and Murænoids of the temperate zones. They do not fall into a condition of complete torpidity, as Reptiles and Mammals, but their vital functions are simply lowered, and they hide in sheltered holes, and cease to go abroad in search of their food. Between the tropics a great number of fishes (especially Siluroids, Labyrinthici, Ophiocephaloids, the Dipnoi), are known to survive long-continued droughts by passing the dry season in a perfectly torpid state, imbedded in the hardened mud. Protopterus, and probably many of the other fishes mentioned, prepare for themselves a cavity large enough to hold them, and coated on the inside with a layer of hardened mucus, which preserves them from complete desiccation. It has been stated that in India fishes may survive in this condition for more than one season, and that ponds known to have been dry for several years, and to the depth of many feet, have swarmed with fishes as soon as the accumulation of water released them from their hardened bed.

The principal *use* derived by man from the class of Fishes consists in the abundance of wholesome and nourishing food which they yield. In the Polar regions especially, whole tribes are entirely dependent on this class for subsistence; and in almost all nations fishes form a more or less essential part of food, many being, in a preserved condition, most important articles of trade. The use derived by man from them in other respects is of but secondary importance. Cod-liver oil is prepared from the liver of some of the Gadoids of the Northern Hemisphere, and of Sharks; isinglass from the swim-bladder of Sturgeons, Sciænoids, and Polynemoids; shagreen from the skin of Sharks and Rays.

The flesh of some fishes is at times, or constantly, *poisonous.* When eaten, it causes symptoms of more or less intense irritation of the stomach and intestines, inflammation of the mucous membranes, and not rarely death. The fishes, the flesh of which appears always to have poisonous properties, are *Clupea thrissa*, *Clupea venenosa*, and some species of *Scarus*, *Tetrodon*, and *Diodon*. There are many others which have occasionally or frequently caused symptoms of poisoning Poey enumerates not less than seventy-two different kinds from Cuba; and various species of *Sphyræna*, *Balistes*, *Ostracion*, *Caranx*, *Lachnolæmus*, *Tetragonurus*, *Thynnus*, have been found to be poisonous in all seas between the tropics. All or nearly all these fishes acquire their poisonous properties from their food which consists of poisonous Medusæ, Corals, or decomposing substances. Frequently the fishes are found to be eatable if the head and intestines be removed immediately after capture. In the West Indies it has been ascertained that all the fishes living and feeding on certain coral banks are poisonous. In other fishes the poisonous properties are developed at certain seasons of the year only, especially the season of propagation: as the Barbel, Pike, and Burbot, whose roe causes violent diarrhœas when eaten during the season of spawning.

Poison-organs are more common in the class of Fishes than was formerly believed, but they seem to have exclusively the function of defence, and are not auxiliary in procuring food, as in venomous Snakes. Such organs are found in the Sting-rays, the tail of which is armed with one or more powerful barbed

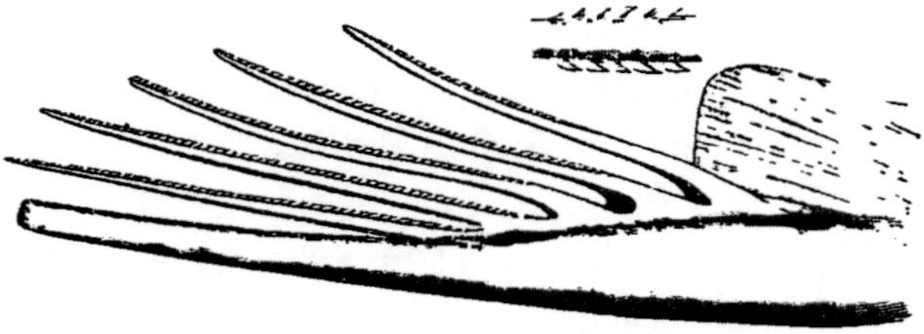

Fig. 98.—Portion of tail, with spines, of *Aëtobatis narinari*, a Sting-ray from the Indian Ocean. *a*, nat. size.

spines. Although they lack a special organ secreting poison, or a canal in or on the spine by which the venomous fluid is conducted, the symptoms caused by a wound from the spine of a Sting-ray are such as cannot be accounted for merely by the mechanical laceration, the pain being intense, and the subsequent inflammation and swelling of the wounded part terminating not rarely in gangrene. The mucus secreted from the surface of the fish and inoculated by the jagged spine evidently possesses venomous properties. This is also the case in many Scorpænoids, and in the Weaver (*Trachinus*), in which the dorsal and opercular spines have the same function as the caudal spines of the Sting-rays; however, in the Weavers the spines are deeply grooved, the groove being charged with a fluid mucus. In *Synanceia* the poison-organ (Fig. 99,) is still more developed: each dorsal spine is in its terminal half provided with a deep groove on each side, at the lower end of which lies a pear-shaped bag containing the milky poison; it is prolonged into a membranous duct, lying in the groove of the spine, and open at its point. The native fishermen, well acquainted with the dangerous nature of these fishes, carefully avoid handling them; but it often happens that persons wading with naked feet in the sea, step upon the fish, which generally lies hidden in the sand. One or more of the erected spines penetrate the skin, and the poison is injected into the wound by the pressure of the foot on the poison-bags. Death has not rarely been the result.

Fig. 99.—A dorsal spine, with poison-bags, of *Synanceia verrucosa*. Indian Ocean.

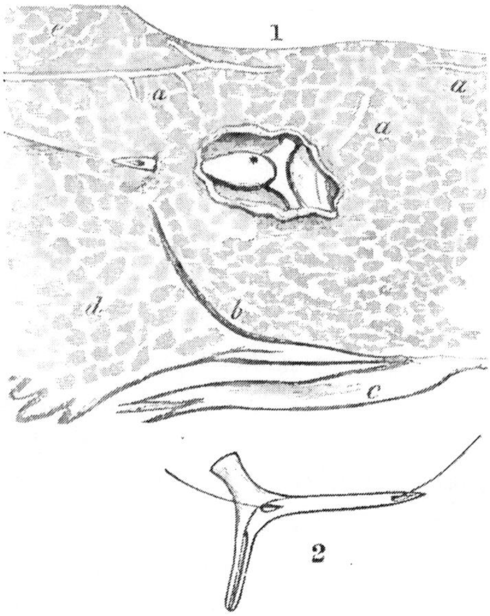

Fig. 100.—Opercular part of the Poison-apparatus of *Thalassophryne* (Panama).

1. Hinder half of the head, with the venom-sac* *in situ*. *a*, Lateral line and its branches; *b*, Gill-opening; *c*, Ventral fin; *d*, Base of Pectoral fin; *e*, Base of dorsal.
2. Operculum with the perforated spine.

The most perfect poison-organs hitherto discovered in fishes are those of *Thalassophryne*, a Batrachoid genus of fishes from the coasts of Central America. In these fishes the operculum again and the two dorsal spines are the weapons. The former (Fig. 100, [2]) is very narrow, vertically styliform and very mobile; it is armed behind with a spine, eight lines long, and of the same form as the hollow venom-fang of a snake, being perforated at its base and at its extremity. A sac covering the base of the spine discharges its contents through the apertures and the canal in the interior of the spine. The structure of the dorsal spines is similar. There are no secretory glands imbedded in the membranes of the sacs; and the fluid must be secreted by their mucous membrane. The sacs are without an external muscular layer, and situated immediately below the thick loose skin which envelops the spines to their extremity; the ejection of the poison into a living animal, therefore, can only be effected, as in *Synanceia*, by the pressure to which the sac is subjected the moment the spine enters another body.

Finally, a singular apparatus found in many Siluroids may be mentioned in connection with the poison-organs, although its function is still problematical. Some of these fishes are armed with powerful pectoral spines and justly feared on account of the dangerous wounds they inflict; not a few of them possess, in addition to the pectoral spines, a sac with a more or less wide opening in the axil of the pectoral fin; and it does not seem improbable that it contains a fluid which may be introduced into a wound by means of the pectoral spine, which would be covered with it, like the barbed arrow-head of an Indian. However, whether this secretion is equally poisonous in all the species provided with that axillary sac, or whether it has poisonous qualities at all, is a question which can be decided by experiments only made with the living fishes.

CHAPTER XV.

DISTRIBUTION OF FISHES IN TIME.

Of what kind the fishes were which were the first to make their appearance on the globe; whether or not they were identical with, or similar to, any of the principal types existing at present; are questions which probably will for ever remain hidden in mystery and uncertainty. The supposition that the Leptocardii and Cyclostomes, the lowest of the vertebrate series, must have preceded the other sub-classes, is an idea which has been held by many Zoologists: and as the horny teeth of the Cyclostomes are the only parts of their body which under favourable circumstances might have been preserved, Palæontologists have ever been searching for this evidence.

Fig. 101. Right dental plate of Myxine affinis.

Indeed, in deposits belonging to the Lower Silurian and Devonian, in Russia, England, and North America, minute, slender, pointed horny bodies, bent like a hook, with sharp opposite margins, have been found and described under the name of *Conodonts*. More frequently they possess an elongated basal portion, in which there is generally a larger tooth with rows of similar but smaller denticles on one or both sides of the larger tooth, according as this is central or at one end of the base. In other examples there is no prominent central tooth, but a series of more or less similar teeth is implanted on a straight or curved base. Modifications of these arrangements are very numerous, and many Palæontologists entertain still doubts whether the origin of these remains is not rather from Annelids and Mollusks than from Fishes.

[See G. J. Hinde, in "Quarterly Journal of the Geological Society," 1879.]

The first undeniable evidence of a fish, or, indeed, of a vertebrate animal, occurs in the *Upper Silurian* Rocks, in a bone-bed of the Downton sandstone, near Ludlow. It consists of compressed, slightly curved, ribbed spines, of less than two inches in length (*Onchus*); of small shagreen-scales (*Thelodus*); the fragment of a jaw-like bar with pluricuspid teeth (*Plectrodus*); the cephalic bucklers of what seems to be a species of *Pteraspis;* and, finally, the coprolitic bodies of phosphate and carbonate of lime, including recognisable remains of the Mollusks and Crinoids inhabiting the same waters. But no vertebra or other part of the skeleton has been found. The spines and scales seem to have belonged to the same kind of fish, which probably was a Plagiostome. It is quite uncertain whether or not the jaw (if it be the jaw of a fish[1]) belonged to the buckler-bearing *Pteraspis*, the position of which among Ganoids, with which it is generally associated, is open to doubt.

No detached undoubted tooth of a Plagiostome or Ganoid scale has been discovered in the Ludlow deposits: but so much is certain that those earliest remains in Palæozoic rocks belonged to fishes closely allied to forms occurring in greater abundance in the succeeding formation, the Devonian, where they are associated with undoubted Palæichthyes, Plagiostomes as well as Ganoids.

These fish-remains of the *Devonian* or *Old Red Sandstone*, can be determined with greater certainty. They consist of spines or the so-called *Ichthyodorulites*, which show sufficiently distinctive characters to be referred to several genera, one of

[1] Ray Lankester considers it to be a portion of the long denticulated cornua of a genus *Eukeraspis* allied to *Cephalaspis*.

them, *Onchus*, still surviving from the Silurian epoch. All these spines are believed to be those of Chondropterygians, to which order some pluricuspid teeth (*Cladodus*) from the Old Red Sandstone in the vicinity of St. Petersburg have been referred likewise.

The remains of the Ganoid fishes are in a much more perfect state of preservation, so that it is even possible to obtain a tolerably certain idea of the general appearance and habits of some of them, especially of such as were provided with hard carapaces, solid scales, and ordinary or bony fin-rays. A certain proportion of them, as might have been expected, remind us, with regard to external form, of Teleosteous fishes rather than of any of the few still existing Ganoid types; but it is contrary to all analogy and to all palæontological evidence to suppose that those fishes were, with regard to their internal structure, more nearly allied to Teleosteans than to Ganoids. If they were not true Ganoids, they may be justly supposed to have had the essential characters of Palæichthyes. Other forms exhibit even at that remote geological epoch so unmistakably the characteristics of existing Ganoids, that no one can entertain any doubt with regard to their place in the system. In none of these fishes is there any trace of vertebral segmentation.

The Palæichthyes of the Old Red Sandstone, the systematic position of which is still obscure, are the *Cephalaspidæ* from the Lower Old Red Sandstone of Great Britain and Eastern Canada; *Pterichthys, Coccosteus*, and *Dinichthys*: genera which have been combined in one group—*Placodermi*; and *Acanthodes* and allied genera, which combined numerous branchiostegals with chondropterygian spines and a shagreen-like dermal covering.

Among the other Devonian fishes (and they formed the majority) two types may be recognised, both of which are unmistakably Ganoids. The first approaches the still living

Polypterus, with which some of the genera like *Diplopterus* singularly agree in the form and armature of the head, the lepidosis of the body, the lobate pectoral fins, and the termination of the vertebral column. Other genera, as *Holoptychius*, have cycloid scales; many have two dorsal fins (*Holoptychius*), and, instead of branchiostegals, jugular scutes; others one long dorsal confluent with the caudal (*Phaneropleuron*).

In the second type the principal characters of the *Dipnoi* are manifest, and some of them, for example *Dipterus, Palædaphus, Holodus*, approach so closely the Dipnoi which still survive, that the differences existing between them warrant a separation into families only.

Devonian fishes are frequently found under peculiar circumstances, enclosed in the so-called *nodules*. These bodies are elliptical flattened pebbles, which have resisted the action of water in consequence of their greater hardness, whilst the surrounding rock has been reduced to detritus by that agency. Their greater density is due to the dispersion in their substance of the fat of the animal which decomposed in them. Frequently, on cleaving one of these nodules with the stroke of the hammer, a fish is found embedded in the centre. At certain localities of the Devonian, fossil fishes are so abundant that the whole of the stratum is affected by the decomposing remains emitting a peculiar smell when newly opened, and acquiring a density and durability not possessed by strata without fishes. The flagstones of Caithness are a remarkable instance of this.

The fish-remains of the *Carboniferous* formation show a great similarity to those of the preceding. They occur throughout the series, but are very irregularly distributed, being extremely scarce in some countries, whilst in others entire beds (the so-called bone-beds) are composed of ichthyolites. In the ironstones they frequently form the nuclei of nodules, as in the Devonian.

Of Chondropterygians the spines of *Onchus* and others still occur, with the addition of teeth indicative of the existence of fishes allied to the Cestracion-type (*Cochliodus, Psammodus*): a type which henceforth plays an important part in the composition of the extinct marine fish faunæ. Another extinct Selachian family, that of Hybodontes, makes its appearance, but is known from the teeth only.

Of the Ganoid fishes, the family *Palæoniscidæ* (Traquair) is numerously represented; others are Cœlacanths (*Cœlacanthus, Rhizodus*), and *Saurodipteridæ* (*Megalichthys*). None of these fishes have an ossified vertebral column, but in some (*Megalichthys*) the outer surface of the vertebræ is ossified into a ring; the termination of their tail is heterocercal. The carboniferous *Uronemus* and the Devonian *Phaneropleuron* are probably generically the same; and the Devonian *Dipnoi* are continued as, and well represented by, *Ctenodus*.

The fishes of the *Permian* group are very similar to those of the Carboniferous. A type which in the latter was but very scantily represented, namely the *Platysomidæ*, is much developed. They were deep-bodied fish, covered with hard rhomboid scales possessing a strong anterior rib, and provided with a heterocercal caudal, long dorsal and anal, short non-lobate paired fins (when present), and branchiostegals. The *Palæoniscidæ* are represented by many species of *Palæoniscus, Pygopterus* and *Acrolepis*, and Cestracionts by *Janassa* and *Strophodus*.

The passage from the Palæozoic into the *Mesozoic* era is not indicated by any marked change as far as fishes are concerned. The more remarkable forms of the Trias are Shark-like fishes represented by ichthyodorulithes like *Nemacanthus, Liacanthus,* and *Hybodus;* and Cestracionts represented by species of *Acrodus* and *Strophodus.* Of the Ganoid genera *Cœlacanthus, Amblypterus* (*Palæoniscidæ*), *Saurichthys* persist from the Carboniferous epoch. *Ceratodus* appears for the first time (Muschel-Kalk of Germany).

Thanks to the researches of Agassiz, and especially Sir P. Egerton, the ichthyological fauna of the Lias is, perhaps, the best known of the Mesozoic era, 152 species having been described. Of the various localities, Lyme Regis has yielded more than any other, nearly all the Liassic genera being represented there by not less than seventy-nine species. The Hybodonts and Cestracionts continue in their fullest development. Holocephales (*Ischyodus*), true Sharks (*Palæoscyllium*), Rays (*Squaloraja, Arthropterus*), and Sturgeons (*Chondrosteus*) make their first appearance; but they are sufficiently distinct from living types to be classed in separate genera, or even families. The Ganoids, especially Lepidosteoids, predominate over all the other fishes: *Lepidotus, Semionotus, Pholidophorus, Pachycormus, Eugnathus, Tetragonolepis,* are represented by numerous species; other remarkable genera are *Aspidorhynchus, Belonostomus, Saurostomus, Sauropsis, Thrissonotus, Conodus, Ptycholepis, Endactis, Centrolepis, Legnonotus, Oxygnathus, Heterolepidotus, Isocolum, Osteorhachis, Mesodon.* These genera offer evidence of a great change since the preceding period, the majority not being represented in older strata, whilst, on the other hand, many are continued into the succeeding oolithic formations. The homocercal termination of the vertebral column commences to supersede the heterocercal, and many of the genera have well ossified and distinctly segmented spinal columns. Also the cycloid form of scales becomes more common: one genus (*Leptolepis*) being, with regard to the preserved hard portions of its organisation, so similar to the Teleosteous type that some Palæontologists refer it (with much reason) to that sub-class.

[See *E. Sauvage*, Essai sur la Faune Ichthyologique de la période Liasique. In "Bibl. de l'école des hautes études," xiii. art. 5. Paris 1875. 8º.]

As already mentioned, the *Oolithic* formations show a great similarity of their fish-fauna to that of the Lias; but still more apparent is its approach to the existing fauna. Teeth have been found which cannot even generically be distinguished from *Notidanus*. The Rays are represented by genera like *Spathobatis, Belemnobatis, Thaumas;* the *Holocephali* are more numerous than in the Lias (*Ischyodus, Ganodus*). The most common Ganoid genera are *Caturus, Pycnodus, Pholidophorus Lepidotus, Leptolepis,* all of which had been more or less fully represented in the Lias. Also *Ceratodus* is continued into it.

The *Cretaceous* group offers clear evidence of the further advance towards the existing fauna. Teeth of Sharks of existing genera *Carcharias* (*Corax*), *Scyllium, Notidanus,* and *Galeocerdo,* are common in some of the marine strata, whilst Hybodonts and Cestracionts are represented by a small number of species only; of the latter one new genus, *Ptychodus,* appears and disappears. A very characteristic Ganoid genus, *Macropoma,* comprises homocercal fishes with rounded ganoid scales sculptured externally and pierced by prominent mucous tubes. *Caturus* becomes extinct. Teeth and scales of *Lepidotus* (with *Sphærodus* as sub-genus), clearly a fresh-water fish, are widely distributed in the Wealden, and finally disappear in the chalk; its body was covered with large rhomboidal ganoid scales. *Gyrodus* and *Aspidorhynchus* occur in the beds of Voirons, *Coelodus* and *Amiopsis* (allied to Amra), in those of Comen, in Istria. But the Palæichthyes are now in the minority; undoubted Teleosteans have appeared, for the first time, on the stage of life in numerous genera, many of which are identical with still existing fishes. The majority are Acanthopterygians, but Physostomes and Plectognaths are likewise well represented, most of them being marine. Of Acanthopterygian families the first to appear are the *Bery-*

cidæ, represented by several very distinct genera: *Beryx*; *Pseudoberyx* with abdominal ventral fins; *Beryeopsis* with cycloid scales; *Homonotus*, *Stenostoma*, *Sphenocephalus*, *Acanus*, *Hoplopteryx*, *Platycornus* with granular scales; *Podocys* with a dorsal extending to the neck; *Acrogaster*, *Macrolepis*, *Rhacolepis* from the chalk of Brazil. The position of *Pycnosterynx* is uncertain, it approaches certain Pharyngognaths. True *Percidæ* are absent, whilst the *Carangidæ*, *Sphyrænidæ*, *Cataphracti*, *Gobiidæ*, *Cottidæ*, and *Sparidæ* are represented by one or more genera. Somewhat less diversified are the Physostomes, which belong principally to the *Clupeidæ* and *Dercetidæ*, most of the genera being extinct; Clupea is abundant in some localities. *Scopelidæ* (*Hemisaurida* and *Saurocephalus*) occur in the chalk of Comen in Istria, and of Mæstricht. Of all cretaceous deposits none surpass those of the Lebanon with regard to the number of genera, species, and individuals; the forms are exclusively marine, and the remains in the most perfect condition.

In the *Tertiary* epoch the Teleosteans have almost entirely replaced the Ganoids; a few species only of the latter make their appearance, and they belong to existing genera, or, at least, very closely allied forms (*Lepidosteus*, *Amia*, *Hypamia*, *Acipenser*). The Chondropterygians merge more and more into recent forms; Holocephali continue, and still are better represented than in the present fauna. The Teleosteans show even in the Eocene a large proportion of existing genera, and the fauna of some localities of the Miocene (Oeningen) is almost wholly composed of them. On the whole, hitherto more than one-half have been found to belong to existing genera, and there is no doubt that the number of seemingly distinct extinct genera will be lessened as the fossils will be examined with a better knowledge of the living forms. The distribution of the fishes differed widely from that of our period, many of our tropical genera occurring in localities which are now included within our temperate zone, and being mixed with others, which nowadays are restricted to a colder climate: a mixture which continues throughout the Pliocene.

A few families of fishes, like the freshwater Salmonidæ, seem to have put in their appearance in *Post-pliocene* times; however, not much attention has been paid to fish-remains of these deposits; and such as have been incidentally examined offer evidence of the fact that the distribution of fishes has not undergone any further essential change down to the present period.

[See *E. Sauvage*, Mémoire sur la Faune Ichthyologique de la période Tertiaire. Paris 1873. 8°.]

Fig. 102.—*Pycnodus rhombus*, a Ganoid from the Upper Oolite.

CHAPTER XVI.

THE DISTRIBUTION OF EXISTING FISHES OVER THE EARTH'S SURFACE—GENERAL REMARKS.

In an account of the geographical distribution of fishes the *Freshwater* forms are to be kept separate from the *Marine*. However, when we attempt to draw a line between these two kinds of fishes, we meet with a great number of species and of facts which would seem to render that distinction very vague. There are not only species which can gradually accommodate themselves to a sojourn in either salt or fresh water, but there are also such as seem to be quite indifferent to a rapid change from one into the other: so that individuals of one and the same species (Gastrosteus, Gobius, Blennius, Osmerus, Retropinna, Clupea, Syngnathus, etc.), may be found at some distance out at sea, whilst others live in rivers far beyond the influence of the tide, or even in inland fresh waters without outlet to the sea. The majority of these fishes belong to forms of the fauna of the *brackish* water, and as they are not an insignificant portion of the fauna of almost every coast, we shall have to treat of them in a separate chapter.

Almost every large river offers instances of truly marine fishes (such as *Serranus, Sciænidæ, Pleuronectes, Clupeidæ, Tetrodon, Carcharias, Trygonidæ*), ascending for hundreds of miles of their course; and not periodically, or from any apparent physiological necessity, but sporadically throughout the year, just like the various kinds of marine Porpoises which are found all along the lower course of the Ganges, Yang-tse-

Kiang, the Amazons, the Congo, etc. This is evidently the commencement of a change in a fish's habits, and, indeed, not a few of such fishes have actually taken up their permanent residence in fresh waters (as species of Ambassis, Apogon Dules, Therapon, Sciæna, Blennius, Gobius, Atherina, Mugil, Myxus, Hemirhamphus, Clupea, Anguilla, Tetrodon, Trygon): all forms *originally marine.*

On the other hand, we find fishes belonging to fresh-water genera descending rivers and sojourning in the sea for a more or less limited period ; but these instances are much less in number than those in which the reverse obtains. We may mention species of *Salmo* (the Common Trout, the Northern Charr), and Siluroids (as *Arius, Plotosus*). *Coregonus*, a genus so characteristic of the inland lakes of Europe, Northern Asia, and North America, nevertheless offers some instances of species wandering by the effluents into the sea, and taking up their residence in salt water, apparently by preference, as *Coregonus oxyrhynchus.* But of all the Fresh-water families none exhibit so great a capability of surviving the change from fresh into salt water, as the *Gastrosteidæ* (Stickle-backs), of the northern Hemisphere, and the equally diminutive *Cyprinodontidæ* of the tropics ; not only do they enter into, and live freely in, the sea, but many species of the latter family inhabit inland waters, which, not having an outlet, have become briny, or impregnated with a larger proportion of salts than pure sea water. During the voyage of the "Challenger" a species of *Fundulus, F. nigrofasciatus,* which inhabits the fresh and brackish waters of the Atlantic States of North America, was obtained, with Scopelids and other pelagic forms, in the tow-net, midway between St. Thomas and Teneriffe.

Some fishes annually or periodically ascend rivers for the purpose of spawning, passing the rest of the year in the sea, as Sturgeons, many Salmonoids, some Clupeoids, Lampreys, etc. The two former evidently belonged originally to the freshwater series, and it was only in the course of their existence that they acquired the habit of descending to the sea, perhaps because their freshwater home did not offer a sufficient supply of food. These migrations of freshwater fishes have been compared with the migrations of birds; but they are much more limited in extent, and do not impart an additional element to the fauna of the place to which they migrate, as is the case with the distant countries to which birds migrate.

The distinction between freshwater and marine fishes is further obscured by geological changes, in consequence of which the salt water is gradually being changed into fresh, or *vice versa*. These changes are so gradual and spread over so long a time, that many of the fishes inhabiting such localities accommodate themselves to the new conditions. One of the most remarkable and best studied instances of such an alteration is the Baltic, which, during the second half of the Glacial period, was in open and wide communication with the Arctic Ocean, and evidently had the same marine fauna as the White Sea. Since then, by the rising of the land of Northern Scandinavia and Finland, this great gulf of the Arctic Ocean has become an inland sea, with a narrow outlet into the North Sea, and its water, in consequence of the excess of the fresh water pouring into it over the loss by evaporation, has been so much diluted as to be nearly fresh at its northern extremities: and yet nine species, the origin of which from the Arctic Ocean can be proved, have survived the changes, propagating their species, agreeing with their brethren in the Arctic Ocean in every point, but remaining comparatively smaller. On the other hand, fishes which we must regard as true freshwater fishes, like the Rudd, Roach, Pike, Perch, enter freely the brackish water of the Baltic.[1] Instances of marine fishes

[1] Ekström, Fische in den Scheeren von Mörkö.

being permanently retained in fresh water in consequence of geological changes are well known: thus *Cottus quadricornis* in the large lakes of Scandinavia; species of *Gobius, Blennius,* and *Atherina* in the lakes of Northern Italy; *Comephorus,* of the depths of the Lake of Baikal, which seems to be a dwarfed Gadoid. *Carcharias gangeticus* in inland lakes of the Fiji Islands, is another instance of a marine fish which has permanently established itself in fresh water.

In the miocene formation of Licata in Sicily, in which fish remains abound, numerous Cyprinoids are mixed with littoral and pelagic forms. Sauvage found in 450 specimens from that locality, not less than 266, which were Leucisci, Alburni, or Rhodei. Now, although it is quite possible that in consequence of a sudden catastrophe the bodies of those Cyprinoids were carried by a freshwater current into, and deposited on the bottom of, the sea, the surmise that they lived together with the littoral fishes in the brackish water of a large estuary, which was not rarely entered by pelagic forms, is equally admissible. And, if confirmed by other similar observations, this instance of a mixture of forms which are now strictly freshwater or marine, may have an important bearing on the question to what extent fishes have in time changed their original habitat.

Thus there is a constant exchange of species in progress between the freshwater and marine faunæ, and in not a few cases it would seem almost arbitrary to refer a genus or even larger group of fishes to one or the other; yet there are certain groups of fishes which entirely, or with but few exceptions are, and, apparently, during the whole period of their existence have been, inhabitants either of the sea or of fresh water; and as the agencies operating upon the distribution of marine fishes differ greatly from those influencing the dispersal of freshwater fishes, the two series must be treated separately. The most obvious fact that dry land, which intervenes between river systems, offers to the rapid spreading of a freshwater fish an obstacle which can be surmounted only exceptionally or by a most circuitous route, whilst marine fishes may readily and voluntarily extend their original limits, could be illustrated by a great number of instances. Without entering into details, it may suffice to state as the general result, that no species or genus of freshwater fishes has anything like the immense range of the corresponding categories of marine fishes; and that, with the exception of the Siluroids, no other freshwater family is so widely spread as the families of marine fishes. Surface temperature or climate which is, if not the most, one of the most important physical factors in the limitation of freshwater fishes, similarly affects the distribution of marine fishes, but in a less degree, and only those which live near to the shore or the surface of the ocean; whilst it ceases to exercise its influence in proportion to the depth, the true deep-sea forms being entirely exempt from its operation. Light, which is pretty equally distributed over the localities inhabited by freshwater fishes, cannot be considered as an important factor in their distribution, but it contributes towards constituting the impassable barrier between the surface and abyssal forms of marine fishes. Altitude has stamped the fishes of the various Alpine provinces of the globe with a certain character, and limited their distribution; but the number of these Alpine forms is comparatively small, ichthyic life being extinguished at great elevations even before the mean temperature equals that of the high latitudes of the Arctic region, in which some freshwater fishes flourish. On the other hand, the depths of the ocean, far exceeding the altitude of the highest mountains, still swarm with forms specially adapted for abyssal life. That other physical conditions of minor and local importance, under which fresh water fishes live, and by which their dispersal is regulated, are more complicated than similar ones of the ocean, is probable, though perhaps less so than is generally supposed: for the fact is that the former are more accessible to observation than the latter, and are, therefore, more generally and more readily comprehended and acknowledged. Thus, not only because many of the most characteristic forms of the marine and freshwater series are found, on taking a broader view of the subject, to be sufficiently distinct, but also because their distribution depends on causes different in their nature as well as the degree of their action, it will be necessary to treat of the two series separately. Whether the oceanic areas correspond in any way to the terrestrial will be seen in the sequel.

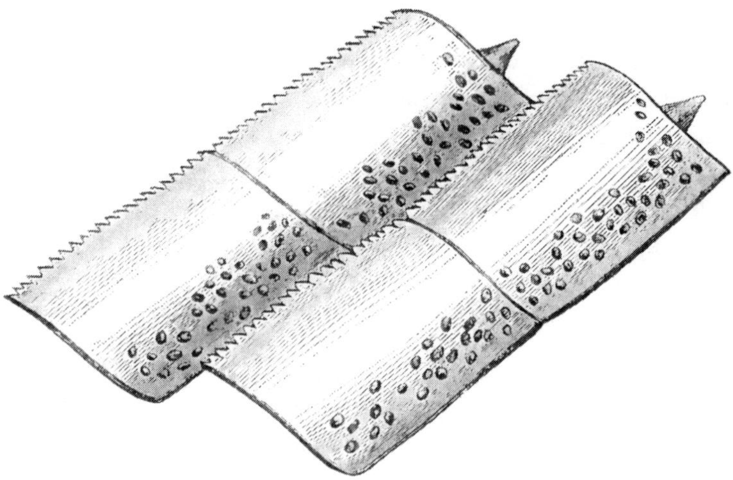

Fig. 103.—Ganoid scales of *Tetragonolepis.*

CHAPTER XVII.

THE DISTRIBUTION OF FRESHWATER FISHES.

HAVING shown above that numerous marine fishes enter fresh waters, and that some of them have permanently established themselves therein, we have to eliminate from the category of freshwater fishes all such adventitious elements. They are derived from forms, the distribution of which is regulated by other agencies, and which, therefore, would obscure the relations of the faunæ of terrestrial regions if they were included in them. They will be mentioned with greater propriety along with the fishes constituting the fauna of the brackish water.

True freshwater fishes are the following families and groups only:—

Dipnoi . .	with	4	species.	Labyrinthici .	with	30	species.
Acipenseridæ and Polyodontidæ	,,	26	,,	Luciocephalidæ	,,	1	,,
Amiidæ .	,,	1	,,	Gastrosteus .	,,	10	,,
Polypteridæ .	,,	2	,,	Ophiocephalidæ	,,	31	,,
Lepidosteidæ .	,,	3	,,	Mastacembelidæ	,,	13	,,
Percina .	,,	46	,,	Chromides .	,,	105	,,
Grystina .	,,	11	,,	Comephoridæ	,,	1	,,
Aphredoderidæ	,,	1	,,	Gadopsidæ .	,,	1	,,
Centrarchina .	,,	26	,,	Siluridæ .	,,	572	,,
Dules . .	,,	10	,,	Characinidæ .	,,	261	,,
Nandidæ .	,,	7	,,	Haplochitonidæ	,,	3	,,
Polycentridæ .	,,	3	,,	Salmonidæ (3 genera excepted) .	,,	135	,,

Percopsidæ	with	1 species.	Kneriidæ	with	2	species.
Galaxiidæ	„	15 „	Hyodontidæ	„	1	„
Mormyridæ (and Gymnarchidæ)	„	52 „	Osteoglossidæ	„	5	„
Esocidæ	„	8 „	Notopteridæ	„	5	„
Umbridæ	„	2 „	Gymnotidæ	„	20	„
Cyprinodontidæ	„	112 „	Symbranchidæ	„	5	„
Heteropygii	„	2 „	Petromyzontidæ	„	12	„
Cyprinidæ	„	724 „	Total		2269	species.

As in every other class of animals, these freshwater genera and families vary greatly with regard to the extent of their geographical range; some extend over the greater half of the continental areas, whilst others are limited to one continent only, or even to a very small portion of it. As a general rule, a genus or family of freshwater fishes is regularly dispersed and most developed within a certain district, the species and individuals becoming scarcer towards the periphery as the type recedes more from its central home, some outposts being frequently pushed far beyond the outskirts of the area occupied by it. But there are not wanting those remarkable instances of closely allied forms occurring, almost isolated, at most distant points, without being connected by allied species in the intervening space; or of members of the same family, genus, or species inhabiting the opposite shores of an ocean, and separated by many degrees of abyssal depths. We mention of a multitude of such instances the following only:—

A. Species identical in distant continents—

1. A number of species inhabiting Europe and the temperate parts of eastern North America, as *Perca fluviatilis, Gastrosteus pungitius, Lota vulgaris, Salmo salar, Esox lucius, Acipenser sturio, Acipenser maculosus,* and several Petromyzonts.

2. *Lates calcarifer* is common in India as well as in Queensland.

3. *Galaxias attenuatus* inhabits Tasmania, New Zealand, the Falkland Islands, and the southernmost part of the South American continent.

4. Several Petromyzonts enter the fresh waters of Tasmania, South Australia, New Zealand, and Chili.

B. Genera identical in distant continents—

1. The genus *Umbra*, so peculiar a form as to be the type of a distinct family consisting of two most closely allied species only, one of which is found in the Atlantic States of North America, the other in the system of the Danube.

2. A very distinct genus of Sturgeons, *Scaphirhynchus*, consisting of two species only, one inhabiting fresh waters of Central Asia, the other the system of the Mississippi.

3. A second most peculiar genus of Sturgeons, *Polyodon*, consists likewise of two species only, one inhabiting the Mississippi, the other the Yang-tse-kiang.

4. *Amiurus*, a Siluroid, and *Catostomus*, a Cyprinoid genus, both well represented in North America, occur in a single species in temperate China.

5. *Lepidosiren* is represented by one species in tropical America, and by the second in tropical Africa (*Protopterus*).

6. *Notopterus* consists of three Indian and two West African species.

7. *Mastacembelus* and *Ophiocephalus*, genera characteristic of the Indian region, emerge severally by a single species in West and Central Africa.

8. *Symbranchus* has two Indian and one South American species.

9. *Prototroetes*, the singular antarctic analogue of *Coregonus*, consists of two species, one in the south of Australia the other in New Zealand.

10. *Galaxias* is equally represented in Southern Australia, New Zealand, and the southern parts of South America.

C. Families identical in distant continents—

1. The *Labyrinthici*, represented in Africa by 5, and in India by 25 species.

2. The *Chromides*, represented in Africa by 25, and in South America by 80 species.

3. The *Characinidæ*, represented in Africa by 35, and in South America by 226 species.

4. The *Haplochitonidæ*, represented in Southern Australia by one, in New Zealand by one, and in Patagonia by a third species.

This list could be much increased from the families of *Siluridæ* and *Cyprinidæ*, but as these have a greater range than the other Freshwater fishes, they do not illustrate with equal force the object for which the list has been composed.

The ways in which the dispersal of Freshwater fishes has been effected were various; they are probably all still in operation, but most work so slowly and imperceptibly as to escape direct observation; perhaps, they will be more conspicuous, after science and scientific inquiry shall have reached to a somewhat greater age. From the great number of freshwater forms which we see at this present day acclimatised in, gradually acclimatising themselves in, or periodically or sporadically migrating into, the sea, we must conclude that, under certain circumstances, salt water may cease to be an impassable barrier at some period of the existence of freshwater species, and that many of them have passed from one river through salt water into another. Secondly, the headwaters of some of the grandest rivers, the mouths of which are at opposite ends of the continents which they drain, are sometimes distant from each other a few miles only; the intervening space may have been easily bridged over for the passage of fishes by a slight geological change affecting the level of the watershed, or even by temporary floods; and a communication of this kind, if existing for a limited period only, would afford the ready means of an exchange of a number of species previously peculiar to one or the other of those river or lake systems. Some fishes, provided with gill-openings so narrow that the water moistening the gills cannot readily evaporate; and endowed, besides, with an extraordinary degree of vitality, like many Siluroids (*Clarias, Callichthys*), Eels, etc., are enabled to wander for some distance over land, and may thus reach a watercourse leading them thousands of miles from their original home. Finally, fishes or their ova may be accidentally carried by waterspouts, by aquatic birds or insects, to considerable distances.

Freshwater fishes of the present fauna were already in existence when the great changes of the distribution of land and water took place in the tertiary epoch; and having stated that salt water is not an absolute barrier to the spreading of Freshwater fishes, we can now more easily account for those instances of singular disconnection of certain families or genera. It is not necessary to assume that there was a continuity of land stretching from the present coast of Africa to South America, or from South America to New Zealand and Australia, to explain the presence of identical forms at so distant localities; it suffices to assume that the distances were lessened by intervening archipelagoes, or that an oscillation has taken place in the level of the land area.

Dispersal of a type over several distant continental areas may be evidence of its great antiquity, but it does not prove that it is of greater antiquity than another limited to one region only. Geological evidence is the only proof of the antiquity of a type. Thus, although the *Dipnoi* occur on the continents of Africa, South America, and Australia, and their present distribution is evidently the consequence of their wide range in palæozoic and secondary epochs; the proof of their high antiquity can be found in their fossil remains only. For, though the Siluroids have a still greater range, their wide distribution is of comparatively recent date, as the few fossil remains that have been found belong to the tertiary epoch. The rapidity of dispersal of a type depends entirely on its facility to accommodate itself to a variety of physical conditions, and on the degree of vitality by which it is enabled to survive more or less sudden changes under unfavourable conditions; proof of this is afforded by the family of Siluroids, many of which can suspend for some time the energy of their respiratory functions, and readily survive a change of water.

To trace the geological sequence of the distribution of an ichthyic type, and to recognise the various laws which have governed, and are still governing its dispersal, is one of the ultimate tasks of Ichthyology. But the endeavour to establish by means of our present fragmentary geological knowledge the divisions of the fauna of the globe, leads us into a maze of conflicting evidence; or, as Mr. Wallace truly observes, "any attempt to exhibit the regions of former geological ages in combination with those of our own period must lead to confusion." Nevertheless, as the different types of animals found at the present day within a particular area have made their appearance therein at distant periods, we should endeavour to decide as far as we can, in an account of the several zoo-geographical divisions, the following questions:—

1. Which of the fishes of an area should be considered to be the remnants of *ancient* types, probably spread over much larger areas in preceding epochs?

2. Which of them are to be considered to be *autochthont* species, that is, forms which came in the tertiary epoch or later into existence within the area to which they are still limited, or from which they have since spread?

3. Which are the forms which must be considered to be *immigrants* from some other region?

The mode of division of the earth's surface into zoological regions or areas now generally adopted, is that proposed by Mr. Sclater, which recommends itself as most nearly agreeing with the geographical divisions. These regions are as follows:—

I. PALÆOGÆA.

1. The *Palæarctic* region; including Europe, temperate Asia, and North Africa.
2. The *Ethiopian* region; including Africa, south of the Sahara, Madagascar, and the Mascarene Islands; also Southern Arabia.
3. The *Indian* region; including India south of the Himalayas, to Southern China, Borneo, and Java.
4. The *Australian* region; including Australia, the Pacific Islands, Celebes, and Lombock.

II. NEOGÆA.

5. The *Nearctic* region; including North America to Northern Mexico.
6. The *Neotropical* region; including South America, the West Indies, and Southern Mexico.

Comparatively few classes and orders of animals have been carefully studied with regard to their geographical distribution, but the majority of those which have been examined show that the difference of latitude is accompanied by a greater dissimilarity of indigenous species than that of longitude, and that a main division into an old world and new world fauna is untenable. More especially the Freshwater fishes, with which we are here solely concerned, have been spread in *circumpolar* zones, and in a but limited degree from north to south. No family, much less a genus, ranges from the north to the south, whilst a number of families and genera make the entire circuit, and some species more than half of the circuit round the globe within the zone to which they belong. Not even the Cyprinoids and Siluroids, which are most characteristic of the freshwater fauna of our period, are an exception to this. Temperature and climate, indeed, are the principal factors by which the character of the freshwater fauna is determined; they form the barriers which interfere with the unlimited dispersal of an ichthyic type, much more than mountain ranges, deserts, or oceans. Hence the tropical zone is an impassable barrier to the northern Freshwater fish in its progress towards the south; where a similarly temperate climate obtains in the southern hemisphere, fish-forms appear *analogous* to those of the north, but *genetically* and *structurally distinct.*

The similarity which obtains in fishes at somewhat distant points of the same degree of longitude, rarely extends far, and is due to the natural tendency of every animal to spread as far as physical conditions will permit. Between two regions situated north and south of each other there is always a debateable border ground, in parts of which sometimes the fishes of the one, sometimes those of the other, predominate, and which is, in fact, a *band* of demarcation. Within this band the regions overlap each other; therefore, their border *lines* are rarely identical, and should be determined by the northern and southernmost extent of the most characteristic types of each region. Thus, for instance, in China, a broad band intervenes between temperate and tropical Asia, in which these two faunæ mix, and the actual northern border line of the tropical fauna is north of the southern border line of temperate Asia.

It is the aim of every philosophical classification to indicate the degree of affinity which obtains between the various divisions; but the mode of division into six equivalent regions, as given above, does not fulfil this aim with regard to Freshwater fishes, the distribution of which allows of further generalisation and subdivision. The two families, *Cyprinidæ* and *Siluridæ*, of which the former yields a contingent of one-third, and the latter of one-fourth of all the freshwater species known of our period, afford most valuable guidance for the valuation of the degrees of affinity between the various divisions. The Cyprinoids may be assumed to have taken their origin in the Alpine region, dividing the temperate and tropical parts of Asia; endowed with a greater capability of acclimatising themselves in a temperate as well as tropical climate than any other family of freshwater fishes, they spread north and south as well as east and west; in the preglacial epoch they reached North America, but they have not had time to penetrate into South America, Australia, or the islands of the Pacific. The Siluroids, principally fishes of the sluggish waters of the plains, and well adapted for surviving changes of the water in which they live, for living in mud or sea-water, flourish most in the tropical climate, in which this type evidently had its origin. They came into existence after the Cyprinoids, fossil remains being known only from tertiary deposits in India, none from Europe. They rapidly spread over the areas of land within the tropical zone, reaching northern Australia from India, and one species even immigrated into the Sandwich Islands, probably from South America. The Coral Islands of the Pacific still remain untenanted by them. Their progress into temperate regions was evidently slow, only very few species penetrating into the temperate parts of Asia and Europe; and the North American species, although more numerous, showing no great variety of structure, all belonging to the same group (*Amiurina*). Towards the south their progress was still slower, Tasmania, New Zealand, and Patagonia being without representatives, whilst the streams of the Andes of Chili are inhabited by a few dwarfed forms identical with such as are characteristic of similar localities in the more northern and warmer parts of the South American continent.

After these preliminary remarks we propose the following division of the fauna of Freshwater fishes :—

I. THE NORTHERN ZONE.—Characterised by Acipenseridæ. Few Siluridæ. Numerous Cyprinidæ. Salmonidæ, Esocidæ.

1. *Europo-Asiatic or Palæarctic Region.*—Characterised by absence of osseous Ganoidei ; Cobitidæ and Barbus numerous.

2. *North American Region.*—Characterised by osseous Ganoidei, Amiurina, and Catostomina ; but no Cobitidæ or Barbus.

II. THE EQUATORIAL ZONE.—Characterised by the development of Siluridæ.

A. *Cyprinoid Division.*—Characterised by presence of Cyprinidæ and Labyrinthici.

1. *Indian Region.*—Characterised by [absence of Dipnoi [1]] Ophiocephalidæ, Mastacembelidæ. Cobitidæ numerous.

2. *African Region.*—Characterised by presence of Dipnoi and Polypteridæ. Chromides and Characinidæ numerous. Mormyridæ. Cobitidæ absent.

[1] Will probably be found.

B. *Acyprinoid Division.*—Characterised by absence of Cyprinidæ and Labyrinthici.

1. *Tropical American Region.*—Characterised by presence of Dipnoi. Chromides and Characinidæ numerous. Gymnotidæ.

2. *Tropical Pacific Region.*—Characterised by presence of Dipnoi. Chromides and Characinidæ absent.

III. THE SOUTHERN ZONE.—Characterised by absence of Cyprinidæ, and scarcity of Siluridæ. Haplochitonidæ and Galaxiidæ represent the Salmonoids and Esoces of the Northern zone. One region only.

1. *Antarctic Region.*—Characterised by the small number of species; the fishes of—

   a. The Tasmanian sub-region;
   b. The New Zealand sub-region;
   c. The Patagonian sub-region;

   being almost identical.[1]

In the following detailed account we begin with a description of the equatorial zone, this being the one from which the two principal families of freshwater fishes seem to have spread.

I. EQUATORIAL ZONE.

Roughly speaking, the borders of this zoological zone coincide with the geographical limits of the tropical zone, the tropics of the Cancer and Capricorn; its characteristic forms, however, extend in undulating lines several degrees north and southwards. Commencing from the west coast of

[1] We distinguish these sub-regions, because their distinction is justified by other classes of animals; as regards freshwater fishes their distinctness is even less than that between Europe and Northern Asia.

Africa the desert of the Sahara forms a well-marked boundary between the equatorial and northern zones; as the boundary approaches the Nile it makes a sudden sweep towards the north as far as Northern Syria (*Mastacembelus*, near Aleppo, and in the Tigris; *Clarias* and *Chromides*, in the lake of Galilee); crosses through Persia and Afghanistan (*Ophiocephalus*), to the southern ranges of the Himalayas, and follows the course of the Yang-tse-Kiang, which receives its contingent of equatorial fishes through its southern tributaries. Its continuation through the North Pacific may be considered to be indicated by the tropic which strikes the coast of Mexico at the southern end of the Gulf of California. Equatorial types of South America are known to extend so far northwards; and by following the same line the West India Islands are naturally included in this zone.

Towards the south the equatorial zone embraces the whole of Africa and Madagascar, and seems to extend still farther south in Australia, its boundary probably following the southern coast of that continent; the detailed distribution of the freshwater fishes of South-Western Australia has been but little studied, but the few facts which we know show that the tropical fishes of Queensland follow the principal water-course of that country, the Murray River, far towards the south and probably to its mouth. The boundary-line then stretches northwards of Tasmania and New Zealand, coinciding with the tropic until it strikes the western slope of the Andes, on the South American Continent, where it again bends southwards to embrace the system of the Rio de la Plata.

The equatorial zone is divided into four regions:—

A. The Indian region.
B. The African region.
C. The Tropical American region.
D. The Tropical Pacific region.

These four regions diverge into two well-marked divisions, one of which is characterised by the presence of Cyprinoid fishes, combined with the development of Labyrinthici; whilst in the other both these types are absent. The boundary between the Cyprinoid and Acyprinoid division seems to follow Wallace's line, a line drawn from the south of the Philippines between Borneo and Celebes, and farther south between Bali and Lombock. Borneo abounds in Cyprinoids; from the Philippine Islands a few only are known at present, and in Bali two species have been found; but none are known from Celebes or Lombock, or from islands situated farther east of them.[1]

Taking into consideration the manner in which Cyprinoids and Siluroids have been dispersed, we are obliged to place the Indian region as the first in the order of our treatment; and indeed the number of its freshwater fishes, which appear to have spread from it into the neighbouring regions, far exceeds that of the species which it has received from them.

A. The Indian Region comprises the whole continent of Asia south of the Himalayas and the Yang-tse-kiang; it includes the islands to the west of Wallace's line. Towards the north-east the island of Formosa, which also by other parts of its fauna leans more towards the equatorial zone, has received some characteristic Japanese Freshwater fishes, for instance, the singular Salmonoid *Plecoglossus*. Within the geographical boundaries of China the Freshwater fishes of the tropics pass gradually into those of the northern zone, both being separated by a broad debateable ground. The affluents of the great river tra-

[1] Martens (Preuss. Exped. Ostas. Zool. i. p. 356), has already drawn attention that a Barbel, said to have been obtained by Ida Pfeiffer in Amboyna (Günth. Fish. vii. p. 123), cannot have come from that locality.

versing this district are more numerous from the south than from the north, and carry the southern fishes far into the temperate zone. The boundary of this region towards the north-west is scarcely better defined. Before Persia passed through the geological changes by which its waters were converted into brine and finally dried up, it seems to have been inhabited by many characteristic Indian forms, of which a few still survive in the tract intervening between Afghanistan and Syria; *Ophiocephalus* and *Discognathus* have each at least one representative, *Macroncs* has survived in the Tigris, and Mastacembelus has penetrated as far as Aleppo. Thus, Freshwater fishes belonging to India, Africa, and Europe, are intermingled in a district which forms the connecting link between the three continents. Of the freshwater fishes of Arabia we are perfectly ignorant; so much only being known that the Indian *Discognathus lamta* occurs in the reservoirs of Aden, having, moreover, found its way to the opposite African coast; and that the ubiquitous Cyprinodonts flourish in brackish pools of Northern Arabia.

The following is the list of the forms of freshwater fishes inhabiting this region:[1]—

Percina—		
Lates[2] [Africa, Australia] . .	1	species.
Nandina	7	,,
Labyrinthici [Africa]	25	,,
Luciocephalidæ	1	,,
Ophiocephalidæ [1 species in Africa] . .	30	,,
Mastacembelidæ [3 species in Africa] . .	10	,,
Chromides [Africa, South America]		
Etroplus	2	,,

[1] In the following and succeeding lists, those forms which are peculiar to and exclusively characteristic of, the region, are printed in italics; the other regions, in which the non-peculiar forms occur, are mentioned within brackets [ ].

[2] Lates calcarifer in India as well as Australia.

Siluridæ—	
Clariina [Africa]	12 species.
Chacina	3 ,,
Silurina [Africa, Palæarct.]	72 ,,
Bagrina [Africa]	50 ,,
Ariina [Africa, Australia, South America]	40 ,,
Bagariina	20 ,,
Rhinoglanina [Africa]	1 ,,
Hypostomatina [South America]	5 ,,
Cyprinodontidæ—	
Carnivoræ [Palæarct., North America, Africa, South America]	
Haplochilus [Africa, South America, North America, Japan]	4 ,,
Cyprinidæ [Palæarct., N. America, Africa]—	
Cyprinina [Palæarct., N. America, Africa]	190 ,,
Rasborina [Africa, 1 species]	20 ,,
Semiplotina	4 ,,
Danionina [Africa]	30 ,,
Abramidina [Palæarct., N. Amer., Africa]	30 ,,
Homalopterina	10 ,,
Cobitidina [Palæarct.]	50 ,,
Osteoglossidæ [Africa, Australia, S. America]	1 ,,
Notopteridæ [Africa]	3 ,,
Symbranchidæ—	
Amphipnous	1 ,,
Monopterus	1 ,,
Symbranchus [1 species in S. America]	2 ,,
	625 species.

In analysing this list we find that out of 39 families or groups of freshwater fishes 12 are represented in this region, and that 625 species are known to occur in it; a number equal to two-sevenths of the entire number of freshwater fishes known. This large proportion is principally due to the development of numerous local forms of Siluroids and

Cyprinoids, of which the former show a contingent of about 200, and the latter of about 330 species. The combined development of those two families, and their undue preponderance over the other freshwater types, is therefore the principal characteristic of the Indian region. The second important character of its fauna is the apparently total absence of Ganoid and Cyclostomous fishes. Every other region has representatives of either Ganoids or Cyclostomes, some of both. However, attention has been directed to the remarkable coincidence of the geographical distribution of the *Sirenidæ* and *Osteoglossidæ*, and as the latter family is represented in Sumatra and Borneo, it may be reasonably expected that a Dipnoous form will be found to accompany it. The distribution of the *Sirenidæ* and *Osteoglossidæ* is as follows :—

Tropical America.

Lepidosiren paradoxa.	Osteoglossum bicirrhosum. Arapaima gigas.

Tropical Australia.

Ceratodus forsteri. Ceratodus miolepis.	Osteoglossum leichardti.

East Indian Archipelago.

?	Osteoglossum formosum.

Tropical Africa.

Protopterus annectens.	Heterotis niloticus.

Not only are the corresponding species found within the same region, but also in the same river systems ; and although such a connection may and must be partly due to a similarity of habit, yet the identity of this singular distribution is so striking that it can only be accounted for by assuming that the *Osteoglossidæ* are one of the earliest Teleosteous types which have been contemporaries of and have accompanied the present Dipnoi since or even before the beginning of the tertiary epoch.

Of the *autochthont* freshwater fishes of the Indian region, some are still limited to it, viz., the *Nandina*, the *Luciocephalidæ* (of which one species only exists in the Archipelago), of Siluroids the *Chacina* and *Bagariina*, of Cyprinoids the *Semiplotina* and *Homalopterina;* others very nearly so, like the *Labyrinthici, Ophiocephalidæ, Mastacembelidæ,* of Siluroids the *Silurina,* of Cyprinoids the *Rasborina* and *Danionina,* and *Symbranchidæ.*

The regions with which the Indian has least similarity are the North American and Antarctic, as they are the most distant. Its affinity to the other regions is of a very different degree:—

1. Its affinity to the Europo-Asiatic region is indicated almost solely by three groups of Cyprinoids, viz., the *Cyprinina, Abramidina,* and *Cobitidina.* The development of these groups north and south of the Himalayas is due to their common origin in the highlands of Asia; but the forms which descended into the tropical climate of the south are now so distinct from their northern brethren that most of them are referred to distinct genera. The genera which are still common to both regions are only the true Barbels (*Barbus*), a genus which, of all Cyprinoids, has the largest range over the old world, and of which some 160 species have been described; and, secondly, the Mountain Barbels (*Schizothorax,* etc.), which, peculiar to the Alpine waters of Central Asia, descend a short distance only towards the tropical plains, but extend farther into rivers within the northern temperate districts. The origin and the laws of the distribution of the *Cobitidina* appear to have been identical with those of *Barbus,* but they have not spread into Africa.

If, in determining the degree of affinity between two

regions, we take into consideration the extent in which an exchange has taken place of the faunæ originally peculiar to each, we must estimate that obtaining between the fresh-water fishes of the Europo-Asiatic and Indian regions as very slight indeed.

2. There exists a great affinity between the Indian and African regions; seventeen out of the twenty-six families or groups found in the former are represented by one or more species in Africa, and many of the African species are not even generically different from the Indian. As the majority of these groups have many more representatives in India than in Africa, we may reasonably assume that the African species have been derived from the Indian stock; but this is probably not the case with the Siluroid group of *Clariina*, which with regard to species is nearly equally distributed between the two regions, the African species being referable to three genera (*Clarias, Heterobranchus, Gymnallabes*, with the sub-genus *Channallabes*), whilst the Indian species belong to two genera only, viz. *Clarias* and *Heterobranchus*. On the other hand, the Indian region has derived from Africa one fresh-water form only, viz. *Etroplus*, a member of the family of *Chromides*, so well represented in tropical Africa and South America. *Etroplus* inhabits Southern and Western India and Ceylon, and has its nearest ally in a Madegasse Freshwater fish, *Paretroplus*. Considering that other African Chromides have acclimatised themselves at the present day in saline water, we think it more probable that Etroplus should have found its way to India through the ocean than over the connecting land area; where, besides, it does not occur.

3. A closer affinity between the Indian and Tropical American regions than is indicated by the character of the equatorial zone generally, does not exist. No genus of Fresh-water fishes occurs in India and South America without being found in the intermediate African region, with two exceptions.

Four small Indian Siluroids (*Sisor*, *Erethistes*, *Pseudecheneis*, and *Exostoma*) have been referred to the South American *Hypostomatina*; but it remains to be seen whether this combination is based upon a sufficient agreement of their internal structure, or whether it is not rather artificial. On the other hand, the occurrence and wide distribution in tropical America of a fish of the Indian family Symbranchidæ (*Symbranchus marmoratus*), which is not only congeneric with, but also most closely allied to, the Indian *Symbranchus bengalensis*, offers one of those extraordinary anomalies in the distribution of animals of which no satisfactory explanation can be given at present.

4. The relation of the Indian region to the Tropical Pacific region consists only in its having contributed a few species to the poor fauna of the latter. This immigration must have taken place within a recent period, because some species now inhabit fresh waters of tropical Australia and the South Sea Islands without having in any way changed their specific characters, as *Lates calcarifer*, species of *Dules*, *Plotosus anguillaris*; others (species of *Arius*) are but little different from Indian congeners. All these fishes must have migrated by the sea; a supposition which is supported by what we know of their habits. We need not add that India has not received a single addition to its freshwater fish-fauna from the Pacific region.

Before concluding these remarks on the Indian region, we must mention that peculiar genera of Cyprinoids and Siluroids inhabit the streams and lakes of its alpine ranges in the north. Some of them, like the Siluroid genera *Glyptosternum*, *Euglyptosternum*, *Pseudecheneis*, have a folded disk on the thorax between their horizontally spread pectoral fins; by means of this they adhere to stones at the bottom of the mountain torrents, and without it they would be swept away into the lower courses of the rivers. The Cyprinoid genera inhabiting similar localities, and the lakes into which the alpine rivers pass, such as *Oreinus, Schizothorax, Ptychobarbus, Schizopygopsis, Diptychus, Gymnocypris,* are distinguished by peculiarly enlarged scales near the vent, the physiological use of which has not yet been ascertained. These alpine genera extend far into the Europo-Asiatic region, where the climate is similar to that of their southern home. No observations have been made by which the altitudinal limits of fish life in the Himalayas can be fixed, but it is probable that it reaches the line of perpetual snow, as in the European Alps which are inhabited by Salmonoids. Griffith found an *Oreinus* and a Loach, the former in abundance, in the Helmund at Gridun Dewar, altitude 10,500 feet; and another Loach at Kaloo at 11,000 feet.

B. The African Region comprises the whole of the African continent south of the Atlas and the Sahara. It might have been conjectured that the more temperate climate of its southern extremity would have been accompanied by a conspicuous difference of the fish-fauna. But this is not the case; the difference between the tropical and southern parts of Africa consists simply in the gradual disappearance of specifically tropical forms, whilst Siluroids, Cyprinoids, and even Labyrinthici penetrate to its southern coast; no new form has entered to impart to South Africa a character distinct from the central portion of the continent. In the north-east the African fauna passes the Isthmus of Suez and penetrates into Syria; the system of the Jordan presenting so many African types that it has to be included in a description of the African region as well as of the Europo-Asiatic. This river is inhabited by three species of *Chromis,* one of *Hemichromis,* and *Clarias macracanthus,* a common fish of the Upper Nile. This infusion of African forms cannot be accounted for by any one of those accidental means of dispersal, as *Hemichromis* is not represented in the north-eastern parts of Africa proper, but chiefly on the west coast and in the Central African lakes.

Madagascar clearly belongs to this region. Besides some Gobies and *Dules*, which are not true freshwater fishes, four *Chromides* are known. To judge from general accounts, its Freshwater fauna is poorer than might be expected; but, singular as it may appear, collectors have hitherto paid but little attention to the Freshwater fishes of this island. The fishes found in the freshwaters of the Seychelles and Mascarenes are brackish-water fishes, such as *Fundulus*, *Haplochilus*, *Elops*, *Mugil*, etc.

The following is the list of the forms of Freshwater fishes inhabiting this region:—

Dipnoi [Australia, Neotrop.]—		
Lepidosiren annectens	1	species.
Polypteridæ	2	,,
Percina (Cosmopol.)—		
Lates [India, Australia]	1	,,
Labyrinthici [India]	5	,,
Ophiocephalidæ [India]	1	,,
Mastacembelidæ [India]	3	,,
Chromides [South America]—		
Chromis	23	,,
Hemichromis	5	,,
Paretroplus	1	,,
Siluridæ—		
Clariina [India]	14	,,
Silurina [India, Palæarct.]	11	,,
Bagrina [India]	10	,,
Pimelodina [South America]	2	,,
Ariina[1] [India, Australia, S. Amer., Patagonia]	4	,,
Doradina [South America]—		
Synodontis	15	,,

[1] One species (*Arius thalassinus*) found in Indian and African rivers.

Rhinoglanina [India]	2 species.
Malapterurina	3 „
Characinidæ [South America]—	
Citharinina	2 „
Nannocharacina	2 „
Tetragonopterina—	
Alestes	14 „
Crenuchina—	
Xenocharax	1 „
Hydrocyonina—	
Hydrocyon	4 „
Distichodontina	10 „
Ichthyborina	2 „
Mormyridæ (*Gymnarchidæ*)	51 „
Cyprinodontidæ—	
Carnivoræ [Palæarct., India, S. America—	
Haplochilus [India, South America] .	7 „
Fundulus [Palæarct., Nearct.] . .	1 „
Cyprinidæ [Palæarct., India, North America]—	
Cyprinina [Palæarct., India, N. America—	
Labeo [India]	6 „
Barynotus [India]	2 „
Abrostomus	2 „
Discognathus lamta[1] [India] . .	1 „
Barbus [Palæarct., India] . .	35 „
Rasborina [India]	1 „
Danionina [India]—	
Barilius [India]	3 „
Abramidina [Palæarct., India, N. America]—	
Pelotrophus	2 „
Kneriidæ	2 „
Osteoglossidæ [India, Australia, South America]—	
Heterotis	1 „
Pantodontidæ	1 „
Notopteridæ [India]	2 „
	255 species.

[1] This species extends from India into East Africa.

Out of the 39 families or groups of freshwater fishes 15 are represented in the African region, or three more than in the Indian region; however of two of them, viz., the *Ophiocephalidæ* and *Mastacembelidæ*, a few species only have found their way into Africa. On the other hand, the number of species is much less, viz. 255, which is only two-fifths of that of the known Indian species. The small degree of specialisation and localisation is principally due to the greater uniformity of the physical conditions of this continent, and to the almost perfect continuity of the great river systems, which take their origin from the lakes in its centre. This is best shown by a comparison of the fauna of the Upper Nile with that of the West African rivers. The number of species known from the Upper Nile amounts to 56, and of these not less than 25 are absolutely identical with West African species. There is an uninterrupted continuity of the fish-fauna from west to the north-east, and the species known to be common to both extremities may be reasonably assumed to inhabit also the great reservoirs of water in the centre of the continent. A greater dissimilarity is noticeable between the west and north-east fauna on the one hand, and that of the Zambezi on the other; the affinity between them is merely generic; and all the fishes hitherto collected in Lake Nyassa have proved to be distinct from those of the Nile, and even from those of other parts of the system of the Zambezi.

Africa, unlike India, does not possess either alpine ranges or outlying archipelagoes, the fresh waters of which would swell the number of its indigenous species; but at a future time, when its fauna is better known than at present, it is possible that the great difference in the number of species between this and the Indian regions may be somewhat lessened.

The most numerously-represented families are the Siluroids, with 61 species; the Cyprinoids, with 52; the Mormyridæ, with 51; the Characinidæ, with 35; and the Chromides, with 29. There is not, therefore, that great preponderance of the two first families over the remaining, which we noticed in the Indian region; in Africa there is a comparatively greater variety of distinct Freshwater types, imparting to the study of its fauna an unflagging pleasure such as is scarcely gained by the study of the other region. With the forms peculiar to it there are combined those of India as well as South America.

In Tropical Africa there are still remnants of Ganoids: *Protopterus* (*Lepidosiren*) *annectens* and *Polypterus bichir*, with the singularly modified *Calamoichthys*. The two former range from east to west, and are accompanied by an Osteoglossoid (*Heterotis*) which has hitherto been found in the Nile and on the West Coast only.

Autochthont and limited to this region are the *Mormyridæ*, *Pantodontidæ*, and *Kneriidæ*, a singular type, somewhat akin to the Loaches. Of Siluroid genera the most characteristic are *Synodontis*, *Rhinoglanis*, and the electric *Malapterurus*; of Characinoids, *Citharinus*, *Alestes*, *Xenocharax*, *Hydrocyon*, *Distichodon*, *Ichthyborus*.

The regions with which Africa (like India) has least similarity are, again, the North American and Antarctic. Its affinity with the Europo-Asiatic region consists only in having received, like this latter, a branch of the Cyprinoids, the African Carps and Barbels resembling, on the whole, more Indian than Europo-Asiatic forms. Its similarity to Australia is limited to the two regions possessing Dipnoous and Osteoglossoid types. But its relations to the two other regions of the equatorial zone are near and of great interest.

1. Africa has, in common with India, the Siluroid group of *Clariina*, the *Silurina*, and *Bagrina*; and more especially the small but very natural family of *Notopteridæ*, represented by three species in India, and by two on the *west coast* of Africa. It would be hazardous to state at present in which

of the two regions these fishes first made their appearance, but the discovery of remains of *Notopteridæ* and *Silurina* in tertiary deposits of Sumatra points to the Indian region as their original home. We can have less doubt about the other fishes common to both regions; they are clearly immigrants into Africa from the East, and it is a remarkable fact that these immigrants have penetrated to the most distant limits of Africa in the west as well as in the south,—viz. the *Labyrinthici*, represented by two genera closely allied to the Indian *Anabas;* the *Ophiocephalidæ* and *Mastacembelidæ*, a few species of which have penetrated to the west coast, singularly enough being absent from the eastern rivers; the *Ariina*, represented by several species, of which one or two are identical with Indian, having extended their range along the intervening coasts to the east coast of Africa. The Cyprinoids also afford an instance of an Indian species ranging into Africa, viz. *Discognathus lamta*, which seems to have crossed at the southern extremity of the Red Sea, as it is found in the reservoirs at Aden and the hill-streams of the opposite coast-region of Abyssinia.

2. No such direct influx of species and genera has occurred from South America into Africa. Yet the affinity of their Freshwater fishes is striking. Two of the most natural families of fishes, the *Chromides* and *Characinidæ*, are peculiar, and (with the exception of *Etroplus*) restricted to them. The African and South American Dipnoi are closely allied to each other. The *Pimelodina*, so characteristic of Tropical America, have three representatives in Africa, viz., *Pimelodus platychir*, *P. balayi*, and *Auchenoglanis biscutatus;* the *Doradina* are another Siluroid group restricted to these two continents.[1] Yet, with all these points of close resem-

[1] We have left out from these considerations the Ariina and Cyprinodonts, which can pass with impunity through salt water, and are spread over much larger areas.

blance, the African and South American series are, with the exception of the two species of Pimelodus, *generically* distinct; which shows that the separation of the continents must have been of an old date. On the other hand, the existence of so many similar forms on both sides of the Atlantic affords much support to the supposition that at a former period the distance between the present Atlantic continents was much less, and that the fishes which have diverged towards the East and West are descendants of a common stock which had its home in a region now submerged under some intervening part of that ocean. Be this as it may, it is evident that the physical conditions of Africa and South America have remained unchanged for a considerable period, and are still sufficiently alike to preserve the identity of a number of peculiar freshwater forms on both sides of the Atlantic. Africa and South America are, moreover, the only continents which have produced in Freshwater fishes, though in very different families, one of the most extraordinary modifications of an organ—the conversion, that is, of muscle into an apparatus creating electric force.

C. The boundaries of the TROPICAL AMERICAN (*Neotropical*) REGION have been sufficiently indicated in the definition of the Equatorial zone. A broad and most irregular band of country, in which the South and North American forms are mixed, exists in the north; offering some peculiarities which deserve fuller attention in the subsequent description of the relations between the South and North American faunæ. The following Freshwater fishes inhabit this region:—

Dipnoi [Australia, Africa]—
Lepidosiren paradoxa . . . 1 species.
Polycentridæ . . . . . 3 „
Chromides [Africa]—
Heros, Acara, Cichla, etc. . . 80 „
(*Lucifuga* . . . . . 2 „)

Siluridæ—		
Hypophthalmina	5	species.
Pimelodina [Africa, 2 species]	70	„
Ariina [Africa, India, Australia, Fuegian]	35	„
Doradina [Africa]	60	„
Hypostomatina [India]	90	„
Aspredinina	9	„
Nematogenyina [Fuegian]—	2	„
Trichomycterina [Fuegian]	2	„
Stegophilina	3	„
Characinidæ [Africa]—		
Erythrinina	15	„
Curimatina	40	„
Anastomatina	25	„
Tetragonopterina	80	„
Hydrocyonina	30	„
Crenuchina	1	„
Serrasalmonina	35	„
Cyprinodontidæ—		
Carnivoræ [Europe, Asia, N. America, India, Africa]	30	„
Limnophagæ	31	„
Osteoglossidæ [Africa, India, Australia]	2	„
Gymnotidæ	20	„
Symbranchidæ [India]	1	„
	672	species.

Out of the 39 families or groups of Freshwater fishes, 9 only are represented in the Tropical American region. This may be accounted for by the fact that South America is too much isolated from the other regions of the Equatorial zone to have received recent additions to its fauna. On the other hand, the number of species exceeds that of every other region, even of the Indian, with which, in regard to the comparative development of families, the Neotropical region shows great analogy, as will be seen from the following Table :—

INDIAN.		NEOTROPICAL.	
Siluridæ .	200 sp.	Siluridæ .	276 sp.
Cyprinidæ .	330 sp.	Characinidæ .	226 sp.
Labyrinthici .	25 sp.	Chromides .	80 sp.
Ophiocephalidæ	30 sp.	Cyprinodontidæ	60 sp.
Mastacembelidæ	10 sp.	Gymnotidæ .	20 sp.

In both regions the great number of species is due to the development of numerous local forms of two families, the *Characinidæ* taking in the New World the place of the *Cyprinidæ* of the Old World. Thereto are added a few smaller families with a moderately large number of species, which, however, is only a fraction of that of the leading families, the remainder of the families being represented by a few species only. The number of genera within each of the two regions of the two principal families is also singularly alike; the Indian region having produced about 45 Siluroid and as many Cyprinoid genera, whilst the Neotropical region is tenanted by 54 Siluroid and 40 Characinoid genera. These points of similarity between the two regions cannot be accidental; they indicate that agreement in their physical and hydrographical features which in reality exists.

Of Ganoids, we find in Tropical America one species only, *Lepidosiren paradoxa*, accompanied by two Osteoglossoids (*Osteoglossum bicirrhosum* and *Arapaima gigas*).

Autochthont and limited to this region are the *Polycentridæ*, all the non-African genera of *Chromides* and *Characinidæ;* of Siluroids, the *Hypophthalmina*, *Aspredinina*, and *Stegophilina*, and the majority of *Pimelodina*, *Hypostomatina*, and *Doradina;* the herbivorous Cyprinodonts or *Limnophagæ*, and numerous insectivorous Cyprinodonts or *Carnivoræ;* and the *Gymnotidæ* (Electric eel).

The relations to the other regions are as follows:—

1. The resemblances to the Indian and Tropical Pacific regions partly date from remote geological epochs, or are partly

due to that similarity of physical conditions to which we have already referred. We have again to draw attention to the unexplained presence in South America of a representative of a truly Indian type (not found in Africa), viz. *Symbranchus marmoratus.* On the other hand, a direct genetic affinity exists between the Neotropical and African regions, as has been noticed in the description of the latter, a great part of their freshwater fauna consisting of descendants from a common stock.

2. A comparison of the specifically Neotropical with the specifically North American types shows that no two regions can be more dissimilar. It is only in the intervening borderland, and in the large West Indian Islands, that the two faunæ mix with each other. We need not enter into the details of the physical features of Central America and Mexico —the broken ground, the variety of climate (produced by different altitudes) within limited districts, the hot and moist alluvial plains surrounding the Mexican Gulf, offer a diversity of conditions most favourable to the intermixture of the types from the north and the south. But yet the exchange of peculiar forms appears to be only beginning; none have yet penetrated beyond the debateable ground, and it is evident that the land-connection between the two continents is of comparatively recent date: a view which is confirmed by the identity of the marine fishes on both sides of Central America.

Cuba—and this is the only island in the West Indies which has a number of freshwater fishes sufficient for the determination of its zoogeographical relations—is inhabited by several kinds of a perch (*Centropomus*), freshwater mullets, Cyprinodonts, one species of Chromid (an *Acara*), and *Symbranchus marmoratus.* All these fishes are found in Central America, and as they belong to forms known to enter brackish water more or less freely, it is evident that they have crossed from the mainland of South America or from Central

America. But with them there came a remarkable North American type, *Lepidosteus*. *Lepidosteus viridis*, which is found in the United States, has penetrated on the mainland to the Pacific coast of Guatemala, where it is common at the mouth of the rivers and in brack-water lakes along the coast; it probably crossed into Cuba from Florida. A perfectly isolated type of fishes inhabits the subterranean waters of the caves of Cuba (two species of *Lucifuga*). The eyes are absent or quite rudimentary, as in most other cave animals. Singularly, it belongs to a family (*Ophidiidæ*), the members of which are strictly marine; and its nearest ally is a genus, *Brotula*, the species of which are distributed over the Indo-Pacific Ocean, one only occurring in the Caribbean Sea. This type must have witnessed all the geological changes which have taken place since Cuba rose above the surface of the sea.

A similar mixture of forms of the Tropical and Temperate types of Freshwater fishes takes place in the south of South America; its details have not yet been so well studied as in the north; but this much is evident that, whilst in the East Tropical forms follow the Plate river far into the Temperate region, in the West the Temperate Fauna finds still a congenial climate in ranges of the Andes, situated close to, or even north of, the Tropic.

Like the Indian region, the Tropical American has a peculiar Alpine Fauna, the Freshwater fishes of which, however, belong to the Siluroids and Cyprinodonts. The former are small, dwarfed forms (*Arges, Stygogenes, Brontes, Astroblepus, Trichomycterus, Eremophilus*), and have a perfectly naked body, whilst the representatives in the lowlands of, at least, the first four genera are mailed. The Alpine Cyprinodonts, on the other hand, (*Orestias*) exceed the usual small size of the other members of this family, are covered with thick scales, but have lost their ventral fins. Some of these Alpine forms, like *Trichomycterus*, follow the range of the Andes far into the southern temperate region. The majority reach to a height of 15,000 feet above the level of the sea, and a few are found still higher.

D. The Tropical Pacific Region includes all the islands east of Wallace's Line, New Guinea, Australia—with the exception of its south-eastern portion,—and all the islands of the Tropical Pacific to the Sandwich group. Comparing the area of this region with that of the others, we find it to be not only the poorest in point of the number of its species generally, but also in that of the possession of peculiar forms, as will appear from the following list:—

Dipnoi [Neotrop., Africa.]		
Ceratodus	2	species.
Percidæ [Cosmopol.]—		
Lates (calcarifer) [India]	1	„
Nannoperca	1	„
Oligorus [New Zealand]	1	„
Dules [India]	8	„
(*Macquaria*)	1	„
Labyrinthici—		
Anabas (scandens) [India]	1	„
Ophiocephalidæ—		
Ophiocephalus (striatus) [India]	1	„
Atherinidæ [Brack-water]—		
Atherinichthys	2	„
Osteoglossidæ [India, Africa, Neotrop.]	1	„
Siluridæ—		
Plotosina [India]	9	„
Ariina [India, Africa, Neotrop.]	7	„
Symbranchidæ—		
Monopterus (javanicus) [India]	1	„
Total	36	species.

The paucity of freshwater fishes is due, in the first place, to the arid climate and the deficiency of water in the Australian continent, as well as to the insignificant size of the fresh-water courses in the smaller islands. Still this cannot be the only cause: the large island of Celebes, which, by its mountainous portions, as well as by its extensive plains and lowlands, would seem to offer a favourable variety of conditions for the development of a freshwater Fauna,is, as far as has been ascertained, tenanted by seven Freshwater fishes only, viz. 2 *Arius*, 2 *Plotosus*, 1 *Anabas*, 1 *Ophiocephalus*, 1 *Monopterus*, all of which are the commonest species of the Indian region. New Guinea has not yet been explored, but, from the faunæ nearest to this island, we expect its fresh-water fishes will prove to be equally few in number, and identical with those of Celebes and North Australia; a supposition confirmed by the few small collections which have reached Europe. Finding, then, that even those parts of this region, which are favourable to the development of Freshwater fishes, have not produced any distinct forms, and that the few species which inhabit them, are unchanged, or but slightly modified Indian species, we must conclude that the whole of this area has remained geologically isolated from the other regions of this zone since the commencement of the existence of Teleostei; and that, with the exception of *Ceratodus* and *Osteoglossum*, the immigration of the other species is of very recent date.

Fossil remains of *Ceratodus* have been found in Liassic and Triassic formations of North America, England, Germany, and India; and it is, therefore, a type which was widely spread in the Mesozoic epoch. Although it would be rash to conclude that its occupation of Australia dates equally far back, for it may have reached that continent long afterwards; yet it is evident that, as it is one of the most ancient of the existing types, so it is certainly the first of the Freshwater fishes which appeared in Australia. *Osteoglossum*, of which no fossil remains yet have been found, is proved by its distribution to be one of the oldest Teleosteous types. There must have been a long gap of time before these ancient types were joined by the other Teleostei. All of them migrated through the intervening parts of the ocean from India. Most of the *Plotosina*, some of the *Arii*, *Dules*, and the *Athcrinichthys*, also *Nannoperca* (allied to *Apogon*), were among the earliest arrivals, being sufficiently differentiated to be specifically or even generically (*Cnidoglanis*, *Nannoperea*) distinguished; but some others, like *Anabas seandens*, *Latcs calcarifcr*, *Dules marginatus*, must have reached the Australian continent quite recently, for they are indistinguishable from Indian specimens.

In South-western Australia a mingling of the scanty fauna with that of the southern temperate parts takes place. *Oligorus macquariensis* (The Murray Cod), which has a congener on the coast of New Zealand, ascends high up the Murray river, so that we cannot decide whether this Percoid should be located in the Tropical or Temperate part of Australia. Several *Galaxias* also extend to the confines of Queensland, and will probably some day be found members of this region.

In the smaller Pacific islands the Freshwater fishes exhibit a remarkable sameness: two or three species of *Dules*, several Eels, an Atherine, or some Gobies, Mullets, and other fishes which with equal readiness exchange fresh for salt water, and which would at once reach and occupy any streams or freshwater lakes that may be formed on an island.

The Sandwich Islands are the only group among the smaller islands which are tenanted by a Siluroid, a species of *Arius*, which is closely allied to Central American species, and, therefore, probably immigrated from Tropical America.

II. Northern Zone.

The boundaries of the Northern Zone coincide in the

main with the northern limit of the Equatorial Zone; but at three different points they overlap the latter, as has been already indicated. This happens in, and east of, Syria, where the mixed faunæ of the Jordan and the rivers of Mesopotamia demand the inclusion of this territory into the Northern Zone as well as the Equatorial; in the island of Formosa, where a Salmonoid and several Japanese Cyprinoids flourish; and in Central America, where a *Lepidosteus*, a Cyprinoid (*Sclerognathus meridionalis*), and an Amiurus (*A. meridionalis*) represent the North American fauna in the midst of a host of tropical forms.

A separate *Arctic* Zone does not exist for Freshwater fishes; ichthyic life becomes extinct towards the pole as soon as the fresh water remains frozen throughout the year, or thaws for a few weeks only; and the few fishes which extend into high latitudes, in which lakes are open for two or three months in the year, belong to types in no wise differing from those of the more temperate south. The highest latitude at which fishes have been obtained is 82° lat. N., whence the late Arctic Expedition brought back specimens of Charr (*Salmo arcturus* and *Salmo narcsii*).

The ichthyological features of this zone are well marked: the Chondrosteous Ganoids or Sturgeons, and the families of *Salmonidæ* and *Esocidæ* are limited to, and characteristic of, it; Cyprinoids flourish with the Salmonoids, both families preponderating in numbers over the others, whilst the Siluroids are few in number and in variety.

The two regions in which this zone is divided are very closely related to one another, and their affinity is not unlike that which obtains between the sub-regions of the Southern Zone. The subjoined list will show their close agreement with regard to families as well as species. Several of the latter are common to both, viz.—*Acipenser sturio, A. maculatus, Perca fluviatilis, Gastrosteus pungitius, Salmo salar,*

Esox lucius, Lota vulgaris, Petromyzon marinus, P. fluviatilis, and *P. branchialis*; and all recent investigations have resulted in giving additional evidence of the affinity, and not of the diversity of the two regions.

In Europe and temperate Asia, as well as in North America, mountain ranges elevated beyond the line of perpetual snow would seem to offer physical conditions favourable for the development of a distinct alpine fauna. But this is not the case, because the difference of climate between the mountain districts and the lowlands is much less in this zone than in the Equatorial. Consequently the alpine freshwater fishes do not essentially differ from those of the plains; they are principally Salmonoids; and in Asia, besides, mountain-barbels and Loaches. *Salmo orientalis* was found by Griffith to abound in the tributaries of the Bamean river at an altitude of about 11,000 feet.

	Europo-Asiatic.	N. American.
Acipenseridæ—		
Acipenser	9 species.	12 species.
Scaphirhynchus . .	2 „	1 „
Polyodon . . .	1 „	1 „
Lepidosteidæ	0 „	3 „
Amiidæ	0 „	1 „
Percina [Cosmopol.] . .	10 „	30 „
Grystina [Australia, New Zealand]	0 „	2 „
Centrarchina . . .	0 „	26 „
Aphredoderidæ . . .	0 „	1 „
Cottidæ [partly marine]—		
Cottus	3 „	8 „
Ptyonotus . . .	0 „	1 „
Gastrosteidæ	5 „	5 „
Comephoridæ	1 „	0 „
Gadidæ [marine]—		
Lota	1 „	1 „
Siluridæ—		
Silurina [India, Africa] .	5 „	0 „

	Europo-Asiatic.	N. American.
Bagrina	2 species.	0 species.
Amiurina	1 "	17 "
Salmonidæ	90 "	45 "
Percopsidæ	0 "	1 "
Esocidæ	1 "	7 "
Umbridæ	1 "	1 "
Cyprinodontidæ Carnivoræ [India, Africa, Neotrop.] . .	9 "	30 "
Heteropygii	0 "	2 "
Cyprinidæ—		
Catostomina	1 "	25 "
Cyprinina [India, Africa] .	80 "	30 "
Leuciscina	60 "	70 "
Rhodeina	10 "	0 "
Abramidina [India, Africa]	44 "	10 "
Cobitidina [India] . .	20 "	0 "
Hyodontidæ	0 "	1 "
Petromyzontidæ [Southern Zone]	4 "	8 "
	360 species.	339 species.

A. The Europo-Asiatic (Palæarctic) Region.—Its western and southern boundaries coincide with those of the Northern Zone, so that only those which divide it from North America have to be indicated. Behring's Strait and the Kamtschatka Sea have been conventionally taken as the boundary, but this is shown to be artificial by the fact that the animals of both coasts, as far as they are known at present, are not sufficiently distinct to be referred to two distinct regions. As to the freshwater fishes those of North-western America and of Kamtschatka are but imperfectly known, but there can be little doubt that the same agreement exists between them as is the case with other classes of animals. The Japanese islands exhibit a decided Palæarctic fish-fauna, which includes Barbus and Cobitioids, forms strange to the North American fauna. A slight influx of tropical forms is perceived in the south of Japan, where two Bagrina (*Pseudobagrus aurantiacus* and *Liocassis longirostris*) have established themselves for a considerable period, for both are peculiar to the island, and have not been found elsewhere.

In the east, as well as in the west, the distinction between the Europo-Asiatic and North American regions disappears almost entirely the farther we advance towards the north. Of four species of the genus Salmo known from Iceland, one (*S. salar*) is common to both regions, two are European (*S. fario* and *S. alpinus*), and one is a peculiarly Icelandic race (*S. nivalis*). As far as we know the Salmonoids of Greenland and Baffin's Land they are all most closely allied to European species, though they may be distinguished as local races.

Finally, as we have seen above, the Europo-Asiatic fauna mingles with African and Indian forms in Syria, Persia, and Afghanistan. *Capoëta*, a Cyprinoid genus, is characteristic of this district, and well represented in the Jordan and rivers of Mèsopotamia.

Assuming that the distribution of Cyprinoids has taken its origin from the alpine tract of country dividing the Indian and Palæarctic regions, we find that this type has found in the temperate region as equally favourable conditions for its development as in the tropical. Out of the 360 species known to exist in the Palæarctic regions, no less than 215 are Cyprinoids. In the countries and on the plateaus immediately joining the Himalayan ranges those mountain forms which we mentioned as peculiar to the Indian Alps abound and extend for a considerable distance towards the west and east, mixed with other *Cyprinina* and *Cobitidina.* The representatives of these two groups are more numerous in Central and Eastern Asia than in Europe and the northern parts of Asia, where the *Leuciscina* predominate. *Abramidina* or Breams are more numerous in the south and east of Asia, but they spread to the extreme north-western and northern limits, to which the Cyprinoid type reaches. The *Rhodeina* are a small family especially characteristic of the East, but with one or two off-shoots in Central Europe. Very significant is the appearance in China of a species of the *Catostomina*, a group otherwise limited to North America.

The Cyprinoids, in their dispersal from the south northwards, are met from the opposite direction by the Salmonoids. These fishes are, without doubt, one of the youngest families of *Teleostei*, for they did not appear before the Pliocene era; they flourished at any rate during the glacial period, and, as is testified by the remnants which we find in isolated elevated positions, like the Trout of the Atlas, of the mountains of Asia-Minor, and of the Hindu Kush, they spread to the extreme south of this region. At the present day they are most numerously represented in its northern temperate parts; towards the south they become scarcer, but increase again in numbers and species, wherever a great elevation offers them the snow-fed waters which they affect. In the rivers of the Mediterranean Salmonoids are by no means scarce, but they prefer the upper courses of those rivers, and do not migrate to the sea.

The Pike, Umbra, several species of Perch and Stickleback, are also clearly autochthont species of this region. Others belong to marine types, and seem to have been retained in fresh water at various epochs: thus the freshwater Cottus (Miller's Thumb); *Cottus quadricornis*, which inhabits lakes of Scandinavia, whilst other individuals of the same species are strictly marine; the Burbot (*Lota vulgaris*); and the singular *Comephorus*, a dwarfed and much-changed Gadoid which inhabits the greatest depths of Lake Baikal.

Remnants of the Palæichthyic fauna are the *Sturgeons* and *Lampreys*. The former inhabit in abundance the great rivers of Eastern Europe and Asia, periodically ascending them from the sea; their southernmost limits are the Yang-tse-kiang in the east, and rivers flowing into the Adriatic, Black and Caspian Seas, and Lake Aral, towards the centre of this region. None are known to have gone beyond the boundaries of the Northern Zone. If the Lampreys are justly reckoned among Freshwater fishes, their distribution is unique and exceptional. In the Palæarctic region some of the species descend periodically to the sea, whilst others remain stationary in the rivers; the same has been observed in the Lampreys of North America. They are entirely absent in the Equatorial Zone, but reappear in the Temperate Zone of the Southern Hemisphere. Many points of the organisation of the Cyclostomes indicate that they are a type of great antiquity.

The remaining Palæarctic fishes are clearly immigrants from neighbouring regions: thus *Silurus, Macrones,* and *Pseudobagrus* from the Indian region; *Amiurus* (and, as mentioned above, *Catostomus*) from North America. The Cyprinodonts are restricted to the southern and warmer parts; all belong to the carnivorous division. The facility with which these fishes accommodate themselves to a sojourn in fresh, brackish, or salt water, and even in thermal springs, renders their general distribution easily comprehensible, but it is impossible to decide to which region they originally belonged; their remains in tertiary deposits round the Mediterranean are not rare.

B. The boundaries of the North American or Nearctic Region have been sufficiently indicated. The main features and the distribution of this fauna are identical with those of the preceding region. The proportion of Cyprinoid species to the total number of North-American fishes (135 : 339) appears to be considerably less than in the Palæarctic region, but we cannot admit that these figures approach the truth, as the Cyprinoids of North America have been much less studied than those of Europe; of many scarcely more than the name is known. This also applies in a great measure to the Salmonoids, of which only half as many as are found in the Palæarctic region have been sufficiently described to be worthy of consideration. North America will, without doubt, in the end show as many distinct races as Europe and Asia.

Cyprinoids, belonging to genera living as well as extinct, existed in North America in the tertiary period. At present, *Cyprinina, Leuciscina,* and *Abramidina* are well represented, but there is no representative of the Old World genus *Barbus,* or of the *Cobitidina*[1]; *Rhodeina* are also absent. On the other hand, a well-marked Cyprinoid type is developed—the *Catostomina,* of which one species has, as it were, returned into Asia. Very characteristic is the group of *Centrarchina,* allied to the Perch, of which there are some thirty species; two *Grystina.* Of the Sticklebacks there are as many species as in Europe, and of Pike not less than seven species have been distinguished. *Umbra* appears to be as local as in Europe. Some very remarkable forms, types of distinct families, though represented by one or two species only, complete the number of North American autochthont fishes—viz., *Aphredoderus, Percopsis, Hyodon,* and the *Heteropygii* (*Amblyopsis* and *Chologaster*). The last are allied to the Cyprinodonts, differing from them in some points of the structure of their intestines. The two genera are extremely similar, but *Chologaster,* which is found in ditches of the rice-fields of South Carolina, is provided with eyes, and lacks the ventral fins. *Amblyopsis* is the celebrated Blind Fish of the Mammoth Cave of Kentucky: colourless, eyeless, with rudimentary ventral fins, which may be occasionally entirely absent.

[1] Cope has discovered in a tertiary freshwater-deposit at Idaho an extinct genus of this group, *Diastichus.* He considers this interesting fact to be strongly suggestive of continuity of territory of Asia and North America.—"Proc. Am. Phil. Soc. 1873," p. 55.

A peculiar feature of the North American Fish Fauna is that it has retained, besides the Sturgeons and Lampreys, representatives of two Ganoid families, *Lepidosteus* and *Amia*. Both these genera existed in tertiary times: the former occurs in tertiary deposits of Europe as well as North America, whilst fossil remains of *Amia* have been found in the Western Hemisphere only.

It is difficult to account for the presence of the *Amiurina* in North America. They form a well-marked division of the *Bagrina*, which are well represented in Africa and the East Indies, but absent in South America; it is evident, therefore, they should not be regarded as immigrants from the south, as is the case with the Palæarctic Siluroids. Nor, again, has the connection between South and North America been established sufficiently long to admit of the supposition that these Siluroids could have spread in the interval from the south to the northern parts of the continent, for some of the species are found as far north as Pine Islands Lake (54° lat. N.)[1]

III. Southern Zone.

The boundaries of this zone have been indicated in the description of the Equatorial Zone; they overlap the southern boundaries of the latter in South Australia and South America, but we have not at present the means of exactly defining the limits to which southern types extend northwards. This zone includes Tasmania with at least a portion of South-eastern Australia (*Tasmanian sub-region*), New Zealand and the Auckland Islands (*New Zealand sub-region*), and Chili, Patagonia, Terra del Fuego, and the Falkland Islands (*Fuegian sub-region*). No freshwater fishes are known

[1] Leidy describes a Siluroid (*Pimelodus*) from tertiary deposits of Wyoming Territory. "Contrib. to the Extinct Vert. Fauna of the Western Territ. 1873," p. 193.

from Kerguelen's Land, or from islands beyond 55° lat. S. The southern extremity of Africa has to be excluded from this zone so far as Freshwater fishes are concerned.

This zone is, with regard to its extent as well as to the number of species, the smallest of the three; yet its ichthyological features are well marked; they consist in the presence of two peculiar families, each of which is analogous to a northern type, viz. the *Haplochitonidæ*, which represent the *Salmonidæ*, *Haplochiton* being the analogue of *Salmo*, and *Prototroctes* that of *Coregonus*; and the *Galaxiidæ*, which are the Pikes of the Southern Hemisphere.

Although geographically widely separate from each other, the Freshwater fishes of the three divisions are nevertheless so closely allied that conclusions drawn from this group of animals alone would hardly justify us in regarding these divisions as sub-regions. One species of *Galaxias* (*G. attenuatus*) and the three Lampreys are found in all three, or at least two, sub-regions.

Freshwater Fishes of the Southern Zone.

	Tasmanian.	N. Zealand.	Fuegian.
Perciehthys	...	...	3
Siluridæ—			
Diplomystax	...	...	1
Nematogenys	...	...	1
Trichomycterina [Neotrop.]	...	...	5
Gadopsidæ	1	...	...
(*Retropinna*	...	1	...)
Haplochitonidæ	1	1	1
Galaxiidæ	6	5	4
Petromyzontidæ . . .	3	1	3
	11	8	18

But little remains to be added in explanation of this list; *Percichthys* is in Chili the autochthont form of the cosmopolitan group of *Percina*. *Diplomystax*, an Arioid fish of Chili, and *Nematogenys* seem to have crossed the Andes from Tropical America at a comparatively early period, as these genera are not represented on the eastern side of South America; the *Trichomycterina* occur on both sides of the Andes, which they ascend to a considerable height. *Retropinna* is a true Salmonoid, allied to, and representing in the Southern Hemisphere the Northern Smelt, *Osmerus*. In both these genera a part of the specimens live in the sea, and ascend rivers periodically to spawn; another part remain in rivers and lakes, where they propagate, never descending to the sea, this freshwater race being constantly smaller than their marine brethren. That this small Teleostean of the Northern Hemisphere should reappear, though in a generically modified form, in New Zealand, without having spread over other parts of the Southern Zone, is one of the most remarkable, and at present inexplicable facts of the geographical distribution of freshwater fishes.

Fig. 104.—*Haplochiton zebra*, Straits of Magelhæn.

CHAPTER XVIII.

THE FISHES OF THE BRACKISH WATER.

On such parts of a coast at which there is a mixture of fresh and salt water, either in consequence of some river emptying its water into the sea or from an accumulation of land surface water forming lagunes, which are in uninterrupted or temporary communication with the sea, there flourishes a peculiar brackish water fauna which is characterised by the presence of fishes found sometimes in sea-, sometimes in pure fresh-water.

This fauna can be rather sharply defined if a limited district only is taken into consideration; thus, the species of the brackish water fauna of Great Britain, the Pacific coast of Central America, of the larger East India Islands, etc., can be enumerated without much hesitation. But difficulties arise when we attempt to generalise in the enumeration of the forms referable to the brackish water fauna; because the genera and families enumerated include certain species and genera which have habituated themselves exclusively either to a freshwater or marine existence; and, besides, because a species of fish may be at one locality an inhabitant of brackish water, at another of the sea, and at a third of fresh water. The circumstance that these fishes can live in sea and fresh water has enabled them to spread readily over the globe, a few only being limited to particular regions; therefore, for the purposes of dividing the earth's surface into natural zoological regions the brackish water forms are useless. The following fishes may be referred to this Fauna :—

1. Species of *Rajidæ* (*Raja*, *Trygon*) prefer the mouths of rivers, probably because the muddy or sandy bottom offers the most suitable conditions for fishes which can feed on the bottom only; such brackish water species belong chiefly to the Equatorial Zone, some having taken up their abode entirely in fresh water (South American Trygons).

2. *Ambassis*, a Percoid genus, consisting of numerous small species, inhabiting the shores of the tropical parts of the Indian Ocean and the coasts of Tropical Australia. Many species enter, and all seek the neighbourhood of, fresh water; hence they disappear in the islands of the Pacific, and are scarce in the Red Sea.

3. *Therapon*, with the same distribution as the former.

4. Numerous *Sciænidæ* of the Equatorial Zone.

5. The *Polynemidæ*, chiefly inhabitants of brackish water of the Equatorial Zone, most developed in the Indian region, and scarce in the Tropical Pacific.

6. Numerous species of *Caranx* (or Horse Mackerels) of the Equatorial Zone.

7. Nearly all species of *Gastrosteus* enter brackish water, *G. spinachia* being almost exclusively confined to it: Northern Zone.

8. The most important genera of the Gobies (*Gobiina*): *Gobius* (nearly cosmopolitan), *Sicydium*, *Boleophthalmus*, *Periophthalmus*, *Eleotris* (equatorial). Many of the species are entirely confined to fresh water.

9. The *Amblyopina*, similar to the Gobies, but with more elongated body: Tropical Indo-Pacific.

10. The *Trypauchenina*: Coasts of the Indian region.

11. Many species of *Blennius*, of which several are found far inland in fresh waters—for instance in North Italy, in the Lake of Galilee, in the eastern parts of Asia Minor.

12. The majority of *Atherinidæ*, and

13. The *Mugilidæ*: both families being most numerous and abundant in brackish water, and almost cosmopolitan.

14. Many *Pleuronectidæ* prefer the mouths of rivers for the same reason as the Rays; some ascend rivers, as the Flounder, *Cynoglossus*, etc.

15. Several *Siluridæ*, as especially the genera *Plotosus*, *Cnidoglanis*, *Arius*, which attain their greatest development in brackish water.

16. The *Cyprinodontidæ* are frequently found in brackish water.

17. Species of *Clupea*, some of which ascend rivers, and become acclimatized in fresh water, as *Clupea finta*, which has established itself in the lakes of northern Italy.

18. *Chatoessus*, a genus of Clupeoid fishes of the Equatorial Zone, of which some species have spread into the Northern Zone.

19. *Megalops*: Equatorial Zone.

20. *Anguilla*. The distribution, no less than the mode of propagation, and the habits generally, of the so-called Freshwater-eels still present us with many difficult problems. As far as we know at present their birthplace seems to be the coast in the immediate neighbourhood of the mouths of rivers. They are much more frequently found in fresh water than in brackish water, but the distribution of some species proves that they at times migrate by sea as well as by land and river. Thus *Anguilla mauritiana* is found in almost all the fresh and brackish waters of the islands of the Tropical Indian Ocean and Western Pacific, from the Comoros to the South Sea; *Anguilla vulgaris* is spread over temperate Europe (exclusive of the system of the Danube, the Black and Caspian Seas), in the Mediterranean district (including the Nile and rivers of Syria), and on the Atlantic coast of North America; *Anguilla bostoniensis*, in Eastern North America, China, and Japan; *Anguilla lati-*

rostris, in Temperate Europe, the whole Mediterranean district, the West Indies, China, and New Zealand. The other more local species are found, in addition to localities already mentioned, on the East Coast of Africa, South Africa, on the continent of India, various East Indian Islands, Australia, Tasmania, Auckland Islands; but none have ever been found in South America, the West Coast of North America, and the West Coast of Africa: surely one of the most striking instances of irregular geographical distribution.

21. Numerous *Syngnathidæ* have established themselves in the Northern Zone as well as in the Equatorial, in the vegetation which flourishes in brackish water.

This list could be considerably increased if an enumeration of species, especially of certain localities, were attempted; but this is more a subject of local interest, and would carry us beyond the scope of a general account of the distribution of Fishes.

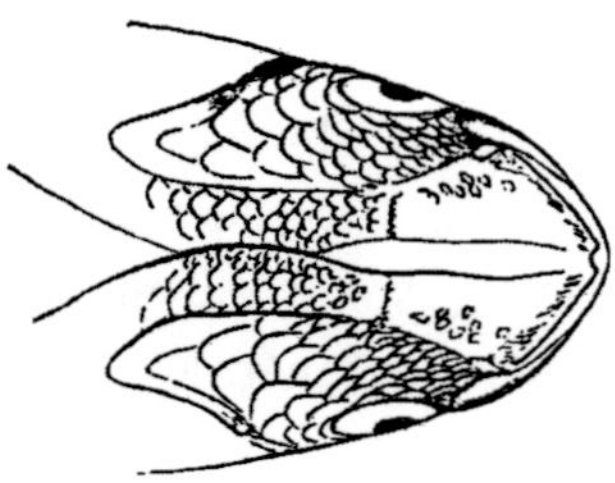

Fig. 105.—*Mugil octo-radiatus.*

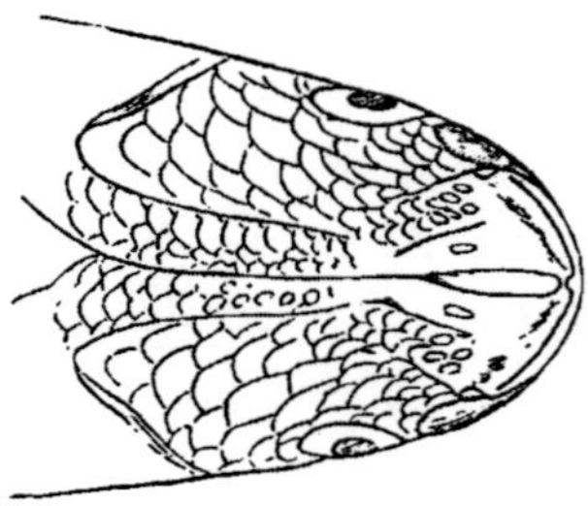

Fig. 106.—*Mugil auratus.*

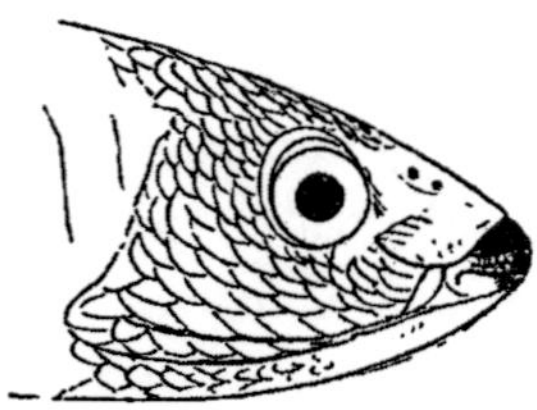

Fig. 107.—*Mugil septentrionalis.*

Heads of Grey Mullets, fishes of Brackish water.

CHAPTER XIX.

THE DISTRIBUTION OF MARINE FISHES.

MARINE fishes fall, with regard to their mode of life and distribution, into three distinct categories :—

1. *Shore Fishes*—That is, fishes which inhabit chiefly parts of the sea in the immediate neighbourhood of land either actually raised above, or at least but little submerged below, the surface of the water. They do not descend to any great depth,—very few to 300 fathoms, and the majority live close to the surface. The distribution of these fishes is determined not only by the temperature of the surface water but also by the nature of the adjacent land, and its animal and vegetable products; some of these fishes being confined to flat coasts with soft or sandy bottoms, others to rocky and fissured coasts, others to living coral formations. If it were not for the frequent mechanical and involuntary removals to which these fishes are exposed, their distribution within certain limits, as it no doubt originally existed, would resemble still more that of freshwater fishes than we find it actually does at the present period.

2. *Pelagic Fishes*—that is, fishes which inhabit the surface and uppermost strata of the open ocean, which approach the shores only accidentally, or occasionally (in search of prey), or periodically (for the purpose of spawning). The majority spawn in the open sea, their ova and young being always found at great distance from the shore. With regard to their distribution, they are still subject to the influences of light and the temperature of the surface water; but they are

independent of the variable local conditions which tie the shore fish to its original home, and therefore roam freely over a space which would take a freshwater or shore fish thousands of years to cover in its gradual dispersal. Such as are devoid of rapidity of motion are dispersed over similarly large areas by the oceanic currents, more slowly than, but as surely as, the strong swimmers. Therefore, an accurate definition of their distribution within certain areas equivalent to the terrestrial regions is much less feasible than in the case of shore fishes.

3. *Deep-sea Fishes*—that is, fishes which inhabit such depths of the ocean as to be but little or not influenced by light or the surface temperature; and which, by their organisation are prevented from reaching the surface stratum in a healthy condition. Living almost under identical tellurian conditions, the same type, the same species, may inhabit an abyssal depth under the equator as well as one near the arctic or antarctic circle; and all we know of these fishes points to the conclusion that no separate horizontal regions can be distinguished in the abyssal fauna, and that no division into bathymetrical strata can be attempted on the base of generic much less of family characters.

It must not be imagined that these three categories are more sharply defined than Freshwater and Marine Fishes. They gradually pass into each other, and there are numerous fishes about which uncertainty exists whether they should be placed in the Shore or Pelagic series, or in the Pelagic or Deep-sea series; nay, many facts favour the view that changes in the mode of life and distribution of fishes are still in progress.

The change in habitat of numerous fishes is regulated by the distribution of their favourite food. At certain seasons the surface of the sea in the vicinity of land swarms with mollusks, larval Crustaceans, Medusæ, attracting shoals of fishes from the open ocean to the shores; and these are again pursued by fishes of larger size and predacious habits, so that all these fishes might be included, with equal propriety, in the littoral or pelagic series. However, species which are known to normally spawn in the open ocean must be always referred to the latter division.

Chondropterygii, Acanthopterygii, Anacanths, Myxinoids, and Pharyngobranchii furnish the principal contingents to the Marine Fauna; whilst the majority of Malacopterygians, the Ganoids, and Cyclostomes are Freshwater Fishes.

I.—Distribution of Shore Fishes.

The principal types of Shore-fishes are the following:—

CHONDROPTERYGII—

HOLOCEPHALA	4	species.
PLAGIOSTOMATA—		
Carchariidæ (part.) . . .	12	„
Scylliidæ	30	„
Cestraciontidæ . . .	4	„
Spinacidæ (part.) . . .	8	„
Rhinidæ	1	„
Pristiophoridæ . . .	4	„
Pristidæ	5	„
Rhinobatidæ	14	„
Torpedinidæ	15	„
Rajidæ	34	„
Trygonidæ	47	„

ACANTHOPTERYGII—

Percidæ (part. incl. Pristipomatidæ) .	625	„
Mullidæ	35	„
Sparidæ	130	„
Squamipinnes	130	„
Cirrhitidæ	40	„

Heterolepidina	12	species.
Scorpænidæ	120	"
Cottidæ (part.)	100	"
Cataphracti (part.)	20	"
Trachinidæ	100	"
Sciænidæ	100	"
Sphyrænidæ	15	"
Trichiuridæ	17	"
Elacate	1	"
Nomeidæ (part.)	5	"
Cyttidæ	8	"
Stromateus	9	"
Mene	1	"
Carangidæ (part.)	130	"
Kurtidæ	7	"
Gobiodon	7	"
Callionymina	30	"
Discoboli	11	"
Batrachidæ	14	"
Pediculati (part.)	11	"
Blenniidæ	90	"
Acanthoclinidæ	1	"
Teuthididæ	30	"
Acronuridæ	60	"
Hoplognathidæ	3	"
Malacanthidæ	3	"
Plesiopina	4	"
Trichonotidæ	2	"
Cepolidæ	7	"
Gobiesocidæ	21	"
Psychrolutidæ	2	"
Centriscidæ	7	"
Fistulariidæ	4	"

ACANTHOPTERYGII PHARYNGOGNATHI—

Pomacentridæ	150	"
Labridæ	400	"
Embiotocidæ	17	"

ANACANTHINI—	
Gadopsidæ	1 species.
Lycodidæ	15 „
Gadidæ (part.)	50 „
Ophidiidæ (part.) . . .	40 „
Pleuronectidæ	160 „
PHYSOSTOMI—	
Saurina (part.)	16 „
Salmonidæ (part.) . . .	7 „
Clupeidæ (part.)	130 „
Chirocentridæ	1 „
Chilobranchus	1 „
Murænidæ (part.) . . .	200 „
Pegasidæ	4 „
LOPHOBRANCHII	120 „
PLECTOGNATHI—	
Sclerodermi	95 „
Gymnodontes	83 „
CYCLOSTOMATA—	
MYXINIDÆ	5 „
LEPTOCARDII	2 „
	3587 species.

These types of Shore fishes are divided among the following oceanic areæ :—

I. The Arctic Ocean.

II. The Northern Temperate Zone.

A. The Temperate North Atlantic.
1. The British district.
2. The Mediterranean district.
3. The North American district.

B. The Temperate North Pacific.
1. The Kamtschatkan district.
2. The Japanese district.
3. The Californian district.

III. The Equatorial Zone.

A. The Tropical Atlantic.
B. The Tropical Indo-Pacific.
C. The Pacific Coast of Tropical America.

1. The Central American district.
2. The Galapagoes district.
3. The Peruvian district.

IV. The Southern Temperate Zone.

1. The Cape of Good Hope district.
2. The South Australian district.
3. The Chilian district.
4. The Patagonian district.

V. The Antarctic Ocean.

As with freshwater fishes, the main divisions of the Shore-fish faunæ are determined by their distance from the equator, the equatorial zone of the Freshwater series corresponding entirely to that of the Shore-fish series. But as Marine fishes extend farther towards the Poles than Freshwater fishes, and as the polar types are more specialised, a distinct Arctic and Antarctic fauna may be separated from the faunæ of the temperate zones. The two subdivisions of the Northern temperate zone in the Freshwater series are quite analogous to the corresponding divisions in the Coast series. In the Southern Hemisphere the Shore-fishes of the extremity of Africa form a separate district of the temperate zone, whilst the Freshwater fishes of South Africa were found to be tropical types. The Marine series of the Southern temperate zone is also much more diversified than the Freshwater series, and admits of further subdivision, which, although in some degree indicated in the Freshwater series, does not entirely correspond to that proposed for the latter.

I. SHORE FISHES OF THE ARCTIC OCEAN.

The Shore fishes clearly prove a continuity of the Arctic circumpolar fauna, as the southern limit of which we may indicate the southern extremity of Greenland and the Aleutian Archipelago, or 60° of lat. N.

Towards the North, fishes become less in variety of species and fewer in number of individuals, and only very few genera are restricted to this fauna.

The highest latitude at which Shore fishes have been observed is 83° N. lat. The late Arctic Expedition collected at and near that latitude specimens of *Cottus quadricornis, Icelus hamatus, Cyclopterus spinosus, Liparis fabricii, Gymnelis viridis,* and *Gadus fabricii.* This number probably would have been larger if the difficulties of collecting fishes in those high latitudes were not almost insuperable for the greater part of the year.

As far as we know, the fishes north and south of Behring's Straits belong to the same generic or family types as those of the corresponding latitudes of the Eastern Hemisphere, though the majority are specifically distinct. But the information we possess of the fishes of the northernmost extremity of the Pacific is extremely scanty and vague. Farther south, whence now and then a collection reaches Europe, we meet with some European species, as the Herring, Holibut, Hake.

The Chondropterygians are very scarce, and it is doubtful whether another Chondropterygian, beside the pelagic *Læmargus* or Greenland Shark, crosses the Arctic circle. In the more temperate latitudes of South Greenland, Iceland, and Northern Scandinavia, *Acanthias, Centroscyllium,* and a species of *Raja,* also *Chimæra,* are met with.

Of Acanthopterygians the families of *Cottidæ, Cataphracti, Discoboli,* and *Blenniidæ* are well represented, and several of the genera are characteristic of the Arctic fauna: marine species of *Cottus; Centridermichthys, Icelus, Triglops; Agonus, Aspidophoroides; Anarrhichas, Centronotus, Stichæus; Cyclopterus* and *Liparis.* Two species of *Sebastes* are rather common.

Characteristic is also the development of Gadoid fishes, of which some thirteen species, belonging to *Gadus, Merluccius,* and *Molva,* form one of the principal articles of food to the inhabitants of the coasts of the Arctic Ocean. The Blennioid Anacanthini or *Lycodidæ,* are limited to the Arctic and Antarctic coasts. *Ammodytes* and a few Flat-fishes (*Hippoglossoides* and *Pleuronectes*) are common in the more temperate parts.

Labroids only exceptionally penetrate so far towards the north.

Physostomes are very scarce, and represented only by a few species of *Clupea* and by *Mallotus;* the latter is an ancient inhabitant of the Greenland coasts, fossil remains, indistinguishable from the species of the present day, being frequently found in nodules of clay of comparatively recent formation.

The Arctic climate is still less favourable to the existence of Lophobranchs, only a few *Syngnathus* and *Nerophis* being present in the more southern latitudes, to which they have been carried by oceanic currents from their more congenial home in the south. Sclerodermes and Plectognaths are entirely absent.

The Gadoids are accompanied by *Myxine,* which parasitically thrives in them.

II. The Northern Temperate Zone.

A. *Shore Fishes of the Temperate North Atlantic.*

This part of the fauna may be subdivided into three districts:—

1. The fishes of the north-eastern shores, viz. of the British islands, of Scandinavia so far as it is not included in the Arctic fauna, and of the continent of Europe southwards to about 40° of lat. N.—*British* district.

2. The fishes of the Mediterranean shores and of the adjoining shores of the Atlantic, including the Azores, Madeira, and the Canary Islands—*Mediterranean* district.

3. The fishes of the western shores, from 60° lat. N. to about 30° lat. N.—the *North American* district.

1. The *British* district shows scarcely any marked distinctive features; the character of its fauna is simply intermediate between that of the Arctic Ocean and the Mediterranean district; truly Arctic forms disappear, while such as are also found in the Mediterranean make their appearance. Also with regard to the abundance of individuals and variety of fishes this district forms a transition from the north towards the south.

Besides the few Arctic Chondropterygians, all of which extend into this district, the small shore Dog-fishes are well represented (*Mustelus, Galeus, Scyllium, Pristiurus*); the ubiquitous *Rhina* or Monk-fish is common; of Rays, *Raja* predominates in a variety of species over *Torpedo* and *Trygon*, which are still scarce.

Of Acanthopterygians, *Centridermichthys, Icelus, Triglops*, and *Aspidophoroides*, do not extend from the north into this district; and *Cottus, Anarrhichas, Centronotus, Stichæus*, the *Discoboli* disappear within its limits. Nearly all the remainder are genera which are also found in the Mediterranean districts. The following are the principal forms, and known to propagate on these shores: *Labrax; Serranus, Polyprion, Dentex; Mullus; Cantharus, Pagrus, Pagellus; Sebastes; Cottus, Trigla, Agonus; Trachinus; Sciæna* (?); *Zeus; Trachurus, Capros; Callionymus; Discoboli; Lophius; Anarrhichas, Centronotus, Stich-*

œus; Blenniops, Zoarces (not in Mediterranean); *Cepola; Lepadogaster.*

Of the Anacanthini the Gadoids are as numerous as in the Arctic Ocean, most being common to both districts; they are represented by *Gadus, Gadiculus, Merluccius, Phycis, Molva, Motella, Raniceps,* and *Brosmius;* but, whilst the majority show their northern origin by not extending into the Mediterranean, *Ammodytes* and most *Pleuronectidæ* prove themselves to be the more southern representatives of this order. In the British district we find *Hippoglossus, Hippoglossoides, Rhombus, Phrynorhombus, Pleuronectes, Solea,* and only the two first are not met with in the Mediterranean.

Labroids are common; with the exception of the North American *Tautoga,* all the other genera are met with.

Physostomes are not well represented, viz. by one species of *Osmerus,* one of *Engraulis,* one of *Conger,* and about five of *Clupea.*

Syngnathus and *Nerophis* become more common as we proceed southwards; but the existence of Sclerodermis and Plectognaths is indicated by single individuals only, stragglers from their southern home, and unable to establish themselves in a climate ungenial to them.

The Gadoids are accompanied by *Myxine;* and *Branchiostoma* may be found in all suitable localities.

2. The *Mediterranean* district is distinguished by a great variety of forms; yet, with the exception of a few genera established for single species, none of the forms can be considered peculiar to it; and even that small number of peculiar genera is more and more diminished as our knowledge of the distribution of fishes advances. Some genera are identical with those found on the western coasts of the Atlantic and in the West Indies; but a most remarkable and unexpected affinity obtains with another very distant

fauna, viz. that of Japan. The number of genera common to the Mediterranean district and the Japanese coasts is larger than that of the genera common to the Mediterranean and the opposite American coasts.

The Chondropterygians found in the British district continue in the Mediterranean, their number being increased by *Centrina*, *Spinax*, *Pteroplatea*, and some species of *Rhinobatus*, a genus more numerously represented in the Tropics. *Torpedo* and *Trygon* are common.

The greatest variety belong to the Acanthopterygians, as will be seen from the following list:—*Labrax*; *Anthias*, *Serranus*, *Polyprion*, *Apogon*, *Pomatomus*, *Pristipoma*, *Diagramma* (an Indian genus with two Mediterranean species, and otherwise not represented in the Atlantic), *Dentex*, *Mæna*, *Smaris*; *Mullus*; *Cantharus*, *Box*, *Scatharus*, *Oblata*, *Sargus*, *Pagrus*, *Pagellus*, *Chrysophrys*; *Sebastes*, *Scorpæna*; *Hoplostethus*, *Beryx*, *Polymixia*; *Trigla*, *Lepidotrigla*, *Agonus*, *Peristethus*; *Trachinus*, *Uranoscopus*; *Umbrina*, *Sciæna*; *Sphyræna*; *Aphanopus*, *Lepidopus*, *Nesiarchus*, *Trichiurus*, *Thyrsites*; *Cubiceps*; *Zeus*, *Cyttus*; *Stromateus*; *Trachurus*, *Caranx*, *Capros*, *Diretmus*, *Antigonia*; *Callionymus*; *Batrachus*; *Lophius*; *Cristiceps*, *Tripterygium*; *Cepola*; *Lepadogaster*; *Centriscus*; *Notacanthus*.

The *Labridæ* are as common as, or even more so than, in the British district, and represented by the same genera. But, besides, some other Pharyngognaths, properly belonging to the Tropical Atlantic, have fully established themselves, though only by a few species, viz. *Glyphidodon* and *Heliastes*; *Cossyphus*, *Novacula*, *Julis*, *Coris*, and *Scarus*.

The Gadoids show a marked decrease of development; and the species of *Gadus*, *Gadiculus*, *Mora*, *Strinsia*, *Phycis*, and *Molva*, which are peculiar to the Mediterranean, seem to inhabit rather the colder water of moderate depths, than the surface near the shore. *Motella*, however, proves to be a true

Shore fish also in the Mediterranean, at least in its adult state. *Ophidium* and *Fierasfer* appear now besides *Ammodytes*. As the Gadoids decrease, so the *Pleuronectidæ* increase, the genera of the Mediterranean district being *Rhombus*, *Phrynorhombus*, *Arnoglossus*, *Citharus*, *Rhomboidichthys*, *Pleuronectes* (a northern genus not extending farther southwards), *Solea*, *Synaptura*, and *Ammopleurops*.

The variety of Physostomes is small; the following only being superadded to those of the British district:—*Saurus* (a tropical genus), *Aulopus*; *Congromuræna*, *Heteroconger*, *Myrus*, *Ophichthys*, *Muræna*.

The Lophobranchs are more numerous in species and individuals than in the British district; and, besides *Syngnathus* and *Nerophis*, several species of *Hippocampus* are common. Also a few species of *Balistes* occur.

Myxine is lost in this district; whilst *Branchiostoma* is abundant.

3. The shore fishes of the *North American* district consist, as on the eastern coasts of the North Atlantic, of northern and southern elements; but they are still more mixed with each other than on the European coasts, so that a boundary line cannot be drawn between them. The affinity to the fauna of the eastern shores is great, but almost entirely limited to the genera composing the fauna of the British district. British genera not found on the American coasts are—*Galeus*, *Scyllium*, *Chimæra*, *Mullus*, *Pagellus*, *Trigla*, *Trachinus*, *Zeus*, *Callionymus*. The southern elements of North America are rather derived from the West Indies, and have no special affinity to Mediterranean forms; very few of the non-British Mediterranean forms extend across the Atlantic; instead of a Mediterranean we find a West Indian element. Many of the British *species* range across the Atlantic, and inhabit in an unchanged condition the northern parts of

this district; and from the frequent occurrence of isolated specimens of other British species on the North American coast, we may presume that many more occasionally cross the Atlantic, but without being able to obtain a permanent footing.

The genera peculiar to this district are few in number, and composed of very few species, viz. *Hemitripterus, Pammelas, Chasmodes, Cryptacanthodes*, and *Tautoga*.

The close resemblance of what must be considered northern forms to those of Europe will be evident from the following list:—

Mustelus, Rhina, Torpedo, Raja, Trygon.

Labrax, Centropristis, Serranus; Pagrus, Chrysophrys; Sebastes, Hemitripterus; Cottus, Aspidophoroides; Uranoscopus; Micropogon, Pogonias, Sciæna; Trachurus, Pammelas; Cyclopterus, Liparis; Lophius; Anarrhichas, Chasmodes, Stichæus, Centronotus, Cryptacanthodes, Zoarces.

Tautoga, Ctenolabrus.

Gadus, Merluccius, Phycis, Molva, Motella, Brosmius; Ophidium (one species, perhaps identical with a Mediterranean species); *Ammodytes; Hippoglossus, Hippoglossoides, Rhombus, Pleuronectes.*

Osmerus, Mallotus; Engraulis, Clupea; Conger.

Syngnathus—Myxine—Branchiostoma.

West Indian genera, or at least genera which are more developed within the tropics, and which extend more or less northwards in the North American district, are:—

Pteroplatea (also in the Mediterranean).

Gerres, Dules (*auriga*), *Lobotes, Ephippus; Sargus; Prionotus; Umbrina, Otolithus, Larimus; Sphyræna* (Mediterr.); *Trichiurus* (Mediterr.); *Elacate; Cybium, Trachynotus; Stromateus* (Mediterr.); *Caranx; Batrachus* (Mediterr.); *Malthe.*

Pseudorhombus, Solea (Mediterr.)

Saurus (Mediterr.); *Etrumeus, Albula, Elops, Megalops.*

Hippocampus (Mediterr.)
Balistes, Monacanthus.

B. *Shore Fishes of the Temperate North Pacific.*

This fauna shows a great affinity to that of the temperate North Atlantic, not only in including a considerable proportion of identical genera, and even of species, but also in having its constituent parts similarly distributed. However, our knowledge of the ichthyology of this fauna is by no means complete. Very few collections have been made in Northern Japan, and on the coasts farther north of it; and, again, the ichthyology of the coasts of Southern California is but little known. Southern Japan has been well searched, but very little attention has been paid to the extent of the northward range of the species. In collections made by Mr. Swinhoe at Chefoo, in lat. 37° N., the proportions of temperate and tropical fishes were found to be about equal. Thus, the details of the distribution of the fishes of these shores have still to be worked out; nevertheless, three divisions may be recognised which, for the present, may be defined as follows:—

1. The fishes of the north-western shores, to about 37° lat. N., including the corresponding northern parts of Japan—*Kamtschatkan* district; this corresponds to the British district of the Atlantic.

2. The fishes of Southern Japan and the corresponding shores of the continent of Asia, between 37° and 30° lat. N.—*Japanese* district, which corresponds to the Mediterranean.

3. The fishes of the eastern shores southwards to the latitude of San Francisco—*Californian* district; this corresponds to the North American district of the Atlantic.

Too little is known of the shore fishes of the coasts between San Francisco and the tropic to enable us to treat of it as a separate division.

The Shore fishes of the North Pacific generally are composed of the following elements:—

a. Arctic forms which extend into the Arctic Ocean, and the majority of which are also found in the British district.

b. Peculiar forms limited to the North Pacific, like the *Heterolepidina, Embiotocidæ,* and certain Cottoid and Blennioid genera.

c. Forms identical with fishes of the Mediterranean.

d. Peculiar forms limited to the southern parts of Japan.

e. Tropical forms which have entered the North Pacific from the south.

1. The small list of fishes which we can assign to the *Kamtschatkan* district is due rather to the imperfect manner in which its fauna has been explored than to its actual poverty of fishes; thus, although we may be sure that sooner or later the small kinds of Dog-fishes of the British district will be found there also, at present we have positive knowledge of the occurrence of only two Chondropterygians, viz. *Chimæra* and *Raja.* The species of the latter genus seem to be much less numerous than in the Atlantic.

Of Acanthopterygians the following are known:—*Sebastes; Chirus, Agrammus; Podabrus, Blepsias, Cottus, Centridermichthys, Hemilepidotus, Agonus; Trichodon; Callionymus; Liparis; Dictyosoma, Stichæus, Centronotus.*

Labroids are absent; they are clearly a type unable to endure great cold; of the Embiotocoids which represent them in the Pacific, one species only (a species of *Ditrema*) is known from this district.

The Gadoids are, so far as we know at present, sparsely represented, viz. by isolated species of *Gadus, Motella,* and *Lotella,* the latter being an inhabitant of moderate depths rather than of the surface. *Hippoglossus, Pleuronectes,* and *Parophrys,* seem to occur everywhere at suitable localities.

The Physostomes are nearly the same as in the British district, viz. a Smelt (*Hypomesus*), probably also the Arctic *Mallotus*, an Anchovy, several species of *Clupea*, and the Conger-eel. A very singular Salmonoid fish, *Salanx*, which is limited to the north-western Pacific, occurs in great abundance.

Also, the Lophobranchs correspond in their development to those of the British district, *Nerophis* being replaced by *Urocampus*.

Neither Myxinoids nor *Branchiostoma* have as yet been found.

2. The *Japanese* district is, like the Mediterranean, distinguished by a great variety of forms; some of them are peculiar to it (marked *J.* in the following list); others occur in the Mediterranean, though also in other districts (*M.*) The resemblance to the Mediterranean is even greater than would appear from the following list of genera, inasmuch as a considerable number of species are identical in both districts. Three of the Berycoid genera have hitherto been found in the Japanese and Mediterranean districts only, and nowhere else. Another very singular fact is that some of the most characteristic genera, like *Mullus, Zeus, Callionymus, Centriscus,* inhabit the Mediterranean and Japanese districts, *but have never reached the opposite American coasts, either in the Atlantic or Pacific;* although, at least in the latter, the oceanic currents would rather favour than obstruct their dispersal in the direction towards America. Bold as the hypothesis may appear, we can only account for the singular distribution of these shore fishes by assuming that the Mediterranean and Japanese seas were in direct and open communication with each other within the period of the existence of the present Teleosteous Fauna.

Gadoids have disappeared, or are represented by forms

inhabiting moderate depths. Neither *Myxine* nor *Branchiostoma* are known to have as yet been found.

List of Japanese Shore Fishes.

Chimæra (M.)

Galeus (M.), *Mustelus* (M.), *Triacis*, *Scyllium* (M.), *Crossorhinus*, *Pristiophorus*, *Cestracion*; *Rhina* (M.); *Rhinobatus* (M.), *Narcine*, *Raja* (M.), *Trygon* (M.), *Pteroplatea* (M.)

Percalabrax (J.), *Niphon* (J.), *Centropristis*, *Anthias* (M.), *Serranus* (M.), *Apogon* (M.), *Scombrops* (J.), *Acropoma*, *Anoplus* (J.), *Pristipoma* (M.), *Hapalogenys* (J.), *Histiopterus*, *Velifer* (J.), *Dentex* (M.), *Erythrichthys*—*Mullidæ* (M.)—*Girella*, *Pagrus* (M.), *Chrysophrys* (M.)—*Chilodactylus*—*Sebastes* (M.), *Scorpæna* (M.), *Aploactis*, *Trichopleura*, *Pelor*—*Monocentris* (J.), *Hoplostethus* (M.), *Beryx* (M.), *Polymixia* (M.)—*Platycephalus*, *Hoplichthys* (J.), *Bembras* (J.), *Prionotus*, *Lepidotrigla* (M.), *Trigla* (M.), *Peristethus* (M.)—*Uranoscopus* (M.), *Percis*, *Sillago*, *Latilus*.—*Sciæna* (M.), *Otolithus*—*Sphyræna* (M.)—*Lepidopus* (M.), *Trichiurus* (M.)—*Zeus* (M.)—*Caranx*, *Trachurus* (M.)—*Callionymus* (M.)—*Lophius* (M.), *Halicuthæa* (J.)—*Hoplognathus*—*Cepola* (M.)—*Centriscus* (M.), *Fistularia*.

Heliastes (M.)—*Labrichthys*, *Duymæria*, *Platyglossus*, *Novacula* (M.), *Julis* (M.), *Coris* (M.)

Sirembo (J.)—*Motella* (M.)—*Atelcopus* (J.)

Pseudorhombus, *Pleuronectes* (M.), *Solea* (M.), *Synaptura* (M.)

Saurus (M.), *Harpodon*.—*Salanx* (J.)—*Engraulis* (M.), *Clupea* (M.), *Etrumeus*—*Conger* (M.), *Congromuræna* (M.), *Muræncsox* (M.), *Oxyconger*, *Myrus* (M.), *Ophichthys* (M.), *Muræna* (M.)

Syngnathus (M.), *Hippocampus* (M.), *Solenognathus*.

Triacanthus, *Monacanthus*, *Ostracion*.

3. The *Californian* district includes a marked northern element, the principal constituents of which are identical with types occurring in the corresponding district of the Atlantic, viz. the North American, as exemplified by *Discoboli, Anarrhichas, Centronotus, Cottus, Hippoglossus, Clupea* (*harengus*), etc. But it possesses also, in the greatest degree of development, some types almost peculiar to itself, as the *Heterolepidina*, some remarkable Cottoid and Blennioid genera, and more especially the Embiotocoids—viviparous Pharyngognaths—which replace the Labroids of the other hemisphere. Gadoids are much less numerous than in the North American district. The southern forms are but little known, but it may be anticipated that, owing to the partial identity of the Faunæ of the two coasts of the Isthmus of Panama, a fair proportion of West Indian forms will be found to have entered this district from the south. The following are the principal genera :—

Chimæra, Galeus, Mustelus, Triacis, Cestracion, Rhina, Raja.

Serranus ; Chirus, Ophiodon, Zaniolepis ; Sebastes ; Nautichthys, Scorpænichthys, Cottus, Centridermichthys, Hemilepidotus, Artedius, Prionotus, Agonus ; Cyclopterus, Liparis ; Anarrhichas, Neoclinus, Cebidichthys, Stichæus, Centronotus, Apodichthys ; Psychrolutes ; Auliscops.

Embiotocidæ.

Gadus. Hippoglossus, Psettichthys, Citharichthys, Paralichthys, Pleuronectes, Parophrys.

Osmerus, Thalcichthys, Hypomesus ; Engraulis, Clupea.

Syngnathus.

III.—The Equatorial Zone.

As we approach the Tropic from the north, the tribes characteristic of the Arctic and Temperate zones become scarcer, and disappear altogether : to be replaced by the greater variety of Tropical types. Of Chondropterygians, the *Chimæridæ, Spinacidæ, Mustelus,* and *Raja,* do not pass the

Tropic, or appear in single species only; and of Teleosteans, the *Berycidæ*, *Pagrus*, the *Heterolepidina*, *Cottus* and allied genera, *Lophius*, *Anarrhichas*, *Stichæus*, *Lepadogaster*, *Psychrolutes*, *Centriscus*, *Notacanthus*, the *Labridæ* and *Embiotocidæ*, the *Lycodidæ*, *Gadidæ*, and marine *Salmonidæ* disappear either entirely, or retire from the shores and surface into the depths of the ocean.

With regard to variety of forms, as well as to number of individuals, this zone far surpasses either of the temperate zones; in this respect, the life in the sea is as that on the land. Coast fishes are not confined to the actual coast-line, but abound on the coral reefs, with which some parts of the Atlantic and Pacific are studded, and many of which are submerged below the water. The abundance of animal and vegetable life which flourishes on them renders them the favourite pasture-grounds for the endless variety of coral-fishes (*Squamipinnes*, *Acronuridæ*, *Pomacentridæ*, *Julidæ*, *Plectognathi*, etc.), and for the larger predatory kinds. The colours and grotesque forms of the Fishes of the Tropics have justly excited the admiration of the earliest observers. Scarlet, black, blue, pink, red, yellow, etc., are arranged in patterns of the most bizarre fashion, mingling in spots, lines, bands; and reminding us of the words of Captain Cook when describing the coral-reefs of Palmerston Island: "The glowing appearance of the Mollusks was still inferior to that of the multitude of fishes that glided gently along, seemingly with the most perfect security. The colours of the different sorts were the most beautiful that can be imagined—the yellow, blue, red, black, etc., far exceeding anything that art can produce. Their various forms, also, contributed to increase the richness of this sub-marine grotto, which could not be surveyed without a pleasing transport."

Of Chondropterygians the *Scylliidæ*, *Pristis* (Saw-fishes). *Rhinobatidæ*, and *Trygonidæ* attain to the greatest development. Of Acanthopterygians *Centropristis, Serranus, Plectropoma, Mesoprion, Priacanthus, Apogon, Pristipoma, Hæmulon, Diagramma, Gerres, Scolopsis, Synagris, Cæsio, Mullidæ, Lethrinus, Squamipinnes, Cirrhites,* some genera of *Scorpænidæ, Platycephalus, Sciænidæ, Sphyræna, Caranx Equula, Callionymus, Teuthis, Acanthurus, Naseus,* are represented by numerous species; and the majority of these genera and families are limited to this zone. Of Pharyngognaths the *Pomacentridæ, Julidina,* and *Scarina,* are met with near every coral formation in a living condition. Of Gadoids, a singular minute form, *Bregmaceros,* is almost the only representative, the other forms belonging to deep water, and rarely ascending to the surface. Flat-fishes (*Pleuronectidæ*) are common on sandy coasts, and the majority of the genera are peculiar to the Tropics. Of *Physostomi* only the *Saurina, Clupeidæ,* and *Murænidæ* are represented, the *Clupeidæ* being exceedingly numerous in individuals, whilst the *Murænidæ* live more isolated, but show a still greater variety of species. Lophobranchii and Sclerodermi are generally distributed. Branchiostoma has been found on several coasts.

Geographically it is convenient to describe the Coast fauna of the tropical Atlantic separately from that of the Indo-Pacific ocean. The differences between them, however, are far less numerous and important than between the freshwater or terrestrial faunæ of continental regions. The majority of the principal types are found in both, many of the species being even identical; but the species are far more abundant in the Indo-Pacific than in the Atlantic, owing to the greater extent of the archipelagoes in the former. But for the broken and varied character of the coasts of the West Indies, the shores of the tropical Atlantic would, by their general uniformity, afford but a limited variety of conditions to the development of specific and generic forms, whilst the deep inlets of the

Indian ocean, with the varying configuration of their coasts, and the different nature of their bottom, its long peninsulas, and its archipelagoes, and the scattered islands of the tropical Pacific, render this part of the globe the most perfect for the development of fish-life. The fishes of the Indian and Pacific oceans (between the Tropics) are almost identical, and the number of species ranging from the Red Sea and east coast of Africa to Polynesia, even to its westernmost islands, is very great indeed. However, this Indo-Pacific fauna does not reach the Pacific coast of South America. The wide space devoid of islands, east of the Sandwich Islands and the Marquesas group, together with the current of cold water which sweeps northwards along the South American coast, has proved to be a very effectual barrier to the eastward extension of the Indo-Pacific fauna of coast fishes; and, consequently, we find an assemblage of fishes on the American coast and at the Galapagoes Islands, sufficiently distinct to constitute a distinct zoological division.

The following list, which contains only the principal genera and groups of coast fishes, will give an idea of the affinity of the tropical Atlantic and Indo-Pacific :—[1]

	Trop.-Atl.	Indo-Pac.
Scylliidæ . . .	—	13
Pristis . . .	3	4
Rhinobatidæ . .	4	8
Torpedinidæ . .	1	8
Trygonidæ . . .	14	24
Etelis . . .	1	1
Aprion . . .	—	1
Apsilus . . .	1	—
Centropristis . .	15	—
Anthias . . .	4	5
Serranus . . .	30	85
Plectropoma . .	11	5

[1] The genera peculiar to the Equatorial zone are printed in italics.

	Trop.-Atl.	Indo-Pac.
Grammistes . .	—	2
Rhypticus . . .	3	—
Diploprion . . .	—	1
Myriodon . . .	—	1
Mesoprion . . .	15	50
Priacanthus . .	4	12
Apogon and *Chilodipterus* .	2	75
Pristipoma . . .	12	14
Hæmulon . . .	15	—
Diagramma . .	—	30
Gerres . . .	12	16
Scolopsis . . .	—	20
Dentex and *Symphorus* .	—	7
Synagris and *Pentapus* .	—	24
Cæsio . . .	—	12
Mullidæ . . .	5	22
Sargus . . .	7	2
Lethrinus . . .	1	18
Chrysophrys . .	1	7
Pimelepterus . .	1	5
Squamipinnes . .	13	110
Toxotes . . .	—	2
Cirrhites . . .	—	20
Scorpænidæ . .	2	65
Myripristis . .	3	15
Holocentrum . .	6	25
Platycephalus . .	—	25
Prionotus . . .	1	—
Trigla . . .	—	4
Peristethus . .	2	6
Uranoscopina . .	2	8
Champsodon . .	—	1
Percis . . .	—	10
Sillago . . .	—	5
Latilus . . .	1	2
Opisthognathus . .	2	5
Pseudochromis . .	—	8
Cichlops and *Pseudoplesiops*	—	2

	Trop. Atl.	Ind.-Pac.
Sciænidæ	44	43
Sphyræna	1	10
Trichiuridæ	6	5
Caranx	20	60
Chorinemus	4	7
Trachynotus	6	4
Psettus	1	2
Platax	—	7
Zanclus	—	1
Equula and *Gazza*	—	20
Teuthis	—	30
Acanthurus	3	42
Naseus	—	12
Kurtidæ	1	6
Gobiodon	—	7
Callionymus	—	17
Batrachidæ	5	4
Tetrabrachium	—	1
Malthe	1	—
Petroscirtes	—	30
Clinus	6	—
Dactyloscopus	1	—
Malacanthus	1	2
Cepola	—	1
Gobiesocidæ	5	1
Amphisile	—	3
Fistulariidæ	3	3
Pomacentridæ	17	120
Lachnolæmus	1	—
Julidina	36	190
Pseudodax	—	1
Scarina	21	65
Pseudophycis	—	1
Bregmaceros	—	1
Ophidiidæ	3	7
Fierasfer	—	6
Pleuronectidæ	21	56
Saurina	5	9

	Trop.-Atl.	Ind.-Pac.
Clupeidæ	33	84
Chirocentrus	—	1
Murænidæ	47	130
Pegasus	—	3
Solenostoma	—	2
Syngnathidæ	7	41
Sclerodermi	16	67
Gymnodontes	23	40

A. *Shore Fishes of the Tropical Atlantic.*

The boundaries of the tropical Atlantic extend zoologically a few degrees beyond the Northern and Southern Tropics, but as the mixture with the types of the temperate zone is very gradual, no distinct boundary line can be drawn between the tropical and temperate faunæ.

Types, almost exclusively limited to it, and not found in the Indo-Pacific, are few in nmmber, as *Centropristis, Rhypticus, Hæmulon, Malthe.* A few others preponderate with regard to the number of species, as *Plectropoma, Sargus, Trachynotus, Batrachidæ,* and *Gobiesocidæ.* The Sciænoids are equally represented in both oceans. All the remainder are found in both; but in the minority in the Atlantic, where they are sometimes represented by one or two species only (for instance, *Lethrinus*).

B. *Shore Fishes of the Tropical Indo-Pacific Ocean.*

The ichthyological boundaries of this part of the tropical zone may be approximately given as 30° of lat. N. and S.; on the Australian coasts it should probably be placed still farther south, viz., to 34°; it includes, as mentioned above, the Sandwich Islands, and all the islands of the South Sea, but not the American coasts.

Some eighty genera of Shore fishes are peculiar to the Indo-Pacific, but the majority consists of one or a few species only; comparatively few have a plurality of species, as *Diagramma, Lethrinus, Equula, Teuthis, Amphiprion, Dascyllus, Choerops, Chilinus, Anampses, Stethojulis, Coris, Coilia.*

The Sea-perches, large and small, which feed on Crustaceans and other small fishes, and the coral-feeding Pharyngognaths are the types which show the greatest generic and specific variety in the Indo-Pacific. Then follow the *Squamipinnes* and *Muraenidae*, the *Clupeidae* and *Carangidae* families in which the variety is more that of species than of genus. The *Scorpaenidae, Pleuronectidae, Acronuridae, Sciaenidae, Syngnathidae*, and *Teuthyes*, are those which contribute the next largest contingents. Of shore-loving Chondropterygians the *Scylliidae* and *Trygonidae* only are represented in moderate numbers, though they are more numerous in this ocean than in any other.

C. *Shore Fishes of the Pacific Coasts of Tropical America.*

As boundaries within which this fauna is comprised, may be indicated 30° lat. N. and S., as in the Indo-Pacific. Its distinction from the Indo-Pacific lies in the almost entire absence of coral-feeding fishes. There are scarcely any Squamipinnes, Pharyngognaths or Acronuridae, and the Teuthyes are entirely absent. The genera that remain are such as are found in the tropical zone generally, but the species are entirely different from those of the Indo-Pacific. They are mixed with a sprinkling of peculiar genera, consisting of one or two species, like *Discopyge, Hoplopagrus, Doydixodon*, but they are too few in number to give a strikingly peculiar character to this fauna.

Three districts are distinguishable:—

a. The *Central American district*, in which we include, for the present, Lower California, shows so near an affinity to the tropical Atlantic that, if it were not separated from it by the neck of land uniting the two American Continents, it would most assuredly be regarded as a portion of the Fauna of the tropical Atlantic. With scarcely any exceptions the genera are identical, and of the species found on the Pacific side nearly one-half have proved to be the same as those of the Atlantic. The explanation of this fact has been found in the existence of communications between the two oceans by channels and straits which must have been open till within a recent period. The isthmus of Central America was then partially submerged, and appeared as a chain of islands similar to that of the Antilles ; but as the reef-building corals flourished chiefly north and east of those islands, and were absent south and west of them, reef-fishes were excluded from the Pacific shores when the communications were destroyed by the upheaval of the land.

b. The *Galapagoes district* received its coast fauna principally from the Central American district, a part of the species being absolutely the same as on the coast of the Isthmus of Panama, or as in the West Indies. Yet the isolation of this group has continued a sufficiently long period to allow of the development of a number of distinct species of either peculiarly Atlantic genera (such as *Centropristis, Rhypticus, Gobiesox, Prionotus*), or at least tropical genera (such as *Chrysophrys, Pristipoma, Holacanthus, Caranx, Balistes*). A few other types from the Peruvian coast (*Doydixodon*), or even from Japan (*Prionurus*), have established themselves in this group of islands. A species of *Cestracion* has also reached the Galapagoes, but whether from the south, north, or west, cannot be determined.

The presence of the Atlantic fauna on the Pacific side is felt still farther west than the Galapagoes, some Atlantic species having reached the Sandwich Islands, as *Chætodon humeralis* and *Blennius brevipinnis.*

c. The *Peruvian district* possesses a very limited variety of shore fishes, which belong, with few exceptions, like *Discopyge, Hoplognathus, Doydixodon,* to genera distributed throughout the tropical zone, or even beyond it. But the species, so far as they are known at present, are distinct from those of the Indo-Pacific, as well as of the tropical Atlantic; and therefore this district cannot be joined either to the Central American or the Galapagoes.

IV.—The Southern Temperate Zone.

This zone includes the coasts of the southern extremity of Africa, from about 30° lat. S., of the south of Australia with Tasmania, of New Zealand, and the Pacific and Atlantic coasts of South America between 30° and 50° lat. S.

The most striking character of this fauna is the reappearance of types inhabiting the corresponding latitudes of the Northern Hemisphere, and not found in the intervening tropical zone. This interruption of the continuity in the geographical distribution of Shore-fishes is exemplified by species as well as genera, for instance—*Chimæra monstrosa, Galeus canis, Acanthias vulgaris, Acanthias blainvillii, Rhina squatina, Zeus faber, Lophius piscatorius, Centriscus scolopax, Engraulis encrasicholus, Clupea sprattus, Conger vulgaris.* Instances of genera are still more numerous—*Cestracion, Spinax, Pristiophorus, Raja; Callanthias, Polyprion, Histiopterus, Cantharus, Box, Girella, Pagellus, Chilodactylus, Sebastes, Aploactis, Agonus, Lepidopus, Cyttus, Psychrolutidæ, Notacanthus; Lycodes, Merluccius, Lotella, Phycis, Motella; Aulopus; Urocampus, Solenognathus; Myxine.*

Naturally, where the coasts of the tropical zone are continuous with those of the temperate, a number of tropical genera enter the latter, and genera which we have found between the tropics as well as in the temperate zone of the Northern Hemisphere, extend in a similar manner towards the south. But the truly tropical forms are absent; there are no *Squamipinnes*, scarcely any *Mullidæ*, no *Acronuri*, no *Teuthyes*, no *Pomacentridæ* (with a single exception on the coast of Chili), only one genus of *Julidina*, no *Scarina*, which are replaced by another group of Pharyngognaths, the *Odacina*. The *Labrina*, so characteristic of the temperate zone of the Northern Hemisphere, reappear in a distinct genus (*Malacopterus*) on the coast of Juan Fernandez.

The family of *Berycidæ*, equally interesting with regard to their distribution in time and in space, consists of temperate and tropical genera. The genus by which this family is represented in the southern temperate zone (*Trachichthys*) is much more nearly allied to the northern than to the tropical genera.

The true *Cottina* and *Heterolepidina* (forms with a bony stay of the præoperculum, which is generally armed) have not crossed the tropical zone; they are replaced by fishes extremely similar in general form, and having the same habits, but lacking that osteological peculiarity. Their southern analogues belong chiefly to the family *Trachinidæ*, and are types of genera peculiar to the Southern Hemisphere.

The *Discoboli* of the Northern Hemisphere have likewise not penetrated to the south, where they are represented by *Gobiesocidæ*. These two families replace each other in their distribution over the globe.

Nearly all the *Pleuronectidæ* (but they are not numerous) belong to distinct genera, some, however, being remarkably similar in general form to the northern *Pleuronectes*.

With Gadoids *Myxinidæ* reappear, one species being extremely similar to the European Myxine. *Bdellostoma* is a genus peculiar to the southern temperate zone.

As in the northern temperate zone, so in the southern, the number of individuals and the variety of forms is much less than between the tropics. This is especially apparent on comparing the numbers of species constituting a genus. In this zone genera composed of more than ten species are the exception, the majority having only from one to five.

The proportion of genera limited to this zone is rather high; they will be indicated under the several districts, which we distinguish on geographical rather than zoological grounds.

1. The *Cape of Good Hope* district.

The principal genera found in this district are the following (those limited to the entire zone being marked with a single (*) and those peculiar to this district with a double (**) asterisk) :—

Chimæra, **Callorhynchus*, *Galeus*, ***Leptocarcharias*, *Scyllium*, *Acanthias*, *Rhinobatus*, *Torpedo*, *Narcine*, *Astrape*, *Raja*.

Serranus, *Dentex*, *Pristipoma*; *Cantharus*, *Box*, ***Dipterodon*, *Sagrus*, *Pagrus*, *Pagellus*, *Chrysophrys*; **Chilodactylus*; *Sebastes*, **Agriopus*; *Trigla*; *Sphyræna*; *Lepidopus*, *Thyrsites*; *Zeus*; *Caranx*; *Lophius*; *Clinus* (10 species), *Cristiceps*; ***Chorisochismus*.

***Halidesmus*, **Genypterus*, *Motella*.

Syngnathus.—**Bdellostoma*.

This list contains many northern forms, which in conjunction with the peculiarly southern types (*Callorhynchus*, *Chilodactylus*, *Agriopus*, *Clinus*, *Genypterus*, *Bdellostoma*) leave no doubt that this district belongs to the southern temperate zone, whilst the Freshwater fishes of South Africa are members of the tropical fauna. Only a few (*Rhinobatus*, *Narcine*, *Astrape*, and *Sphyræna*) have entered from the neighbouring tropical coasts. The development of Sparoids is greater than in any of the other districts of this zone, and may be regarded as one of its distinguishing features.

2. The *South Australian* district comprises the southern coasts of Australia (northwards, about to the latitude of

Sydney), Tasmania, and New Zealand. It is the richest in the southern temperate zone, partly in consequence of a considerable influx of tropical forms on the eastern coast of Australia, where they penetrate farther southwards than should have been expected from merely geographical considerations; partly in consequence of the thorough manner in which the ichthyology of New South Wales and New Zealand has been explored. On the other hand, the western half of the south coast of Australia is still almost a *terra incognita*.

The shore-fishes of New Zealand are not so distinct from those of south-eastern Australia as to deserve to be placed in a separate district. Beside the genera which enter this zone from the Tropics, and which are more numerous on the Australian coast than on that of New Zealand, and beside a few very local genera, the remainder are identical. Many of the South Australian species, besides, are found also on the coasts of New Zealand. The principal points of difference are the extraordinary development of Monacanthus on the coast of South Australia, and the apparently total absence in Australia of Gadoids, which in the New Zealand Fauna are represented by six genera.

Shore-fishes of the South Australian district.

	South Australia and Tasmania.	New Zealand.
*Callorhynchus (antarcticus) .	1	1
Galeus (canis) . .	1	1
Scyllium . . .	2	1
**Parascyllium . .	1	—
Crossorhinus . .	1	—
Cestracion . . .	2	1
Mustelus (antarcticus) .	1	1
Acanthias (vulgaris and blainvillii)	2	1
Rhina . . .	1	—
Pristiophorus . .	1	—

	South Australia and Tasmania.	New Zealand.
**Trygonorhina (fasciata) .	1	1
Rhinobatus . .	1	1
Torpedo . . .	—	1
Narcine . . .	1	—
Raja . . .	3	1
Trygon (Urolophus) .	3	2
**Enoplosus . . .	1	—
Anthias (richardsonii) .	1	1
Callanthias . .	1	—
Serranus . . .	x [1]	—
Plectropoma . .	4	—
**Lanioperca . . .	1	—
**Arripis . . .	3	1
Histiopterus . .	1	—
Erythrichthys . .	—	1
*Haplodactylus . .	2	2
Girella . . .	4	—
**Tephræops . . .	1	—
Pagrus . . .	1	1
*Scorpis . . .	2	1
**Atypichthys . .	1	—
**Trachichthys . .	—	1
**Chironemus . .	1	1
**Holoxenus . . .	1	—
Chilodactylus . .	9	4
**Nemadactylus . .	1	—
**Latris . . .	2	2
Scorpæna . . .	4	2
**Glyptauchen . .	1	—
Centropogon . .	2	—
*Agriopus . . .	1	1
*Aploactis . . .	1	—
**Pentaroge . . .	1	—
Platycephalus . .	5	—
Lepidotrigla . .	3	1
Trigla . . .	3	1

[1] Number of species uncertain.

	South Australia and Tasmania.	New Zealand.
**Anema . . .	—	1
**Crapatalus . . .	—	1
**Kathetostoma . .	1	2
**Leptoscopus . .	1	3
Percis . . .	2	1
*Aphritis . . .	1	—
Sillago . . .	2	—
*Bovichthys . . .	1	1
*Notothenia . . .	—	1
Sphyræna . . .	1	—
Lepidopus . . .	—	1
Trichiurus . . .	1	—
Thyrsites . . .	1	1
**Platystethus . .	—	2
Zeus (faber) . .	1	1
Cyttus . . .	1	1
Trachurus (trachurus) .	1	1
Caranx . . .	x	2
*Seriolella . . .	—	1
Pempheris . . .	1	—
Callionymus . .	3	—
Batrachus . . .	1	—
**Brachionichthys . .	2	—
**Saccarius . . .	—	1
Clinus . . .	1	1
**Lepidoblennius . .	1	—
Cristiceps and Tripterygium	4	5
**Patæcus . . .	3	—
**Acanthoclinus . .	—	1
**Diplocrepis . . .	—	1
**Crepidogaster . .	3	1
**Trachelochismus . .	—	1
**Neophrynichthys . .	—	1
Centriscus . . .	2	1
Notacanthus (sexspinis)	1	1
**Labrichthys . .	8	2
**Odax . . .	5	1

	South Australia and Tasmania.	New Zealand.
**Coridodax . . .	—	1
**Olistherops . . .	1	—
**Siphonognathus . .	1	—
Gadus . . .	—	1
Merluccius . . .	—	1
Lotella . . .	—	1
**Pseudophycis . .	—	1
Motella . . .	—	1
Bregmaceros . .	—	1
*Genypterus . . .	1	1
**Lophonectes . .	1	—
**Brachypleura . .	—	1
Pseudorhombus . .	—	1
**Ammotretis . .	1	1
**Rhombosolea . .	3	3
**Peltorhamphus . .	—	1
Solea . . .	1	—
Aulopus . . .	1	—
Gonorhynchus (greyi) .	1	1
Engraulis (encrasicholus) .	1	1
Clupea . . .	1	1
**Chilobranchus . .	1	—
Conger (vulgaris) . .	1	1
Ophichthys . .	1	1
Muræuichthys . .	1	—
Congromuræna . .	—	1
Syngnathus . .	5	2
Ichthyocampus . .	—	1
**Nannocampus . .	1	—
Urocampus . .	1	—
**Stigmatophora . .	2	1
Solenognathus . .	2	1
**Phyllopteryx . .	2	—
Monacanthus . .	15	1
Ostracion . . .	3	1
*Bdellostoma . .	—	1
Branchiostoma . .	1	1

3. The coast-line of the *Chilian district* extends over 20 degrees of latitude only, and is nearly straight. In its northern and warmer parts it is of a very uniform character, and exposed to high and irregular tides, and to remarkable and sudden changes of the levels of land and water, which must seriously interfere with fishes living and propagating near the shore. No river of considerable size interrupts the monotony of the physical conditions, to offer an additional element in favour of the development of littoral animals. In the southern parts, where the coast is lined with archipelagoes, the climate is too severe for the majority of fishes. All these conditions combine to render this district comparatively poor as regards variety of Shore fishes, as will be seen from the following list:—

*Callorhynchus; Scyllium, Acanthias, Spinax; Urolophus.

Serranus, Plectropoma, Polyprion, Pristipoma, Erythrichthys; *Haplodactylus; *Scorpis; Chilodactylus, **Mendosoma; Sebastes, *Agriopus; Trigla, Agonus; *Aphritis, *Eleginus, Pinguipes, Latilus, Notothenia (1 sp.) Umbrina; Thyrsites; Trachurus, Caranx, *Seriolella; Porichthys; **Myxodes, Clinus; Sicyases, Gobiesox.

Heliastes; **Malacopterus; *Labrichthys.

Merluccius; *Genypterus; Pseudorhombus.

Engraulis, Clupea; Ophichthys, Muræna.

Syngnathus.—*Bdellostoma.

Of these genera six only are not found in other districts of this zone. Three are peculiar to the Chilian district; *Porichthys* and *Agonus* have penetrated so far southwards from the Peruvian and Californian districts; and *Polyprion* is one of those extraordinary instances in which a very specialised form occurs at almost opposite points of the globe, without having left a trace of its previous existence in, or of its passage through, the intermediate space.

4. The *Patagonian district* is, with the exception of the neighbourhood of the mouth of the Rio de la Plata, almost unknown. In that estuary occur *Mustelus vulgaris,* two *Raja,* two *Trygon,* several Sciænoids, *Paropsis signata* and *Percophis brasilianus* (two fishes peculiar to this coast), *Prionotus punctatus, Læmonema longifilis* (a Gadoid), a *Pseudorhombus,* two Soles, *Engraulis olidus,* a *Syngnathus, Conger vulgaris,* and *Ophichthys ocellatus;* and if we notice the occurrence of a *Serranus* and *Caranx,* of *Aphritis* and *Pinguipes,* and of two or three *Clupea,* we shall have enumerated all that is known of this fauna. The fishes of the southern part, viz. the coast of Patagonia proper, southwards to Magelhæn's Straits, are unknown; which is the more to be regretted, as it is most probably the part in which the characteristic types of this district are most developed.

V.—Shore Fishes of the Antarctic Ocean.

To this fauna we refer the shore fishes of the southernmost extremity of South America, from 50° lat. S., with Terra del Fuego and the Falkland Islands, and those of Kerguelen's Land, with Prince Edward's Island. No fishes are known from the other oceanic islands of these latitudes.

In the Southern Hemisphere surface fishes do not extend so far towards the Pole as in the Northern; none are known from beyond 60° lat. S., and the Antarctic Fauna, which is analogous to the Arctic Fauna, inhabits coasts more than ten degrees nearer to the equator. It is very probable that the shores between 60° and the Antarctic circle are inhabited by fishes sufficiently numerous to supply part of the means of subsistence for the large Seals which pass there at least some season of the year, but hitherto none have been obtained by naturalists; all that the present state of our knowledge justifies us in saying is, that the general character of the Fauna of Magelhæn's Straits and Kerguelen's Land is extremely similar to that of Iceland and Greenland.

As in the arctic Fauna, Chondropterygians are scarce, and represented by *Acanthias vulgaris* and species of *Raja*. *Holocephali* have not yet been found so far south, but *Callorhynchus*, which is not uncommon near the northern boundary of this fauna, will prove to extend into it.

As to Acanthopterygians, *Cataphracti* and *Scorpænidæ* are represented as in the arctic Fauna, two of the genera (*Sebastes* and *Agonus*) being identical. The *Cottidæ* are replaced by six genera of *Trachinidæ*, remarkably similar in form to arctic types; but *Discoboli* and the characteristic Arctic Blennioids are absent.

Gadoid Fishes reappear, but are less developed; as usual they are accompanied by *Myxine*. The reappearance of so specialised a genus as *Lycodes* is most remarkable. Flatfishes are scarce as in the North, and belong to peculiar genera.

Physostomes are probably not entirely absent, but hitherto none have been met with so far south. Lophobranchs are scarce, as in the Arctic zone; however, it is noteworthy that a peculiar genus, with persistent embryonic characters (*Protocampus*), is rather common on the shores of the Falkland Islands.

The following are the genera known from this zone. Those with a single asterisk (*) are known to extend into the Temperate zone, but not beyond it; those with a double asterisk (**) are limited to the Antarctic shores:—

	Magelhæn's and Falkland.	Kerguelen.
Acanthias vulgaris . .	1	—
Raja . . .	1	2
Psammobatis . .	1	—
Sebastes . . .	1	—
**Zanclorhynchus . .	—	1

	Magelhæn's and Falkland.	Kerguelen.
*Agriopus . . .	1	—
Agonus . . .	1	—
*Aphritis . . .	1	—
*Eleginus . . .	1	—
**Chænichthys . .	1	1
*Bovichthys . . .	2	—
*Notothenia . . .	8	7
**Harpagifer . . .	1	1
Lycodes . . .	4	—
**Magnea . . .	1	—
Lotella . . .	1	—
Merluccius . . .	1	—
**Lepidopsetta . .	—	1
**Thysanopsetta . .	1	—
Syngnathus . .	1	—
**Protocampus . .	1	—
Myxine . . .	1	—
	31	13

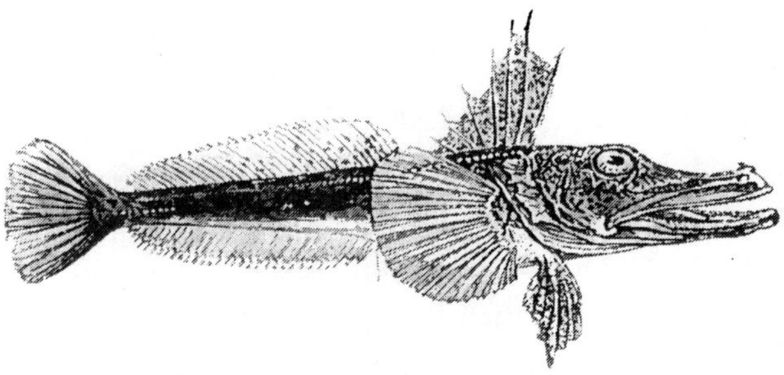

Fig. 108.—*Chænichthys rhinoceratus*, shores of the Antarctic Ocean.

CHAPTER XX.

DISTRIBUTION OF PELAGIC FISHES.

PELAGIC Fishes,—that is, fishes inhabiting the surface of mid-ocean (see p. 255), belong to various orders, viz. Chondropterygians, Acanthopterygians, Physostomes, Lophobranchs, and Plectognaths. But neither Anacanths nor Pharyngognaths contribute to this series of the Marine Fauna. The following genera and families are included in it:—

CHONDROPTERYGII: Carcharias, Galeocerdo, Thalassorhinus, Zygæna, Triænodon, Lamnidæ, Rhinodon, Notidanidæ, Læmargus, Euprotomicrus, Echinorhinus, Isistius; Myliobatidæ.

ACANTHOPTERYGII: Dactylopterus, Micropteryx, Scombrina, Gastrochisma, Nomeus, Centrolophus, Coryphænina, Seriola, Temnodon, Naucrates, Psenes, Xiphiidæ, Antennarius.

PHYSOSTOMI: Sternoptychidæ, Scopelus, Astronesthes, Scombresocidæ (majority).

LOPHOBRANCHII: Hippocampus.

PLECTOGNATHI: Orthagoriscus, and some other Gymnodonts.

Pelagic fishes differ much from one another in their mode of life. The majority are excellent swimmers, which not only can move with great rapidity, but also are possessed of great powers of endurance, and are thus enabled to continue their course for weeks, apparently without the necessity of rest: such are many Sharks, Scombroids, Dolphins, Pilot-fish, Sword-fishes. In some, as in *Dactylopterus* and *Exocoetus,* the ability of taking flying leaps out of the water is superadded to the power of swimming (Flying-fishes). But in others the power of swimming is greatly reduced, as in *Antennarius, Hippocampus*, and Gymnodonts; they frequent places in the ocean covered with floating seaweed, or drift on the surface without resistance, at the mercy of wind and current. The *Echeneis* or Sucking-fishes attach themselves to other large fish, ships, or floating objects, and allow themselves to be carried about, unless change of climate or want of food obliges them to abandon their temporary carrier. Finally, another class of Pelagic fishes come to the surface of the ocean during the night only; in the day time they descend to some depth, where they are undisturbed by the rays of the sun or the agitation of the surface-water: such are *Brama*, the *Sternoptychidæ, Scopelus, Astronesthes;* fishes, the majority of which are provided with those extraordinary luminary organs that we find so much developed in the true Deep-sea fishes. Indeed, this last kind of Pelagic fishes forms a passage to the Deep-sea forms.

Pelagic fishes, like shore fishes, are most numerous in the Tropical Zone; and, with few exceptions (*Echinorhinus, Psenes, Sternoptychidæ, Astronesthes*), the same genera are represented in the tropical Atlantic as well as in the Indo-Pacific. The number of identical species occurring in both these oceans is great, and probably still greater than would appear from systematic lists, in which there are retained many specific names that were given at a time when species were believed to have a very limited range. The Pelagic fauna of the tropics gradually passes into that of the temperate zones, only a few genera, like *Cybium, Psenes, Antennarius*, being almost entirely confined to the tropics. All the other tropical genera range into the temperate zones, but their representatives become scarcer with the increasing distance from the equator. North of 40° lat. N. many genera have disappeared, or are met with in isolated examples only, as *Carcharias, Zygæna, Notidanus, Myliobatidæ, Dactylopterus, Echeneis,*

Nomeus, Coryphæna, Schedophilus, Seriola, Temnodon, Antennarius, Sternoptychidæ, Astronesthes, Exocoetus, Tetrodon, Diodon; and only one genus of Sharks, *Galeocerdo*, approaches the Arctic circle. Some few species, like *Antennarius, Scopelus*, are carried by currents near to the northern confines of the temperate zones; but such occurrences are accidental, and these fishes must be regarded as entirely foreign to the fauna of those latitudes. On the other hand, some Pelagic fishes inhabit the temperate zones, whilst their occurrence within the tropics is very problematical; thus, in the Atlantic, *Thalassorhinus, Selache, Læmargus, Centrolophus, Diana, Ausonia, Lampris* (all genera composed of one or two species only). Beside the Shark mentioned, no other Pelagic fishes are known from the Arctic Ocean.

We possess very little information about the Pelagic fish-fauna of the Southern oceans. So much only is certain, that the tropical forms *gradually* disappear; but it would be hazardous, in the present state of our knowledge, to state even approximately, the limits of the southward range of a single genus. Scarcely more is known about the appearance of types peculiar to the Southern temperate zone; for instance, the gigantic Shark (*Rhinodon*), representing the Northern Selache, near the coasts of South Africa, and the Scombroid genus, *Gastrochisma*, in the South Pacific.

The largest of marine fishes, *Rhinodon, Selache, Carcharodon, Myliobatidæ, Thynnus, Xiphiidæ, Orthagoriscus*, belong to the Pelagic Fauna. Young fishes are frequently found in mid-ocean, which are the offspring of shore-fishes normally depositing their spawn near the coast. The manner, in which this fry passes into the open sea, is unknown; for it has not yet been ascertained whether it is carried by currents from the place where it was deposited originally, or whether shore-fishes sometimes spawn at a distance from the coast. We may remember that shore-fishes inhabit not only coasts but also submerged banks with some depth of water above, and that, by the action of the water, spawn deposited on these latter localities is very liable to be dispersed over wide areas of the ocean. Embryoes of at least some shore-fishes hatched under abnormal conditions seem to have an abnormal growth up to a certain period of their life, when they perish. The *Leptocephali* must be regarded as such abnormally developed fish (see p. 179). Fishes of a similar condition are the so-called Pelagic *Plagusiæ*, young Pleuronectoids, the origin of which is still unknown. As mentioned before, Flat-fishes, like all the other Anacanths, are otherwise not represented in the Pelagic fauna.

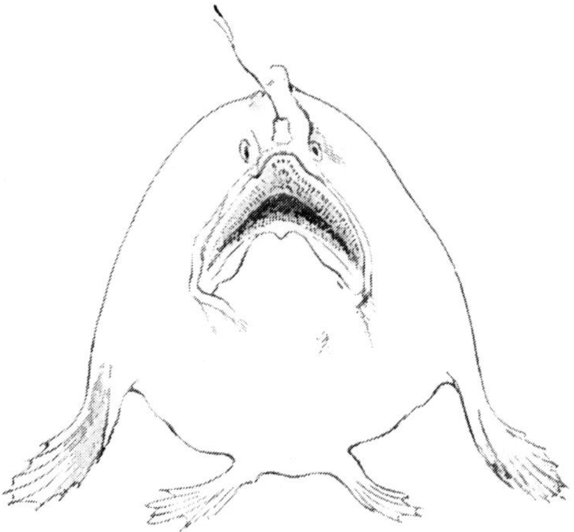

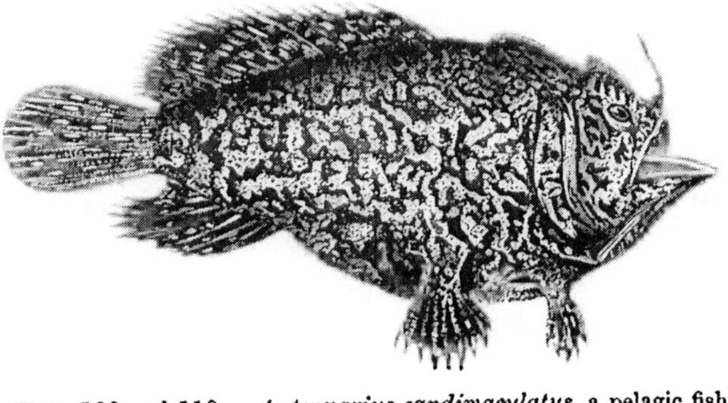

Figs. 109 and 110.—*Antennarius caudimaculatus*, a pelagic fish, from the Indian Ocean.

CHAPTER XXI.

THE FISHES OF THE DEEP SEA.

THE knowledge of the existence of deep-sea fishes is one of the recent discoveries of ichthyology. It is only about twenty years ago that, from the evidence afforded by the anatomical structure of a few singular fishes obtained in the North Atlantic, an opinion was expressed that these fishes inhabited great depths of the ocean, and that their organisation was specially adapted for living under the physical abyssal conditions. These fishes agreed in the character of their connective tissue, which was so extremely weak as to yield to, and to break under, the slightest pressure, so that the greatest difficulty is experienced to preserve their body in its continuity. Another singular circumstance was, that some of the specimens were picked up floating on the surface of the water, having met their deaths whilst engaged in swallowing or digesting another fish not much inferior or even superior in size to themselves.

The first peculiarity was accounted for by the fact that, if those fishes really inhabited the great depths supposed, their removal from the enormous pressure under which they lived would be accompanied by such an expansion of gases within their tissues as to rupture them, and to cause a separation of the parts which had been held together by the pressure. The second circumstance was explained thus:—A raptatorial fish organised to live at a depth of between 500 and 800 fathoms seizes another usually inhabiting a depth of between 300 and 500 fathoms. In its struggles to escape, the fish seized, nearly as large or strong as the attacking fish, carries the latter out of its depth into a higher stratum, where the diminished pressure causes such an expansion of gases as to make the destroyer with its victim rise with increasing rapidity towards the surface, which they reach dead or in a dying condition. Specimens in this condition are not rarely picked up; and as, of course, comparatively few can by accident fall into the hands of naturalists, occurrences of the kind related must happen very often.

Thus, the existence of fishes peculiarly adapted for the deep sea has been a fact maintained and admitted for some time in Ichthyology; and as the same genera and species were found at very distant parts of the ocean, it was further stated that those Deep-sea fishes were not limited in their range, and that, consequently, the physical conditions of the depths of the ocean must be the same or nearly the same over the whole globe. That Deep-sea fishes were not of a peculiar order, but chiefly modified forms of surface types, was another conclusion arrived at from the sporadic evidence collected during the period which preceded systematic deep-sea dredging.

However, nothing was positively known as to the exact depths inhabited by those fishes until observations were made during the voyage of H.M.S. "Challenger." The results obtained by this expedition afforded a surer and more extended basis for our knowledge of Deep-sea fishes.

The physical conditions of the deep sea, which must affect the organisation and distribution of fishes, are the following:—

1. Absence of sunlight. Probably the rays of the sun do not penetrate to, and certainly do not extend beyond, a depth of 200 fathoms, therefore we may consider this to be the depth where the Deep-sea fauna commences. Absence of light is, of necessity, accompanied by modifications of the organs of vision and by simplification of colours.

2. The absence of sunlight is in some measure compensated for by the presence of phosphorescent light, produced by many marine animals, and also by numerous Deep-sea fishes.

3. Depression and equality of the temperature. At a depth of 500 fathoms the temperature of the water is already as low as 40° Fahr., and perfectly independent of the temperature of the surface-water; and from the greatest depths upwards to about 1000 fathoms the temperature is uniformly but a few degrees above freezing-point. Temperature, therefore, ceases to offer an obstacle to the unlimited dispersal of Deep-sea fishes.

4. The increased pressure by the water. The pressure of the atmosphere on the level of the sea amounts to fifteen pounds per square inch of the surface of the body of an animal; but the pressure amounts to a ton weight for every 1000 fathoms of depth.

5. With the sunlight, vegetable life ceases in the depths of the sea. All Deep-sea fishes are therefore carnivorous; the most voracious feeding frequently on their own offspring, and the toothless kinds being nourished by the animalcules which live on the bottom, or which, "like a constant rain," settle down from the upper strata towards the bottom of the sea.

6. The perfect quiet of the water at great depths. The agitation of the water, caused by the disturbances of the air, does not extend beyond the depth of a few fathoms; below this surface-stratum there is no other movement except the quiet flow of ocean-currents, and near the bottom of the deep sea the water is probably in a state of almost entire quiescence.

The effect upon fishes of the physical conditions described is clearly testified by the modification of one or more parts of their organisation, so that every Deep-sea fish may be recognised as such, without the accompanying positive evidence that it has been caught at a great depth; and *vice versa*, fishes reputed to have been obtained at a great depth, and not having any of the characteristics of the dwellers of the deep sea, must be regarded as surface-fishes.

The most striking characteristic, found in many Deep-sea fishes, is in relation to the tremendous pressure under which they live. Their osseous and muscular systems are, as compared with the same parts of surface-fishes, very feebly developed. The bones have a fibrous, fissured, and cavernous texture; are light, with scarcely any calcareous matter, so that the point of a needle will readily penetrate them without breaking. The bones, especially the vertebræ, appear to be most loosely connected with one another; and it requires the most careful handling to prevent the breaking of the connective ligaments. The muscles, especially the great lateral muscles of the trunk and tail, are thin, the fascicles being readily separated from one another or torn, the connective tissue being extremely loose, feeble, or apparently absent. This peculiarity has been observed in the *Trachypteridæ*, *Plagyodus*, *Chiasmodus*, *Melanocetus*, *Saccopharynx*. But we cannot assume that it actually obtains whilst those fishes exist under their natural conditions. Some of them are most rapacious creatures which must be able to execute rapid and powerful movements to catch and overpower their prey; and for that object their muscular system, thin as its layers may be, must be as firm, and the chain of the segments of their vertebral column as firmly linked together as in surface-fishes. Therefore, it is evident that the change which the body of those fishes has undergone on their withdrawal from the pressure under which they live is a much aggravated form of the affection that is experienced by persons reaching great altitudes in their ascent of a mountain or in a balloon. In every living organism with an intestinal tract there are accumulations of free gases; and, moreover, the blood and other fluids, which permeate every part of the body, contain gases in solution. Under greatly diminished pressure these gases expand, so that, if the withdrawal from a depth is not an extremely slow and gradual process, the various tissues must be distended, loosened, ruptured; and what is a vigorous fish at a depth of 500 or more fathoms, appears at the surface as a loosely-jointed body which, if the skin is not of sufficient toughness, can only be kept together with difficulty. At great depths a fibrous osseous structure and a thin layer of muscles suffices to obtain the same results for which, at the surface, thickness of muscle and firm osseous or cartilaginous tissue are necessary.

The muciferous system of many Deep-sea fishes is developed in an extraordinary degree. We find already in fishes which are comparatively little removed from the surface (that is to depths of 100-200 fathoms), the lateral line much wider than in their congeners or nearest allies which live on the surface, as in *Trachichthys*, *Hoplostethus*, many *Scorpænidæ*. But in fishes inhabiting depths of 1000 and more fathoms, the whole muciferous system is dilated; it is especially the surface of the skull which is occupied by large cavities (*Macruridæ*, deep-sea *Ophidiidæ*), and the whole body seems to be covered with a layer of mucus. These cavities collapse and shrink in specimens which have been preserved in spirit for some time, but a re-immersion in water for a short time generally suffices to show the immense quantity of mucus secreted by them. The physiological use of this secretion is unknown; it has been observed to have phosphorescent properties in perfectly fresh specimens.

The colours of Deep-sea fishes are extremely simple, their bodies being either black or silvery; in a few only are some filaments or the fin-rays of a bright scarlet colour. Among the black forms albinoes are not scarce.

The organ of sight is the first to be affected by a sojourn in deep water. Even in fishes which habitually live at a depth of only 80 fathoms, we find the eye of a proportionally larger size than in their representatives at the surface. In such fishes the eyes increase in size with the depth inhabited by them, down to the depth of 200 fathoms, the large eyes being necessary to collect as many rays of light as possible. Beyond that depth small-eyed fishes as well as large-eyed occur, the former having their want of vision compensated for by tentacular organs of touch, whilst the latter have no such accessory organs, and evidently see only by the aid of phosphorescence. In the greatest depths blind fishes occur with rudimentary eyes and without special organs of touch.

Many fishes of the deep sea are provided with more or less numerous, round, shining, mother-of-pearl-coloured bodies, imbedded in the skin. These so-called phosphorescent or luminous organs are either larger bodies of an oval or irregularly elliptical shape placed on the head, in the vicinity of the eye, or smaller round globular bodies arranged symmetrically in series along the side of the body and tail, especially near the abdominal profile, less frequently along the back. The former have not yet been anatomically examined. The number of pairs of the latter is in direct relation to that of the segments of the vertebral column, the muscular system, etc. (metameres); and two kinds may be distinguished differing from each other in their anatomical structure. The organs of one kind consist of an anterior, biconvex, lens-like body, which is transparent during life, simple or composed of rods (*Chauliodus*); and of a posterior chamber which is filled with a transparent fluid, and coated with a dark membrane composed of hexagonal cells, or of rods arranged as in a retina. This structure is found in *Astronesthes*, *Stomias*, *Chauliodus*, etc. In the other kind the organ shows throughout a simply glandular structure, but apparently without an efferent duct (*Gonostoma*, *Scopelus*, *Maurolicus*, *Argyropelecus*). Branches of the spinal nerves run to each organ, and are distributed over the retina-like membrane or the glandular follicles. The former kind of organs are considered by some naturalists true organs of vision (accessory eyes), the function of the latter being left unexplained by them.

Although, thus, these organs morphologically differ from each other, there is no doubt that the functions of all have some relation to the peculiar conditions of light under which the fishes provided with them live; these fishes being either deep-sea forms or nocturnal pelagic kinds. There are three possible hypotheses as to the function of these organs :—

1. All the different kinds of organs are sensory, or, in other words, accessory eyes.

2. Only the organs with a lenticular body are sensory, and those with a glandular structure produce and emit phosphorescent light.

3. All are producers of light.

There are very serious objections to adopting the first view. *Scopelus* and *Argyropelecus* possess not only perfectly developed, but even large eyes, specially adapted for a nocturnal life; and therefore accessory organs of vision must appear to be quite superfluous to them. On the other hand, in Deep-sea fishes without external eyes, which would seem to especially require these metameric organs of sense, they are invariably absent. And, finally, it is quite inconceivable that the glandular structures should have the faculty of conveying impressions of light to the nervous centre. The second supposition seems therefore to be nearer the truth; and is supported by the fact that the glandular organs of Scopeli have actually been observed to gleam with phosphorescent light, and by the obvious morphological similarity of the organs with a lenticular body and retina-like membrane to an organ of vision. We are, moreover, justified, from an *à priori* consideration, in supposing that in depths to which no sunlight descends, and which are illuminated by phosphorescent light only, peculiar organs of vision would have been developed. On the other hand, this supposition is opposed by the fact that many fishes which dwell in those abyssal depths are provided with large ordinary eyes (as the *Trachypteri*, the majority of *Macruridæ*), and, therefore, that the ordinary organ of vision is quite sufficient for seeing by phosphorescent light. Thus, whilst we must admit that those compound organs may prove to be organs of sense, we maintain at the same time that their morphological nature is not opposed to the belief that they too, like the glandular organs, are producers of light. It may be produced at the bottom of the posterior chamber, and emitted through the lenticular body in particular directions, with the same effect as light is sent through the convex glass of a "bull's eye." This hypothesis seems to be less bold than the other, which would require us to believe that vertebrate animals, with a nervous centre specialised for the reception of the impressions of the higher senses, should receive them through the spinal chord.

[See *Ussow*, "Ueber den Bau der sogenannten augenaehnlichen Flecken einiger Knochenfische." St. Petersburg, Bullet. 1879.]

Whenever we find in a fish long delicate filaments, developed in connection with the fins or the extremity of the tail, we may conclude that it is an inhabitant of still water and of quiet habits. Many deep-sea fishes (*Trachypteridæ*, *Macruridæ*, *Ophidiidæ*, *Bathypterois*) are provided with such filamentous prolongations, the development of which is perfectly in accordance with their sojourn in the absolutely quiet waters of abyssal depths.

Some of the raptatorial Deep-sea fishes have a stomach so distensible and capacious that it can receive a fish of twice or thrice the bulk of the destroyer (*Melanocetus*, *Chiasmodus*, *Saccopharynx*). Deglutition is performed in them not by means of the muscles of the pharynx, as in other fishes, but by the independent and alternate action of the jaws, as in Snakes. These fishes cannot be said to swallow their food, but rather draw themselves over their victim, in the fashion of an *Actinia.*

Before the voyage of H.M.S. "Challenger," scarcely thirty Deep-sea fishes were known. This number is now much increased by the discovery of many new species and genera; but, singularly, no new types of families were discovered: nothing but what might have been expected from our previous knowledge of this group of fishes. Modifications of certain organs, perfectly novel, and of the greatest interest, were found, as we shall see in the "Systematic Part;" but the most important results of this voyage are that the general character of the abyssal fish-fauna, the abundance of fishes, and the exact depths to which fishes may descend, have been ascertained.

However, the statements of the depths at which the fishes collected by the "Challenger" were taken cannot be received without some critical examination of each individual species. No precaution was taken to keep the mouth of the dredge closed during its descent or ascent, and therefore it is quite within the limits of probability that sometimes fishes were accidentally enclosed within the dredge, whilst it was traversing the surface strata. And this has happened more than once; for it is quite certain that common surface fishes like *Sternoptyx* and *Astronesthes*, never ranged to a depth of 2500 fathoms. On the other hand, the majority of the fishes obtained offer sufficient evidence from their own organisation that they live on the bottom, and are unable to support themselves in the water at a certain distance from the bottom or surface; and, consequently, that they actually were obtained at the depth to which the dredge descended.

As far as the observations go at present, no distinct bathymetrical regions, which would be characterised by peculiar forms, can be defined. The depths from 200 to

600 fathoms are inhabited by numerous forms, still strongly reminding us of surface types. To this fauna belong the few Chondropterygians of the deep sea, a Sebastes and Setarches, a Beryx and Polymixia, a Cottus, etc.; but they are associated with many others which descend to the greatest depths. And before anything like a division into bathymetrical zones can be attempted, the observations of the "Challenger" expedition must be confirmed and supplemented by other series of similar systematic observations. One of the most startling conclusions at which we would have to arrive from the "Challenger" observations is, that some of the species of Deep-sea fishes would range from a depth of some 300 fathoms down to one of 2000 fathoms; or, in other words, that a fish which has once attained in its organisation to that modification by which it is enabled to exist under the pressure of half a ton, can easily accommodate itself to one of two tons or more,—a conclusion which is not in accordance with anatomical facts, and which must be confirmed by other observations before we can adopt it. But if the vertical range of Deep-sea fishes is actually as it appears from the "Challenger" lists, then there is no more distinct vertical than horizontal distribution of Deep-sea fishes.

The greatest depth reached hitherto by a dredge in which fishes were enclosed is 2900 fathoms. But the specimens thus obtained belong to a species (*Gonostoma microdon*), which seems to be extremely abundant in upper strata of the Atlantic and Pacific, and were therefore most likely caught by the dredge in its ascent. The next greatest depth, viz., 2750 fathoms, must be accepted as one at which fishes undoubtedly do live; the fish obtained from this depth of the Atlantic, *Bathyophis ferox,* showing by its whole habit that it is a form living on the bottom of the ocean.

The fish-fauna of the deep sea is composed chiefly of forms or modifications of forms which we find represented at the surface in the cold and temperate zones, or which appear as nocturnal pelagic forms. The Chondropterygians are few in number, not descending to a depth of more than 600 fathoms. The Acanthopterygians, which form the majority of the coast and surface faunas, are also scantily represented; genera identical with surface types are confined to the same inconsiderable depths as the Chondropterygians, whilst those Acanthopterygians which are so much specialised for the life in the deep sea as to deserve generic separation, range from 200 to 2400 fathoms. Three distinct families of Acanthopterygians belong to the deep-sea fauna, viz. *Trachypteridæ*, *Lophotidæ*, and *Notacanthidæ;* they respectively consist of three, one, or two genera only.

Gadidæ, *Ophidiidæ*, and *Macruridæ* are very numerous, ranging through all depths; they constitute about one-fourth of the whole deep-sea fauna.

Of *Physostomi*, the families of *Sternoptychidæ*, *Scopelidæ*, *Stomiatidæ*, *Salmonidæ*, *Bathythrissidæ*, *Alepocephalidæ*, *Halosauridæ*, and *Murænidæ* are represented. Of these the *Scopeloids* are the most numerous, constituting nearly another fourth of the fauna. *Salmonidæ* are scarce, with three small genera only. *Bathythrissidæ* include one species only, which is probably confined in its vertical as well as horizontal range; it occurs at a depth of about 350 fathoms in the sea of Japan. The *Alepocephalidæ* and *Halosauridæ*, known before the "Challenger" expedition from isolated examples only, prove to be true, widely-spread, deep-sea types. Eels are well represented, and seem to descend to the greatest depths.

Myxine has been obtained from a depth of 345 fathoms.

It will be useful to append a complete list of Deep-sea fishes, with the depths as ascertained by the dredgings of the "Challenger:"—

List of Deep-sea Fishes.

	Fathoms.
CHONDROPTERYGIANS—	
Raja	565
Scyllium	400
Centroscyllium . . .	245
Centrophorus . . .	345-500
ACANTHOPTERYGIANS—	
Pomatomus . . .	(? down to) 200
Sebastes	275
Setarches	215
Beryx	345
Melamphaes. . .	(? beyond) 200
Polymixia	345
Nealotus	
Nesiarchus	
Aphanopus	
Enoxymetopon . . .	
Lepidopus	345
Gempylus	
Anomalops	
? Antigonia	
Diretmus	
Cottus	565
Bathydraco	1260
Oncirodes	
Melanocetus johnsonii . .	1850
,, bispinosus . .	360
Himantolophus . . .	
Chaunax	215
Ceratias	2400
Halieutichthys . . .	
Dibranchus	360
Trachypteridæ . . .	
Lophotes	
Notacanthus rissoanus . .	1875
,, bonapartii . .	400

ANACANTHINI—	Fathoms.
Melanonus	1975
Halargyreus	
Lotella marginata . . .	120-345
Physiculus	345
Uraleptus	
Læmonema	
Haloporphyrus australis . .	55-70
„ lepidion . .	345-600
„ rostratus . .	600-1375
Chiasmodus niger . . .	1500
Sirembo grandis . . .	1875
„ macrops . . .	375
„ messieri . . .	345
„ ocellatus . . .	350
„ brachysoma . . .	350
Acanthonus armatus . . .	1075
Typhlonus nasus . . .	2440
Aphyonus gelatinosus . .	1400
Rhinonus ater . . .	2150
Bathynectes laticeps . . .	2500
„ compressus . .	1075-2500
„ gracilis . . .	1400
Pteridium	
Macrurus (12 species) . .	120-700
Coryphænoides norvegicus . .	
„ serratus . .	
„ nasutus . .	345-565
„ villosus . .	345
„ rudis . .	500-650
„ æqualis . .	600
„ crassiceps . .	520-650
„ microlepis . .	215
„ murrayi . .	1100
„ serrulatus . .	700
„ filicauda . .	1800-2650
„ variabilis . .	1375-2425
„ affinis . .	1900

	Fathoms.
Coryphænoides carinatus	500
„ longifilis	565
„ altipinnis	565-1875
„ asper	500-1875
„ leptolepis	350-2050
„ sclerorhynchus	1090
„ denticulatus	275-520
Malacocephalus	350
Bathygadus cottoides	520-700
„ multifilis	500
STERNOPTYCHIDÆ—	
Argyropelecus	1127 [?]
Sternoptyx	0-2500 [?]
Polyipnus	255
Gonostoma denudatum	
„ microdon	500-2900 [?]
„ elongatum	360-800
„ gracile	345-2425
Chauliodus	565-2560
SCOPELIDÆ—	
Bathysaurus ferox	1100
„ mollis	1875-2385
Bathypterois longifilis	520-630
„ longipes	2650
„ quadrifilis	500-770
„ longicauda	2550
Chlorophthalmus agassizii	215
„ nigripinnis	120
„ gracilis	1100-1425
Scopelus engraulis	255
„ antarcticus	1950
„ mizolepis	800
„ dumerilii	215
„ macrolepidotus	520-630
„ crassiceps	675-1550
„ macrostoma	2350-2425
„ microps	1375
Odontostomus hyalinus	

	Fathoms.
Odontostomus humeralis	500
Nannobrachium nigrum	500
Ipnops murrayi	1600-2150
Paralepis	
Sudis	
Plagyodus	
STOMIATIDÆ—	
Astronesthes niger	2500 [?]
Stomias boa	450-1800
„ barbatus	
„ ferox	
Echiostoma barbatum	
„ micripnus	2150
„ microdon	2440
Malacosteus niger	
„ indicus	500
Bathyophis ferox	2750
SALMONIDÆ—	
Argentina	
Microstoma	
Bathylagus antarcticus	1950
„ atlanticus	2040
BATHYTHRISSIDÆ—	
Bathythrissa dorsalis	345
ALEPOCEPHALIDÆ—	
Alepocephalus rostratus	
„ niger	1400
Platytroctes apus	1500
Bathytroctes microlepis	1090
„ rostratus	675
„ macrolepis	2150
Xenodermichthys	345
HALOSAURIDÆ—	
Halosaurus owenii	
„ affinis	565
„ macrochir	1090-1375
„ mediorostris	700
„ rostratus	2750

MURÆNIDÆ—	Fathoms.
Nemichthys scolopacea . .	
„ infans . . .	500-2500
Cyema atrum . . .	1500-1800
Saccopharynx . . .	
Synaphobranchus pinatus . .	345-1200
„ bathybius . .	1875-2050
„ brevidorsalis .	1075-1375
„ affinis . .	345
Nettastoma parviceps . .	345
CYCLOSTOMATA—	
Myxine australis . . .	345

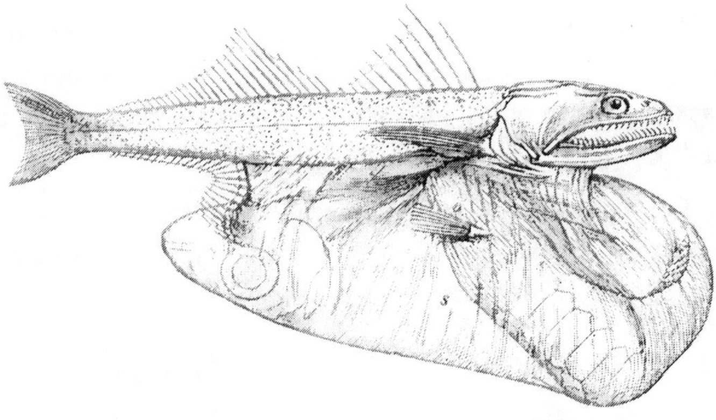

Fig. 111.—*Chiasmodus niger;* obtained in the North Atlantic at a depth of 1500 fathoms; the specimen has swallowed a large Scopelus (*s*); *o*, ventral fin.

SYSTEMATIC AND DESCRIPTIVE PART.

THE Class of Fishes is divided into four sub-classes :—

I. PALÆICHTHYES.—Heart with a contractile conus arteriosus; intestine with a spiral valve; optic nerves non-decussating, or only partially decussating.

II. TELEOSTEI.—Heart with a non-contractile bulbus arteriosus; intestine without spiral valve; optic nerves decussating. Skeleton ossified, with completely separated vertebræ.

III. CYCLOSTOMATA.—Heart without bulbus arteriosus; intestine simple. Skeleton cartilaginous and notochordal. One nasal aperture only. No jaws; mouth surrounded by a circular lip.

IV. LEPTOCARDII.—Heart replaced by pulsating sinuses; intestine simple. Skeleton membrano-cartilaginous and notochordal. No skull; no brain.

FIRST SUB-CLASS: PALÆICHTHYES.

Heart with a contractile conus arteriosus;[1] *intestine with a spiral valve;*[2] *optic nerves non-decussating, or only partially decussating;*[3] *skeleton cartilaginous or osseous.*

This sub-class comprises the Sharks and Rays, and the Ganoid fishes. Although based upon a singular concurrence of most important characters, its members exhibit as great a diversity of form, and as manifold modifications in the re-

[1] See p. 151, Fig. 67. [2] See p. 128, Fig. 55. [3] See p. 104.

mainder of their organisation as the *Teleostei*. The *Palæichthyes* stand to the *Teleostei* in the same relation as the Marsupials to the Placentalia. Geologically, as a sub-class, they were the predecessors of Teleosteous fishes; and it is a remarkable fact that all those modifications which show an approach of the ichthyic type to the Batrachians are found in this sub-class. We divide it into two orders: *Chondropterygii* and *Ganoidei*.

FIRST ORDER: CHONDROPTERYGII.

Skeleton cartilaginous. Body with medial and paired fins, the hinder pair abdominal. Vertebral column generally heterocercal, the upper lobe of the caudal fin produced. Gills attached to the skin by the outer margin, with several intervening gill-openings: rarely one external gill-opening only. No gill-cover. No air-bladder. Two, three, or more series of valves in the conus arteriosus. Ova large and few in number,[1] *impregnated and, in some species, developed within a uterine cavity. Embryo with deciduous external gills.*[2] *Males with intromittent organs attached to the ventral fins.*[3]

This order, for which, also, the name *Elasmobranchii* has been proposed (by Bonaparte), comprises the Sharks and Rays and Chimæras, and is divided into two sub-orders: *Plagiostomata* and *Holocephala*.

FIRST SUB-ORDER: PLAGIOSTOMATA.

From five to seven gill-openings. Skull with a suspensorium and the palatal apparatus detached. Teeth numerous.

The Plagiostomes differ greatly among each other with regard to the general form of their body: in the Sharks or *Selachoidei* the body is elongate, more or less cylindrical, gradually passing into the tail; their gill-openings are lateral. In the Rays, or *Batoidei*, the gill-openings are always placed

[1] See p. 167, Figs. 79, 81. [2] See p. 136, Fig. 58. [3] See p. 167, Fig. 78.

on the abdominal aspect of the fish; the body is depressed, and the trunk, which is surrounded by the immensely developed pectoral fins, forms a broad flat disk, of which the tail appears as a thin and slender appendage. Spiracles are always present; the number of gill-openings is constantly five; no anal fin; dorsal fins, if present, situated on the tail. However, some of the Rays approach the Sharks in having the caudal portion less abruptly contracted behind the trunk.

Fossil Plagiostomes are very numerous in all formations. Some of the earliest determinable fish remains are believed to be, or are, derived from Plagiostomes. Those which can be referred to any of the following families will be mentioned subsequently: but there are others, especially fin-spines, which leave us in doubt to which group of Plagiostomes their owners had any affinity, thus *Onchus* from the upper Silurian, continuing to carboniferous formations; *Dimeracanthus, Homocanthus*, from the Devonian; *Oracanthus, Gyracanthus, Tristychius, Astroptychius, Ptychacanthus, Sphenacanthus*, etc., from carboniferous formations; *Leptacanthus*, from the coal to the Oolite; *Cladacanthus, Cricacanthus, Gyropristis*, and *Lepracanthus*, from the coal measures; *Nemacanthus, Liacanthus*, from the Trias; *Astracanthus, Myriacanthus, Pristacanthus*, from the jurassic group.

A. Selachoidei: Sharks.

The elongate cylindrical body, generally terminating in a more or less pointed snout, and passing into a powerful and flexible tail, blade-like at its extremity, gives to the Sharks a most extraordinary power of swimming, with regard to endurance as well as rapidity of motion. Many, especially the larger kinds, inhabit the open ocean, following ships for weeks, or pursuing shoals of fishes in their periodical migrations. Other large-sized sharks frequent such parts of the coast as offer them abundance of food; whilst the majority of the smaller kinds are shore fishes, rarely leaving the bottom, and sometimes congregating in immense numbers. The movements of sharks resemble in some measure those of snakes, their flexible body being bent in more than one curve when moving.

Sharks are most numerous in the seas between the Tropics, and become scarcer beyond, a few only reaching the Arctic circle; it is not known how far they advance southwards towards the Antarctic region. Some species enter fresh waters, and ascend large rivers, like the Tigris or Ganges, to a considerable distance. The pelagic as well as the shore species have a wide geographical range. Very few descend to a considerable depth, probably not exceeding 500 fathoms. There are about 140 different species known.

Sharks have no scales like those of other fishes; their integuments are covered with calcified papillæ which, under the microscope, show a structure similar to that of teeth. If the papillæ are small, pointed, and close set, the skin is called "shagreen;" rarely they are larger, appearing as bucklers or spines, of various sizes.

These fishes are exclusively carnivorous, and those armed with powerful cutting teeth are the most formidable tyrants of the ocean. They have been known to divide the body of a man in two at one bite, as if by the sweep of a sword. Some of the largest sharks, however, which are provided with very small teeth, are almost harmless, feeding on small fishes only or marine invertebrates. Others, particularly of the smaller kinds, commonly called "Dog-fishes," have short or obtuse teeth, and feed on shells or any other animal substance. Sharks scent their food from a distance, being readily attracted by the smell of blood or decomposing bodies.

In China and Japan, and many other eastern countries, the smaller kinds of sharks are eaten. Sharks' fins form in

India and China a very important article of trade, the Chinese preparing from them gelatine, and using the better sorts for culinary purposes. The fins are obtained not exclusively from Sharks but also from Rays, and assorted in two kinds, viz. "white and black." The white consist exclusively of the dorsal fins, which are on both sides of the same uniform light colour, and reputed to yield more gelatine than the other fins. The pectoral, ventral, and anal fins pass under the denomination of black fins.; the caudal fin is not used. One of the principal places where shark fishery is practised as a profession is Kurrachee. Dr. Buist, writing in 1850 ("Proc. Zool. Soc." 1850, p. 100), states that there are thirteen large boats, with crews of twelve men each, constantly employed in this pursuit; that the value of the fins sent to the market varies from 15,000 to 18,000 rupees; that one boat will sometimes capture at a draught as many as one hundred sharks of various sizes; and that the number total of sharks captured during the year amounts probably to not less than 40,000. Large quantities are imported from the African coast and the Arabian Gulf, and various ports on the coast of India. In the year 1845-46, 8770 cwt. of sharks' fins were exported from Bombay to China.

First Family—Carchariidæ.

Eye with a nictitating membranc. Mouth cresecent-shaped, inferior. Anal fin present. Two dorsal fins, the first opposite to the spaec between pectoral and ventral fins, without spine in front.

Carcharias.—Snout produced in the longitudinal axis of the body; mouth armed with a series of large flat triangular teeth, which have a smooth cutting or serrated edge. Spiracles absent. A transverse pit on the back of the tail, at the root of the caudal fin.

This genus comprises the true Sharks, common in the

tropical, but less so in the temperate seas. Between thirty and forty different species have been distinguished, of which

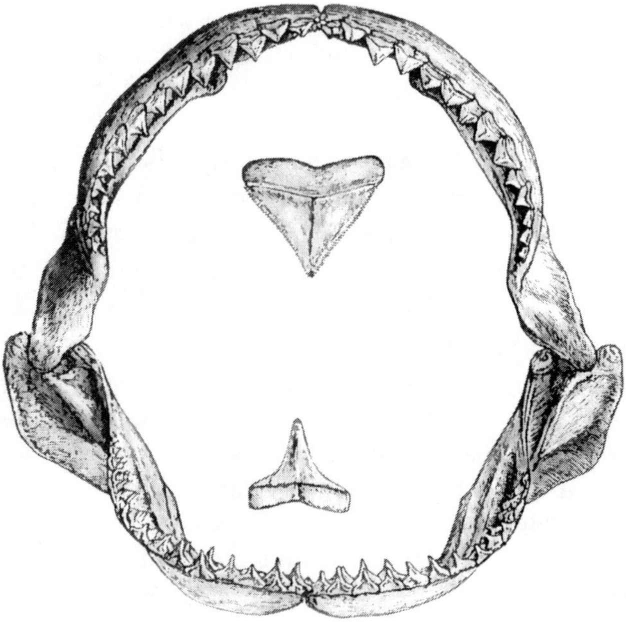

Fig. 112.—Dentition of the Blue Shark (Carcharias glaucus); the single teeth are of the natural size.

one of the most common is the "Blue Shark" (*Carcharias glaucus*). Individuals of from twelve to fifteen feet are of very common occurrence, but some of the species attain a much larger size, and a length of 25 and more feet. Fishes of this genus or of closely allied genera (*Corax, Hemipristis*) are not uncommon in the chalk and tertiary formations.

Galeocerdo.—Teeth large, flat, triangular, oblique, serrated on both edges, with a deep notch on the outer margin. Spiracles small. A pit on the tail, above and below, at the root of the caudal fin. Two notches on the under caudal border, one of them at the end of the spine.

Three species, of which one (*G. arcticus*) is confined to

the arctic and sub-arctic oceans. The others inhabit temperate and tropical seas, and all attain to a very large size.

Galeus.—Snout produced in the longitudinal axis of the body; teeth equal in both jaws, rather small, flat, triangular, oblique, serrated and with a notch. Spiracles small. No pit at the commencement of the caudal fin, which has a single notch on its lower margin.

These are small sharks, commonly called "Tope." The species found on the British coast is spread over nearly all the temperate and tropical seas, and is common in California and Tasmania. It lives on the bottom, and is very troublesome to fishermen by constantly taking away bait or driving away the fishes which they desire to catch.

Zygæna.—The anterior part of the head is broad, flattened, and produced into a lobe on each side, the extremity of which is occupied by the eye. Caudal fin with a single notch at its lower margin. A pit at the root of the caudal fin. Spiracles none. Nostrils situated on the front edge of the head.

The "Hammerheads," or Hammerheaded Sharks, have a dentition very similar to that of *Carcharias*, and although they do not attain to the same large size, they belong to the most formidable fishes of the ocean. The peculiar form of their head is quite unique among fishes; young examples have the lateral extension of the skull much less developed than adults. Five species are known, which are most abundant in the tropics. By far the most common is *Zygæna malleus*, which occurs in nearly all tropical and subtropical seas. Specimens of this species may be often seen ascending from the clear blue depths of the ocean like a great cloud. Cantor found in a female, nearly 11 feet long, thirty-seven embryons.—Hammerheads have lived from the cretaceous epoch.

Mustelus.—The second dorsal fin is not much smaller than the first. No pit at the root of the caudal, which is without distinct lower lobe. Snout produced in the longitudinal axis of the body. Spiracles small, behind the eyes. Teeth small, numerous, similar in both jaws, obtuse, or with very indistinct cusps, arranged like pavement.

The "Hounds" are small Sharks, abundant on the coasts of all the temperate and tropical seas; two of the five species known occur on the coasts of Europe, viz. *M. lævis* and *M. vulgaris*. Closely allied as these two species are, they yet show a most singular difference, viz. that a placenta is developed in the uterus for the attachment of the embryo in *M. lævis* (the Γαλεὸς λεῖος of Aristotle, to whom this fact was already known); whilst the embryons of *M. vulgaris* are developed without such placenta (see *J. Müller*, "Abhandl. Ak. Wiss." Berl. 1840). The Hounds are bottom fish, which feed principally on shells, crustaceans, and decomposing animal substances.

Several other genera belong to the family *Carchariidæ*, but it will be sufficient to mention their names:—*Hemigaleus*, *Loxodon*, *Thalassorhinus*, *Triænodon*, *Leptocarcharias*, and *Triacis*.

Second Family—Lamnidæ.

Eye without nictitating membrane. Anal fin present. Two dorsal fins; the first opposite to the space between pectoral and ventral fins, without spine in front. Nostrils not confluent with the mouth which is inferior. Spiracles absent or minute.

All the fishes of this family attain to a very large size, and are pelagic. But little is known of their reproduction. The first appearance of this family is indicated by *Carcharopsis*, a genus from carboniferous formations, the teeth of which differ from those of *Carcharodon* only by having a broad fold at the base. In the chalk and tertiary formations almost all the existing genera are represented; and, besides, *Oxytes*, *Sphenodus*, *Gomphodus*, and *Ancistrodon*, which are known from teeth only, have been considered generically distinct from the living Porbeagles.

LAMNA (OXYRHINA).—The second dorsal and anal are very small. A pit at the root of the caudal, which has the lower lobe much developed. Side of the tail with a prominent longitudinal keel. Mouth wide. Teeth large, lanceolate, not serrated, sometimes with additional basal cusps. On each side of the upper jaw, at some distance from the symphysis, there is one or two teeth conspicuously smaller than the others. Gill-openings very wide. Spiracles minute.

Fig. 113.—Upper and lower tooth of Lamna.

Of the "Porbeagles," three species have been described, of which the one occurring in the North Atlantic, and frequently straying to the British coasts (*L. cornubica*), is best known. It attains to a length of ten feet, and feeds chiefly on fishes; its lanceolate teeth are not adapted for cutting, but rather for seizing and holding its prey, which it appears to swallow whole. According to Pennant it is viviparous; only two embryoes were found in the female which came under his observation. Haast has found this species also off the coast of New Zealand.

CARCHARODON.—The second dorsal and anal are very small. Pit at the root of the caudal, which has the lower lobe well developed. Side of the tail with a prominent longitudinal keel. Mouth wide. Teeth large, flat, erect, regularly triangular, serrated. On each side of the upper jaw, at some distance from the symphysis, there is one or two teeth conspicuously smaller than the others. Gill-openings wide.

One species only is known (*C. rondeletii*), which is the most formidable of all Sharks. It is strictly pelagic; and appears to occur in all tropical and sub-tropical seas. It is known to attain to a length of 40 feet. The tooth figured here, of the natural size, is taken from a jaw 20 inches wide in its transverse diameter (inside measure), each half of the mandible measuring 22 inches.[1] The whole length of the fish was 36½ feet.

Carcharodon teeth are of very common occurrence in various tertiary strata, and have been referred to several species, affording ample evidence that this type was much more numerously represented in that geological epoch than in the recent fauna. Some individuals attained to an immense size, as we may judge from teeth found in the Crag, which are 4 inches wide at the base, and 5 inches long, measured along their lateral margin. The naturalists of the "Challenger" expedition have made the highly interesting discovery that teeth of similar size are of common occurrence in the ooze of the Pacific, between Polynesia and the west coast of America. As we have no record of living individuals of that bulk having been observed, the gigantic species to which these teeth belonged must have become extinct within a comparatively recent period. Nothing is known of the anatomy, habits, and reproduction of the surviving species, and no opportunity should be lost of obtaining information on this Shark.

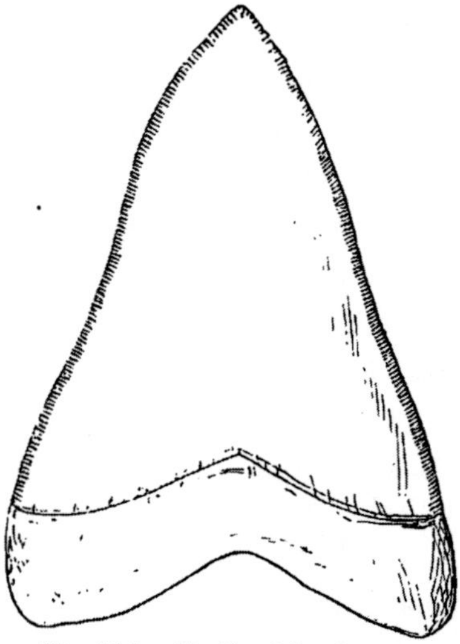

Fig. 114.—Tooth of Carcharodon rondeletii.

ODONTASPIS.—The second dorsal and anal are not much smaller than the first dorsal. No pit at the root of the caudal. Side of the tail without keel. Mouth wide. Teeth large, awl-shaped, with one or two small cusps at the base. Gill-openings of moderate width.

Large Sharks from tropical and temperate seas; two species.

1 The cartilaginous jaws of Sharks shrink at least a third in drying, and, therefore, cannot be kept at full stretch without tearing.

Alopecias.—The second dorsal and anal very small. Caudal fin of extraordinary length, with a pit at its root. No keel on the side of the tail. Mouth and gill-openings of moderate width. Teeth equal in both jaws, of moderate size, flat, triangular, not serrated.

This genus consists of one species only, which is known by the name of "Fox-shark" or "Thresher." It is the most common of the larger kinds of Sharks which occur on the British coasts; and seems to be equally common in other parts of the Atlantic and Mediterranean, as well as on the coasts of California and New Zealand. It attains to a length of fifteen feet, of which the tail takes more than one half; and is quite harmless to man. It follows the shoals of Herrings, Pilchards, and Sprats in their migrations, destroying incredible numbers. When feeding it uses the long tail in splashing the surface of the water, whilst it swims in gradually decreasing circles round a shoal of fishes, which are thus kept crowded together, falling an easy prey to their enemy. Statements that it has been seen to attack Whales and other large Cetaceans, rest upon erroneous observations.

Selache.—The second dorsal and anal very small. A pit at the root of the caudal fin, which is provided with a lower lobe. Side of the tail with a keel. Gill-openings extremely wide. Teeth very small, numerous, conical, without serrature or lateral cusps.

Also this genus consists of one species only, the "Basking Shark" (Pélerin of the French). It is the largest Shark of the North Atlantic, growing to a length of more than thirty feet. It is quite harmless if not attacked; its food consisting of small fishes, and other small marine animals swimming in shoals. On the west coast of Ireland it is chased for the sake of the oil which is extracted from the liver, one fish yielding from a ton to a ton and a-half. Its capture is not unattended with danger, as one blow from the enormously strong tail is sufficient to stave in the sides of a large boat. At certain seasons it is gregarious, and many specimens may be seen in calm weather lying together motionless, with the upper part of the back raised above the surface of the water; a habit from which this Shark has derived its name. The buccal and branchial cavities are of extraordinary width, and, in consequence of the flabby condition of those parts, the head presents a variable and singular appearance in specimens lying dead on the ground. This peculiarity, as well as the circumstance that young specimens have a much longer and more pointed snout than adult ones, has led to the erroneous opinion that several different genera and species of Basking Shark occur in the European seas. The branchial arches of *Selache* are provided with a very broad fringe of long (five to six inches) and thin gill-rakers, possessing the same microscopical structure as the teeth and dermal productions of Sharks. Similar gill-rakers have been found in a fossil state in the Crag of Anvers in Belgium, proving the existence of this Selachian type in the tertiary epoch. Nothing is known of the reproduction of this fish. The latest contributions to its history are by *Steenstrup* in "Overs. Dansk. Vidensk. Selsk., Forhandl." 1873, and by *Pavesi* in "Annal. Mus. Civ. Genova," 1874 and 1878.

THIRD FAMILY—RHINODONTIDÆ.

No nictitating membrane. Anal fin present. Two dorsal fins, the first nearly opposite to the ventrals, without spine in front. Mouth and nostril near the extremity of the snout.

This small family comprises one species only, *Rhinodon typicus*, a gigantic Shark, which is known to exceed a length of fifty feet, but is stated to attain that of seventy. It does not appear to be rare in the western parts of the Indian Ocean, and possibly occurs also in the Pacific. It is one of the most interesting forms, not unlike the Basking Shark of the Northern Seas, having gill-rakers like that species;

but very little is known of its structure and mode of life. It is perfectly harmless, its teeth being extremely small and numerous, placed in broad bands; it has been stated to feed on tang, an observation which requires confirmation. The snout is very broad, short, and flat; the eyes are very small. A pit at the root of the caudal fin which has the lower lobe well developed; side of the tail with a keel. A characteristic figure of this fish has been given by A. Smith in his "Illustrations of the Zoology of South Africa," Plate 26, from a specimen which came ashore at the Cape of Good Hope.

FOURTH FAMILY—NOTIDANIDÆ.

No nictitating membrane. One dorsal fin only, without spine, opposite to the anal.

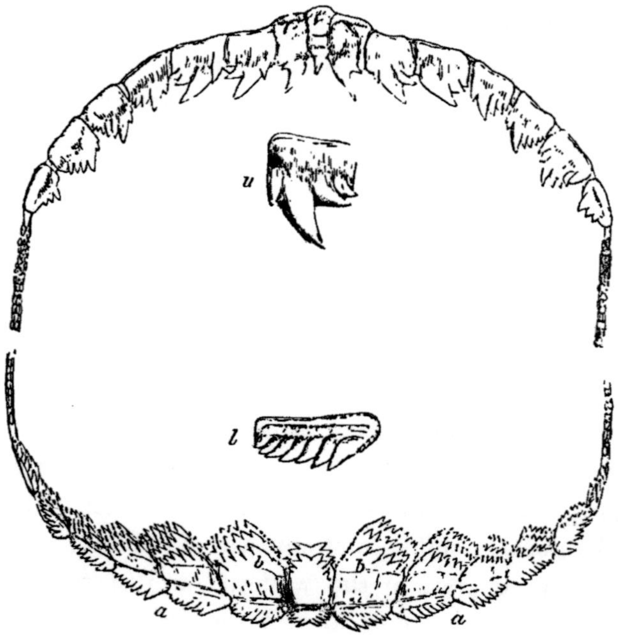

Fig. 115.—Dentition of Notidanus indicus. *a*, teeth in function; *b*, teeth in reserve; *u*, upper, and *l*, lower, tooth, of natural size.

NOTIDANUS.—Dentition unequal in the jaws: in the upper jaw one or two pairs of awl-shaped teeth, the following six being broader, and provided with several cusps, one of which is much the strongest. Lower jaw with six large comb-like teeth on each side, beside the smaller posterior teeth. Spiracles small, on the side of the neck. No pit at the root of the caudal fin. Gill-openings wide, six in number in *Hexanchus*, seven in *Heptanchus*.

Four species are known, distributed over nearly all the tropical and sub-tropical seas; they attain to a length of about fifteen feet. Fossil teeth belonging to this type have been found in jurassic and later formations (*Notidanus* and *Acllopos*).

FIFTH FAMILY—SCYLLIIDÆ.

Two dorsal fins, without spine: the first above or behind the ventrals; anal fin present. No nictitating membrane. Spiracle always distinct. Mouth inferior. Teeth small, several series generally being in function.

SCYLLIUM.—The origin of the anal fin is always in advance of that of the second dorsal. Nasal cavity separate from the mouth. Teeth small, with a middle longer cusp, and generally one or two small lateral cusps arranged in numerous series. Eggs similar to those of the Rays (Fig. 79, p. 167).

The fishes of this genus are of small size, and commonly called "Dog-fishes." They are coast fishes, living on the bottom, and feeding on Crustaceans, dead fishes, etc. None of the eight species known have a very wide distribution, but where they occur they are generally sufficiently abundant to prove troublesome to fishermen. They inhabit most parts of the temperate and tropical seas. On the British coasts two species are found, the "Larger" and "Lesser spotted Dog-fish," *Scyllium canicula* and *Scyllium catulus*, which are said to be more plentiful among the Orkney Islands than elsewhere. They are scarcely ever brought to market; but the fishermen of some localities do not disdain to eat them. Their flesh is remarkably white, a little fibrous, and dry. In the Orkneys they are skinned, split up, cleaned, and then spread out on the rocks to dry for home consumption. The skins are used for smoothing down cabinet-work. It would be worth while to apply the fins of these and other Sharks, which are so extensively used in China for making gelatine soups, to the same purpose in this country, or to dry them for exportation to the East. Most of the species of Dog-fishes are spotted, and those of the allied genera, *Parascyllium* and *Chiloscyllium*, very handsomely ornamented.

Closely allied to *Scyllium* is *Pristiurus*, from the coasts of Europe, which is provided with a series of small flat spines on each side of the upper edge of the caudal fin.

Fossil forms of Dog-fishes are not scarce in the Lias and Chalk: *Scylliodus, Palæoscyllium, Thyellina, Pristiurus.*

GINGLYMOSTOMA.—The second dorsal fin opposite to, and somewhat in advance of, the anal. Eyes very small; spiracle minute and behind the eye. Nasal and buccal cavities confluent. The nasal valves of both sides form one quadrangular flap in front of the mouth, each being provided with a free cylindrical cirrhus. The fourth and fifth gill-openings are close together. The teeth stand either in many series, each having a strong median cusp and one or two smaller ones on each side (*Ginglymostoma*), or they stand in a few (three) series only, the foremost only being in function, and each tooth having a convex, finely and equally serrated margin (*Nebrius*).

Four species from the tropical parts of the Atlantic and Indian Oceans, attaining to a length of some 12 feet. Pelagic.

STEGOSTOMA.—The first dorsal above the ventrals, the second in advance of the anal, which is very close to the caudal. Tail, with the caudal fin, exceedingly long, measuring one-half of the total length. Eyes very small, spiracle as wide as, and situated behind, the orbit. Nasal and buccal cavities confluent. Snout very obtuse; upper lip very thick, like a pad, bent downwards over the mouth, with a free cylindrical cirrhus on each side. Teeth small, trilobed, in many series, occupying in both jaws a transverse flat sub-quadrangular patch. The fourth and fifth gill-openings are close together.

The single species (*St. tigrinum*) for which this genus has been formed, is one of the commonest and handsomest sharks of the Indian Ocean. Young individuals keep generally close to the coasts, whilst the adult, which are from 10 to 15 feet long, are not rarely met in the open ocean. The colour is a brownish yellow, ornamented with black or brown transverse bands, or with snuff-coloured rounded spots; hence this shark is frequently mentioned by the names of "Zebra-Shark" or "Tiger-Shark."

CHILOSCYLLIUM.—The first dorsal fin above or behind the ven-

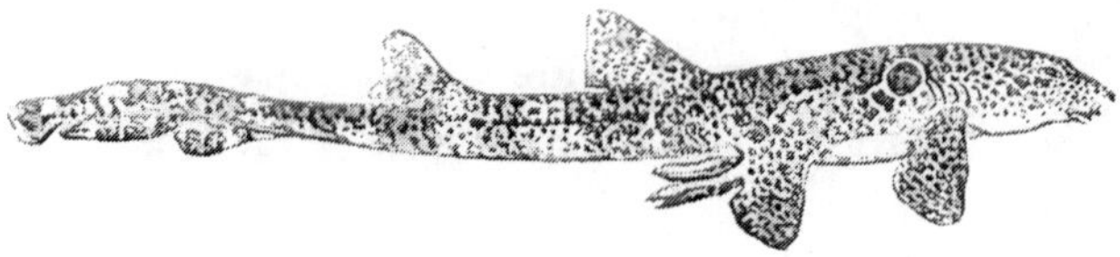

Fig. 116.—Chiloscyllium trispeculare, from North-western Australia.

trals. Anal fin placed far behind the second dorsal, and very close to the caudal. Spiracle very distinct, below the eye. Nasal and buccal cavities confluent. Nasal valve folded, with a cirrhus. Teeth small, triangular, with or without lateral cusps. The two last gill-openings close together.

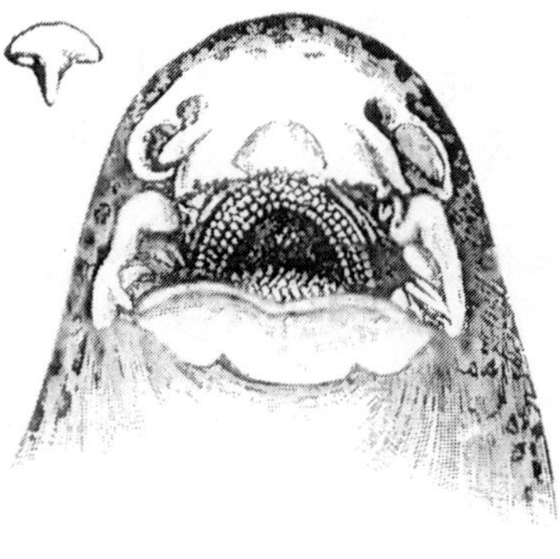

Fig. 117.—Confluent nasal and buccal cavities of the same fish.

"Dog-fishes," from the Indian Ocean, of small size. Four species are known, of which one, *Ch. indicum*, is one of the commonest shore-fishes on the coasts of this region, extending from the southern extremity of the African Continent to Japan.

CROSSORHINUS.—The first dorsal behind the ventrals, the second in advance of the anal, which is very close to the caudal. Tail rather short. Eyes small. Spiracle a wide oblique slit, behind and below the eye. Nasal and buccal cavities confluent. Head broad, flat, with the snout very obtuse; mouth wide, nearly anterior. A free nasal cirrhus; sides of the head with skinny appendages. Anterior teeth rather large, long and slender, without lateral lobes, the lateral tricuspid, smaller, forming a few series only. The fourth and fifth gill-openings close together.

Three species are known from the Australian and Japanese coasts. They are evidently ground-sharks, which lie concealed on the bottom watching for their prey. In accordance with this habit their colour closely assimilates that of a rock or stone covered with short vegetable and coralline growth—a resemblance increased by the frond-like tentacles on the side of the head. This peculiarity of the integuments, which is developed in a yet higher degree in Pediculati and Lophobranchs, is not met with in any other Selachian. These Sharks grow to a length of 10 feet.

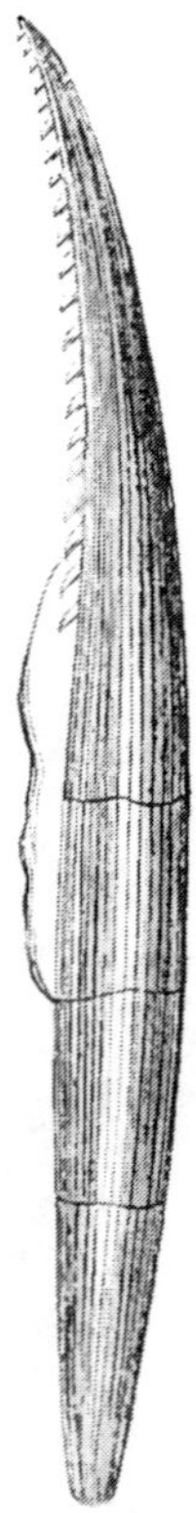

Fig. 118.—Spine of Hybodus subcarinatus.

SIXTH FAMILY—HYBODONTIDÆ.

Two dorsal fins, each with a serrated spine. Teeth rounded, longitudinally striated, with one larger, and from two to four smaller lateral cusps. Skin covered with shagreen.

Extinct. From carboniferous, liassic, and triassic formations. Several genera have been distinguished; and if *Cladodus* belongs to this family, it would have been represented even in the Devonian.

SEVENTH FAMILY—CESTRACIONTIDÆ.

No nictitating membrane. Two dorsal fins, the first

opposite to the space between pectoral and ventral fins; anal fin present. Nasal and buccal cavities confluent. Teeth obtuse, several series being in function.

This family is one of particular interest, because representatives of it occur in numerous modifications in primary and secondary strata. Their dentition is uniformly adapted for the prehension and mastication of crustaceous and hard-shelled animals. The fossil forms far exceeded in size the species of the only surviving genus; they make their appearance with *Ctenoptychius* in the Devonian; this is succeeded in

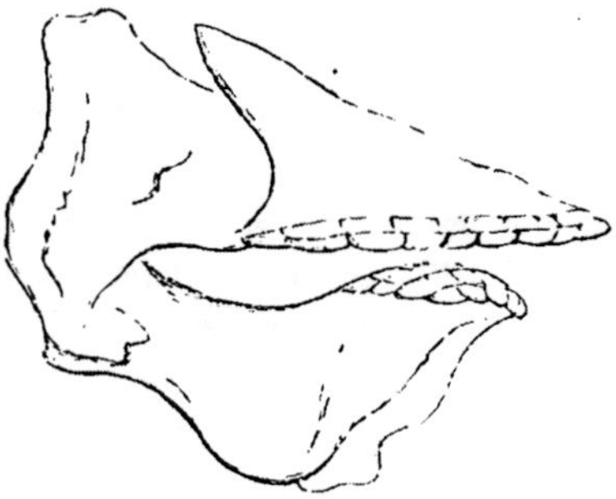

Fig. 119.—Jaws of Port Jackson Shark, Cestracion philippi.

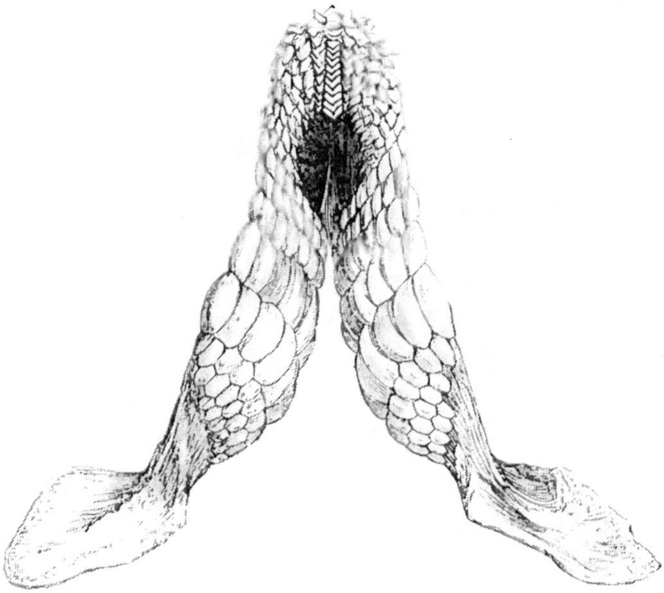

Fig. 120.—Upper jaw of the same, half natural size.

the coal-measures by *Psammodus, Chomatodus, Petrodus, Coch-*

liodus, *Polyrhizodus*, etc.; in the Trias and Chalk by *Strophodus*, *Acrodus*, *Thectodus*, and *Ptychodus*. Of the 25 genera known, 22 have lived in the periods preceding the Oolitic.

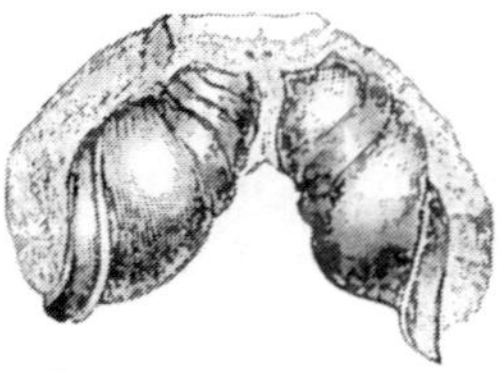

Fig. 121.—Cochliodus contortus.

CESTRACION (HETERODONTUS). — Each dorsal fin armed with a spine in front; the second in advance of the anal. Mouth rather narrow. Spiracles small, below the posterior part of the eye. Gill-openings rather narrow. Dentition similar in both jaws, viz. small obtuse teeth in front, which in young individuals

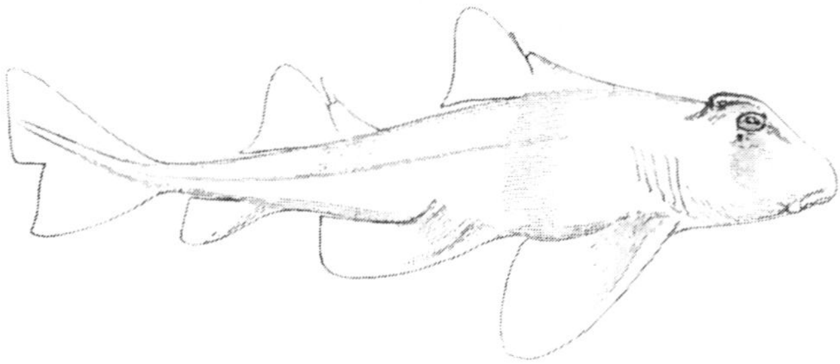

Fig. 122.—Cestracion galeatus, Australia.

are pointed and provided with from three to five cusps. The lateral teeth are large, padlike, twice as broad as long, arranged in oblique series, one series being formed by much larger teeth than those in the other series.

Four species are known from Japan, Amboyna, Australia, the Galapagoes Islands, and California; none exceed a length of 5 feet. The egg has been figured on p. 168 (Fig. 80).

EIGHTH FAMILY—SPINACIDÆ.

No membrana nictitans. Two dorsal fins; no anal. Mouth but slightly arched; a long, deep, straight, oblique groove on each side of the mouth. Spiracles present; gill-openings narrow. Pectoral fins not notched at their origin.

The oldest representative of this family (*Palæospinax*)

occurs at Lyme Regis; its skin is granular; each dorsal fin possesses a spine; the teeth in the jaws are dissimilar—the upper being multicuspid, longitudinally ribbed as in Hybodus, the lower smooth and tricuspid. *Drepanophorus* and *Spinax primævus* occur in Cretaceous formations of England and the Lebanon.

CENTRINA.—Each dorsal fin with a strong spine. Trunk rather elevated, trihedral, with a fold of the skin running along each side of the belly. Teeth of the lower jaw erect, triangular, finely serrated; those of the upper slender, conical, forming a group in front of the jaw. Spiracles wide, behind the eye.

One species, *Centrina salviani*, from the Mediterranean and neighbouring parts of the Atlantic; of small size.

ACANTHIAS.—Each dorsal fin with a spine. Teeth equal in both jaws, rather small; their point is so much turned aside that the inner margin of the tooth forms the cutting edge. Spiracles rather wide, immediately behind the eye.

The two species of "Spiny Dog-fishes," *A. vulgaris* and *A. blainvillii*, have a very remarkable distribution, being found in the temperate seas of the Northern and Southern Hemispheres, but not in the intermediate tropical zone. They are of small size, but occur at times in incredible numbers, 20,000 having been taken in one sean on the Cornish coast. They do much injury to the fishermen by cutting their lines and carrying off their hooks.

CENTROPHORUS.—Each dorsal fin with a spine which, however, is sometimes so small as to be hidden below the skin. Mouth wide. Teeth of the lower jaw with the point more or less inclined backwards and outwards. Upper teeth erect, triangular, or narrow, lanceolate, with a single cusp. Spiracles wide, behind the eye.

Eight species are known from the southern parts of the European seas, and one from the Moluccas; they do not appear to exceed a length of five feet. According to the observations of E. P. Wright, some of the species at least live at a considerable depth, perhaps at a greater depth than any of the other known Sharks. The Portuguese fishermen fish for them in 400 or 500 fathoms with a line of some 600 fathoms in length. The Sharks caught were specimens of *Centrophorus coelolepis,* from three to four feet long. "These sharks, as they were hauled into the boat, fell down into it like so many dead pigs; there was not the smallest motion of their bodies. There can be no reasonable doubt that they were inhabitants of the same great depth as *Hyalonema,*" and that, in fact, they were killed by being dragged to the surface from the pressure of water under which they lived. The dermal productions of some of the species have a very peculiar form, being leaf-shaped, pedunculate, or ribbed, or provided with an impression.

Spinax.—Each dorsal fin with a spine. Teeth of the lower jaw with the point so much turned aside that the inner margin of the tooth forms the cutting edge. Upper teeth erect, each with a long-pointed cusp and one or two small ones on each side. Spiracles wide, superior, behind the eye.

Three small species from the Atlantic and the southern extremity of America. *Centroscyllium* is an allied genus from the coast of Greenland.

Scymnus.—Two short dorsal fins without spine, the first at a considerable distance from the ventrals. Dermal productions uniformly small. Nostrils at the extremity of the snout. Upper teeth small, pointed; lower much larger, dilated, erect, triangular, not very numerous. Spiracles wide.

A single species, *S. lichia,* is rather common in the Mediterranean and the neighbouring parts of the Atlantic.

Læmargus.—All the fins small; two dorsal fins, without spine, the first at a considerable distance from the ventrals. Skin uniformly covered with minute tubercles. Nostrils near the extremity of the snout. The upper teeth small, narrow, conical; the lower teeth numerous, in several series, the point

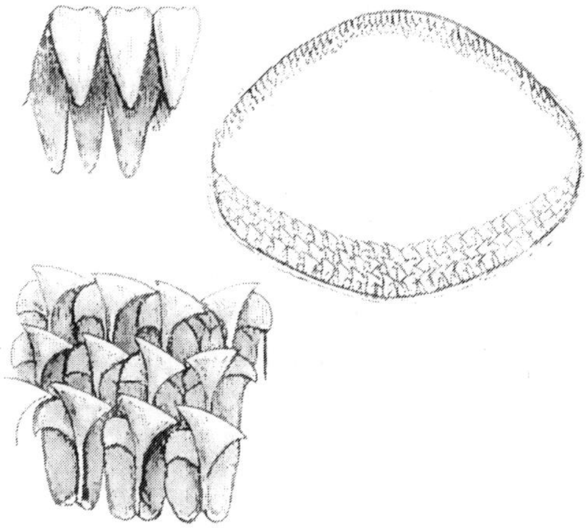

Fig. 123.—Dentition of the Greenland Shark. Some teeth are represented of the natural size; those of the lower jaw in three series.

so much turned aside that the inner margin forms a cutting, non-serrated edge. Jaws feeble. Spiracles of moderate width.

The "Greenland Shark" is an inhabitant of the Arctic regions, but rarely straying to the latitudes of great Britain; it grows to a length of about 15 feet, and, although it never or but rarely attacks man, is one of the greatest enemies of the whale, which is often found with large pieces bitten out

Fig. 124.—*Læmargus borealis*, Greenland Shark.

of the tail by this Shark. Its voracity is so great that, according to Scoresby, it is absolutely fearless of the presence of man whilst engaged in feeding on the carcase of a whale, so that it can be pierced through with a spear or knife without being driven away. It is stated to be viviparous, and to produce about four young at a birth.

Echinorhinus.—Two very small dorsal fins, without spine,

the first opposite to the ventrals. Skin with scattered large round tubercles. Nostrils midway between the mouth and the end of the snout. Teeth equal in both jaws, very oblique, the point being turned outwards; several strong denticulations on each side of the principal point. Spiracles small.

The "Spinous Shark" is readily recognised by the short bulky form of its body, short tail, and large spinous tubercles. It is evidently a ground-shark, which probably lives at some depth and but accidentally comes to the surface. More frequently met with in the Mediterranean, it has been found several times on the south coast of England, and near the Cape of Good Hope.

Euprotomicrus and *Isistius* are two other genera of this family; they are pelagic and but little known.

NINTH FAMILY—RHINIDÆ.

No anal fin; two dorsal fins. Spiracles present. Pectoral fins large, with the basal portion prolonged forwards, but not grown to the head.

RHINA.—Head and body depressed, flat; mouth anterior. Gill-openings rather wide, lateral, partly covered by the base of the pectoral. Spiracles wide, behind the eyes. Teeth conical, pointed, distant. Dorsal fins on the tail.

The "Angel-fish," or "Monk-fish" (*Rh. squatina*), approaches the Rays as regards general form and habits. Within the temperate and tropical zones it is almost cosmopolitan, being well known on the coasts of Europe, eastern North America, California, Japan, South Australia, etc.; it does not seem to exceed a length of five feet; it is viviparous, producing about twenty young at a birth.

Extinct forms, closely allied to the "Angel-fish," are found in the Oolite, and have been described as *Thaumas*. The carboniferous genus, *Orthacanthus*, may have been allied to this family, but it was armed with a spine immediately behind the head.

TENTH FAMILY—PRISTIOPHORIDÆ.

The rostral cartilage is produced into an excess flat lamina, armed along each edge with a series of te

These Sharks resemble so much the common Sa as to be easily confounded with them, but their gill-o are lateral, and not inferior. They are also much sma size, and a pair of long tentacles are inserted at the side of the saw. The four species known (*Pristiophorus*) o in the Australian and Japanese seas.

Squaloraja, from the Lias, is supposed to have its nearest affinities to this family.

B. BATOIDEI—RAYS.

In the typical Rays the body is excessively depressed, and forms, with the expanded pectoral fins, a circular or sub-rhomboidal disk, of which the slender tail appears as a more or less long appendage. In the two families which we shall place first (*Pristidæ* and *Rhinobatidæ*), the general habit of the body still resembles that of the Sharks, but the gill-openings are ventral, as in the true Rays; the anal fin is invariably absent, and the dorsal fins, if developed, are placed on the tail. The mode of life of those fishes is quite in accordance with the form of their body. Whilst the species with a shark-like body and muscular tail swim freely through the water, and are capable of executing rapid and sustained motions, the true Rays lead a sedentary life, moving slowly on the bottom, rarely ascending to the surface. Their tail has almost entirely lost the function of an organ of locomotion, acting in some merely as a rudder. They progress solely by means of the pectoral fins, the broad and thin margins of which are set in an undulating motion, entirely identical with that of the dorsal and anal fins of the *Pleuronectidæ*. They are exclusively carnivorous, like the Sharks, but unable

and catch rapidly-moving animals; therefore they
…efly on molluscous and crustaceous animals. How-
…he colour of their integuments assimilates so closely
of their surroundings, that other fishes approach them
…r enough to be captured by them. The mouth of Rays
…eing entirely at the lower surface of the head, the prey is not directly seized with the jaws; but the fish darts over its victim so as to cover and hold it down with its body, when it is conveyed by some rapid motions to the mouth.

Rays do not descend to the same depth as Sharks; with one exception,[1] at least, none have been known to have been caught by a dredge worked in more than 100 fathoms. The majority are coast fishes, and have a comparatively limited geographical range, none extending from the northern temperate zone into the southern. However, some, if not all the species of the family *Myliobatidæ*, which includes the giants of this division of Plagiostomes, have a claim of being included among the Pelagic fishes, as they are frequently met with in the open ocean at a great distance from the shore. It is probable that the occurrence of such individuals in the open sea indicates the neighbourhood of some bank or other comparatively shallow locality. Many species are exclusively confined to fresh water, and occur far inland, especially in tropical America.

The majority are oviparous. All have five pairs of gill-openings. The number of known species is about the same as that of Sharks, viz. 140.

First Family—Pristidæ.

The snout is produced into an exceedingly long flat lamina, armed with a series of strong teeth along each edge (saw).

[1] This exception is a Ray obtained during the "Challenger" expedition, and said to have been dredged in 565 fathoms.

PRISTIS.—Body depressed and elongate, gradually passing into the strong and muscular tail. Pectoral fins, with the front margins quite free, not extending to the head. No tentacles below the saw. Teeth in the jaws minute, obtuse. Dorsal fins without spine, the first opposite or close to the base of the ventrals.

"Saw-fishes." Abundant in tropical, less so in sub-tropical seas. They attain to a considerable size, specimens with a saw 6 feet long and 1 foot broad at the base not being of uncommon occurrence. The saw, which is their weapon of attack, renders them most dangerous to almost all the other large inhabitants of the ocean. Its endo-skeleton consists of three, sometimes five, rarely four, hollow cylindrical tubes, placed side by side, tapering towards the end, and incrusted with an osseous deposit. These tubes are the rostral processes of the cranial cartilage, and exist in all Rays, though in them they are shorter and much less developed. The teeth of the saw are implanted in deep sockets of the hardened integument. The teeth proper, with which the jaws are armed, are much too small for inflicting wounds or seizing other animals. Saw-fishes use this weapon in tearing pieces of flesh off an animal's body or ripping open its abdomen. The detached fragments or protruding soft parts are then seized by them and swallowed. Five distinct species of Saw-fishes are known.

Saws of extinct species have been found in the London clay of Sheppey and in the Bagshot sands.

SECOND FAMILY—RHINOBATIDÆ.

Tail strong and long, with two well-developed dorsal fins, and a longitudinal fold on each side; caudal developed. Disk not excessively dilated, the rayed portion of the pectoral fins not being continued to the snout.

RHYNCHOBATUS.—Dorsal fins without spine, the first opposite to the ventrals. Caudal fin with the lower lobe well developed.

Teeth obtuse, granular, the dental surfaces of the jaws being undulated.

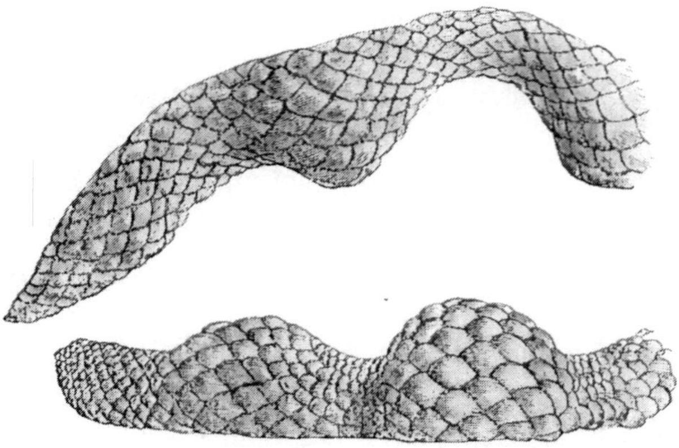

Fig. 125.—Dentition of *Rhynchobatus*.

Two species, *Rh. ancylostomus* and *Rh. djeddensis*, are very common on the tropical coasts of the Indian Ocean. They feed on hard-shelled animals, and attain scarcely a length of 8 feet.

RHINOBATUS.—Cranial cartilage produced into a long rostral process, the space between the process and pectoral fin being filled by a membrane. Teeth obtuse, with an indistinct transverse ridge. Dorsal fins without spine, both at a great distance behind the ventral fins. Caudal fin without lower lobe.

Numerous on the coasts of tropical and sub-tropical seas; about twelve species. *Trygonorhina* is an allied genus from South Australia.

The oolitic genus *Spathobatis* is scarcely distinct from *Rhinobatus*; and another fossil from Mount Lebanon has been actually referred to this latter genus. *Trigorhina* from Monte Postale must be placed here.

THIRD FAMILY—TORPEDINIDÆ.

The trunk is a broad, smooth disk. Tail with a longitudinal fold on each side; a rayed dorsal fin is generally, and a caudal always, present. Anterior nasal valves confluent into a quadrangular lobe. An electric organ composed of vertical hexagonal prisms between the pectoral fins and the head.

"Electric Rays." The electric organs with which these fishes are armed are large, flat, uniform bodies, lying one on each side of the head, bounded behind by the scapular arch, and laterally by the anterior crescentic tips of the pectoral fins. They consist of an assemblage of vertical hexagonal prisms, whose ends are in contact with the integuments above and below; and each prism is subdivided by delicate transverse septa, forming cells, filled with a clear, trembling, jelly-like fluid, and lined within by an epithelium of nucleated corpuscles. Between this epithelium and the transverse septa and walls of the prism there is a layer of tissue on which the terminations of the nerves and vessels ramify. Hunter counted 470 prisms in each battery of *Torpedo marmorata*, and demonstrated the enormous supply of nervous matter which they receive. Each organ receives one branch of the Trigeminal nerve and four branches of the Vagus, the former, and the three anterior branches of the latter, being each as thick as the spinal chord (electric lobes). The fish gives the electric shock voluntarily, when it is excited to do so in self-defence or intends to stun or to kill its prey; but to receive the shock the object must complete the galvanic circuit by communicating with the fish at two distinct points, either directly or through the medium of some conducting body. If an insulated frog's leg touches the fish by the end of the nerve only, no muscular contractions ensue on the discharge of the battery, but a second point of contact immediately produces them. It is said that a painful sensation may be produced by a discharge conveyed through the medium of a stream of water. The electric currents created in these fishes exercise all the other known powers of electricity: they render the needle magnetic, decompose chemical compounds, and emit the spark. The dorsal surface of the electric organ is positive, the ventral surface negative.

[The literature on the electric organ of Torpedo is very extensive. Here may be mentioned *Lorenzini*, "Osservazioni intorno alle Torpedini (1678); *Walsh*, "On the Electric Property of the Torpedo," in Philos. Trans., 1773; *Hunter*, "Anatomical Observations on the Torpedo," *ibid.*; *Davy*, "Observations on the Torpedo," in Philos. Trans., 1834; *Matteucci* and *Savi*, "Traité des Phénomènes Electro-Physiologiques," 1844.]

Of the genus *Torpedo* six species are known, distributed over the Atlantic and Indian Oceans; three of them are rather common in the Mediterranean, and one (*T. hebetans*) reaches the south coast of England. They attain to a width of from two to three feet, and specimens of that size are able to disable by a single discharge a full-grown man, and, therefore, may prove dangerous to bathing persons. Other genera, differing from *Torpedo* in the position and structure of some of the fins, are found in other tropical and sub-tropical seas, viz. *Narcine*, *Hypnos*, *Discopyge* (Peru), *Astrape*, and *Temera*. All, like electric fishes generally, have a naked body.

A large fish, of the general appearance of a Torpedo, has been found at Monte Bolca; and *Cyclobatis*, from the upper cretaceous limestone of Lebanon, is probably another extinct representative of this family.

FOURTH FAMILY—RAJIDÆ.

Disk broad, rhombic, generally with asperities or spines; tail with a longitudinal fold on each side. The pectoral fins extend to the snout. No electric organ; no serrated caudal spine.

RAJA.—Two dorsal fins on the tail, without spine; tail with a rudimentary caudal fin, or without caudal. Each ventral fin divided into two by a deep notch. Teeth small, obtuse, or pointed. Pectoral fins not extending forwards to the extremity of the snout. Nasal valves separated in the middle, where they are without a free margin (see Fig. 1, p. 34).

Of all the genera of *Batoidei*, Rays have the widest geographical range; they are chiefly inhabitants of temperate

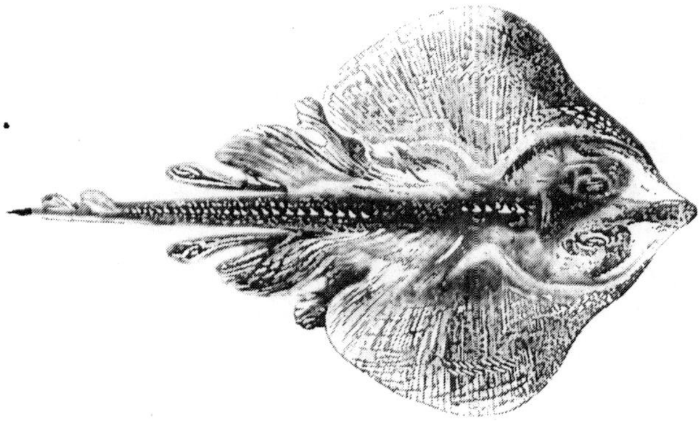

Fig. 126.—Raja lemprieri, from Tasmania.

seas, and much more numerous in those of the Northern than of the Southern Hemisphere. They advance more closely to the Arctic and Antarctic circles than any other member of this group. More than thirty species are known, of which the following are found on the British coast:—The Thornback (*R. clavata*), the Homelyn Ray (*R. maculata*), the Starry Ray (*R. radiata*), the Sandy Ray (*R. circularis*), the common Skate (*R. batis*), the Burton Skate (*R. marginata*), and the Shagreen Skate (*R. fullonica*). Some of these species, especially the Skates, attain a considerable size, the disk measuring six and even seven feet across. All are eatable, and some of them regularly brought to market. In the majority of the species peculiar sexual differences have been observed. In some, as in the Thornback, all or some of the teeth are pointed in the male sex, whilst they are obtuse and flat in the female. The males of all are armed with patches of claw-like spines, retractile in grooves of the integument, and serially arranged occupying a space on the upper side of the pectoral fin

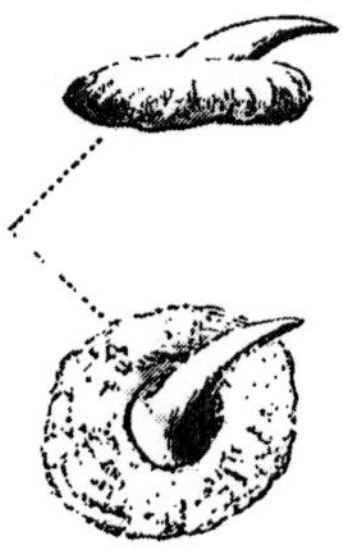

Fig. 127.—Dermal spines of a male Thornback, Raja clavata.

near the angle of the disk, and frequently also the sides of the head. In species which are armed with bucklers or asperities it is the female which is principally provided with these dermal productions, the male being entirely or nearly smooth. Also the colour is frequently different in the two sexes.

Other genera of this family are *Psammobatis*, *Sympterygia*, and *Platyrhina*. Although probably this family was well represented in cretaceous and tertiary formations, the remains found hitherto are comparatively few. *Arthropterus*, from the Lias, seems to have been a true Ray; and dermal spines of a species allied to the Thornback (*Raja antiqua*) are abundant in the crag deposits of Suffolk and Norfolk.

Fifth Family—Trygonidæ.

The pectoral fins are uninterruptedly continued to, and confluent at, the extremity of the snout. Tail long and slender, without lateral longitudinal folds; vertical fins none, or imperfectly developed, often replaced by a strong serrated spine.

The "Sting-Rays" are as numerous as the Rays proper, but they inhabit rather tropical than temperate seas. The species armed with a spine use it as a weapon of defence, and the wounds inflicted by it are, to man, extremely painful, and have frequently occasioned the loss of a limb. We have mentioned above (p. 190) that the danger arises from the lacerated nature of the wound as well as from the poisonous property of the mucus inoculated. The spines (Fig. 98, p. 190) are always barbed on the sides, and may be eight or nine inches long in the larger species. They are shed from time to time, and replaced by others growing behind the one in function, as the teeth of the fishes of this order, or as the fangs of a poisonous snake. Fossil species of *Trygon* and *Urolophus* occur in the tertiary strata of Monte Bolca and Monte Postale. The genera into which the various species have been divided are the following:—

UROGYMNUS.—Tail long, without fin or spine, sometimes with a narrow cutaneous fold below. Body densely covered with osseous tubercles. Teeth flattened.

Only one species is known (*U. asperrimus*), common in the Indian Ocean, and with a body from 4 to 5 feet long; the skin is frequently used for covering shields and the handles of swords and other weapons, its rough surface offering a firm hold to the hand.

TRYGON.—Tail very long, tapering, armed with a long arrow-shaped barbed spine. Body smooth or with tubercles. Nasal valves coalescent into a quadrangular flap. Teeth flattened.

Some twenty-five species are known, one of which (*T. pastinaca*) extends from the south coast of England and the east coast of North America through the Atlantic and Indian Ocean to Japan. The majority of the species belong to the tropical parts of the Indian and Atlantic Oceans; some inhabit exclusively fresh-waters of eastern tropical America. A closely allied genus is *Tæniura*, with six species.

UROLOPHUS.—Tail of moderate length, with a distinct rayed

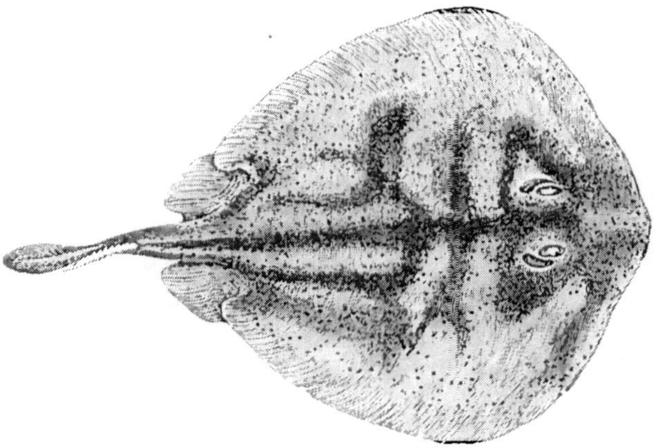

Fig. 128.—Urolophus cruciatus, from Australia.

terminal fin, armed with a barbed spine, without or with a rudimentary dorsal fin. Teeth flattened.

Seven species from tropical seas, apparently of small size.

PTEROPLATEA.—Body at least twice as broad as long; tail very short and thin, without or with a rudimentary fin, and with a serrated spine. Teeth very small, uni- or tri-cuspid.

Six species from temperate and tropical seas.

SIXTH FAMILY—MYLIOBATIDÆ.

The disk is very broad, in consequence of the great development of the pectoral fins, which, however, leave the sides of the head free, and reappear at the extremity of the snout as a pair of detached (cephalic) fins. Viviparous.

"Devil-fishes," "Sea-devils," or "Eagle-rays." Generally of large size, inhabiting temperate and tropical seas. Some genera possess a pair of singular cephalic processes, which generally project in a direction parallel to the longitudinal axis of the body, but are said to be flexible in the living fish, and used for scooping food from the bottom and conveying it to the mouth. In all the species the dentition consists of perfectly flat molars, forming a kind of mosaic pavement in both the upper and lower jaws: a most perfect mechanical arrangement for crushing alimentary substances.

MYLIOBATIS.—Teeth sexangular, large, flat, tessellated, those in the middle much broader than long; several narrower series on

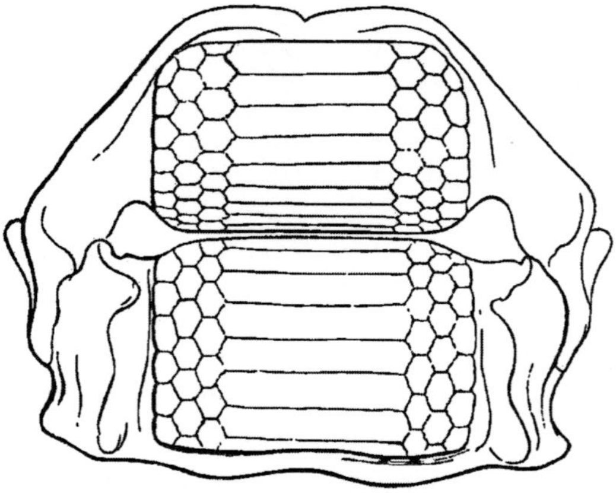

Fig. 129.—Jaws of an Eagle-Ray, *Myliobatis aquila.*

each side. Tail very long and thin, with a dorsal fin near its root; generally a serrated spine behind the fin.

Seven species are known, two of which are European, one (*M. aquila*) being almost cosmopolitan, and occasionally found on the British coast. The young differ much from the adult, having no median series of larger teeth, but all the teeth of equal size and regularly sexangular. Also the tail is much longer in young examples than in old ones, and the coloration more ornamental. Teeth of species very closely allied to, or perhaps even identical with, existing species, are found in tertiary formations.

AËTOBATIS.—Form of the head, body, and tail as in *Myliobatis*. The nasal valves remain separate, each forming a long flap. The lower dental lamina projects beyond the upper. Teeth flat, broad, forming a single series, equivalent to the median series of *Myliobatis*, there being no small lateral teeth.

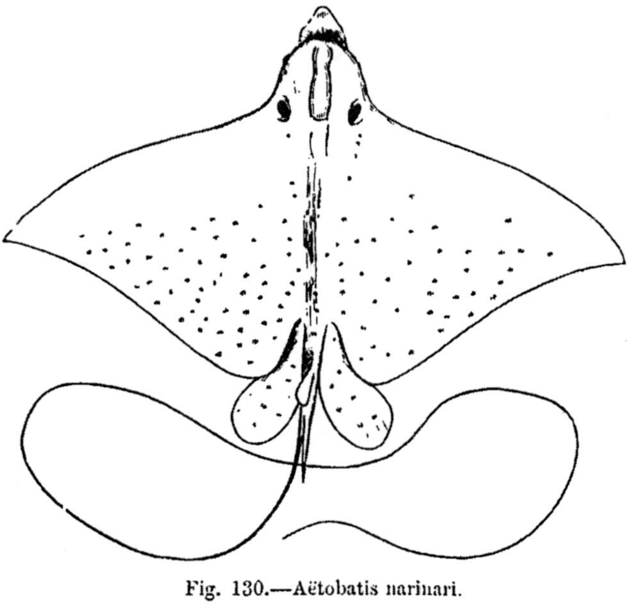

Fig. 130.—Aëtobatis narinari.

One species only (*A. narinari*) which is found in almost

all tropical seas, and of exceedingly common occurrence in the Atlantic and Indian Oceans; it does not seem to grow to a very large size (perhaps not exceeding 5 feet in width), and is readily recognised by numerous round bluish-white spots, with which the back is ornamented. Fossils of this genus occur in the English Eocenes and the Swiss Molasse.

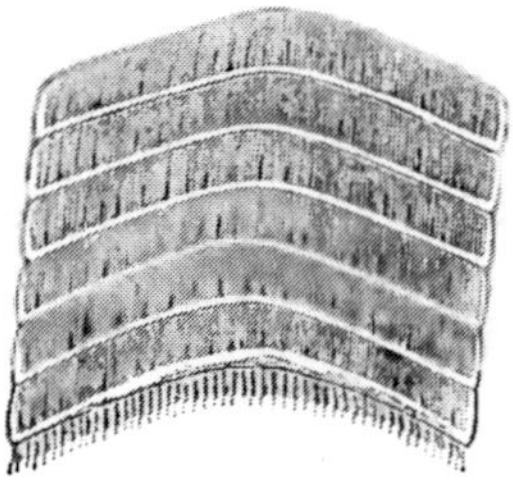

Fig. 131.—Aëtobatis subarcuatus, from Bracklesham.

RHINOPTERA.—The cephalic appendages are bent inwards, and situated at the lower side of the snout. Nasal valves confluent into a broad flap, with free margin. Teeth broad, flat, tessellated, in five or more series, the middle being the broadest, and the others decreasing in width outwards. Tail very slender, with a dorsal fin before the serrated spine.

Seven species from tropical and sub-tropical seas are known; of *Rhinoptera polyodon* nothing is known except the jaws; and as its dentition is very peculiar, no opportunity should be lost of obtaining and preserving entire animals. Teeth very similar to those of existing species, and described as *Zygobatis*, oc-

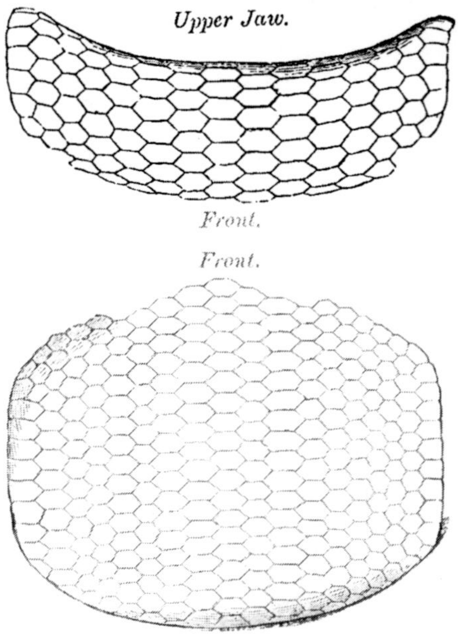

Fig. 133.—Rhinoptera polyodon.

Fig. 132.—Rhinoptera woodwardi; fossil.

cur in the Norwich Crag and in Miocene formations of Switzerland.

DICEROBATIS (CEPHALOPTERA).—Cephalic appendages pointing straight forwards or inwards. Nostrils widely separated from each other. Mouth inferior, wide. Both jaws with very numerous and very small flat or tubercular teeth. Tail very slender, with a dorsal fin between the ventrals, and with or without a serrated spine.

CERATOPTERA.—Cephalic appendages pointing forwards or inwards. Mouth anterior; wide. Teeth in the lower jaw only, very small. Tail very slender, with a dorsal fin between the ventrals and without spine.

The species of these two last genera are not yet well distinguished; about five of *Dicerobatis* and two of *Ceratoptera*

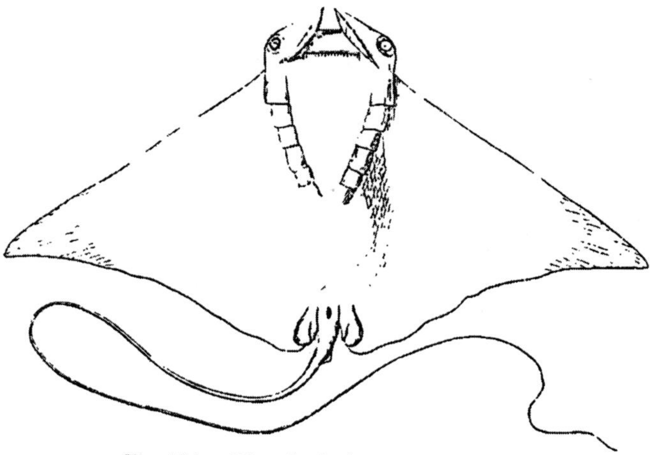

Fig. 134.—Dicerobatis draco, from Misol.

are known from tropical and temperate seas, but their occurrence in the latter is rather sporadic. Some of them, if not all, attain an enormous size. One mentioned by Risso, taken off Messina, weighed 1250 pounds. Several observers speak of having seen them in pairs, the male being usually the smaller. Of a pair mentioned by Risso the female was first taken, and the male remained hovering about the boat for three days, and was afterwards found floating dead on the surface. Still larger individuals, but of uncertain species, are mentioned by Lacépède, who says that one taken at Barbadoes required seven yoke of oxen to draw it. A sketch of another, which was said to be twenty feet long, was sent to Lacépède; and Sonnini speaks of one which appeared to him to be longer and wider than the ship in which he was sailing. A fœtus taken from the uterus of the mother captured at Jamaica, and preserved in the British Museum, is five feet broad, and weighed twenty pounds. The mother measured fifteen feet in width as well as in length, and was between three and four feet thick. The capture of "Devil-fishes" of such large size is attended with danger, as they not rarely attack and capsize the boat. They are said to be especially dangerous when they accompany their young, of which they bring forth one only at a time.

SECOND SUB-ORDER—HOLOCEPHALA.

One external gill-opening only, covered by a fold of the skin, which encloses a rudimentary cartilaginous gill-cover; four branchial clefts within the gill-cavity. The maxillary and palatal apparatus coalescent with the skull.

This sub-order is represented in the living fauna by one family only, *Chimæridæ;* it forms a passage to the following order of fishes, the Ganoids. In external appearance, and with regard to the structure of their organs of propagation, the Chimæras are Sharks (See Fig. 96, p. 184). The males are provided with "claspers" in connection with the ventral fins, and the ova are large, encased in a horny capsule, and few in number; and there is no doubt that they are impregnated within the oviduct, as in Sharks. Chimæras are naked, but, as in *Scylliidæ*, very young individuals possess a series of small "placoid" spines, which occupy the median line of the back, and remind us of similar dermal productions in the Rays. The males, besides, are provided with a singular erectile appendage, spiny at its extremity, and received in a groove on the top of the head. On the other hand, the relations of the Chimæras to the Ganoid, and, more especially, Dipnoous type become manifest in their notochordal skeleton and continuity of cranial cartilage. The spine in front of the first dorsal fin is articulated to the neural apophysis, and not merely implanted in the soft parts, and immovable as in Sharks. A cartilaginous operculum makes its appearance, and the external gill-opening is single. The dentition is that of a Dipnoid, each "jaw" being armed with a pair of broad dental plates, with the addition of a pair of smaller cutting teeth in the upper "jaw." Fossils of similar dental combination are not rare in strata, commencing from the Lias and the bottom of the Oolitic series; but it is impossible to decide in every case whether the fossil should be referred to the Holocephalous or Dipnoous type. According to Newberry, Chimæroid fishes commence in the Devonian with *Rhynchodus*, the remains of which were discovered by him in Devonian rocks of Ohio. Undoubted Chimæroids are *Elasmodus, Psaliodus, Ganodus, Ischyodus, Edaphodon*, and *Elasmognathus*, principally from mesozoic and tertiary formations. Very similar fossils occur in the corresponding strata of North America. A single species of *Callorhynchus* has been discovered by H. Hector in the Lower Greensand of New Zealand.

The living Chimæras are few in number, and remain within very moderate dimensions, probably not exceeding a length of five feet, inclusive of their long filamentous, diphycercal tail. They are referred to two genera.

CHIMÆRA.—Snout soft, prominent, without appendage. The dorsal fins occupying the greater part of the back, anterior with a very strong and long spine. Longitudinal axis of the tail nearly the same as that of the trunk, its extremity being provided with a low fin above and below, similar in form to a dorsal and anal fin. Anal fin very low.

Three species are known: *Ch. monstrosa*, from the coasts of Europe and Japan and the Cape of Good Hope; *Ch. colliei* from the west coast of North America; and *Ch. affinis* from the coast of Portugal. (See Fig. 96, p. 184.)

CALLORHYNCHUS.—Snout with a cartilaginous prominence, terminating in a cutaneous flap. Two dorsal fins, the anterior with a very strong and long spine. Extremity of the tail distinctly turned upwards, with a fin along its lower edge, but without one above. Anal fin close to the caudal, short and deep.

One species (*C. antarcticus*) is common in the Southern temperate zone. Cunningham describes the egg (see Fig. 81, p. 169), as being of a dark greenish-black colour, and, in general, measuring from eight to nine or even ten inches in length, by about three in breadth. It consists of a central, somewhat spindle-shaped convex area (between the horny walls of which the young fish lies), surrounded by a broad plicated margin, which is fringed at the edge, and covered on the under surface with fine light brownish-yellow hairs.

SECOND ORDER—GANOIDEI.

Skeleton cartilaginous or ossified. Body with medial and paired fins, the hinder pair abdominal. Gills free, rarely partially attached to the walls of the gill-cavity. One external gill-opening only on each side; a gill-cover. Air-bladder with a pneumatic duct. Ova small, impregnated after exclusion. Embryo sometimes with external gills.

To this order belong the majority of the fossil fish remains of palæozoic and mesozoic age, whilst it is very scantily represented in the recent fauna, and evidently verging towards total extinction. The knowledge of the fossil forms, based on mere fragments of the hard parts of the body only, is very incomplete, and therefore their classification is in a most unsatisfactory state. In the following pages only the most important groups will be mentioned.

[For a study of details we have to refer to *Agassiz*, "Poissons Fossiles;" *Owen*, "Palæontology," Edinb. 1861, 8vo; *Huxley*, "Preliminary Essay upon the Systematic Arrangement of the Fishes of the Devonian Epoch," in Mem. Geolog. Survey, Dec. 10; Lond. 1861, and "Illustrations of the Structure of Crossopterygian Ganoids," *ibid.* December 12, 1866; *Traquair*, "The Ganoids of the British Carboniferous Formations," part I. "Palæoniscidæ." Palæontogr. Soc. Lond. 1877.]

Eight sub-orders may be distinguished at present.

FIRST SUB-ORDER—PLACODERMI.

Extinct. The head and pectoral region of the body encased in great bony, sculptured plates, with dots of enamel; the remainder of the body naked, or with ganoid scales; skeleton notochordal.

Comprises the oldest vertebrate remains, from Devonian and Carboniferous formations. *Pterichthys:* (Figs. 135 and 136), tail tapering, covered with small ganoid scales, without caudal fin; the cephalic shield was probably moveably joined to the cuirass of the trunk, and both were composed of several pieces; the abdominal shield consisted of one single median plate, and two pairs of lateral plates, a third small pair being sometimes observed detached in front of the anterior pair; pectoral exceedingly long, consisting of two pieces movably connected with each other; tail scaly, and short; a small dorsal fin placed on the tail; a pair of small ventrals; jaws small, with confluent denticles. Several species have been distinguished in remains found in the strata of Caithness and other localities in Scotland. *Coccosteus* (Fig. 137, p. 354): all the bony plates are firmly united, no pectoral spines; tail naked and long; a dorsal and anal fin supported by interneural and interhæmal spines. Dentition unknown. *Dinichthys:* a gigantic fish from the Devonian of North America (estimated at from 15 to 18 feet in length),

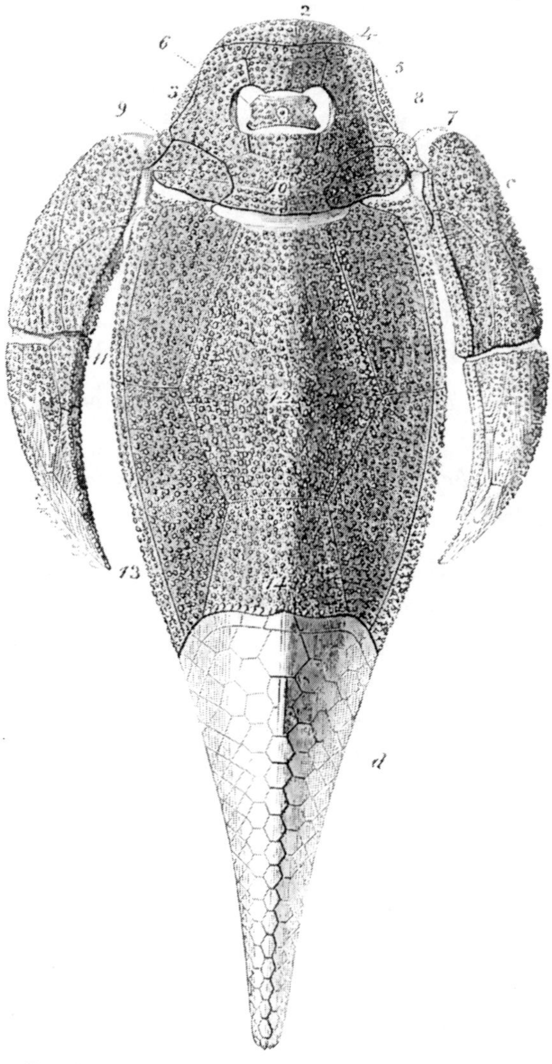

Fig. 135.—Dorsal surface of Pterichthys, after Pander. *d*, Dorsal fin ; *c*, pectoral member ; 2-10, head-bucklers ; 11-13, dorsal bucklers.

with the dermal covering very similar to that of Coccosteus, but with a simple arched dorsal shield. As in this latter genus the caudal extremity does not possess external or internal bony parts, and the ventral plastron of both genera corresponds in every particular; the dentition is so

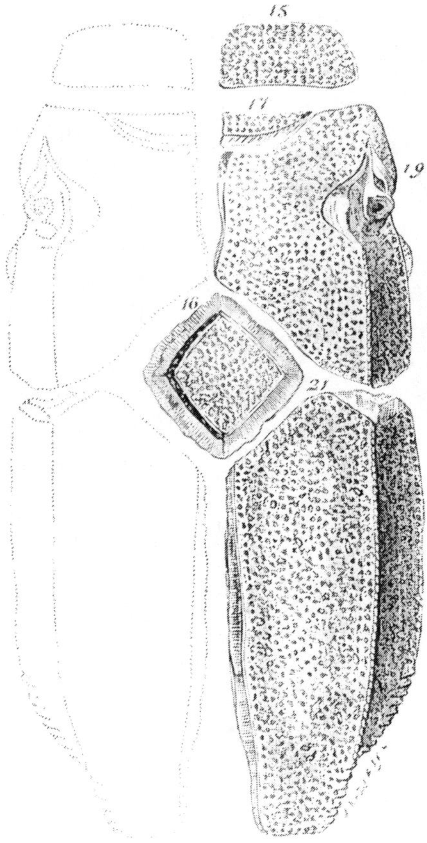

Fig. 136.—Ventral aspect of Pterichthys, after Pander. 15, mandible (?); 16-21, ventral bucklers.

singularly like that of Lepidosiren, that Newberry (Geolog. Survey of Ohio, vol. ii. part 2) considers this genus to be in genetic relation to the Dipnoi. The following genera have been united in a separate family, *Cephalaspidæ;* viz. *Cephal-*

aspis: head covered by a continuous shield with tubercular surface, produced into a horn at each posterior corner; a median dorsal backward prolongation bears a spine; heterocercal. *Auchenaspis* and *Didymaspis:* allied to the preceding, but with the cephalic shield divided into a larger anterior and smaller posterior piece. *Pteraspis:* with the cephalic shield finely striated or grooved, composed of seven pieces. *Scaphaspis* and *Cyathaspis:* with the surface of the head-shield similarly sculptured as in *Pteraspis*, but simple in the former, and composed of four pieces in the latter. *Astrolepis:* attained to the gigantic size of between twenty and thirty feet; its mouth was furnished with two rows of teeth, of which the outer ones were small, the inner much larger.

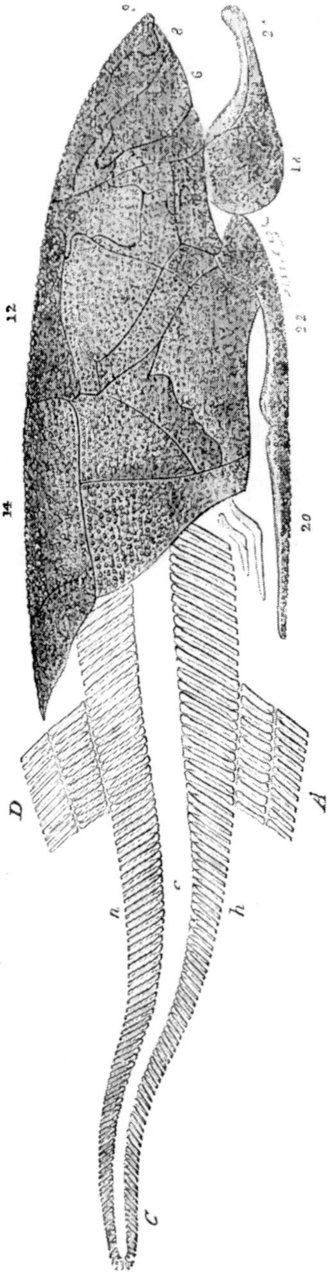

Fig. 137.—Coccosteus, after Pander. *A*, Anal fin; *D*, Dorsal fin; *C*, Heterocercal tail; *c*, notochord; *n*, neurals; *h*, hæmals; 6-24, bucklers.

[See Ray Lankester, A Monograph of the Fishes of the Old Red Sandstone of Britain. Part I. Cephalaspidæ. Lond. 1868 and 1870. 4to.]

SECOND SUB-ORDER—ACANTHODINI.

Extinct. *Body oblong, compressed, covered with shagreen; skull not ossified; caudal heterocercal. Large spines, similar to those of Chondropterygians, in front of some of the median and paired fins. The spines are imbedded between the muscles, and not provided with a proximal joint.*

Acanthodes, Chiracanthus, from Devonian and Carboniferous formations.

THIRD SUB-ORDER—DIPNOI.

Nostrils two pairs, more or less within the mouth; limbs with an axial skeleton. Lungs and gills. Skeleton notochordal. No branchiostegals.[1]

FIRST FAMILY—SIRENIDÆ.

Caudal fin diphycercal; no gular plates; scales cycloid. A pair of molars, above and below, and one pair of vomerine teeth.

LEPIDOSIREN.—Body eel-shaped, with one continuous vertical fin. Limbs reduced to cylindrical filaments, without fringe. Vomerine teeth conical, pointed. Each dental lamina or molar with strong cusps, supported by vertical ridges. No external branchial appendages; five branchial arches, with four intervening clefts. Conus arteriosus with two longitudinal valves. Ovaries closed sacs.

One species only is known from the system of the River Amazons (*L. paradoxa*). It must be very locally distributed, as but a few specimens have been brought to Europe, and all recent endeavours to obtain others have been unsuccessful. Natterer, by whom this most interesting fish was discovered, states that he obtained two specimens, one on the Madeira River, near Borba; the other in a backwater of the Amazons, above Villa Nova. The natives of the former place called it Carámurú, and considered it very scarce. The larger indivi-

[1] See pp. 73 and 74, Figs. 35 and 36.

dual was nearly four feet long. It is said to produce a sound not unlike that of a cat, and to feed on the roots of mandioca and other vegetables. But, to judge from the dentition, this fish is much more likely to be carnivorous, like the following. It is one of the greatest desiderata of Natural History Collections.

[Natterer, "Annalen des Wiener Museum's," 1839, ii.; Bischoff. "Annales des Sciences Naturelles," 1840. xiv.]

PROTOPTERUS.—Very similar in the general form of the body and dentition to *Lepidosiren*. Pectoral and ventral filaments with a fringe containing rays. Three small branchial appendages above the small gill-opening; six branchial arches, with five intervening clefts. Conus arteriosus with two longitudinal valves. Ovaries closed sacs.

Protopterus annectens is the "Lepidosiren" which is commonly found in Zoological collections. It is usually imported

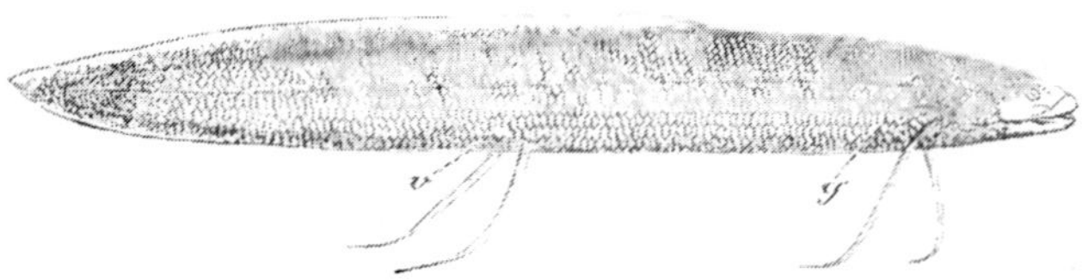

Fig. 138.—Protopterus annectens. *g*, Branchial filaments; *v*, vent.

from the west coast of Africa, where it abounds in many localities; but it is spread over the whole of tropical Africa, and forms in many districts of the central parts a regular article of food.

During the dry season, specimens living in shallow waters which periodically dry up, form a cavity in the mud, the inside of which they line with a protecting capsule of mucus, and from which they emerge again when the rains refill the pools inhabited by them. Whilst they remain in this torpid state of existence, the clay-balls containing them are frequently dug out, and if the capsules are not broken, the fishes imbedded in them can be transported to Europe, and released by being immersed in slightly tepid water. *Protopterus* is exclusively carnivorous, feeding on water-insects, frogs, and fishes, and attains a length of six feet.

[Owen, "Trans. Linn. Soc." 1841, xviii.]

CERATODUS.—Body elongate, compressed, with one continuous vertical fin. Limbs paddle-shaped, with broad, rayed fringe. Vomerine teeth incisor-like; molars with flat, undulated surface, and lateral prongs. No external branchial appendages. Conus arteriosus with transverse series of valves. Ovaries transversely lamellated.[1]

Fig. 139.—*Ceratodus miolepis.*

Two species, *C. forsteri* and *C. miolepis*, are known from fresh waters of Queensland. The specimens hitherto obtained have come from the Burnett, Dawson, and Mary rivers, some from the fresh waters of the upper parts, others from the lower brackish portions. The fish is said to attain to a weight of twenty pounds and to a length of 6 feet. Locally, the settlers call it "Flat-head," "Burnett- or Dawson-Salmon," and the aborigines "Barramunda," a name which they appear to apply also to other large-scaled freshwater fishes, as the Osteoglossum leichardti. In the stomach there is generally found an enormous quantity of the leaves of plants growing on the banks of rivers, evidently eaten after they had fallen into the water and when in a decomposing condition. The flesh of the fish is salmon-coloured, and much esteemed as food.

The Barramunda is said to be in the habit of going on land, or at least on mud-flats; and this assertion appears to

[1] For other illustrations see p. 73, Fig. 35 (palatal view of head); p. 74, Fig. 36 (pectoral skeleton); p. 141, Fig. 60 (gills); p. 148, Fig. 65 (lung); p. 151, Fig. 67 (heart); p. 134, Fig. 57 (intestine); p. 165, Fig. 77 (ovary).

be borne out by the fact that it is provided with a lung. However, it is much more probable that it rises now and then to the surface of the water in order to fill its lung with air, and then descends again until the air is so much deoxygenised as to render a renewal of it necessary. It is also said to make a grunting noise, which may be heard at night for some distance. This noise is probably produced by the passage of the air through the œsophagus when it is expelled for the purpose of renewal. As the Barramunda has perfectly developed gills, beside the lung, we can hardly doubt that, when it is in water of normal composition, and sufficiently pure to yield the necessary supply of oxygen, these organs are sufficient for the purpose of breathing, and that the respiratory function rests with them alone. But when the fish is compelled to sojourn in thick muddy water charged with gases, which are the products of decomposing organic matter (and this must be the case very frequently during the droughts which annually exhaust the creeks of tropical Australia), it commences to breathe air with its lung in the way indicated above. If the medium in which it happens to be is perfectly unfit for breathing the gills cease to have any function; if only in a less degree the gills may still continue to assist in respiration. The Barramunda, in fact, can breathe by either gills or lungs alone, or by both simultaneously. It is not probable that it lives freely out of the water, its limbs being much too flexible for supporting the heavy and unwieldy body, and too feeble generally to be of much use in locomotion on land. However, it is quite possible that it is occasionally compelled to leave the water, although we cannot believe that it can exist without it in a lively condition for any length of time.

Of its propagation or development we know nothing, except that it deposits a great number of eggs of the size of those of a newt, and enveloped in a gelatinous case. We may infer that the young are provided with external gills, as in Protopterus and Polypterus.

The discovery of *Ceratodus* does not date farther back than the year 1870, and proved to be of the greatest interest, not only on account of the relation of this creature to the other

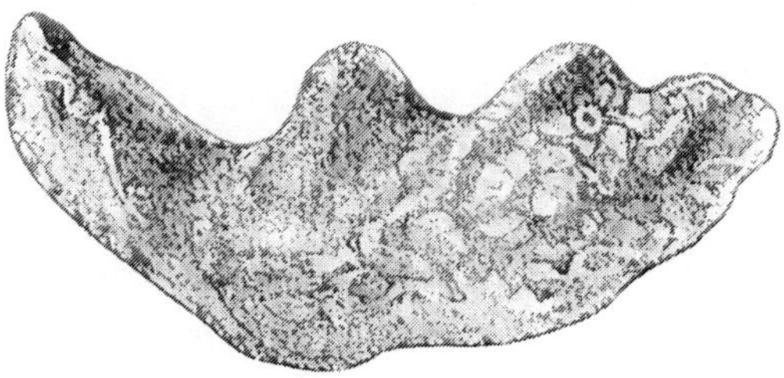

Fig. 140.—Tooth of fossil *Ceratodus* from Aust, near Bristol, natural size.

living *Dipnoi* and *Ganoidei*, but also because it threw fresh light on those singular fossil teeth which are found in strata of Triassic and Jurassic formations in various parts of Europe, India, and America. These teeth, of which there is a great variety with regard to general shape and size, are sometimes two inches long, much longer than broad, depressed, with a flat or slightly undulated, always punctated crown, with one margin convex, and with from three to seven prongs projecting on the opposite margin.

Second Family—Ctenododipteridæ.

Caudal fin heterocercal. Gular plates. Scales cycloid. Two pairs of molars and one pair of vomerine teeth.

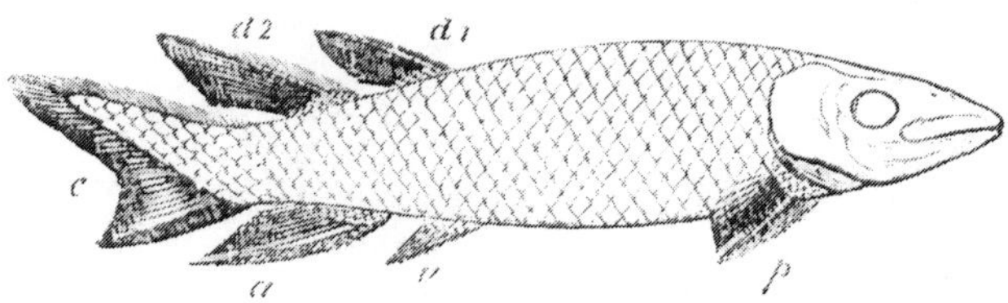

Fig. 141.—Dipterus macrolepidotus.

Extinct. *Dipterus* (*Ctenodus*), *Heliodus* from Devonian strata.

THIRD FAMILY—PHANEROPLEURIDÆ.

Caudal fin diphycercal; vertical fin continuous. Gular plates. Scales cycloid. Jaws with a series of minute conical teeth on the margin.

Extinct. *Phaneropleuron* from Devonian formations, and the carboniferous *Uronemus* are probably generically identical.

FOURTH SUB-ORDER—CHONDROSTEI.

Skeleton notochordal; skull cartilaginous, with dermal ossifications; branchiostegals few in number or absent. Teeth minute or absent. Integuments naked or with bucklers. Caudal fin heterocercal, with fulcra. Nostrils double, in front of the eyes.

FIRST FAMILY—ACIPENSERIDÆ.

Body elongate, sub-cylindrical, with five rows of osseous bucklers. Snout produced, subspatulate or conical, with the mouth at its lower surface, small, transverse, protractile, toothless. Four barbels in a transverse series on the lower side of the snout. Vertical fins with a single series of fulcra in front. Dorsal and anal fins approximate to the caudal. Gill-membranes confluent at the throat and attached to the isthmus. Branchiostegals none. Gills four; two accessory gills. Air-bladder large, simple, communicating with the dorsal wall of the œsophagus.

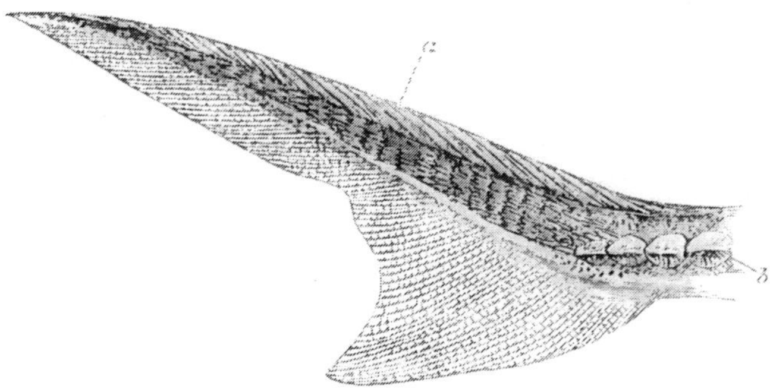

Fig. 142.—Tail of Acipenser. *a*, Fulcra; *b*, osseous bucklers.

Sturgeons are, perhaps, the geologically youngest Ganoids, evidence of their existence not having been met with hitherto in formations of older date than the Eocene clay of Sheppey. They are exclusively inhabitants of the temperate zone of the Northern Hemisphere, being either entirely confined to fresh water, or passing, for the purpose of spawning, a part of the year in rivers. They grow to a large size, and are the largest fishes of the fresh waters of the Northern Hemisphere, specimens 10 feet long being of common occurrence. The ova are very small, and so numerous that one female has been calculated to produce about three millions at one season; therefore their propagation, as well as their growth, must be very rapid; and although in many rivers their number is annually considerably thinned by the systematic manner in which they are caught when they ascend the rivers in shoals from the sea, no diminution has been observed. Wherever they occur they prove to be most valuable on account of their wholesome flesh. In Russia, besides, two not unimportant articles of trade are obtained from them, viz. Caviare, which is prepared from their ovaries, and Isinglass, which is made from the inner coats of their air-bladder. True Sturgeons are divided into two genera, *Acipenser* and *Scaphirhynchus*.

Acipenser.—The rows of osseous bucklers are not confluent on the tail. Spiracles present. Caudal rays surrounding the extremity of the tail.

About twenty different species of Sturgeons may be distinguished from European, Asiatic, and American rivers. The best known are the Sterlet (*A. ruthenus*) from Russian rivers, celebrated for the excellency of its flesh, but rarely exceeding a length of three feet; the Californian Short-snouted Sturgeon (*A. brachyrhynchus*); the Hausen (*A. huso*), from rivers, falling into the Black Sea and the Sea of Azow (rare in Mediterranean), sometimes 12 feet long, and yielding an inferior kind of isinglass; the Chinese Sturgeon (*A. sinensis*);

the Common Sturgeon of the United States (*A. maculosus*), which sometimes crosses the Atlantic to the coasts of Great Britain; Güldenstædt's Sturgeon (*A. güldenstædtii*), common in European and Asiatic rivers, which yields more than one-fourth of the caviare and isinglass exported from Russia; the Common Sturgeon of Western Europe (*A. sturio*), which attains to a length of 18 feet, and has established itself also on the coasts of Eastern North America.

SCAPHIRHYNCHUS.—Snout spatulate; posterior part of the tail attenuated and depressed, so that it is entirely enveloped by the osseous scutes. Spiracles none. The caudal rays do not extend to the extremity of the tail, which terminates in a filament.

Four species are known: one (*S. platyrhynchus*) from the river-system of the Mississippi, and the three others from Central Asia; all are exclusively freshwater fishes; their occurrence in so widely distant rivers is one of the most striking instances by which the close affinity of the North American and North Asiatic faunas is proved.

SECOND FAMILY—POLYODONTIDÆ.

Body naked, or with minute stellate ossifications. Mouth lateral, very wide, with minute teeth in both jaws. Barbels none. Caudal fin with fulcra. Dorsal and anal fins approximate to the caudal. Four gills and a half; no opercular gill or pseudobranchia.

POLYODON (SPATULARIA).—The snout is produced into an exceedingly long, shovel-like process, thin and flexible on the sides. Spiracles present. Gill-cover terminating in a very long tapering flap. One broad branchiostegal. Each branchial arch with a double series of very long, fine, and numerous gill-rakers, the two series being divided by a broad membrane. Air-bladder cellular. Upper caudal fulcra narrow, numerous.

The single species, *P. folium*, occurs in the Mississippi,

and grows to a length of about six feet, of which the rostral shovel takes about one-fourth; in young examples it is comparatively still longer.

PSEPHURUS differs from *Polyodon* in having the rostral process less depressed and more conical. The gill-rakers are comparatively short, in moderate number, and distant from one another. Upper caudal fulcra enormously developed, and in small number (six).

Psephurus gladius inhabits the Yan-tse-Kiang and Hoangho, the distribution of the *Polyodontidæ* being perfectly analogous to that of *Scaphirhynchus*. It grows to an immense size,

Fig. 143.—Psephurus gladius.

specimens of 20 feet in length being mentioned by Basilewsky. The function of the rostral process in the economy of these fishes is not yet sufficiently explained. Martens believes that it serves as an organ of feeling, the water of those large Asiatic and American rivers being too turbid to admit of the Sturgeon seeing its prey, which consists of other fishes. The eyes of *Psephurus*, as well as *Polyodon*, are remarkably small. Both fishes are used as food.

Allied to the *Polyodontidæ*, and likewise provided with a paddle-shaped production of the fore part of the head, is the fossil genus *Chondrosteus*, remains of which occur in the Lias.

FIFTH SUB-ORDER—POLYPTEROIDEI.

Paired fins with axial skeleton, fringed; dorsal fins two or more. Branchiostegals absent, but generally gular plates. Vertebral column diphycercal or heterocercal. Body scaly.

First Family—Polypteridæ.

Scales ganoid; fins without fulcra. A series of dorsal spines, to each of which an articulated finlet is attached; anal placed close to the caudal fin, the vent being near the end of the tail. Abdominal portion of the vertebral column much longer than the caudal.

Polypterus.—Teeth rasp-like, in broad bands in the jaws, on the vomer and palatine bones; jaws with an outer series of closely-set, larger, pointed teeth. Caudal fin surrounding the extremity of the vertebral column; ventral fins well developed. A spiracle on each side of the parietal, covered with an osseous plate. A single large gular plate. Air-bladder double, communicating with the ventral wall of the pharynx.

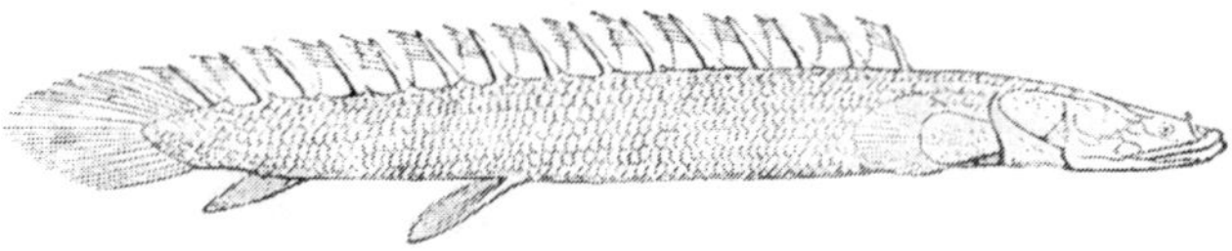

Fig. 144.—Polypterus bichir.

This Ganoid is confined to tropical Africa, occurring in abundance in the rivers of the west coast and in the Upper Nile; but it has not been found in the river-systems belonging to the Indian Ocean. It is scarce in the Middle and Lower Nile, and the specimens found below the Cataracts have been carried down, from southern latitudes, and do not propagate their species in that part of the river. There is only one species known, *Polypterus bichir* ("Bichir" being its vernacular name in Egypt), which varies in the number of the dorsal finlets, the lowest being eight, the highest eighteen. It attains to a length of four feet. Nothing is known of its mode of life, and observations thereon are very desirable.

Calamoichthys.—Distinguished from *Polypterus* by its greatly elongate form, and the absence of ventral fins.

C. calabaricus, a dwarf form from Old Calabar.

Second Family—Saurodipteridæ.

Scales ganoid, smooth like the surface of the skull. Two dorsal fins; paired fins obtusely lobate. Teeth conical. Caudal heterocercal.

Extinct. *Diplopterus, Megalichthys,* and *Osteolepis* from Devonian and Carboniferous formations.

Third Family—Coelacanthidæ.

Scales cycloid. Two dorsal fins, each supported by a single two-pronged interspinous bone; paired fins obtusely lobate. Air-bladder ossified; notochord persistent, diphycercal.

Extinct. *Coelacanthus* from carboniferous strata; several other genera, from the coal formations to the chalk, have been associated with it—*Undina, Graphiurus, Macropoma, Holophagus, Hoplopygus, Rhizodus.*

Fourth Family—Holoptychiidæ.

Scales cycloid or ganoid, sculptured. Two dorsal fins; pectorals narrow, acutely lobate; dentition dendrodont.

Extinct. In this family a peculiar type of dentition is found—the jaws are armed with two kinds of teeth, small ones serially arranged, and much larger fang-like teeth disposed at long intervals. Both kinds show at their base in transverse sections a labyrinthic complexity of structure, numerous fissures radiating from the central mass of vasodentine which fills up the pulp cavity, and sending off small ramifying branches. Genera belonging to this family are *Holoptychius, Saurichthys, Glyptolepis, Dendrodus, Glyptolaemus, Glyptopomus, Tristichopterus, Gyroptychius, Strepsodus,* from Devonian and Carboniferous strata.

SIXTH SUB-ORDER—PYCNODONTOIDEI.

Body compressed, high and short or oval, covered with rhombic scales arranged in decussating pleurolepidal lines. Notochord persistent. Paired fins without axial skeleton. Teeth on the palate and hinder part of the lower jaw molar-like. Branchiostegals, but no gular plates.

Extinct. The regular lozenge-shaped pattern of the integuments of these fishes is described by Sir P. Egerton thus: "Each scale bears upon its inner anterior margin a thick solid bony rib, extending upwards beyond the margin of the scale, and sliced off obliquely, above and below, on opposite sides, for forming splices with the corresponding processes of the adjoining scales. These splices are so closely adjusted that, without a magnifying power or an accidental dislocation, they are not perceptible. When *in situ*, and seen internally, these continuous lines decussate with the true vertebral apophyses." In some genera the "pleurolepidal" lines are confined to the anterior part of the side.

FIRST FAMILY—PLEUROLEPIDÆ.

Homocereal. Body less high. Fins with fulcra.
Pleurolepis and *Homocolepis* from the Lias.

SECOND FAMILY—PYCNODONTIDÆ.

Homocercal. The neural arches and ribs are ossified; the roots of the ribs are but little expanded in the older genera, but enlarged in the tertiary forms, so as to simulate vertebræ. Paired fins not lobate. Obtuse teeth on the palate and the sides of the mandible; maxilla toothless; incisor-like teeth in the intermaxillary and front of the mandible. Fulcra absent in all the fins.

These fishes abound in Mesozoic and Tertiary formations.

Gyrodus, Mesturus, Microdon, Coelodus, Pycnodus, Mesodon, are some of the genera distinguished by palæontologists. (See Fig. 102, p. 201.)

SEVENTH SUB-ORDER—LEPIDOSTEOIDEI.

Scales ganoid, rhombic; fins generally with fulcra; paired fins not lobate. Præ- and inter- operculum developed; generally numerous branchiostegals, but no gular plate.

FIRST FAMILY—LEPIDOSTEIDÆ.

Scales ganoid, lozenge-shaped. Skeleton completely ossified; vertebræ convex in front and concave behind. Fins with fulcra; dorsal and anal composed of articulated rays only, placed far backwards, close to the caudal. Abdominal part of the vertebral column much longer than caudal. Branchiostegals not numerous, without enamelled surface. Heterocercal.

Lepidosteus.—Body elongate, sub-cylindrical; snout elongate, spatulate, or beak-shaped; cleft of the mouth wide; both jaws and palate armed with bands of rasp-like teeth and series of larger conical teeth. Four gills; no spiracles; three branchiostegals. Air-bladder cellular, communicating with the pharynx.

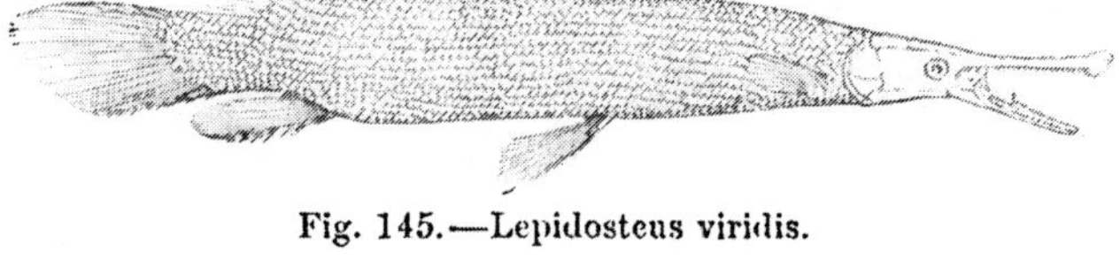

Fig. 145.—Lepidosteus viridis.

Fishes of this genus existed already in Tertiary times; their remains have been found in Europe as well as North America. In our period they are limited to the temperate parts of North America, Central America, and Cuba. Three species can be distinguished which attain to a length of about six feet. They feed on other fishes, and their general resemblance to a pike has given to them the vernacular names of "Gar-Pike," or "Bony Pike."

SECOND FAMILY—SAURIDÆ.

Body oblong, with ganoid scales; vertebræ not completely ossified; termination of the vertebral column homocercal; fins generally with fulcra. Maxillary composed of a single piece; jaws with a single series of conical pointed teeth. Branchiostegals numerous, enamelled, the anterior broad gular plates.

Extinct. Numerous genera occur in Mesozoic formations; one with the widest range is *Semionotus*, with distichous fulcra, from the Lias and Jura; *Eugnathus*, with large posteriorly serrated scales, and fulcra on nearly all fins; *Cephenoplosus* from the Upper Lias; *Macrosemius* from the Oolite; *Propterus, Ophiopsis, Pholidophorus, Pleuropholis, Pachycormus, Oxygnathus, Ptycholepis, Conodus, Eulepidotus, Lophiostomus*, etc.

THIRD FAMILY—STYLODONTIDÆ.

Body rhombic or ovate, with ganoid scales; vertebræ not completely ossified; termination of the vertebral column homocercal; fins with fulcra. Maxillary composed of a single piece; jaws with several series of teeth, the outer ones equal, styliform. Dorsal fin very long, extending to the caudal. Branchiostegals numerous.

Extinct. *Tetragonolepis* from the Lias (see Fig. 103, p. 207).

FOURTH FAMILY—SPHÆRODONTIDÆ.

Body oblong, with rhombic ganoid scales; vertebræ ossified, but not completely closed; homocercal; fins with fulcra. Maxillary composed of a single piece; teeth in several series, obtuse; those on the palate globular. Dorsal and anal fins short. Branchiostegals.

Extinct. The type genus of this family is *Lepidotus*, so named from its large rhombic, dense, and polished scales.

The dorsal is opposite to the anal, and all the fins are provided with a double row of fulcra. This genus ranges from the Lias to the Chalk; one species would seem to have survived into tertiary times, if it should not prove to be a *Lepidosteus.*

FIFTH FAMILY—ASPIDORHYNCHIDÆ.

Body elongate, with ganoid scales; jaws prolonged into a beak; termination of the vertebral column homocercal. Fins with fulcra; a series of enlarged scales along the side of the body. Dorsal fin opposite to the anal.

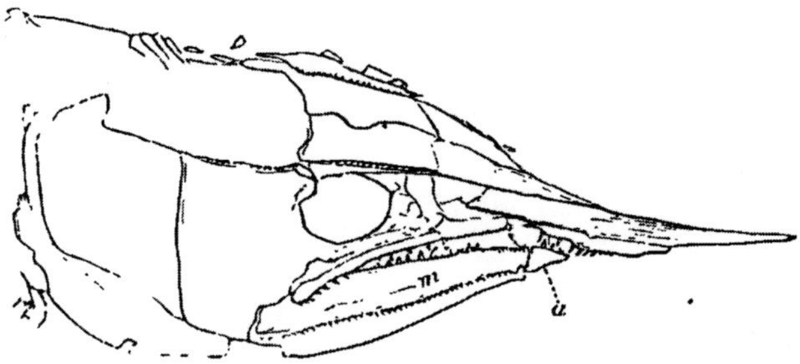

Fig. 146.—Aspidorhynchus fisheri, from the Purbeck beds; *m*, mandible; *a*, presymphyseal bone.

Extinct; mesozoic. *Aspidorhynchus* has the upper jaw longer than the lower; very peculiar is the occurrence of a single, solid, conical bone, situated in front of the symphysis of the lower jaw, to which it is joined by a suture. *Belonostomus* with both jaws of equal length.

SIXTH FAMILY—PALÆONISCIDÆ.

Body fusiform, with rhombic ganoid scales. Notochord persistent, with the vertebral arches ossified. Heterocercal. All the fins with fulcra; dorsal short. Branchiostegals numerous, the foremost pair forming broad gulars. Teeth small, conical, or cylindrical.

Extinct. Many genera are known; from the Old Red

Sandstone—*Chirolepis* and *Acrolepis;* from Carboniferous rocks—*Cosmoptychius, Elonichthys, Nematoptychius, Cycloptychius, Microconodus, Gonatodus, Rhadinichthys, Myriolepis, Urosthenes;* from the Permian—*Rhabdolepis, Palæoniscus, Amblypterus* and *Pygopterus;* from the Lias—*Centrolepis, Oxygnathus, Cosmolepis,* and *Thrissonotus.*

[See Traquair, "The Ganoid Fishes of the British Carboniferous Formations." Part I. *Palæoniscidæ.*]

SEVENTH FAMILY—PLATYSOMIDÆ.

Body generally high, compressed, covered with rhombic ganoid scales arranged in dorso-ventral bands. Notochord persistent, with the vertebral arches ossified. Heterocercal; fins with fulera; dorsal fin long, occupying the posterior half of the back. Branchiostegals numerous: Teeth tubercular or obtuse.

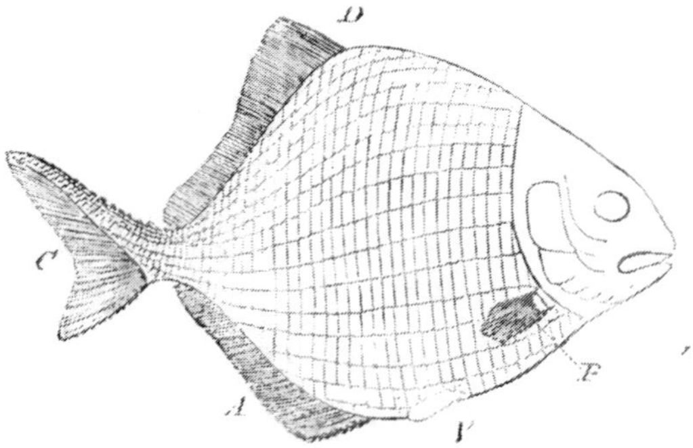

Fig. 147.—Platysomus gibbosus.

Extinct. From Carboniferous and Permian formations—*Eurynotus, Benedenius, Mesolepis, Eurysomus, Wardichthys, Chirodus* (M'Coy), *Platysomus.*

[See Traquair, "On the Structure and Affinities of the Platysomidæ, in "Trans. Roy. Soc.," Edinb., vol. xxix.]

EIGHTH SUB-ORDER—AMIOIDEI.

Vertebral column more or less completely ossified, heterocercal. Body covered with cycloid scales. Branchiostegals present.

First Family—Caturidæ.

Notochord persistent, with partially ossified vertebræ; homocercal; fins with fulcra. Teeth in a single series, small, pointed.

Fig. 148.—Caturus furcatus (Solenhofen).

Extinct. *Caturus* from the Oolite to the Chalk.

Second Family—Leptolepidæ.

Scales cycloid. Vertebræ ossified; homocercal; fins without fulcra; dorsal short. Teeth minute, in bands, with canines in front.

Extinct, and leading to the living representative of this suborder. *Thrissops* with the dorsal fin placed far backwards, and opposite to the long anal. *Leptolepis* with the dorsal fin

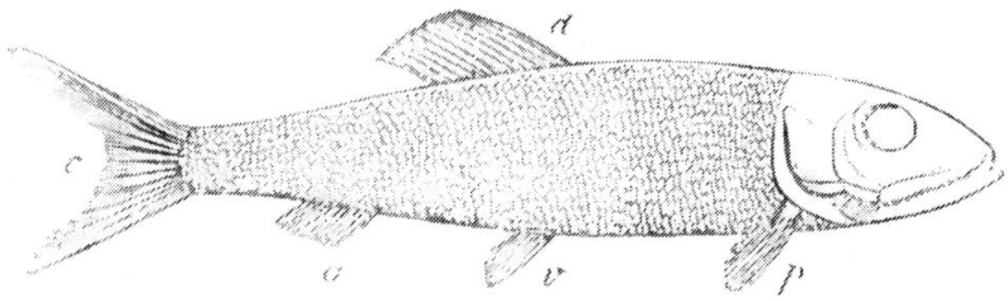

Fig. 149.—Leptolepis sprattiformis.

opposite to the ventrals, from the Lias and Oolite. These fishes, as far as the preserved parts are concerned, cannot be distinguished from Teleosteous fishes, to which they are referred by some Palæontologists.

Third Family—Amiidæ.

Skeleton entirely ossified; a single large gular plate; homo-

cercal; fins without fulcra; a long soft dorsal fin. Abdominal and caudal parts of the vertebral column subequal in extent. Branchiostegals numerous.

AMIA.—Body rather elongate, sub-cylindrical, compressed behind. Snout short, cleft of the mouth of moderate width. Jaws

Fig. 150.—Amia calva; *g*, gular plate.

with an outer series of closely-set pointed teeth, and with a band of rasp-like teeth; similar teeth on the vomer, palatine, and pterygoid bones. Anal short; caudal fin rounded. Gills four; air-bladder bifurcate in front, cellular, communicating with the pharynx.

The "Bow-fin" or "Mud-fish" (*A. calva*) is not uncommon in many of the fresh waters of the United States; it grows to a length of two feet. Little is known about its habits; small fish, crustaceans, and aquatic insects, have been found in its stomach. Wilder has observed its respiratory actions; it rises to the surface, and, without emitting any air-bubble whatever, opens the jaws widely, and apparently gulps in a large quantity of air; these acts of respiration are more frequently performed when the water is foul or has not been changed; and there is no doubt that an exchange of oxygen and carbonic acid is effected, as in the lungs of aërial vertebrates. The flesh of this fish is not esteemed.

Fossil remains occur in tertiary deposits of North America, for instance in the Wyoming territory; they have been distinguished as *Protamia* and *Hypamia*.

SECOND SUB-CLASS—TELEOSTEI.

Heart with a non-contractile bulbus arteriosus; intestine without spiral valve; optic nerves decussating; skeleton ossified, with completely formed vertebræ; vertebral column diphycercal or homocercal; branchiæ free.[1]

The Teleostei form the majority of the fishes of the present fauna, and are the geological successors of the Palæichthyes, undoubted Teleostei not ranging farther back than the Chalk. This sub-class comprises an infinite variety of forms; and as, naturally, many Ganoid fishes lived under similar external conditions, and led a similar mode of life as certain Teleostei, we find not a few analogous forms in both series: some Ganoids resembling externally the Teleosteous Siluroids, others the Clupeoids, others the Chætodonts, others the Scombresoces, etc. But there is no direct genetic relation between those fishes, as some Naturalists were inclined to believe.

The Teleostei are divided into six orders:—

A. Acanthopterygii.—Part of the rays of the dorsal, anal, and ventral fins not articulated, spines. The lower pharyngeals separate. Air-bladder, if present, without pneumatic duct in the adult.

B. Acanthopterygii pharyngognathi.—Part of the rays of the dorsal, anal, and ventral fins not articulated, spines. The lower pharyngeals coalesced. Air-bladder without pneumatic duct.

[1] See p. 97, Fig. 41; and p. 152, Fig. 68.

C. Anacanthini.—Vertical and ventral fins without spinous rays. Ventral fins, if present, jugular or thoracic. Air-bladder, if present, without pneumatic duct. Lower pharyngeals separate.

D. Physostomi.—All the fin rays articulated ; only the first of the dorsal and pectoral fins is sometimes ossified. Ventral fins, if present, abdominal, without spine. Air-bladder, if present, with a pneumatic duct.

E. Lophobranchii.—Gills not laminated, but composed of small rounded lobes, attached to the branchial arches. Gill-cover reduced to a large simple plate. A dermal skeleton replaces more or less soft integuments.

F. Plectognathi.—A soft dorsal fin opposite to the anal; sometimes elements of a spinous dorsal. Ventral fins none, or reduced to spines. Gills pectinate; air-bladder without pneumatic duct. Skin with rough scutes, or with spines, or naked.

FIRST ORDER—ACANTHOPTERYGII.

Part of the rays of the dorsal, anal, and ventral fins are not articulated, more or less pungent spines. The lower pharyngeals are generally separate. Air-bladder, if present, without pneumatic duct in the adult.[1]

First Division—Acanthopterygii Perciformes.

Body more or less compressed, elevated or oblong, but not elongate; the vent is remote from the extremity of the tail, behind the ventral fins if they are present. No prominent anal papilla. No superbranchial organ. Dorsal fin or fins occupying the

[1] The Acanthopterygians do not form a perfectly natural group, some heterogeneous elements being mixed up with it. Neither are the characters, by which it is circumscribed, absolutely distinctive. In some forms (certain Blennioids) the structure of the fins is almost the same as in Anacanths; there are some Acanthopterygians, as *Gerres*, *Pogonias*, which possess coalesced pharyngeals ; and, finally, the presence or absence of a pneumatic duct loses much of its value as a taxonomic character when we consider that probably in all fishes a communication between pharynx and air-bladder exists at an early stage of development.

greater portion of the back; spinous dorsal well developed, generally with stiff spines, of moderate extent, rather longer than, or as long as, the soft; the soft anal similar to the soft dorsal, of moderate extent or rather short. Ventrals thoracic, with one spine and with four or five rays.

First Family—Percidæ.

The scales extend but rarely over the vertical fins, and the lateral line is generally present, continuous from the head to the caudal fin. All the teeth simple and conical; no barbels. No bony stay for the præoperculum.

A large family, represented by numerous genera and species in fresh waters, and on all the coasts of the temperate and tropical regions. Carnivorous.

Fossil Percoids abound in some formations, for instance, at Monte Bolca, where species of *Labrax*, *Lates*, *Smerdis* and *Cyclopoma* (both extinct), *Dules*, *Serranus*, *Apogon*, *Therapon*, and *Pristipoma* have been recognised. *Paraperca* is a genus recently discovered in the Marles of Aix-en-Provence. A species of *Perca* is known from the freshwater deposit of Oeningen.

Perca.—All the teeth are villiform, without canines; teeth on the palatine bones and vomer; tongue toothless. Two dorsal fins, the first with thirteen or fourteen spines; anal fin with two spines. Præoperculum and præorbital serrated. Scales small; head naked above. Branchiostegals seven. Vertebræ more than twenty-four.

The "Freshwater Perch" (*Perca fluviatilis*) is too familiarly known to require description. It is generally distributed over Europe and Northern Asia; and equally common in North America, there being no sufficient ground for separating specifically the specimens of the Western Hemisphere. It frequents especially still waters, and sometimes descends into brackish water. Its weight does not seem to exceed

5 lbs. The female deposits her ova, united together by a viscid matter, in lengthened or net-shaped bands, on water-plants. The number of the eggs of one spawn may exceed a

Fig. 151.—Perca fluviatilis, the Perch.

million. Two other species, *P. gracilis*, from Canada, and *P. schrenckii*, from Turkestan, have been distinguished, but they are very imperfectly known. An allied genus is *Siniperca*, from Northern China.

PERCICHTHYS.—Differing from *Perca* especially in the number of the fin-spines, which are nine or ten in the first dorsal, and three in the anal fin. The upper surface of the head scaly.

These fishes represent the Freshwater Perches of the Northern Hemisphere in the fresh waters of the temperate parts of South America. Two species have been described from Patagonia, and one or two from Chili and Peru.

LABRAX.—All the teeth are villiform, without canines; teeth on the palatine bones, vomer, and the tongue. Two dorsal fins, the first with nine spines; anal fin generally with three. Præoperculum serrated, and with denticulations along its lower limb; præorbital with the margin entire. Scales rather small. Branchiostegals seven; well developed pseudobranchiæ.

The "Bass" are fishes common on the coasts of Europe and the Atlantic coasts, and in the fresh waters of the United States and Canada. The three European species are almost exclusively inhabitants of the sea, entering brackish, but never fresh waters, whilst the American species, the number of which is still uncertain, seem to affect principally fresh water, although some are also found in the sea. The best known European species is *Labrax lupus* (see p. 41, Fig. 4), common on the British coasts. It is a voracious fish, with a remarkably large stomach, and received from the ancient Romans the appropriate name of *lupus*. By the Greeks it was so highly esteemed that Archestratus termed this or one of the two other closely-allied species, taken near Milet, "offspring of the gods." They attributed to it a tender regard for its own safety; and Aristotle says that it is the most cunning of fishes; and that, when surrounded by the net, it digs for itself a channel of escape through the sand. Specimens of from two to three feet are not scarce, but its flesh is nowadays much less esteemed than in ancient times. Of the North American species *Labrax lineatus* and *Labrax rufus* are the most common.

LATES.—All the teeth are villiform, without canines; teeth on the palatine bones and vomer, but none on the tongue. Two dorsal fins—the first with seven or eight, the anal fin with two or three, spines. Præoperculum with strong spines at the angle and the lower limb; also the præorbital is strongly serrated. Scales of moderate size. Branchiostegals seven; pseudobranchiæ present.

Three well-known species belong to this genus. The Perch of the Nile and other rivers of tropical Africa (*Lates niloticus*); the Perch of the Ganges and other East Indian rivers, which enters freely brackish water, and extends to the rivers of Queensland (*Lates calcarifer*). These two species attain to a large size, the Indian species to a length of five feet. Hamilton says that "the vulgar English in Calcutta call it 'Cockup,' and that it is one of the lightest and most esteemed foods brought to table in that city." Specimens two feet in length and caught in salt water are by far the best quality. The third species (*Lates colonorum*) is found in

Australia only, and does not appear to grow to the same large size as its congeners.

Allied to *Lates* is *Psammoperca* from Australia.

PERCALABRAX.—All the teeth villiform, without canines; teeth on the palatine bones and vomer, but none on the tongue. Two dorsal fins—the first with eleven, the anal fin with three spines. Præoperculum serrated along its hinder margin, and with strong spinous teeth below; præorbital not serrated. Scales rather small. Branchiostegals seven; pseudobranchiæ present.

This Perch (*Percalabrax japonicus*) is one of the most common fishes on the coasts of China, Japan, and Formosa; the Japanese name it "Zuzuki," or "Seengo."

ACERINA.—All the teeth villiform, without canines; teeth on the vomer, but none on the palatine bones or the tongue. One continuous dorsal fin, of which the spinous portion consists of from thirteen to nineteen spines; two anal spines. Body rather low, with rather small scales. Bones of the skull with wide muciferous cavities; præoperculum denticulated.

Small freshwater perches, of which *A. cernua*, named "Pope" in England, is the most common, and has the widest distribution in Central Europe and Siberia. The two other species have a more restricted range, *A. schraetzer* being confined to the Danube and other rivers emptying into the Black Sea; and *A. czekanowskii* to Siberian rivers. This genus is not represented in the Western Hemisphere.

LUCIOPERCA.—Teeth in villiform bands, those in the jaws with additional canines; palatine bones toothed. Two dorsal fins—the first with from twelve to fourteen, the anal fin with two spines. Præoperculum serrated; scales small.

The "Pike-Perches" are inhabitants of many lakes and rivers of the temperate northern zone. The European species is confined to the eastern two-thirds of the continent, and one of the most esteemed freshwater fishes; it attains to a length of three or four feet, and to a weight of from 25 to 30 lbs.

It has been recommended for acclimatisation in England, and there is no doubt that in certain localities it might prove a valuable addition to the native fauna; but like all its congeners it is very voracious and destructive to smaller fishes. Two other species inhabit rivers of European and Asiatic Russia, and two or three the fresh waters of North America.

PILEOMA.—All the teeth minute, villiform, without canines; teeth on the vomer and palatine bones. Two dorsal fins—the first with fourteen or fifteen spines. Body rather elongate, with small scales. Præoperculum not serrated.

Small freshwater perches abundant in the United States. Like the following genus, and some others which need not be mentioned here, they can be regarded as small, dwarfed representatives of the preceding genera. The species seem to be numerous, but have not yet been sufficiently well distinguished. The latest and best account of them is by L. Vaillant, "Recherches sur les Poissons d'eaux douces de l'Amérique septentrionale (*Etheostomatidæ*)," in Nouv. Archiv. du Muséum d'Hist. Nat. de Paris, ix., 1873.

BOLEOSOMA.—Allied to *Pileoma*, but with only nine or ten feeble spines in the first dorsal fin. North America.

ASPRO.—Body elongate, cylindrical; snout thick, projecting beyond the mouth, which is situated at its lower side. All the teeth villiform, without canines; teeth on the vomer and palatine bones. Two separate dorsals. Præoperculum serrated; præorbital entire. Scales small.

Two small Perches from the Danube and some other rivers of the continent of Europe, *Aspro vulgaris* and *A. zingel*.

CENTROPOMUS.—Body oblong, compressed, with scales of moderate size. All the teeth villiform, without canines; teeth on the vomer and palatine bones. Two dorsal fins, the first with eight strong spines, the anal with three, the second of which is very strong and long. Præoperculum with a double denticulated edge.

Numerous species are known from the West Indies and Central America. These fishes are found in fresh, brackish, and salt water, and some of the species indiscriminately enter all three kinds of water. They do not grow to any large size, but are esteemed as food.

Enoplosus.—Body much elevated, the depth being still more increased by the high vertical fins. All the teeth are villiform, without canines; teeth on the vomer, palatine bones, and the tongue. Two dorsal fins, the first with seven spines. Præoperculum serrated, with spinous teeth at the angle. Scales of moderate size.

A small and very common marine species (*E. armatus*) on the coast of Australia, especially New South Wales. It is readily recognised by the peculiar shape of its body, and eight black transverse bands on a whitish ground.

This, and the preceding genus, leads to the true "Sea perches," which never, or but rarely, enter fresh water:—

Centropristis.—Body oblong, with scales of small or moderate size. Teeth villiform, with small canines in both jaws; vomerine teeth placed in an angular band, or a short triangular patch; teeth on the palatine bones, but none on the tongue. One dorsal, with the formula $\frac{10}{12 \text{ or less}}$; anal fin $\frac{3}{7\ (6)}$. Præoperculum serrated; sometimes with the angle projecting, and armed with long spines.

About twenty species of small size are known from temperate and tropical seas.

Anthias. — Body rather short, compressed, with scales of moderate size. Teeth villiform, with small canines in both jaws; teeth on the vomer, and palatine teeth. One dorsal, generally with ten spines; anal fin with three; caudal forked. The rays of one or more fins may be prolonged. Præoperculum serrated.

About twenty species are known from temperate and tropical seas; they are mostly of small size, and agreeably coloured, pink and yellow being the predominant colours. *Anthias sacer* is common in the Mediterranean, and was well known to the ancients. Aristotle says that the fishers of Sponges call it sacred, because no voracious fishes came to the places which it frequented, and the diver might descend with safety.—*Callanthias* is a genus closely allied to *Anthias*.

SERRANUS.—Body oblong, compressed, with small scales. Teeth villiform, with very distinct canines in both jaws; teeth on the vomer and palatine bones, none on the tongue. One dorsal, mostly with nine or eleven, rarely with eight, ten or twelve spines; anal fin with three: all the spines being stout. Præoperculum serrated behind and at the angle, but not below.

The "Sea perches proper" are found on the shores of all temperate and tropical seas, most abundantly in the latter. A few species enter brackish and even fresh water, one having been found as high up the Ganges as the confines of Nepal. However, all spawn in the sea. The variety of species is almost infinite, about 150 being tolerably well known, and many more having been described. The distinction of the species is most difficult, and nearly impossible to those who have no opportunity of closely and long observing them in nature, as they are not only subject to great variation of colour, but also to considerable changes dependent on age.

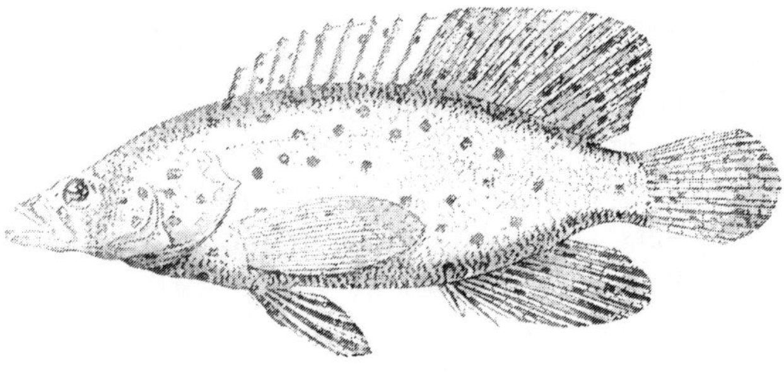

Fig. 152.—Serranus altivelis.

Many are most agreeably coloured, and ornamented with spots, or cross bands or longitudinal stripes; colours which become more uniform with age in those species which attain to a large size. The majority remain of rather small size, growing to a length of one or two feet; but not a few reach more than twice that length, and may even become dangerous to man. Instances of bathers having been attacked by a gigantic species not uncommon at the Seychelles and Aden are on record, the persons having died from the injuries received. Almost all the species are eatable, and many are esteemed as food. One species is common on the British coasts (*S. cabrilla*), and probably some of the more southern species (*S. scriba* and *S. gigas*) occasionally wander as far northwards as the British Channel. The species figured, *S. altivelis*, is locally distributed over nearly all the tropical parts of the Indian Ocean, and distinguished by particularly high dorsal and anal fins. *Anyperodon* and *Prionodes* are two genera closely allied to *Serranus*.

Plectropoma.—Form of the body and dentition (see p. 127, Fig. 54) similar to that of *Serranus*, with a præoperculum serrated behind, and armed with spinous teeth below, which are directed forwards. Dorsal fin with from seven to thirteen spines.

About thirty species from tropical seas are known. *Trachypoma* is allied to this genus.

Polyprion.—Body oblong, rather compressed, covered with small scales. All the teeth are villiform; teeth on the vomer, palatine bones, and the tongue. One dorsal with eleven or twelve spines; anal with three. Præoperculum denticulated; a strong, rough, longitudinal ridge on the operculum.

Two species are known: one from the European coasts (*P. cernium*), and one from Juan Fernandez (*P. kneri*). They attain to a weight of 80 lbs. and more. The European species has the habit of accompanying floating wood, to which they are attracted by the small marine species generally surrounding such objects, and affording a supply of food. It is known by the name of "Stone-bass," and excellent eating.

Grammistes.—Body rather short, compressed, covered with

minute scales imbedded in the thick skin. All the teeth are villiform; teeth on the vomer and palatine bones. Two dorsal fins, the first with seven spines. Præoperculum without serrature, but with three short spines. A short skinny barbel is frequently developed at the chin.

Three species are known from the Indo-Pacific; they are of small size. One, *G. orientalis*, is black, with six or seven white longitudinal stripes, and one of the most common coast-fishes of that ocean.

Rhypticus.—Body oblong, compressed, covered with minute scales imbedded in the thick skin. All the teeth are villiform; teeth on the vomer and palatine bones. The spines of the vertical fins are but little developed, always in small number and short, and may disappear entirely. Præoperculum not serrated, with some obtuse spines.

Four species are known: three from the West Indies and one from the Galapagoes Islands.

Other genera allied to the two preceding are *Aulacocephalus* from Mauritius, Reunion, and Japan; and *Myriodon* from the coasts of Australia.

Diploprion.—Body rather elevated, compressed, with small scales. All the teeth villiform; teeth on the vomer and palatine bones. Two dorsal fins, the first with eight spines; anal with two. Præoperculum with a double denticulated limb.

The single species known (*D. bifasciatum*), is very com-

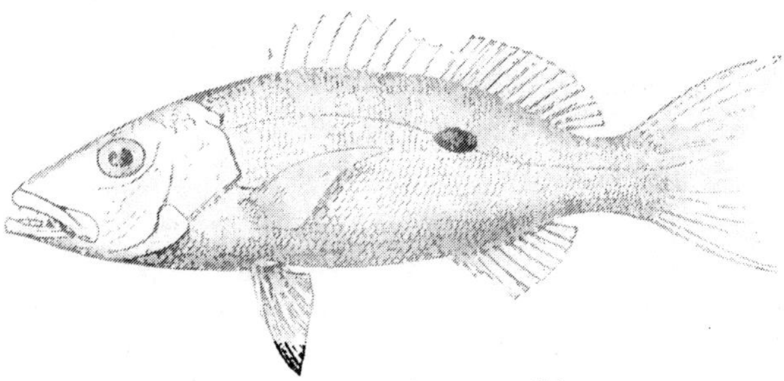

Fig. 153.—Mesoprion monostigma.

mon in the East Indian Archipelago, and on the coasts of

Southern China and Japan. It is of small size, and ornamented with two broad black cross-bands.

Mesoprion.—Body oblong, compressed, covered with scales of moderate size. Teeth villiform, with canines in both jaws; teeth on the vomer and palatine bones. One dorsal fin, with ten or eleven, rarely with more spines; anal fin with three. Præoperculum serrated; in some species (*Genyoroge*) a more or less distinct spinous knob projects from the surface of the interoperculum, and is received in a more or less deep notch of the præopercular margin.

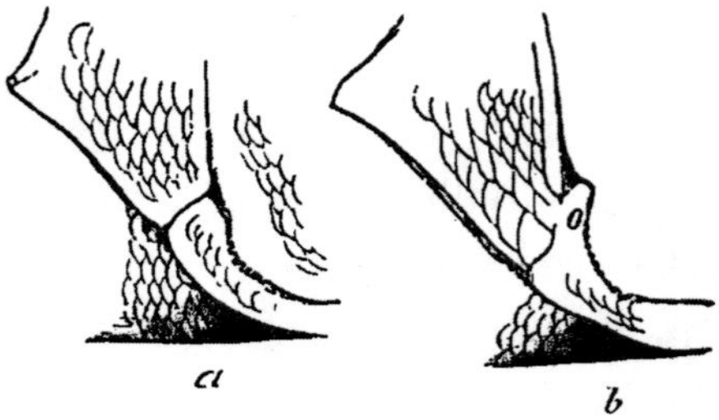

Fig. 154.—Opercles *a*, of *Mesoprion*; *b*, of *Genyoroge*; *o*, knob received in a notch of the præoperculum.

About seventy species are known from tropical seas in both hemispheres, but it is noteworthy that the species with the peculiar protuberance of the interoperculum are confined to the Indo-Pacific. The coloration is much more simple than in the small-scaled Serrani, a uniform hue of greenish, pink, or red prevailing; species with longitudinal bands are scarce, but not rarely dark cross-bands or a large spot on the side occur. The majority of the species remain within very moderate dimensions, specimens exceeding three feet in length being scarce. They are generally eaten, and some of the species belong to the commonest fishes of the tropics, as *M. bengalensis*, *chrysurus*, *gembra*, *griseus*, and others.

Glaucosoma from Japan and Australia is allied to *Mesoprion*.

Dules.—Body oblong, compressed, with scales of moderate size, and very indistinctly ctenoid. All the teeth are villiform; teeth on the vomer and palatine bones. One dorsal with ten spines; anal fin with three. Præoperculum serrated. Six branchiostegals only.

About ten species are known, inhabiting fresh waters of the coasts of the Indo-Pacific, and being especially common in the islands of this region, and also in Tropical Australia. Some live also in brackish water. Though of small size they are esteemed as food.

THERAPON.—Body oblong, compressed, with scales of moderate size. All the teeth are villiform, those of the vomer and palatine bones being rudimentary, and frequently absent. One dorsal, with a depression in its upper margin, and twelve or thirteen spines; anal fin with three. Præoperculum serrated. Air-bladder with two divisions, an anterior and posterior. Six branchiostegals.

About twenty species are known, the distribution of which nearly coincides with *Dules*, but as some of the species are more or less marine, the genus is spread over the whole area of the tropical Indo-Pacific. Other species, especially those of Australian rivers, are entirely limited to fresh water. *Th. theraps, Th. servus*, and *Th. cuvieri* belong to the most common fishes of that area, extending from the east coast of Africa to Polynesia. They are readily recognised by the blackish longitudinal bands with which the body is ornamented. All the species are of small size. *Helotes* is closely allied to this genus.

PRISTIPOMA.—Body oblong, compressed, with ctenoid scales of moderate size. Cleft of the mouth horizontal, not very wide, with the jaws nearly equal in length anteriorly; a central pit below the chin; villiform teeth in the jaws without canines; palate toothless. One dorsal, with eleven to fourteen spines; anal with three. Vertical fins not scaly, or with scales along the base only. Præoperculum serrated. Branchiostegals, seven.

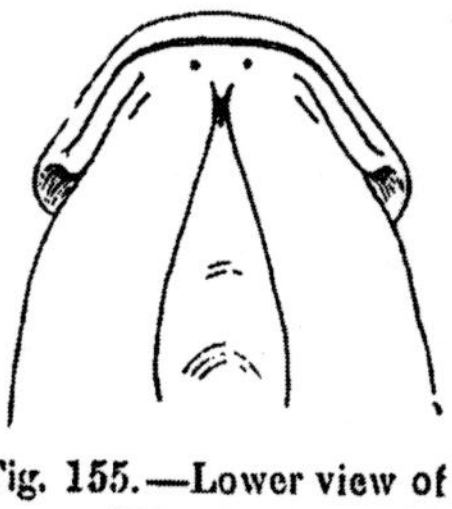

Fig. 155.—Lower view of mandible of Pristipoma manadense.

About forty species are known, all from the sea. They are extremely common between the tropics, some of the species extending into the neighbouring sub-tropical parts. They do not attain a large size, and generally have a plain coloration. *Conodon* is an allied genus.

HÆMULON.—Body oblong, compressed, with ctenoid scales of moderate size. Cleft of the mouth horizontal, generally wide,

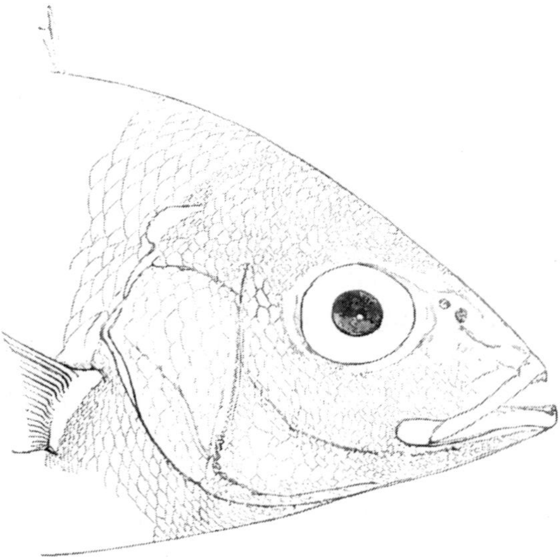

Fig. 156.—Hæmulon brevirostrum.

with the jaws equal in length anteriorly; a central pit below the chin; villiform teeth in the jaws, without canines; palate toothless. One dorsal, with twelve or thirteen spines; anal with three; the soft portions of the vertical fins scaly to their margins. Præoperculum serrated. Branchiostegals, seven.

Marine; sixteen species are known from the coasts of Tropical America; they are of rather small size. The species figured occurs on both sides of Central America. *Hapalogenys* is an allied genus.

DIAGRAMMA.—Body oblong, compressed, covered with rather small ctenoid scales. Upper profile of the head parabolic; cleft of the mouth small, horizontal; from four to six pores under the mandible, but without central pit. Teeth villiform, without canines; palate toothless. One dorsal fin, with from nine to fourteen spines; anal with three. Vertical fins not scaly. Præoperculum serrated; infraorbital not armed. Branchiostegals, six or seven.

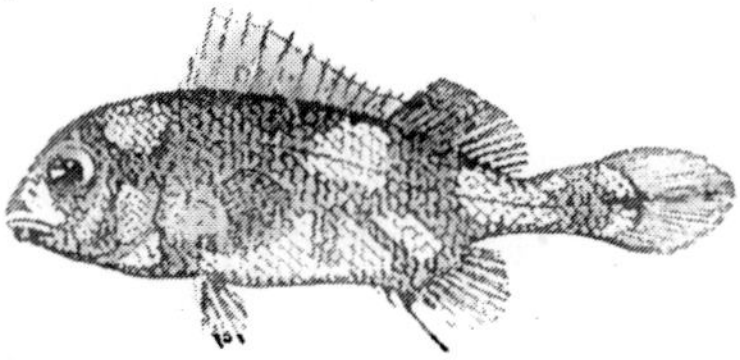

Fig. 157.—Diagramma orientale, from the Indo-Pacific.

Forty species are known, which, with very few exceptions, belong to the tropical parts of the Indo-Pacific. Some attain to a size not very common among Sea-Perches, viz. to a length of from three to four feet. Many are agreeably coloured with black bands or spots. All appear to be esteemed as food. *Hyperoglyphe* from Australia is allied to this genus.

LOBOTES.—Body rather elevated, compressed, with ctenoid scales of moderate size. Eye rather small. Snout obtuse, with oblique cleft of the mouth, and with the lower jaw longest. Teeth villiform, without canines; palate toothless. One dorsal fin with twelve spines; anal with three. Præoperculum denticulated. Branchiostegals, six.

A remarkable fish (*L. auctorum*) on account of its extraordinary range. Common in many localities, scarcer in others, it occurs in the East Indies, and on all the Atlantic coasts of tropical and temperate America. Döderlein found it on the coast of Sicily in 1875. It lives in salt and brackish water, and is known to attain to a length of two feet.

HISTIOPTERUS.—Body rather elevated, strongly compressed, with very small scales. Snout much produced, the anterior profile of the head being concave. Mouth small, at the end of the snout. Teeth villiform, without canines; palate toothless. Some of the spines and rays of the vertical and pectoral fins very long. One dorsal, with about ten spines; anal with three. Præopercular margin partly serrated. Branchiostegals, six.

Marine. Four species are known from Japan and South Australia. The species figured attains to a length of 20 inches, and is esteemed as food. It is known at Melbourne by the names of "Boar Fish" or "Bastard Dorey."

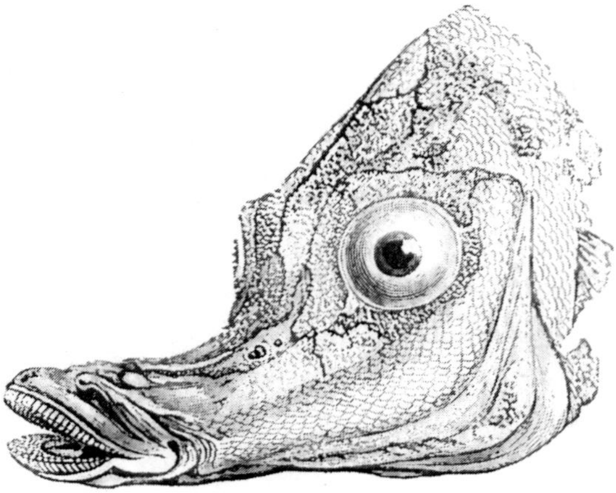

Fig. 158.—Histiopterus recurvirostris.

GERRES.—Body oblong, or rather elevated, covered with scales of moderate size, which are either entirely smooth or minutely ciliated. Mouth very protractile, and descending when thrust out. Eye rather large. No canine teeth; dentition feeble, and

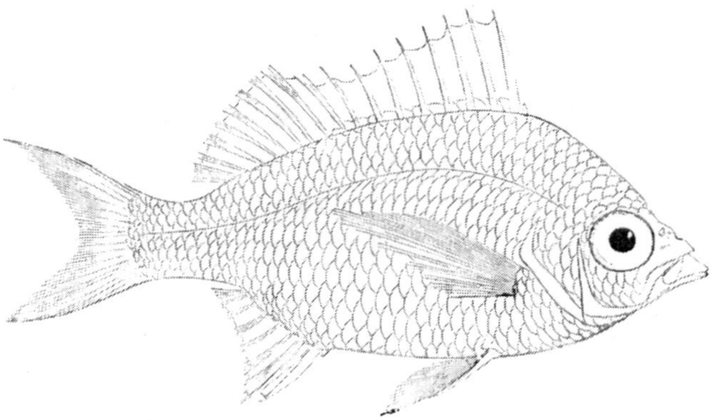

Fig. 159.—Gerres altispinis, from the mouth of the Ganges.

palate toothless. The two divisions of the dorsal fin are nearly separated by a deep incision. Formula of the vertical fins—D. $\frac{9}{10}$ A. $\frac{2\cdot3}{7\cdot9}$. Caudal fin forked. Præoperculum generally without denticulation. Lower pharyngeal bones coalescent.

More than thirty species are known of this genus, which bear so close a resemblance to one another that their distinction is rather difficult. They live in the seas between the Tropics, and some, perhaps all, of the species enter fresh water. Very rarely they exceed a length of ten inches; nearly all have a plain silvery coloration. The coalescence of their lower pharyngeals renders their systematic position rather uncertain, and, indeed, some Ichthyologists have referred them to the Pharyngognaths.

SCOLOPSIS.—Body oblong, covered with scales finely serrated and of moderate size. Jaws nearly equal in length anteriorly; cleft of the mouth horizontal. Teeth villiform, without canines; palate toothless. One dorsal fin. Formula of the vertical fins: D. $\frac{10}{9}$ A. $\frac{3}{7}$. Caudal fin forked. Præoperculum distinctly denticulated; infraorbital ring with a spine directed backwards. Branchiostegals, five.

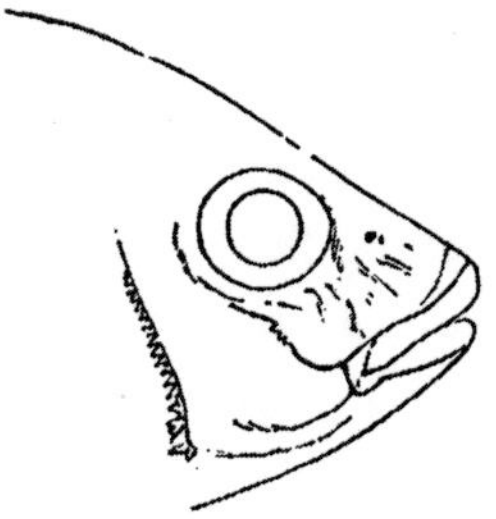

Fig. 160.—Infraorbital spine of Scolopsis monogramma.

Marine, and of small size. Twenty-five species are known from the tropical parts of the Indo-Pacific. *Heterognathodon* is an allied genus, but without the infraorbital spine.

DENTEX.—Body oblong, covered with ctenoid scales of moderate size. Cleft of the mouth nearly horizontal, with the jaws equal in length anteriorly. Canine teeth in both jaws; palate toothless. One dorsal fin. Formula of the vertical fins: D. $\frac{10\text{-}13}{10\text{-}12}$ A. $\frac{3}{8\text{-}9}$. Caudal fin forked. Præoperculum without serrature; præorbital unarmed and broad, there being a wide space between the eye and the cleft of the mouth. Cheek covered by more than three series of scales. Branchiostegals, six.

Marine Fishes, rather locally distributed in the Mediterranean, on the south coast of Africa, in the Red Sea, East Indian Archipelago, and on the coasts of China and Japan. About fourteen species are known, some of which attain a weight of 30 lbs. and more. They form a not unimportant article of food where they are found in any number, as on the Cape of Good Hope. The species found in the Mediterranean (*D. vulgaris*) wanders sometimes to the south coast of England, and is one of the larger species. The coloration of these fishes is rather uniform, silvery, or pink, or greenish. *Symphorus* is an allied genus from the Indo-Pacific.

SYNAGRIS.—Body rather elongate, covered with ciliated scales of moderate size. Cleft of the mouth horizontal, with the jaws equal in length anteriorly. One continuous dorsal, with feeble spines; dorsal $\frac{10}{9}$, anal $\frac{3}{7}$. Caudal deeply forked. Teeth villiform, with canines at least in the upper jaw. Infraorbital not armed; præoperculum without, or with a very indistinct serrature. Cheek with three series of scales. Branchiostegals six.

Marine fishes of small size; about twenty species are known from the tropical parts of the Indo-Pacific. *Pentapus*, *Chætopterus*, and *Aphareus* are allied genera from the same area.

MAENA.—Body oblong, compressed, covered with ciliated scales of moderate size. Mouth very protractile, the intermaxillary pedicles extending backwards to the occiput. Teeth villiform; minute teeth on the vomer. One dorsal, scaleless, with feeble spines. D. $\frac{11}{11}$, A. $\frac{3}{9}$. Caudal fin forked. Præoperculum without serrature. Branchiostegals six.

Small fishes from the Mediterranean, known to the ancients; valueless as food. Three species.

SMARIS.—Body oblong or cylindrical, covered with rather small ciliated scales. Mouth very protractile, the intermaxillary pedicles extending backwards to the occiput. Teeth villiform. Palate toothless. One dorsal, scaleless, with eleven or more very feeble spines; anal with three. Caudal fin forked. Præoperculum without serrature. Branchiostegals six.

Small fishes from the Mediterranean. Six species.

CÆSIO.—Body oblong, covered with ciliated scales of moderate size. Cleft of the mouth more or less oblique, with the jaws equal in length anteriorly, or with the lower somewhat projecting. Teeth villiform; palate generally toothless. One dorsal, with from nine to thirteen very feeble spines, with the anterior part highest, and the posterior covered with minute scales. Caudal fin deeply forked. Præoperculum without, or with minute, serrature.

Small fishes from the Indo-Pacific. Twelve species.

ERYTHRICHTHYS.—Body elongate, covered with small ciliated scales. Mouth very protractile, the pedicles of the intermaxillary extending to the occiput. Dentition quite rudimentary or entirely absent. Two dorsal fins connected by a series of very feeble spines; also the anterior spines are feeble. Præoperculum not serrated.

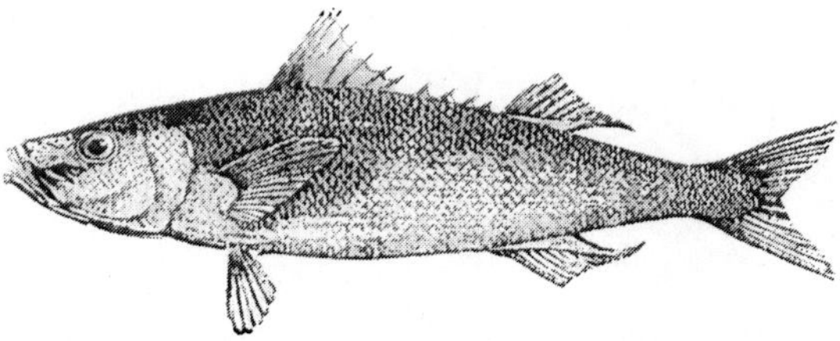

Fig. 161.—Erythrichthys nitidus.

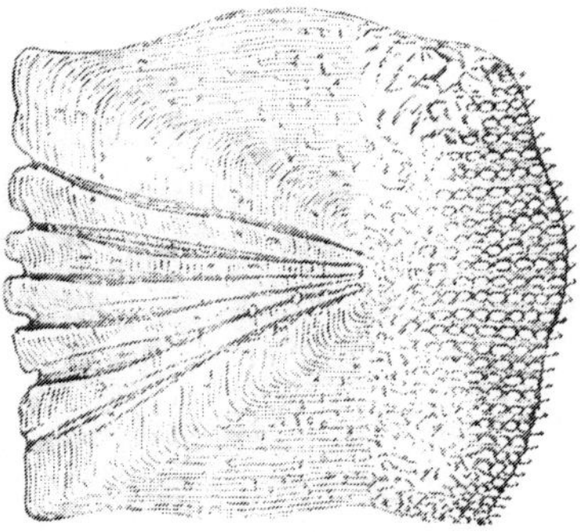

Fig. 162.—Enlarged scale.

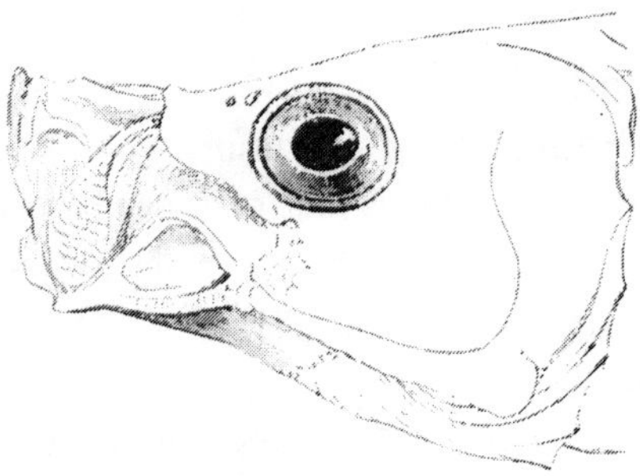

Fig. 163.—Protractile mouth.

Small fishes from various tropical and temperate seas. Four species: the species figured occurs, but is not common, on the coasts of Western Australia, Tasmania, and New Zealand.

OLIGORUS.—Body oblong, covered with small scales. Cleft of the mouth rather oblique, the lower jaw being the longer. Teeth villiform, without canines; teeth on the vomer and palatine bones. One dorsal, with eleven spines; anal with three; caudal fin rounded. Præoperculum with a single smooth or obtusely denticulated margin.

To this genus belong two fishes well known on account of the excellent flavour of their flesh. The first (*O. macquariensis*) is called by the colonists "Murray-Cod," being plentiful in the

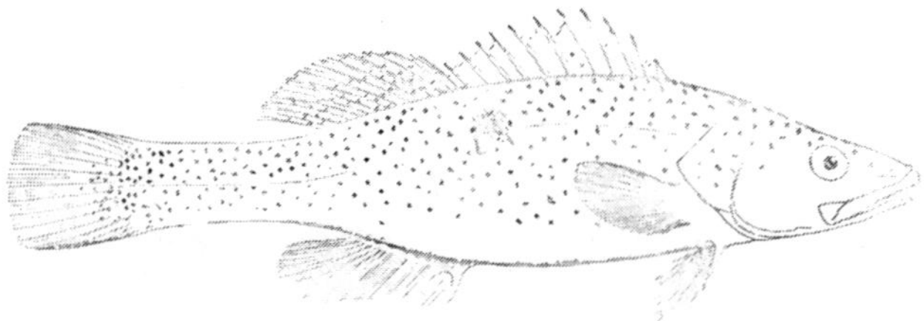

Fig. 164.—The Murray-Cod, *Oligorus macquariensis*.

Murray River and other rivers of South Australia. It attains to a length of more than three feet, and to a weight of nearly 100 lbs. The second (*O. gigas*) is found in the sea, on the coast of New Zealand, and called by the Maoris and colonists "Hapuku." Its average weight is about 45 lbs., but occasionally large specimens of more than a hundredweight are caught. At certain localities it is so plentiful that it may form an important article of trade. Dr. Hector, who has had opportunity of examining it in a fresh state, has pointed out anatomical differences from the Murray-Cod, from which it appears that it would be better placed in a distinct genus.

GRYSTES.—Body oblong, covered with scales of moderate size. All the teeth villiform, without canines; teeth on the vomer and palatine bones. One dorsal fin with ten spines; anal with three; caudal fin rounded. Præoperculum with a single smooth margin.

One species, from the fresh waters of the United States (*G. salmonoides*), attains to a length of more than two feet. It is known by the name of "Growler," and eaten.

ARRIPIS.—Body oblong, covered with scales of moderate size. All the teeth villiform, without canines; teeth on the vomer and palatine bones. One dorsal fin, with nine slender spines; anal with three. Præoperculum denticulated.

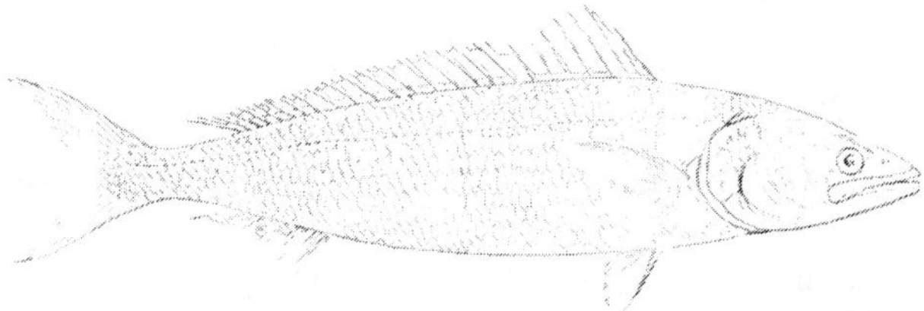

Fig. 165.—*Arripis salar*, South Australia.

Three species are known, from the coasts of Southern Australia and New Zealand. They are named by the colonists Salmon or Trout, from their elegant form and lively habits, and from the sport they afford to the angler. Their usual size is from 1 to 3 lbs., but specimens of double that weight are taken. The smaller specimens are the more delicate and better flavoured. When not fresh, they are liable to assume poisonous properties; and cases of poisoning are not unfrequently caused by them.

HURO.—Body oblong, compressed, covered with scales of moderate size. All the teeth villiform; bones of the head without serrature. Mouth rather oblique, with the lower jaw projecting. Two dorsal fins, the first with six spines.

The "Black Bass" of Lake Huron (*Huro nigricans*).

AMBASSIS.—Body short, strongly compressed, covered with large thin deciduous scales. Mouth oblique, with the lower jaw longest; teeth villiform, without conspicuously larger canines; teeth on the vomer and palatine bones. Two dorsal fins, the first with seven, the anal with three spines; a horizontal spine pointing forwards in front of the dorsal fin. The lower limb of the præoperculum with a double serrated margin.

This genus comprises the smallest of all Percoids, some of the species not much exceeding one inch in length. They are most abundant on the coasts of the tropical Indo-Pacific, and in the fresh waters belonging to that area. The species are numerous (some thirty having been described), and very difficult to distinguish. Their coloration is very plain, a silvery hue prevailing over the whole fish.

APOGON.—Body rather short, covered with large deciduous scales. Mouth oblique, with the lower jaw longest; teeth villiform, without canines; teeth on the vomer and palatine bones. Two dorsal fins, the first with six or seven, the anal with two spines. Præoperculum with a double edge on the margin, one or both edges being serrated. Seven branchiostegals.

Although of similarly small size, the fishes of this genus represent a more highly developed form of the Percoid type than *Ambassis*. Their distribution coincides very much with

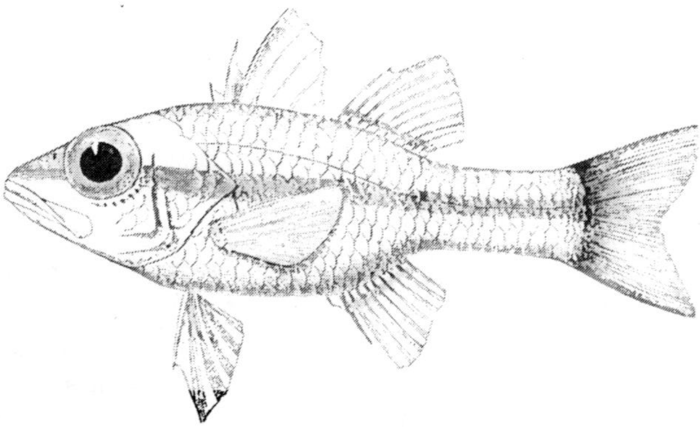

Fig. 166.—Apogon frenatus.

that of *Ambassis*, but they are chiefly marine, comparatively few of the species entering fresh water. They belong to the kind of fishes which, from their habit, are termed "Coral Fishes," being found in greatest abundance on, or in the neighbourhood of, coral reefs, in company with Chætodonts, Pomacentridæ, and others. Their colours also are ornamental and highly diversified, as is generally the case in coral fishes, the majority of the species showing transverse or longitudinal bands or large spots, and numerous other smaller markings which, in the dead fish, soon disappear. Nearly one hundred species have been described, of which a few only occur in the Atlantic, one extending northwards into the Mediterranean.

Chilodipterus, *Acropoma*, and *Scombrops* are allied genera, but with canine teeth in one or both jaws.

POMATOMUS.—Body oblong, covered with scales of moderate size. Eye very large. All the teeth villiform, without canines; teeth on the vomer and palatine bones. Two dorsal fins, the first with seven, the anal with two spines. No serration on any of the bones of the head. Branchiostegals seven.

One species only is known, *P. telescopium*, which grows to a length of nearly two feet. It is not uncommon in the Mediterranean and neighbouring parts of the Atlantic, but only occasionally caught, as it lives habitually at a greater depth than any other Percoid as far as is known at present, probably at depths from 80 to 200 fathoms; a habit sufficiently indicated by its exceedingly large eye.

PRIACANTHUS.—Body short, compressed, covered with small rough scales, which extend also over the short snout. Lower jaw and chin prominent. Eye large. All the teeth villiform, without canines; teeth on the vomer and palatine bones. One dorsal fin with ten spines, anal with three. Præoperculum serrated, with a more or less prominent, flat, triangular spine at the angle.

A very natural genus, easily recognised, and without direct relation to the other Percoid genera. The species, of which seventeen are known, are spread over nearly all the tropical seas, and belong to the more common fishes. They scarcely exceed a length of twelve inches, and are very uniformly coloured, red, pink, and silvery prevailing.

THE following three genera form a group by themselves, which, however, is defined rather by its geographical limits and similarity of general appearance than by distinctive anatomical characters. The species are abundant in the fresh waters of the United States, and well known by the name of "Sun Fishes." They rarely exceed a length of six inches, and are not used as food. The number of species is uncertain.

CENTRARCHUS.—Body short, compressed, with scales of moderate size. All the teeth villiform, without canines; teeth on the vomer, palatines, and on the tongue. One dorsal fin; anal generally with more than three spines. Præoperculum without serrature; operculum not lobed.

BRYTTUS.—Body short, compressed, with scales of moderate size. All the teeth villiform, without canines; teeth on the vomer and palatine bones. One dorsal fin with nine or ten, anal with three spines. Præoperculum not serrated; operculum with a rounded membranaceous coloured lobe behind.

POMOTIS.—Body short, compressed, with scales of moderate length. All the teeth villiform, without canines; teeth on the vomer, but none on the palatine bones. One dorsal, with from nine to eleven spines, anal with three. Præoperculum entire or minutely serrated; operculum with a rounded membranaceous coloured lobe behind.

A NORTH AMERICAN Freshwater genus, *Aphredoderus*, occupies a perfectly isolated position in the system, and is evidently the type of a distinct family. It resembles the "Sun-fishes" of the same country with regard to the structure of the vertical fins, but has the vent situated in front of the ventrals, which are composed of more than five soft rays. The body is oblong, compressed, covered with ctenoid scales. The dorsal fin is single, and has three spines in front. Infraorbital and præoperculum with spinous teeth. Villiform teeth in the jaws, on the vomer and palatine bones. *A. sayanus* from the southern streams and fresh waters of the Atlantic States.

To complete the list of Percoid genera, we have to mention the following:—*Siniperca, Etelis, Niphon, Aprion, Apsilus, Pentaceros, Velifer, Datnioides, Percilia, Lanioperca.*

SECOND FAMILY—SQUAMIPINNES.

Body compressed and elevated, covered with scales, either finely ctenoid or smooth. Lateral line continuous, not continued over the caudal fin. Mouth in front of the snout, generally small, with lateral cleft. Eye lateral, of moderate size. Six or seven branchiostegals. Teeth villiform or setiform, in bands, without canines or incisors. Dorsal fin consisting of a spinous and soft portion of nearly equal development; anal with three or four spines, similarly developed as the soft dorsal, both being many-rayed. The vertical fins more or less densely covered with small scales. The lower rays of the pectoral fin branched, not enlarged; ventrals thoracic, with one spine and five soft rays. Stomach coecal.

The typical forms of this family are readily recognised by the form of their body, and by a peculiarity from which they derive their name *Squamipinnes;* the soft, and frequently also the spinous part of their dorsal and anal fins are so thickly covered with scales that the boundary between fins and body is entirely obliterated. The majority are inhabitants of the tropical seas, and abound chiefly in the neighbourhood of coral-reefs. The beauty and singularity of distribution of the colours of some of the genera, as *Chætodon, Heniochus, Holacanthus,* is scarcely surpassed by any other group of fishes. They remain within small dimensions, and comparatively few are used as food. They are carnivorous, feeding on small invertebrates. Only a few species enter brackish water.

Extinct representatives of this family are not scarce at Monte Bolca and in other tertiary formations. All, at least those admitting of definite determination, belong to existing genera, viz. *Holacanthus, Pomacanthus, Ephippium, Scatophagus.* Very singular is the occurrence of *Toxotes* in the Monte Bolca strata.

The following genera have no teeth on the palate :—

CHÆTODON.—One dorsal fin, without any notch in its upper margin, and with the soft and spinous portions similarly developed; none of the spines elongate. Snout short or of moderate length. Præoperculum without, or with a fine, serration, and without spine at the angle. Scales generally large or of moderate size.

Seventy species are known from the tropical parts of the Atlantic and Indo-Pacific, nearly all being beautifully ornamented with bands or spots. Of the ornamental markings a dark or bicoloured band, passing through the eye and ascending towards the back, is very generally found in these fishes; it frequently occurs again in other marine Acanthopterygians, in which it is not rarely a sign of the immature condition of the individual. The Chætodonts are most

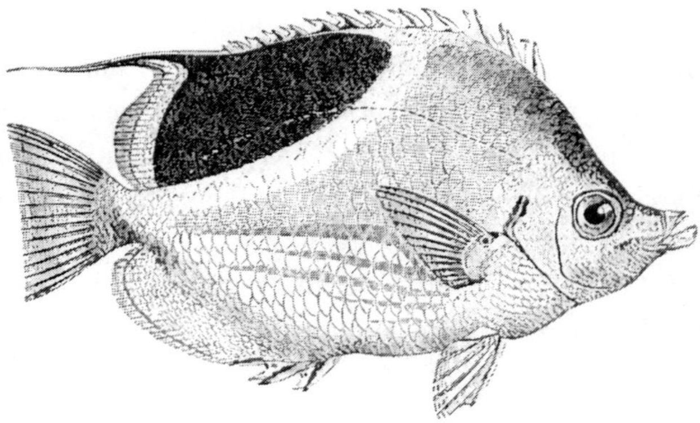

Fig. 167.—Chætodon ephippium.

numerous in the neighbourhood of the coral-reefs of the Indo-Pacific, the species figured (*C. ephippium*) being as common in the East Indian Archipelago as in Polynesia, like many others of its congeners.

CHELMO differs from *Chætodon* only in having the snout produced into a more or less long tube.

Only four species are known, locally distributed in the tropical seas. *Ch. rostratus*, the oldest species known, is

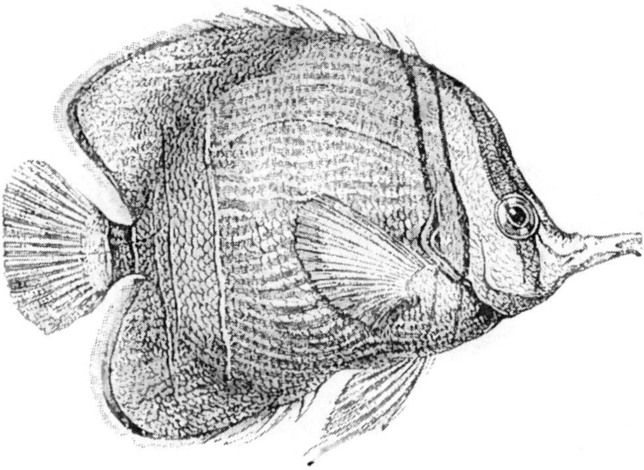

Fig. 168.—Chelmo marginalis, from the coast of Australia.

said to have the instinct of throwing a drop of water from its bill so as to light upon any insect resting on a leaf, and thus make it fall, that it may instantly dart upon it. This statement is erroneous, and probably rests upon the mistaken notion that the long bill is especially adapted for this manœuvre, which, indeed, is practised by another fish of this family (*Toxotes*). The long slender bill of Chelmo (which is a true saltwater fish) rather enables it to draw from holes and crevices animals which otherwise could not be reached by it.

HENIOCHUS.—One dorsal, with from eleven to thirteen spines, the fourth of which is more or less elongate and filiform. Snout rather short or of moderate length. Præoperculum without spine. Scales of moderate size.

Four species are known from the tropical Indo-Pacific. *H. macrolepidotus* is one of the most common fishes of that area; the species figured (*H. varius*) retains in a conspicuous

Fig. 169.—Heniochus varius.

manner horn-like protuberances on bones of the head, with which the young of all the species of this genus seem to be armed.

HOLACANTHUS.—Præoperculum with a strong spine at the angle. One dorsal, with from twelve to fifteen spines. Scales of moderate or small size.

Forty species are known, which, in their geographical distribution accompany, and are quite analogous to, the Chætodonts. One of the most common and most beautiful is called "Emperor of Japan" by the Dutch, which name has been adopted by Bloch for its specific designation, *Holacanthus imperator*. Its body is blue, longitudinally traversed by about thirty yellow bands; the ocular band, and the side behind the head, are black, edged with yellow; the caudal fin is yellow. It is a large species of this genus, sometimes attaining a length of 15 inches, and as an article of food is one of the most esteemed of all the Indian species. With regard to beauty of colours it is surpassed by another allied species, *H. diacanthus*, which likewise ranges from the east coast of Africa to Polynesia.

POMACANTHUS differs from *Holacanthus* in having from eight to ten spines only in the dorsal fin.

The single species (*P. paru*) on which this genus is founded is one of the most common fishes of the West Indies, and offers one of the most remarkable instances of variation of colour within the limits of the same species: some specimens being ornamented with more or less distinct yellowish cross-bands, others with yellow crescent-shaped spots; in others black spots predominate.

SCATOPHAGUS.—Two dorsal fins, united at the base, the first with ten or eleven spines; only the second is scaly. A recum-

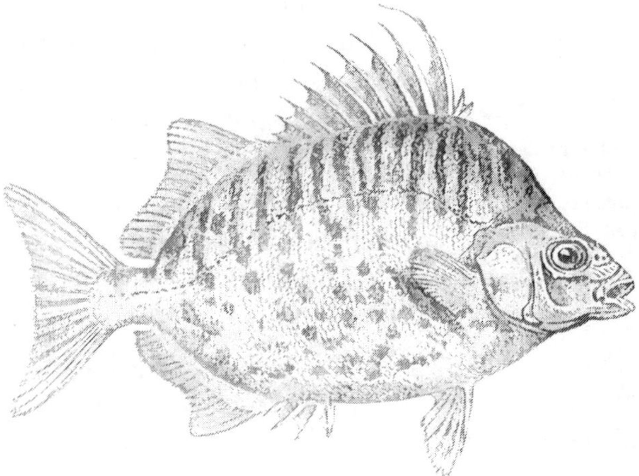

Fig. 170.—Scatophagus multifasciatus.

bent spine before the dorsal, pointing forwards. Anal with four spines. Snout rather short. Præoperculum without spine. Scales very small.

Four species are known, from the Indian Ocean, of which *S. argus* is the most generally known, in fact, one of the most common Indian shore-fishes. It freely enters large rivers, and is said not to be particular in the selection of its food. The species figured (*S. multifasciatus*) represents *S. argus* on the coasts of Australia.

Ephippus.—Snout short, with the upper profile parabolic. Dorsal fin deeply emarginate between the spinous and soft portions, the former with nine spines, the third of which is rather elongate, and flexible; spinous portion not scaly; anal spines three. Pectoral fin short. Præoperculum without spine. Scales of moderate size, or rather small.

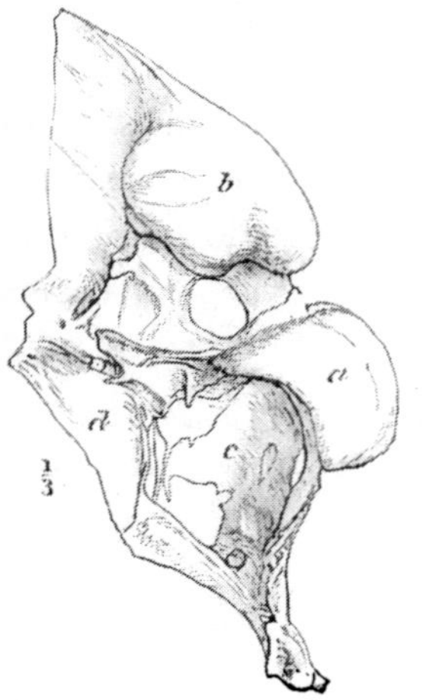

Fig. 171.—Bony enlargement of cranial bones of Ephippus. *a*, Enlargement of the frontal, and *b*, of the supraoccipital bones; *c*, interorbital septum; *d*, basis cranii. ⅓ nat. size.

Two or three species are known from the warmer parts of the Atlantic and Indian Oceans. The Atlantic species (*E. faber*) shows the remarkable peculiarity that in old specimens (12 and more inches long) the occipital crest, and sometimes some of the anterior neural and hæmal spines are enormously enlarged into a globular bony mass. This can hardly be regarded as a pathological change of the bone, as it has been found in all old specimens, without exception.

Drepane is allied to *Ephippus*, but has very long falciform pectoral fins. The single species *D. punctata* is common in the Indian Ocean and on the coasts of Australia. *Hypsinotus*, from Japan, appears to inhabit a greater depth than the other Squamipinnes.

Scorpis and *Atypichthys* are genera distinguished from the preceding by the presence of vomerine teeth. They belong to the coast-fauna of Australia, New Zealand, and Chili.

TOXOTES.—Body short, compressed, covered with scales of moderate size. Snout pointed, with a wide lateral mouth and projecting lower jaw. One dorsal, with five strong spines situated on the posterior part of the back; the soft portion and the anal fin scaly, the latter with three spines. Villiform teeth in the jaws, on the vomer and palatine bones. Scales of moderate size, cycloid.

Two species are known from the East Indies, one (*T. jaculator*), which is the more common, ranging to the north coast of Australia. It has received its name from its habit of throwing a drop of water at an insect which it perceives close

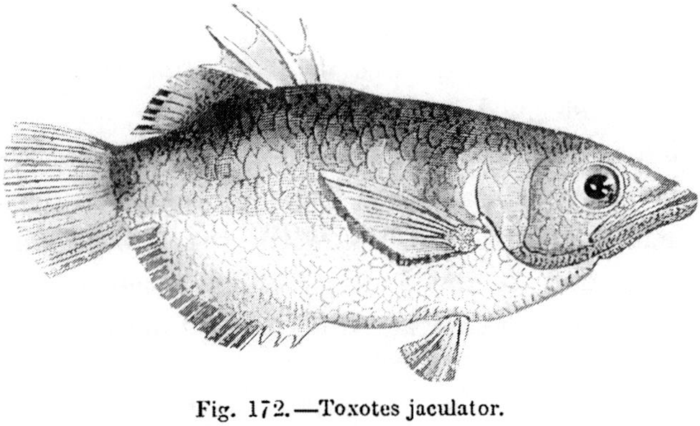

Fig. 172.—Toxotes jaculator.

to the surface, in order to make it fall into it. The Malays, who call it "Ikan sumpit," keep it in a bowl, in order to witness this singular habit, which it continues even in captivity.

THIRD FAMILY—MULLIDÆ.

Body rather low and slightly compressed, covered with large thin scales, without or with an extremely fine serrature. Two long erectile barbels are suspended from the hyoid, and are received between the rami of the lower jaw and opercles. Lateral line continuous. Mouth in front of the snout, with the cleft lateral and rather short; teeth very feeble. Eye lateral, of moderate size. Two short dorsal fins remote from each other, the first with feeble spines; anal similar to the second dorsal. Ventrals with one spine and five rays. Pectorals short. Branchiostegals four; stomach siphonal.

The "Red Mullets" form a very natural family, which, on account of slight modifications of the dentition, has been divided into several sub-genera—*Upeneoides, Upeneichthys, Mullus, Mulloides,* and *Upeneus.* They are marine fishes, but many species enter brackish water to feed on the animalcules abounding in the flora of brackwater. About forty different species are known chiefly from tropical seas, the European species (*M. barbatus,* see p. 43, Fig. 7), extending far northwards into the temperate zone. None attain to a large size, specimens of from two to three lbs. being not common, but all are highly esteemed as food.

The most celebrated is the European species (of which there is one only, *M. surmuletus* being probably the female). The ancient Romans called it *Mullus,* the Greeks τριγλη. The Romans priced it above any other fish; they sought for large specimens far and wide, and paid ruinous prices for them.

> "Mullus tibi quatuor emptus
> Librarum, cœnæ pompa caputque fuit,
> Exclamare libet, non est hic improbe, non est
> Piscis: homo est; hominem, Calliodore, voras."
>
> MARTIAL, x. 31.

Then, as nowadays, it was considered essential for the enjoyment of this delicacy that the fish should exhibit the red colour of its integuments. The Romans brought it, for that purpose, living into the banqueting room, and allowed it to die in the hands of the guests, the red colour appearing in all its brilliancy during the death struggle of the fish. The fishermen of our times attain the same object by scaling the fish immediately after its capture, thus causing a permanent contraction of the chromatophors containing the red pigment (see p. 183).

Fourth Family—Sparidæ.

Body compressed, oblong, covered with scales, the serrature of which is very minute, and sometimes altogether absent. Mouth in front of the snout, with the cleft lateral. Eye lateral, of moderate size. Either cutting teeth in front of the jaws, or molar teeth on the side; palate generally toothless. One dorsal fin, formed by a spinous and soft portion of nearly equal development. Anal fin with three spines. The lower rays of the pectoral fin are generally branched, but in one genus simple. Ventrals thoracic, with one spine and five rays.

The "Sea-breams" are recognised chiefly by their dentition, which is more specialised than in the preceding families, and by which the groups, into which this family has been divided, are characterised. They are inhabitants of the shores of all the tropical and temperate seas. Their coloration is very plain. They do not attain to a large size, but the majority are used as food.

The extinct forms found hitherto are rather numerous; the oldest come from the cretaceous formation of Mount Lebanon; some belong to living genera, as *Sargus, Pagellus;* of others from Eocene and Miocene formations no living representative is known—*Sparnodus, Sargodon, Capitodus, Soricidens, Asima.*

First Group—Cantharina.—More or less broad cutting, sometimes lobate, teeth in front of the jaws; no molars or vomerine teeth; the lower pectoral rays are branched. Partly herbivorous, partly carnivorous. The genera belonging to this group are:—*Cantharus* from the European and South African coasts, of which one species (*C. lineatus*), is common on the coasts of Great Britain, and locally known by the names "Old Wife," "Black Sea-bream;" *Box, Scatharus,* and *Oblata* from the Mediterranean and neighbouring parts of the Atlantic; *Crenidens* and *Tripterodon* from the Indian Ocean; *Pachymetopon, Dipterodon,* and *Gymnocrotaphus* from the Cape

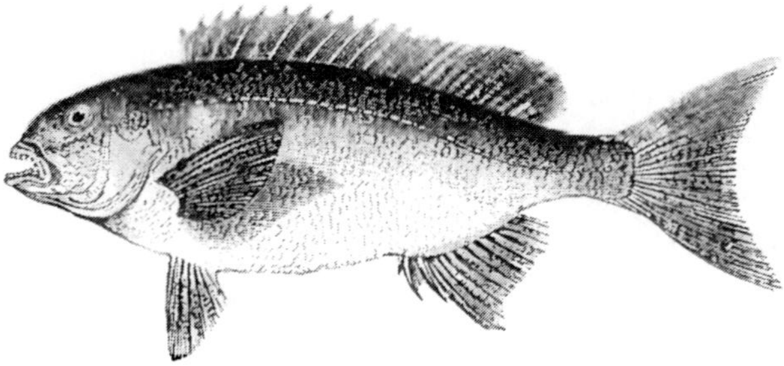

Fig. 173.—Tephræops richardsonii, from King George's Sound.

of Good Hope; *Girella* and *Tephræops* from Chinese, Japanese, and Australian Seas; *Doydixodon* from the Galapagos Islands and the coasts of Peru.

Second Group—Haplodactylina.—In both jaws flat and generally tricuspid teeth; no molars; vomerine teeth. The lower pectoral rays simple, not branched. Vegetable feeders. Only one genus is known, *Haplodactylus,* from the temperate zone of the Southern Pacific.

Third Group—Sargina.—Jaws with a single series of incisors in front, and with several series of rounded molars on the side. One genus is known, *Sargus,* which comprises twenty species; several of them occur in the Mediterranean and the neighbouring parts of the Atlantic, and are popularly called "Sargo," "Sar," "Saragu:" names derived from the word Sargus, by which name these fishes were well known to the ancient Greeks and Romans. One of the largest species is the "Sheep's-head" (*Sargus ovis*), from the coasts of the United States, which attains to a weight of 15 lbs., and is highly esteemed on account of the excellency of its flesh. Singularly enough, this genus occurs also on the east coast of Africa, one of these East-African species being identical with

S. noct from the Mediterranean. These fishes evidently feed

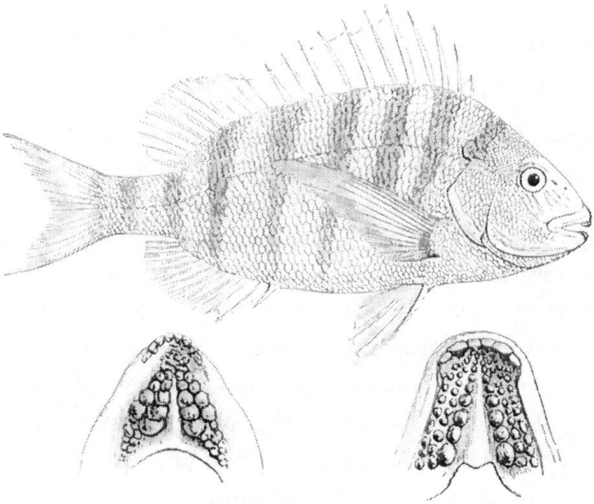

Fig. 174.—The Sheep's-head, *Sargus ovis*, of North America.

on hard-shelled animals, which they crush with their molar teeth.

Fourth Group—Pagrina.—Jaws with conical teeth in front and molar teeth on the sides. Feeding, as the preceding, on hard-shelled animals, like Mollusks and Crustaceans. This group is composed of several genera :—

Lethrinus.—Cheeks scaleless. Body oblong, covered with scales of moderate size (L. lat. 45-50). Canine teeth in front; lateral teeth in a single series, broadly conical or molar-like. Formula of the fins: D. $\frac{10}{9}$, A. $\frac{3}{8}$.

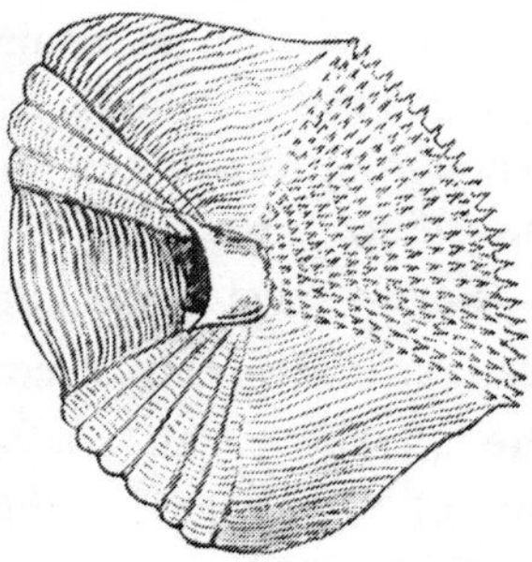

Fig. 175.—Scale of Lethrinus.

More than twenty species are known, all of which, with one exception, occur in the tropical Indo-Pacific. The species, forming this exception, occurs, singularly enough, on the west coast of Africa, where more than one Indian genus reappears in isolated representative species. Some Lethrini attain to a length of three feet.

Sphærodon is closely allied to *Lethrinus*, but has scales on the cheek. One species from the Indo-Pacific.

Pagrus.—Body oblong, compressed, with scales of moderate size. Several pairs of strong canine-like teeth in both jaws; molars arranged in two series. Cheeks scaly. The spines of the dorsal fin, eleven or twelve in number, are sometimes elongate, and can be received in a groove; anal spines three.

Thirteen species are known, chiefly distributed in the warmer parts of the temperate zones, and more scantily represented between the tropics. Several species (*P. vulgaris*, *P. auriga*, *P. bocagii*) occur in the Mediterranean and the neighbouring parts of the Atlantic; one (*P. argyrops*) is well known on the coasts of the United States under the names of "Scup," "Porgy," or "Mishcup," and one of the most important food fishes, growing to a length of 18 inches and a weight of 4 lbs.; another (*P. unicolor*) is one of the best-known sea-fishes of Southern Australia and New Zealand, where it is called "Snapper;" it is considered very good eating, like all the other species of this genus, and attains, like some of them, a length of more than 3 feet and a weight exceeding 20 lbs.

Pagellus.—Body oblong, compressed, with scales of moderate size. Jaws without canines; molars on the sides arranged in several series. Cheeks scaly. The spines of the dorsal fin, from eleven to thirteen in number, can be received in a groove; anal spines three.

Seven species are known, the majority of which are European, as *P. erythrinus*, common in the Mediterranean, and not rare on the south coast of England, where it is generally termed "Becker;" *P. centrodontus*, the common "Sea-bream" of the English coasts, distinguished by a black spot on the origin of the lateral line; in the young, which are called "Chad" by Cornish and Devon fishermen, this spot is absent;

P. owenii, the "Axillary or Spanish Sea-Bream," likewise from the British coasts. *Pagellus lithognathus*, from the coasts of the Cape of Good Hope, attains to a length of four feet, and is one of the fishes which are dried for export and sale to whalers.

CHRYSOPHRYS.—Body oblong, compressed, with scales of moderate size. Jaws with four or six canine teeth in front, and with three or more series of rounded molars on each side. Cheeks scaly. The spines of the dorsal fin, eleven or twelve in number, can be received in a groove; anal spines three.

Some twenty species are known from tropical seas and the warmer parts of the temperate zones. Generally known is *Ch. aurata*, from the Mediterranean, occasionally found on the south coast of England, where it is named "Gilthead." The French call it "Daurade," no doubt from the Latin *Aurata*, a term applied to it by ancient authors. The Greeks named it Chrysophrys (*i.e.* golden eyebrow), in allusion to the brilliant spot of gold which it bears between its eyes. According to Columella, the Aurata was among the number of the fishes brought up by the Romans in their vivaria; and the inventor of those vivaria, one Sergius Orata, is supposed to have derived his surname from this fish. It is said to grow extremely fat in artificial ponds. Duhamel states that it stirs up the sand with the tail, so as to discover the shell-fish concealed in it. It is extremely fond of mussels, and its near presence is sometimes ascertained by the noise which it makes while breaking their shells with its teeth. Several species found on the Cape of Good Hope attain to as large a size as *Pagellus lithagnathus*, and are preserved for sale like that species. *Chrysophrys hasta* is one of the most common species of the East Indian and Chinese coasts, and enters large rivers.

Fifth Group—Pimelepterina.—In both jaws a single anterior series of cutting teeth, implanted by a horizontal posterior process, behind which is a band of villiform teeth.

Villiform teeth on the vomer, palatines and the tongue. Vertical fins densely covered with minute scales. Only one genus is known, *Pimelepterus*, with six species from tropical seas. These fishes are sometimes found at a great distance from the land.

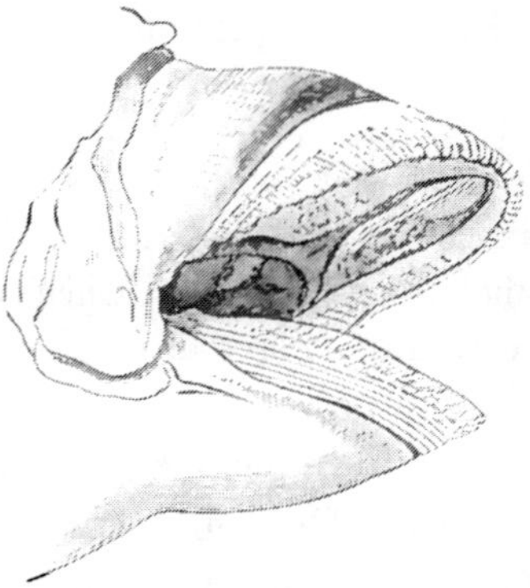

Fig. 176.—Teeth of Hoplognathus.

FIFTH FAMILY—HOPLOGNATHIDÆ.

Body compressed and elevated, covered with very small ctenoid scales. Lateral line continuous. The bones of the jaws have a sharp dentigerous edge, as in Scarus. The teeth, if at all conspicuous, being continuous with the bone, forming a more or less indistinct serrature; no teeth on the palate. The spinous portion of the dorsal fin is rather more developed than the soft; the spines strong; anal with three spines, similar to the soft dorsal. Ventrals thoracic, with one spine and five soft rays.

One genus only is known, *Hoplognathus*, with four species from Australian, Japanese, and Peruvian coasts.

SIXTH FAMILY—CIRRHITIDÆ.

Body oblong, compressed, covered with cycloid scales; lateral line continuous. Mouth in front of the snout, with lateral cleft. Eye lateral, of moderate size. Cheeks without a bony stay for the præoperculum. Generally six, sometimes five or three branchiostegals. Dentition more or less complete, composed of small pointed teeth, sometimes with the addition of canines. One dorsal fin, formed by a spinous and soft portion, of nearly equal development. Anal with three spines, generally less developed than the soft dorsal. The lower rays of the pectoral fins simple and generally enlarged; ventrals thoracic, but remote from the root of the pectorals, with one spine and five rays.

The fishes of this family may be readily recognised by their thickened, undivided lower pectoral rays, which in some are evidently auxiliary organs of locomotion, in others, probably, organs of touch. They differ from the following family, the Scorpænidæ, in lacking the bony connection between the infraorbital ring and the præoperculum. Two groups may be distinguished in this family, which, however, are connected by an intermediate genus (*Chironemus*). The first, distinguished by the presence of vomerine teeth, consists of *Cirrhites* and *Chorinemus*, small prettily coloured fishes. The former genus is peculiar to the Indo-Pacific, and consists of sixteen species; the second, with three species, seems to be confined to the coasts of Australia and New Zealand. The second group lacks the vomerine teeth, and comprises the following genera :—

CHILODACTYLUS.—One dorsal fin, with from sixteen to nineteen spines; anal fin of moderate length; caudal forked. One of the simple pectoral rays more or less prolonged, and projecting beyond the margin of the fin. Teeth in villiform bands; no canines. Præoperculum not serrated. Scales of moderate size. Air-bladder with many lobes.

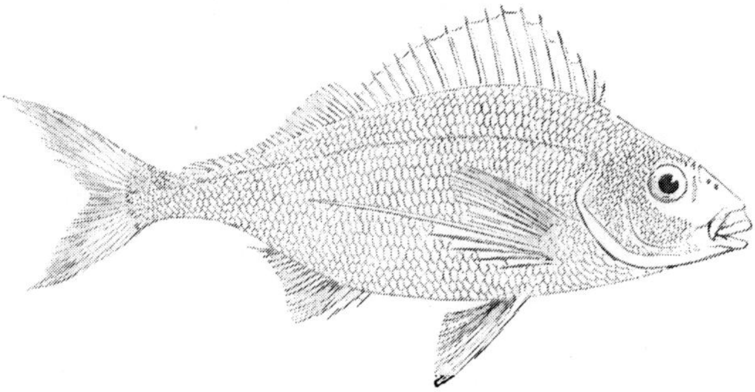

Fig. 177.—*Chilodactylus macropterus*, from Australia.

Seventeen species are known, chiefly from the temperate parts of the Southern Pacific, and also from the coasts of Japan and China. They belong to the most valuable food-fishes, as they grow to a considerable size (from five to twenty-five lbs.), and are easily caught in numbers. At the Cape of Good Hope they are very abundant, and preserved in large quantities for export.

Mendosoma from the coast of Chili, and *Nemadactylus* from Tasmania, are allied genera.

LATRIS.—Dorsal fin deeply notched; the spinous portion with seventeen spines; anal fin many-rayed. None of the simple pectoral rays passes the margin of the fin. Teeth villiform; no canines. Præoperculum minutely serrated. Scales small.

Two species only are known from Tasmania and New Zealand, which belong to the most important food-fishes of the Southern Hemisphere. *Latris hecateia* or the "Trumpeter," ranges from sixty to thirty lbs. in weight, and is considered by the colonists the best flavoured of any of the fishes of South Australia, Tasmania, and New Zealand, and consumed smoked as well as fresh. The second species, *Latris ciliaris*, is smaller, scarcely attaining a weight of twenty lbs., but more abundant; it is confined to the coast of New Zealand.

SEVENTH FAMILY—SCORPÆNIDÆ.

Body oblong, more or less compressed, covered with ordinary scales, or naked. Cleft of the mouth lateral or subvertical. Dentition feeble, consisting of villiform teeth; and generally without canines. Some bones of the head armed, especially the angle of the præoperculum, its armature receiving additional support by a bony stay, connecting it with the infraorbital ring. The spinous portion of the dorsal fin equally

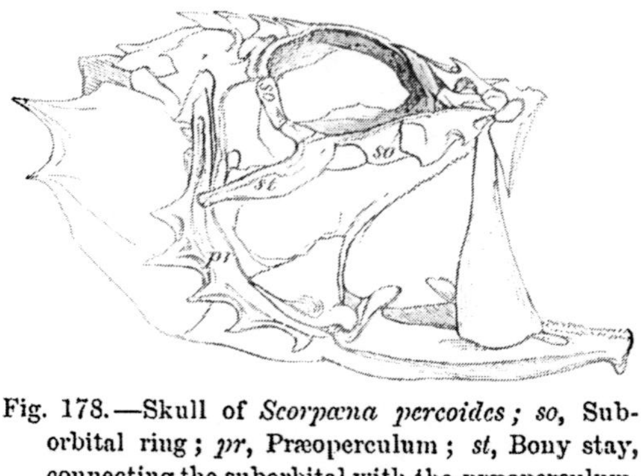

Fig. 178.—Skull of *Scorpæna percoides*; *so*, Suborbital ring; *pr*, Præoperculum; *st*, Bony stay, connecting the suborbital with the præoperculum.

or more developed than the soft and than the anal. Ventrals thoracic, generally with one spine and five soft rays, sometimes rudimentary.

This family consists of carnivorous marine fishes only; some resemble the Sea-Perches in form and habits, as *Sebastes*, *Scorpæna*, etc., whilst others live at the bottom of the sea, and possess in various degrees of development those skinny appendages resembling the fronds of seaweeds, by which they either attract other fishes, or by which they are enabled more effectually to hide themselves. Species provided with those appendages have generally a coloration resembling that of their surroundings, and varying with the change of locality. The habit of living on the bottom has also developed in many Scorpænoids separate pectoral rays, by means of which they move or feel. Some of the genera live at a considerable depth, but apparently not beyond 300 fathoms. Nearly all are distinguished by a powerful armature either of the head, or fin spines, or both; and in some the spines have been developed into poison organs.

The only fossil representative known at present is a species of *Scorpæna* from the Eocene of Oran.

Sebastes.—Head and body compressed; crown of the head scaly to, or even beyond, the orbits; no transverse groove on the occiput. Body covered with scales of moderate or small size, and without skinny tentacles. Fin-rays not elongate; one dorsal, divided by a notch into a spinous and soft portion, with twelve or thirteen spines; the anal with three. No pectoral appendages. Villiform teeth in the jaws, on the vomer, and generally on the palatine bones. Vertebræ more than twenty-four.

About twenty species are known, principally from seas of the temperate zones, as from the coasts of Northern Europe (*S. norvegicus*, *S. viviparus*), of Japan, California, New Zealand, and Van Diemen's Land. All seem to prefer deep water to the surface, and *Sebastes macrochir* has been obtained at a depth of 345 fathoms. In their general form they resemble the Sea-Perches, attain to a weight of from one to four lbs., and are generally esteemed as food.

SCORPÆNA.—Head large, slightly compressed, generally with a transverse naked depression on the occiput; bones of the head armed with spines, and generally with skinny tentacles. Scales of moderate size. Mouth large, oblique. Villiform teeth in the jaws, and at least on the vomer. One dorsal, $\frac{12-13}{9}$, A. $^3/_5$. Pectoral fins without detached rays, large, rounded, with the lower rays simple and thickened. Air-bladder none. Vertebræ twenty-four.

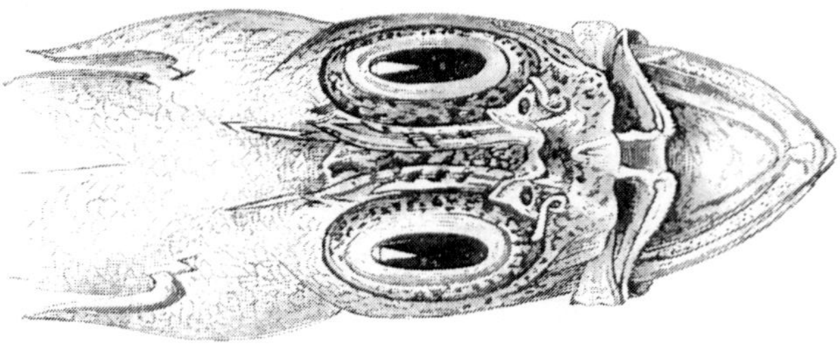

Fig. 179.—Head of Scorpæna percoides, from New Zealand.

About forty species are known from tropical and subtropical seas. They lead a sedentary life, lying hidden in the sand, or between rocks covered with seaweed, watching for their prey, which chiefly consists of small fishes. Their

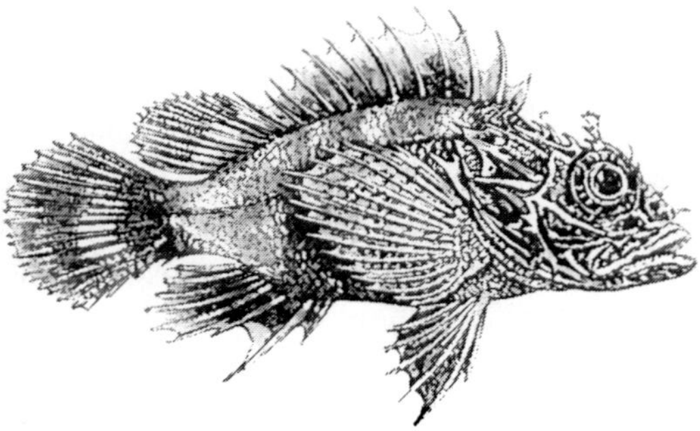

Fig. 180.—Scorpæna bynoensis, from the coasts of Australia.

strong undivided pectoral rays aid them in burrowing in the

sand, and in moving along the bottom. The type of their coloration is very much the same in all the species, viz. an irregular mottling of red, yellow, brown, and black colours, but the distribution of these colours varies exceedingly, not only in the same species but also in the same individuals. They do not attain to any considerable size, probably never exceeding a length of 18 inches. Their flesh is well flavoured. Wounds inflicted by their fin-spines are exceedingly painful, but not followed by serious consequences.

Glyptauchen and *Lioscorpius* are genera closely allied to *Scorpæna*, from Australian seas.

Setarches is also allied to the preceding genera, and provided with very large eyes, in accordance with the depth (215 fathoms) which the two species known at present inhabit; one has been found near Madeira, the other near the Fidji Islands.

Pterois.—Head and body compressed; scales of small or moderate size. Bones of the head armed with numerous spinous projections, between which often skinny tentacles are developed. The dorsal spines and pectoral rays are more or less prolonged, passing beyond the margin of the connecting membrane. Twelve or thirteen dorsal spines. Villiform teeth in the jaws and on the vomer.

Nine species are known from the tropical Indo-Pacific. They belong to the most singularly formed and most beautifully coloured fishes of the Tropics, and formerly were believed to be able to fly, like Dactylopterus. But the membrane connecting their pectoral rays is much too short and feeble to enable them to raise themselves from the surface of the water.

Apistus.—Head and body compressed, covered with ctenoid scales of rather small size. Some bones of the head, and especially the præorbital, are armed with spines. One dorsal with fifteen spines; the anal with three. The pectoral fin is elongate, and one ray is completely detached from the fin.

Villiform teeth in the jaws, on the vomer, and palatine bones. Air-bladder present. A cleft behind the fourth gill.

Two species from the Indian Ocean. These fishes are very small, but of interest on account of the prolongation of their pectoral fins, which indicates that they can take long flying leaps out of the water. However, this requires confirmation by actual observation.

Agriopus.—Head and body compressed, scaleless; head without any, or with very feeble, armature. Cleft of the mouth small, at the end of the produced snout. One dorsal fin, which commences from the head, the spinous portion being formed by from seventeen to twenty-one strong spines; anal short. Villiform teeth in the jaws, generally none on the vomer.

Seven species. This singular genus is peculiar to the temperate parts of the South Pacific, occurring at the Cape, on the coast of South Australia, and Chili. The largest species (*A. torvus*) attains a length of two and a half feet. Nothing is known of its mode of life.

Synanceia.—General appearance of the fish, especially of the head, monstrous. Scales none; skin with numerous soft warty protuberances or filaments. Mouth directed upwards, wide. Eyes small. From thirteen to sixteen dorsal spines; pectoral fins very large. Villiform teeth in the jaws, and sometimes on the vomer.

Four species are known from the Indo-Pacific, of which *S. horrida* and *S. verrucosa* are the most generally distributed, and, unfortunately, the most common. They are justly feared on account of the great danger accompanying wounds which they inflict with their poisoned dorsal spines, as has been already noticed above, p. 191. The greatest length to which they attain does not seem to exceed eighteen inches. They are very voracious fishes, and their stomach is of so great a capacity that they are able to swallow fishes one-third of their own bulk.

Micropus.—Head and body strongly compressed, short, and

deep; no scales, but the skin is covered with minute tubercles. Snout very short, with nearly vertical anterior profile. Præorbital, præ- and inter-operculum with spines on the edge. Dorsal fin with seven or eight, anal with two spines. Pectorals short, ventrals rudimentary. Jaws with villiform teeth.

These fishes belong to the smallest of Acanthopterygians, scarcely exceeding $1\frac{1}{2}$ inches in length. Two species are known, which are rather common on the coral reefs of the Pacific.

CHORISMODACTYLUS.—Head and body rather compressed, scaleless, with skinny flaps. Bones of the head with prominent ridges; the præorbital, præoperculum, and operculum armed;

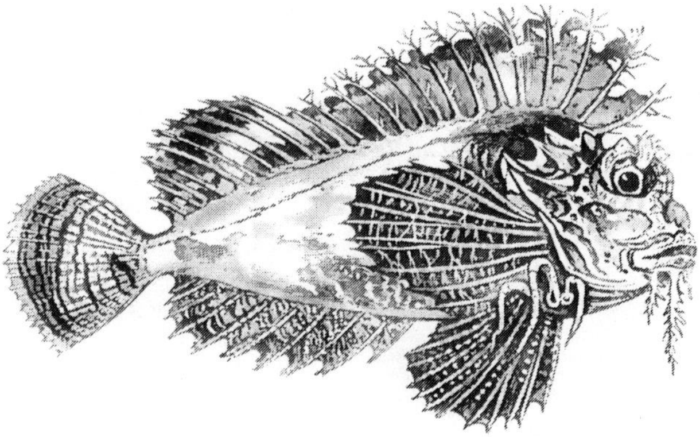

Fig. 181.—Chorismodactylus multibarbis.

a depression on the occiput. One dorsal fin, with thirteen spines; the anal with two. Three free pectoral appendages. Ventral fins with one spine and five rays. Villiform teeth in the jaws only.

Only one small species, *Ch. multibarbis,* is known, from the coasts of India and China.

To complete the list of Scorpænoid genera, we have to mention *Tænianotus, Centropogon, Pentaroge, Tetraroge, Prosopodasys, Aploactis, Trichopleura, Hemitripterus, Minous* and *Pelor*.

EIGHTH FAMILY—NANDIDÆ.

Body oblong, compressed, covered with scales. Lateral line interrupted. Dorsal fin formed by a spinous and soft portion, the number of spines and rays being nearly equal; anal fin with three spines, and with the soft portion similar to the soft dorsal. Ventral fins thoracic, with one spine, and five or four rays. Dentition more or less complete, but feeble.

This small family consists of two very distinct groups.

A. *Plesiopina.* Marine fishes of small size, with pseudobranchiæ and only four ventral rays. *Plesiops* from the coral-reefs of the Indo-Pacific, and *Trachinops* from the coast of New South Wales, belong to this group.

B. *Nandina.* Freshwater fishes of small size from the East Indies, without pseudobranchiæ, and five ventral rays. The genera are *Badis, Nandus,* and *Catopra.*

NINTH FAMILY—POLYCENTRIDÆ.

Body compressed, deep, scaly. Lateral line none. Dorsal and anal fins long, both with numerous spines, the spinous portion being the more developed. Ventrals thoracic, with one spine and five soft rays. Teeth feeble. Pseudobranchiæ hidden.

Only two genera, each represented by one or two species in the Atlantic rivers of Tropical America, *Polycentrus* and *Monocirrhus,* belong to this family. They are small insectivorous fishes.

TENTH FAMILY—TEUTHIDIDÆ.

Body oblong, strongly compressed, covered with very small scales. Lateral line continuous. Eye lateral, of moderate size. A single series of cutting incisors in each jaw; palate toothless. One dorsal fin, the spinous portion being the more deve-

loped; anal with seven spines. Ventral fins thoracic, with an outer and an inner spine, and with three soft rays between.

This family consists of one very natural genus, *Teuthis*, readily recognised by the singular structure of the fins. In all the species the fin-formula is D. $\frac{13}{10}$. A. $\frac{7}{9}$. The incisors are small, narrow, and provided with a serrated edge. The air-bladder is large, and forked anteriorly as well as posteriorly. Their skeleton shows several peculiarities: the number of vertebræ is twenty-three, ten of which belong to

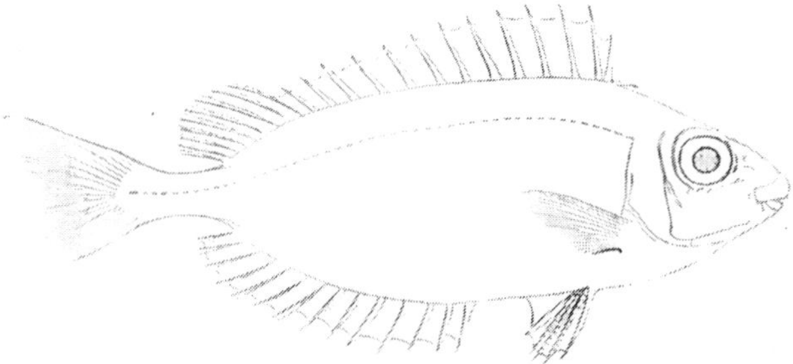

Fig. 182.—Teuthis nebulosa, Indian Ocean.

the abdominal portion. The abdominal cavity is surrounded by a complete ring of bones, the second piece of the coracoid being exceedingly long, and extending along the whole length of the abdomen, where it is joined to a spinous process of the first interhæmal. The pubic bones are slender, long, firmly attached to each other, without leaving a free space between them. They are fastened by a long process which passes the symphysis of the radii, and extends on to that of the humeri.

Thirty species are known, all from the Indo-Pacific; but they do not extend eastwards beyond 140° long., or to the Sandwich Islands. They are herbivorous, and do not exceed a length of fifteen inches.

SECOND DIVISION—ACANTHOPTERYGII BERYCIFORMES.

Body compressed, oblong, or elevated; head with large

muciferous cavities which are covered with a thin skin. Ventral fins thoracic, with one spine and more than five soft rays (in Monocentris *with two only).*

One family only belongs to this division.

FAMILY—BERYCIDÆ.

Body short, with ctenoid scales, which are rarely absent. Eyes lateral, large (except Melamphaës*). Cleft of the mouth lateral, oblique; jaws with villiform teeth; palate generally toothed. Opercular bones more or less armed. Eight (four) branchiostegals.*

This family offers several points of biological interest. All its members are strictly marine; but only two of the genera are surface-forms (*Holocentrum* and *Myripristis*). All the others descend considerably below the surface, and even some of the species of *Myripristis* habitually inhabit depths of from 50 to 100 fathoms. *Polymixia* and *Beryx* have been found in 345 fathoms. *Melamphaës* must live at a still greater depth, as we may infer from the small size of its eye; this fish is not likely to come nearer to the surface than to about 200 fathoms. The other genera named have extremely large eyes, and, therefore, may be assumed to ascend into such superficial strata as are still lit up by a certain proportion of sun-rays. The highly-developed apparatus for the secretion of superficial mucus, with which these fishes are provided, is another sign of their living at a greater depth than any of the preceding families of Acanthopterygians. In accordance with this vertical distribution, Berycoid fishes have a wide horizontal range, and several species occur at Madeira as well as in Japan.

Fossil Berycoids show a still greater diversity of form than living; they belong to the oldest Teleosteous fishes, the majority of the Acanthopterygians found in the chalk being representatives of this family. *Beryx* has been found in several species, with other genera now extinct: *Pseudoberyx*, with abdominal ventrals, from Mount Lebanon; *Berycopsis*, with cycloid scales; *Homonotus*, *Stenostoma*, *Sphenocephalus*, *Acanus*, *Hoplopteryx*, *Platycormus*, with granular scales; *Podocys*, with a dorsal fin extending to the neck; *Acrogaster*, *Macrolepis*, and *Rhacolepis*, from the chalk of Brazil. Species of *Holocentrum* and *Myripristis* occur in the Monte Bolca formation.

MONOCENTRIS.—Snout obtuse, convex, short; eye of moderate size. Villiform teeth on the palatine bones, but none on the vomer. Opercular bones without armature. Scales very large, bony, forming a rigid carapace. Ventrals reduced to a single strong spine and a few rudimentary rays.

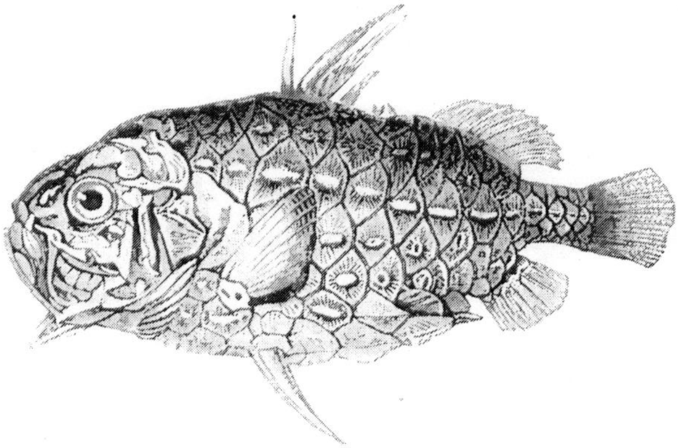

Fig. 183.—Monocentris japonicus.

One species only is known (*M. japonicus*) from the seas off Japan and Mauritius. It does not attain to any size, and is not common.

HOPLOSTETHUS.—Snout very short and obtuse; eye large. Villiform teeth on the palatine bones, but none on the vomer. Operculum unarmed, a strong spine at the scapulary and the angle of the præoperculum. Scales ctenoid, of moderate size; abdominal edge serrated. One dorsal, with six spines; ventrals with six soft rays; caudal deeply forked.

One species only is known (*H. mediterraneus*), which

occurs in the Mediterranean, the neighbouring parts of the Atlantic, and in the sea off Japan.

TRACHICHTHYS.—Snout very short and obtuse, with prominent chin; eye large. Villiform teeth on the palatine bones and on the vomer. A strong spine at the scapulary and at the angle of the præoperculum. Scales rather small; abdomen serrated. One dorsal, with from three to six spines; ventral with six soft rays. Caudal forked.

Four species are known from New Zealand and Madeira.

Anoplogaster is an allied genus from tropical parts of the Atlantic; it is scaleless.

BERYX.—Snout short, with oblique cleft of the mouth and prominent chin; eye large. Villiform teeth on the palatine bones and vomer. Opercular bones serrated; no spine at the angle of the præoperculum. Scales ctenoid, of moderate or large size. One dorsal, with several spines; ventrals with seven or more soft rays. Anal with four spines; caudal forked.

Five species are known from Madeira, the tropical Atlantic, and the seas of Japan and Australia. The species

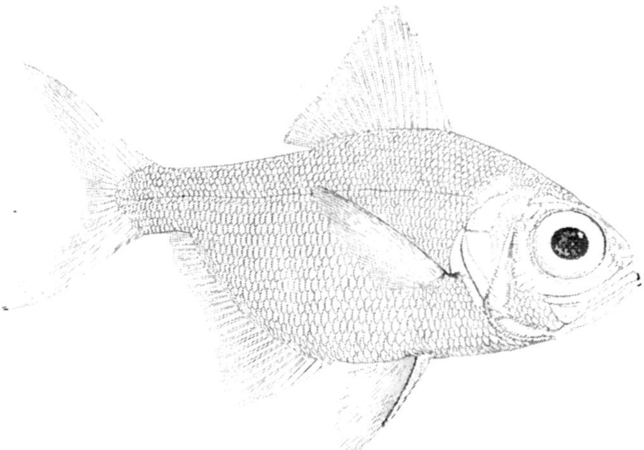

Fig. 184.—Beryx decadactylus.

figured is *B. decadactylus*, common at Madeira, and occurring near Japan at a depth of 345 fathoms; it attains a length of 1½ feet.

MELAMPHAES.—Head large and thick, with very thin bones, nearly all the superficial bones being transformed into wide muciferous channels. Eye small. Palate toothless; no barbels; opercles not armed. Scales large, cycloid. One dorsal, with six spines; anal spines very feeble; caudal forked. Ventrals with seven rays.

Two species, deep-sea fishes of the Atlantic; they are very scarce, as only three or four specimens have been found hitherto.

POLYMIXIA.—Snout short, with the cleft of the mouth nearly horizontal; eye large. Two barbels at the throat. Opercles without armature. Scales of moderate size. One dorsal. Anal with three or four spines; caudal forked; ventrals with six or seven soft rays.

Three species are known: *P. nobilis* from Madeira and St. Helena, *P. lowei* from Cuba, and *P. japonica* from Japan; the latter species from a depth of 345 fathoms. Average size eighteen inches.

MYRIPRISTIS.—Snout short, with oblique cleft of the mouth and prominent chin; eye large or very large. Villiform teeth on the vomer and palatine bones. Opercular bones serrated; præoperculum without spine. Scales large, ctenoid. Two dorsals, the first with ten or eleven spines; anal with four spines; caudal forked; ventrals with seven soft rays. Air-bladder divided by a contraction in two parts, the anterior of which is connected with the organ of hearing.

Eighteen species from the tropical seas of both hemispheres, the majority living near the coast at the surface. The coloration is (principally) red or pink on the back and silvery on the sides. They attain a length of about 15 inches, and are esteemed as food.

HOLOCENTRUM.—Snout somewhat projecting, with the cleft of the mouth nearly horizontal; eye large. Villiform teeth on the vomer and palatine bones. Opercular bones and præorbital serrated; operculum with two spines behind; a large spine at the angle of the præoperculum. Scales ctenoid, of moderate size. Two dorsals, the first with twelve spines; anal with four spines, the third being very long and strong; caudal forked. Ventrals with seven soft rays.

About thirty species are known from the tropical seas of both hemispheres; all are surface fishes, and very common. The young have the upper part of the snout pointed and

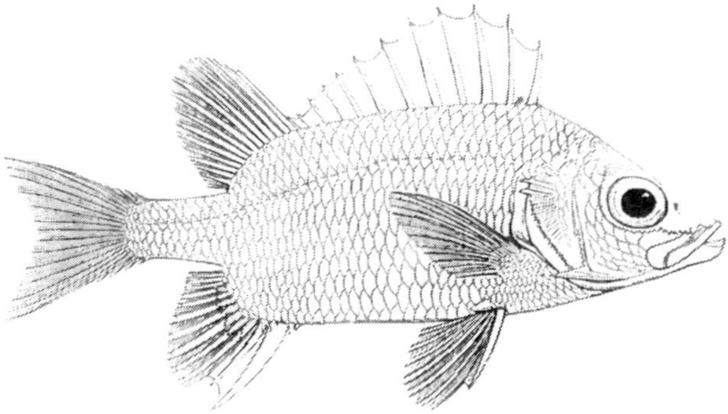

Fig. 185.—*Holocentrum unipunctatum*, from the South Sea.

elongate, and were described as a distinct genus (*Rhynchichthys*). The coloration of the adult is uniform; red, pink, and silvery prevailing. They attain to a length of about 15 inches, and are esteemed as food.

THIRD DIVISION—ACANTHOPTERYGII KURTIFORMES.

One dorsal fin only, much shorter than the anal, which is long and many-rayed. No superbranchial organ.

One family only belongs to this division.

FAMILY—KURTIDÆ.

Body compressed, oblong, deep in front, attenuated behind. Snout short. The spines of the short dorsal are few in number, if developed. Scales small or of moderate size. Villiform teeth in the jaws, on the vomer, and palatine bones.

This family consists of a small number of species only, which form two distinct genera, *Pempheris* and *Kurtus*. They are shore fishes of tropical seas. In both the air-bladder shows some peculiarity: in *Pempheris* it is divided into an anterior and posterior portion; in *Kurtus* it is lodged within the ribs, which are dilated, convex, forming rings. The number of vertebræ is respectively twenty-four and twenty-three.

FOURTH DIVISION—ACANTHOPTERYGII POLYNEMIFORMES.

Two rather short dorsal fins, somewhat remote from each other; free filaments at the humeral arch, below the pectoral fins; muciferous canals of the head well developed.

One family only belongs to this division.

FAMILY—POLYNEMIDÆ.

Body oblong, rather compressed, covered with smooth or very feebly ciliated scales. Lateral line continuous. Snout projecting beyond the mouth, which is inferior, with lateral cleft. Eye lateral, large. Villiform teeth in the jaws and on the palate. Ventrals thoracic, with one spine and five rays.

The fishes of this natural family have been divided, on slight differences, into three genera—*Polynemus*, *Pentanemus*, and *Galeoides*. They are found in rather numerous species on the coasts between the tropics, and the majority enter brackish or even fresh water. Very characteristic are the free filaments which in this family are organs of touch; they are inserted on the humeral arch at some distance from the pectoral fin; but, nevertheless, can be regarded only as a detached portion of that fin; they can be moved quite independently of the fin; their number varies from three to fourteen, according to the species; in some they are exceedingly elongate, twice as long as the fish, in others they are not longer or even

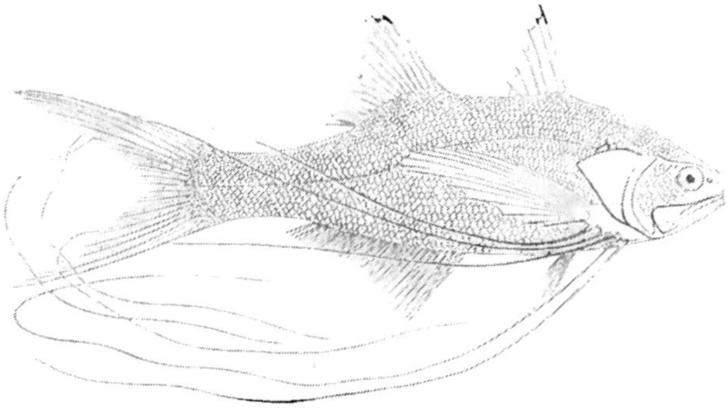

Fig. 186.—*Pentanemus quinquarius*, from the West Coast of Africa and the West Indies.

shorter than the pectoral. It is evident from the whole organisation of these fishes that they live on muddy bottom or in thick water, such as is found near the mouths of great rivers. Their eyes are large, but generally obscured by a filmy skin, so that those feelers must be of great use to them in finding their way and their food. The Polynemoids are very useful to man: their flesh is esteemed, and some of the species are provided with an air-bladder which yields a good sort of isinglass, and forms an article of trade in the East Indies. Some of these fishes attain to a length of four feet.

Fifth Division—Acanthopterygii Sciæniformes.

The soft dorsal is more, generally much more, developed than the spinous, and than the anal. No pectoral filaments; head with the muciferous canals well developed.

Also this division is composed of one family only.

Family—Sciænidæ.

Body rather elongate, compressed, covered with ctenoid scales. Lateral line continuous, and frequently extending over the caudal fin. Mouth in front of the snout. Eye lateral, of moderate size. Teeth in villiform bands, sometimes with the addition of canines; no molars or incisor-like teeth in the jaws; palate toothless. Præoperculum unarmed, and without bony stay. Ventrals thoracic, with one spine and five soft rays. Bones of the head with wide muciferous channels. Stomach cœcal. Air-bladder frequently with numerous appendages (see pp. 144 and *seq.*)

The fishes of the "*Meagre*" family are chiefly coast-fishes of the tropical and sub-tropical Atlantic and Indian Oceans, preferring the neighbourhood of the mouths of large rivers, into which they freely enter, some of the species having become so completely naturalised in fresh water that they are never found now-a-days in the sea. Some of the larger species wander far from their original home, and are not rarely found at distant localities as occasional visitors. In the Pacific and on the coast of Australia, where but a few large rivers enter the ocean, they are extremely rare and, in the Red Sea, they are absent. Many attain a large size, and almost all are eaten.

No fossil species have been as yet discovered.

POGONIAS.—Snout convex, with the upper jaw overlapping the lower. Mandible with numerous small barbels. No canines. The first dorsal with ten stout spines. Two anal spines, the second very strong. Scales of moderate size.

To this fish (*P. chromis*) more especially is given the name of "Drum," from the extraordinary sounds which are produced by it and other allied Sciænoids. These sounds are better expressed by the word drumming than by any other, and are frequently noticed by persons in vessels lying at anchor on the coasts of the United States, where those fishes abound. It is still a matter of uncertainty by what means the "Drum" produces the sounds. Some naturalists believe that it is caused by the clapping together of the pharyngeal teeth, which are very large molar teeth. However, if it be true that the sounds are accompanied by a tremulous motion of the vessel, it seems more probable that they are produced by the fishes beating their tails against the bottom of the vessel in order to get rid of the parasites with which that part of their body is infested. The "Drum" attains to a length of more than four feet, and to a weight exceeding a hundred lbs. Its air-bladder has been figured on p. 146.

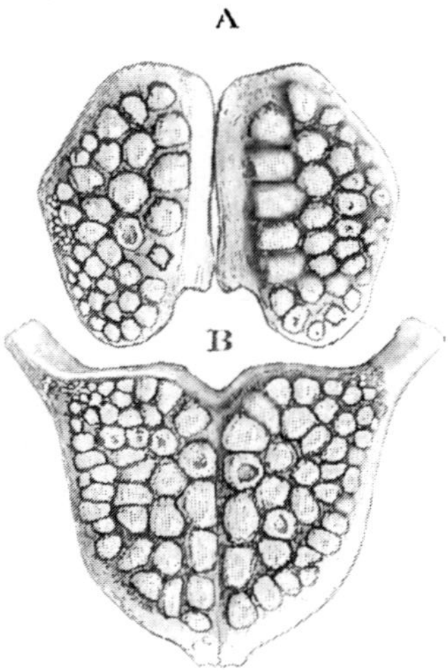

Fig. 187.—Pharyngeal bones and teeth of *Pogonias chromis*. A, Upper; B, Lower pharyngeals.

Micropogon is closely allied to *Pogonias*, but has conical pharyngeal teeth. Two species from the western parts of the Atlantic.

UMBRINA.—Snout convex, with the upper jaw overlapping the lower; a short barbel under the symphysis of the mandible. The first dorsal with nine or ten flexible spines, the anal with one or two. Scales of moderate size.

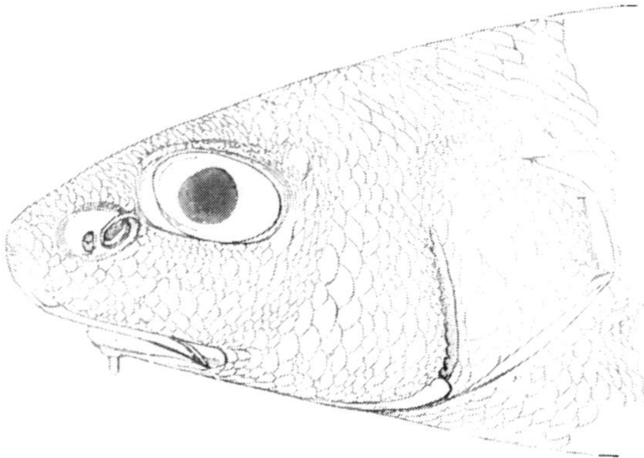

Fig. 188.—*Umbrina nasus*, from Panama.

Twenty species are known from the Mediterranean,

Atlantic, and Indian Ocean. One well known to the ancients,

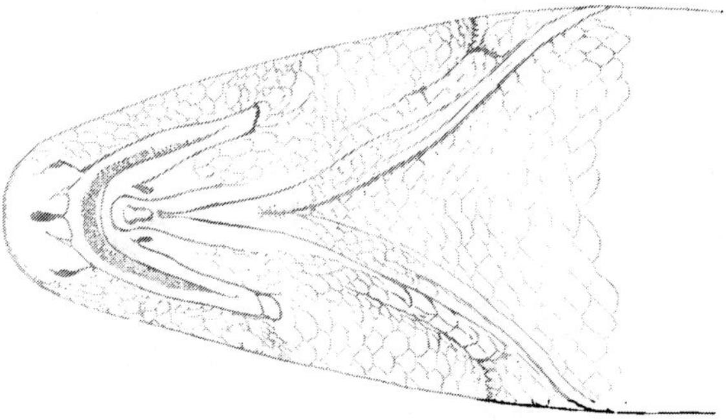

Fig. 189.—*Umbrina nasus*, from Panama.

under the name of *Umbra*, is the *Umbrina cirrhosa* of the Mediterranean, the "Umbrine" or "Ombre" of the French, and the "Corvo" of the Italians. It ranges to the Cape of Good Hope, and attains a length of three feet. Also on the coasts of the United States several species occur, as *U. alburna*, *U. nebulosa*, etc.

Sciæna (including *Corvina*).—The upper jaw overlapping the lower, or both jaws equal in front. Interorbital space moderately broad and slightly convex. Cleft of the mouth horizontal or slightly oblique. The outer series of teeth is generally composed of teeth larger than the rest, but there are no canines. Eye of moderate size, barbel none.

Some fifty species are known, but their distinctive char-

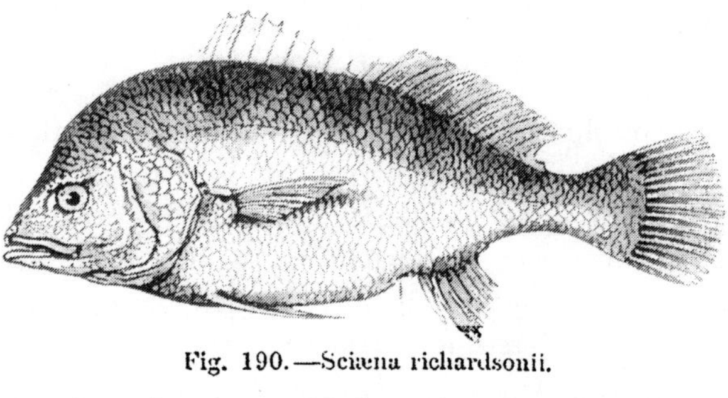

Fig. 190.—Sciæna richardsonii.

acters have been but imperfectly pointed out. They are found in all the seas and rivers in which Sciænoids generally occur, and many are entirely confined to fresh water, for instance the species figured, *Sciæna richardsonii*, from Lake Huron; *Sc. amazonica*; *Sc. obliqua, ocellata, oscula*, etc., from fresh waters of the United States. *Sciæna diacanthus* and *Sc. coitor* belong to the most common fishes of the coasts of the East Indies, ascending the great rivers for a long distance from the sea. One of the European species, *Sciæna aquila*, has an extremely wide range; it not rarely reaches the British coasts, where it is known as "Meagre," and has been found at the Cape of Good Hope and on the coast of southern Australia. Like some of the other species it attains to a length of six feet, but the majority of the species of this genus remain within smaller dimensions. A part of the species have the second anal ray very strong, and have been placed into a distinct genus, *Corvina*,—thus, among others, *Sc. nigra* from the Mediterranean, and *Sc. richardsonii.*

Pachyurus is closely allied to *Sciæna*, but has the vertical fins densely covered with small scales.

Otolithus.—Snout obtuse or somewhat pointed, with the lower jaw longer. The first dorsal with nine or ten feeble spines. Canine teeth more or less distinct. Præoperculum denticulated. Scales of moderate or small size.

About twenty species are known from the tropical and sub-tropical parts of the Atlantic and Indian Oceans. The air-bladder is figured on p. 144.

Ancylodon differs from *Otolithus* in having very long arrow-shaped or lanceolate canine teeth. Coasts of tropical America.

Collichthys.—Body elongate; head very broad, with the upper surface very convex; cleft of the mouth wide and oblique; no large canines. Eye small. No barbel. Scales small, or of moderate size. The second dorsal very long, caudal pointed.

Three species from the East Indian and Chinese coasts. The great development of the muciferous system on the head and the small eye leads one to suppose that these fishes live in muddy water near the mouths of large rivers. The air-bladder has been described on p. 144.

Other genera belonging to this family are *Larimus*, *Eques*, *Nebris*, and *Lonchurus*.

SIXTH DIVISION—ACANTHOPTERYGII XIPHIIFORMES.

The upper jaw is produced into a long cuneiform weapon.

These fishes form one small family only, *Xiphiidæ*.

The "Sword-fishes" are pelagic fishes, occurring in all tropical and sub-tropical seas. Generally found in the open ocean, always vigilant, and endowed with extraordinary strength and velocity, they are but rarely captured, and still more rarely preserved. The species found in the Indian and Pacific Oceans belong to the genus *Histiophorus*, distinguished from the common Mediterranean Sword-fish, or *Xiphias*, by

Fig. 191.—Histiophorus pulchellus.

the presence of ventral fins, which, however, are reduced to two long styliform appendages. The distinction of the species is beset with great difficulties, owing to the circumstance that but few examples exist in museums, and further, because the form of the dorsal fin, the length of the ventrals, the shape and length of the sword, appear to change according to the age of the individuals. Some specimens or species have only the anterior dorsal rays elevated, the remainder of the fin being very low, whilst in others all the rays are exceedingly elongate, so that the fin, when erected, projects beyond the surface of the water. It is stated that Sword-fishes, when quietly floating with the dorsal fin erect, can sail before the wind, like a boat.

Sword-fishes are the largest of Acanthopterygians, and not exceeded in size by any other Teleostean; they attain to a length of from 12 to 15 feet, and swords have been obtained more than three feet long, and with a diameter of at least three inches at the base. The sword is formed by the prolongation and coalescence of the maxillary and intermaxillary bones; it is rough at its lower surface, owing to the development of rudimentary villiform teeth, very hard and strong, and forms a most formidable weapon. Sword-fishes never hesitate to attack whales and other large Cetaceans, and by repeatedly stabbing these animals generally retire from the combat victorious. The cause which excites them to those attacks is unknown; but they follow this instinct so blindly that they not rarely attack boats or large vessels in a similar manner, evidently mistaking them for Cetaceans. Sometimes they actually succeed in piercing the bottom of a ship, endangering its safety; but as they are unable to execute powerful backward movements they cannot always retract their sword, which is broken off by the exertions of the fish to free itself. A piece of a two-inch plank of a whale-boat, thus pierced by a sword-fish, in which the broken sword still remains, is preserved in the British Museum.

The Rev. Wyatt Gill, who has worked as a missionary for many years in the South Sea Islands, communicates that young Sword-fishes are easily caught in strong nets, but no net is strong enough to hold a fish of six feet in length. Specimens of that size are now and then captured by hook and line, a small fish being used as bait. Individuals with the sword broken off are not rarely observed. Larger specimens cannot be captured by the natives, who are in great fear of them. They easily pierce their canoes, and only too often dangerously wound persons sitting in them.

The Mediterranean Sword-fish is constantly caught in the nets of the Tunny-fishers off the coast of Sicily, and brought to market, where its flesh sells as well as that of the Tunny.

The remarkable changes which Sword-fishes undergo at an early stage of their growth have been noticed above, p. 173 and *seq.*

Sword-fishes are as old a type as the Berycoids. Their remains have been found in the chalk of Lewes, and more frequently in the London clay of Sheppy, where an extinct genus, *Coelorhynchus*, has been recognised.

Seventh Division—Acanthopterygii Trichiuriformes.

Body elongate, compressed or band-like; cleft of the mouth wide, with several strong teeth in the jaws or on the palate. The spinous and soft portions of the dorsal fin and the anal are of nearly equal extent, long, many-rayed, sometimes terminating in finlets; caudal fin forked, if present.

Family—Trichiuridæ.

Marine fishes inhabiting the tropical and sub-tropical seas; some of them are surface-fishes, living in the vicinity of the coast, whilst others descend to moderate depths, as the Berycoids. All are powerful rapacious fishes, as is indicated by their dentition.

The oldest of the extinct genera are *Enchodus* and *Anenchelum;* they were formerly referred to the Scombroids, but belong to this family. The former has been found in the chalk of Lewes and Mæstricht; the latter is abundant in the Eocene schists of Glaris. *Anenchelum* is much elongate, and exhibits in the slender structure of its bones the characteristics of a deep-sea fish; it resembles much *Lepidopus*, but has some long rays in the ventrals. Other Eocene genera are *Nemopteryx* and *Xiphopterus*. In the Miocene of Licata in Sicily *Trichiuridæ* are well represented, viz. by a species of *Lepidopus*, and by two genera, *Hemithyrsites* and *Trichiurichthys*, which are allied to *Thyrsites* and *Trichiurus*, but covered with scales.

The following is a complete list of the genera referred to this family:—

NEALOTUS.—Body incompletely clothed with delicate scales. Small teeth in the jaws and on the palatine bones; none on the vomer. Two dorsal fins, the first continuous and extending to the second; finlets behind the second and anal fins. Each ventral fin represented by a single small spine. A dagger-shaped spine behind the vent. Caudal fin well developed.

One specimen only of this fish (*N. tripes*), 10 inches long, has been obtained off Madeira; it evidently lives at a considerable depth, and comes to the surface only by accident.

NESIARCHUS.—Body covered with small scales. Several strong fangs in the jaws; no teeth on the palate. First dorsal not extending to the second. No detached finlets. Ventrals small, but perfectly developed, thoracic. Caudal fin present. A dagger-shaped spine behind the vent.

A rather large fish (*N. nasutus*), very rarely found in the sea off Madeira. The two or three specimens found hitherto measure from three to four feet in length. Probably living at the same depth as the preceding genus.

APHANOPUS.—Scales none. Two very long dorsal fins; caudal well developed; ventrals none. A strong dagger-shaped spine behind the vent. Strong teeth in the jaws; none on the palate.

One species only is known, named *A. carbo* from its coal-black colour; it is evidently a deep-sea fish, very rarely obtained in the sea off Madeira. Upwards of four feet long.

EUOXYMETOPON.—Body naked, very long and thin. Profile

of the head regularly decurved from the nape to the snout, the occiput and forehead being elevated and trenchant. Jaws with fangs; palatine teeth present. One dorsal only, continued from the head to the caudal fin, which is distinct. A dagger-shaped spine behind the vent. Pectoral fins inserted almost horizontally, with the lowest rays longest, and with the posterior border emarginate. Ventral fins rudimentary, scale-like.

This is another deep-sea form of this family, but, at present, no observations have been made as regards the exact depth at which it occurs. A specimen has been known since the year 1812; it was found on the coast of Scotland, and described as *Trichiurus lepturus*. The same species has been re-discovered in the West Indies, where, however, it is also extremely scarce.

Lepidopus.—Body band-like; one single dorsal extends along the whole length of the back; caudal well developed. Ventrals reduced to a pair of scales. Scales none. Several fangs in the jaws; teeth on the palatine bones.

The Scabbard-fish (*L. caudatus*) is rather common in the Mediterranean and warmer parts of the Atlantic, extending

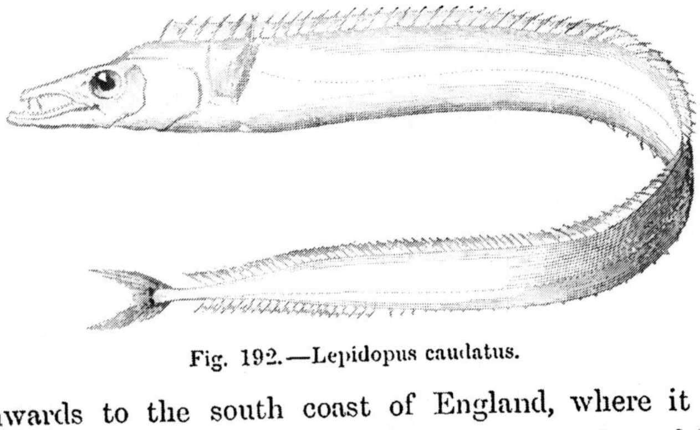

Fig. 192.—Lepidopus caudatus.

northwards to the south coast of England, where it is an occasional visitor, and southwards to the Cape of Good Hope. More recently it has been observed on the coasts of Tasmania and New Zealand. We may, therefore, justly consider it to be a deep-sea fish, which probably descends to the same depth as the preceding allied forms. It grows to a length of five or six feet, but its body is so much compressed that it does not weigh more than as many pounds. It is well known in New Zealand, where it is called "Frost-fish," and esteemed as the most delicious fish of the colony. A still more attenuated species (*L. tenuis*) occurs in the sea off Japan, at a depth of some 340 fathoms.

TRICHIURUS.—Body band-like, tapering into a fine point, without caudal fin. One single dorsal extending the whole length of the back. Ventrals reduced to a pair of scales, or entirely absent. Anal fin rudimentary, with numerous extremely short spines, scarcely projecting beyond the skin. Long fangs in the jaws; teeth on the palatine bones, none on the vomer.

The "Hairtails" belong to the tropical marine fauna, and although generally found in the vicinity of land, they wander frequently out to sea, perhaps merely because they follow some ocean-currents. Therefore they are not rarely found in the temperate zone, the common West Indian species (*T. lepturus*), for instance, on the coast of England. They attain to a length of about four feet. The number of their vertebræ is very large, as many as 160, and more. Six species are known.

EPINNULA—Body rather elongate, covered with minute scales, The first dorsal fin continuous, with spines of moderate strength, and extending on to the second; finlets none; ventrals well developed. Lateral lines two. Teeth of the jaws strong; palatine teeth, none.

The "Domine" of the Havannah, *E. magistralis.*

THYRSITES.—Body rather elongate, for the greater part naked. The first dorsal continuous, with the spines of moderate strength, and extending on to the second. From two to six finlets behind the dorsal and anal. Several strong teeth in the jaws; teeth on the palatine bones.

The species of this genus attain to a considerable size (from four to five feet), and are valuable food fishes; *Th. atun* from the Cape of Good Hope, South Australia, New Zealand, and Chili, is preserved, pickled or smoked. In New Zealand it is called "Barracuda" or "Snoek," and exported from the colony into Mauritius and Batavia as a regular article of commerce, being worth over £17 a ton; *Th. pretiosus*, the "Escholar" of the Havannah, from the Mediterranean, the neighbouring parts of the Atlantic, and the West Indies; *Th. prometheus* from Madeira, Bermuda, St. Helena, and Polynesia; *Th. solandri* from Amboyna and Tasmania is probably the same as *Th. prometheus.*

Young specimens of this (or, perhaps, the following) genus have been described as *Dicrotus.* In them the finlets are not yet detached from the rest of the fin; and the ventral fins, which are entirely obsolete in the adult fish, are represented by a long crenulated spine.

GEMPYLUS.—Body very elongate, scaleless. The first dorsal fin continuous, with thirty and more spines, and extending on to the second. Six finlets behind the dorsal and anal. Several strong teeth in the jaws, none on the palate.

One species (*G. serpens*), inhabiting considerable depths of the Atlantic and Pacific Oceans.

FAMILY—PALÆORHYNCHIDÆ.

This family has been formed for two extinct genera: *Palæorhynchus* from the schists of Glaris, and *Hemirhynchus* from tertiary formations near Paris. These genera resemble much the *Trichiuridæ* in their long, compressed body, and long vertical fins, but their jaws, which are produced into a long beak, are toothless, or provided with very small teeth. The dorsal fin extends the whole length of the back, and the anal reaches from the vent nearly to the caudal, which is forked. The ventrals are composed of several rays and thoracic. The vertebræ long, slender, and numerous, and, like all the bones of the skeleton, thin, indicating that these fishes were inhabitants of considerable depths of the ocean. Both the jaws of *Palæorhynchus* are prolonged into a beak, whilst in *Hemirhynchus* the upper exceeds the lower in length.

Eighth Division—Acanthopterygii Cotto-scombriformes.

Spines developed in one of the fins at least. Dorsal fins either continuous or close together; the spinous dorsal, if present, always short; sometimes modified into tentacles, or into a suctorial disk; soft dorsal always long, if the spinous is absent; anal similarly developed as the soft dorsal, and both generally much longer than the spinous, sometimes terminating in finlets. Ventrals, thoracic or jugular, if present, never modified into an adhesive apparatus. No prominent anal papilla.

Marine fishes, with few exceptions.

First Family—Acronuridæ.

Body compressed, oblong or elevated, covered with minute scales. Tail generally armed with one or more bony plates or spines, which are developed with age, but absent in very young individuals. Eye lateral, of moderate size. Mouth small; a single series of more or less compressed, sometimes denticulated, sometimes pointed incisors in each jaw; palate toothless. One dorsal fin, the spinous portion being less developed than the soft; anal with two or three spines; ventral fins thoracic. Air-bladder forked posteriorly. Intestines with more or less numerous circumvolutions. Nine abdominal, and thirteen caudal vertebræ.

Inhabitants of the tropical seas, and most abundant on coral-reefs. They feed either on vegetable substances or on the superficial animal matter of corals.

Extinct species of *Acanthurus* and *Naseus* have been discovered in the Monte Bolca formation.

ACANTHURUS.—Jaws with a single series of lobate incisors, which are sometimes movable. An erectile spine hidden in a groove on each side of the tail. Ventral fins with one spine and generally five rays. Scales ctenoid, sometimes with minute spines. Branchiostegals five.

The fishes of this genus, which sometimes are termed "Surgeons," are readily recognised by the sharp lancet-shaped spine with which each side of the tail is armed. When at rest the spine is hidden in a sheath; but it can be erected and used by the fish as a very dangerous weapon, by striking with the tail towards the right and left. "Surgeons" occur in all tropical seas, with the exception of the eastern part of the Pacific, where they disappear with the corals. They do

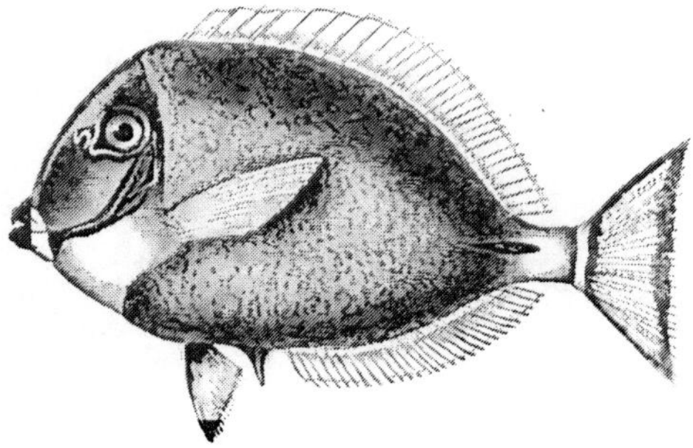

Fig. 193.—Acanthurus leucosternum, Indian Ocean.

not attain to any size, the largest species scarcely exceeding a length of eighteen inches. Many are agreeably or showily coloured, the ornamental colours being distributed in very extraordinary patterns. The larger species are eatable, and some even esteemed as food. It is stated that the fry of some species periodically approaches, in immense numbers, the coasts of some of the South Sea Islands (Caroline Archipelago), and serves as an important article of food to the natives. Nearly fifty species are known.

At an early period of their growth these fishes present so different an aspect that they were considered a distinct genus, *Acronurus*. The form of the body is more circular and exceedingly compressed. No scales are developed, but the skin forms numerous oblique parallel folds. The gill-cover and the breast are shining silvery.

NASEUS.—Tail with two (rarely one or three) bony keeled plates on each side (in the adult). Head sometimes with a bony horn or crest-like prominence directed forwards. Ventral fins composed of one spine and three rays. From four to six spines in the dorsal; two anal spines. Scales minute, rough, forming a sort of fine shagreen. Air-bladder forked behind. Intestinal tract with many circumvolutions.

Twelve species are known from the tropical Indo-Pacific, but none of them extend eastwards beyond the Sandwich Islands. In their mode of life these fishes resemble the *Acanthuri*. Likewise, the young have a very different appearance, and are unarmed, and were described as a distinct

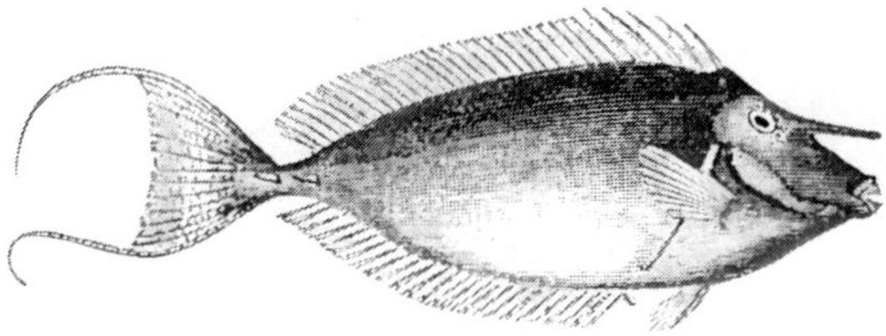

Fig. 194.—Naseus unicornis.

genus, *Keris*. One of the most common species is *N. unicornis*, which, when adult (22 inches long), has a horn about 2 inches long, whilst it is merely a projection in front of the eye in individuals of 7 inches in length.

Prionurus is an allied genus with a series of several keeled bony laminæ on each side of the tail.

SECOND FAMILY—CARANGIDÆ.

Body more or less compressed, oblong or elevated, covered with small scales or naked; eye lateral. Teeth, if present, conical. No bony stay for the præoperculum. The spinous dorsal is less developed than the soft or than the anal, either continuous with, or separated from, the soft portion; sometimes rudimentary.

Ventrals thoracic, sometimes rudimentary or entirely absent. No prominent papilla near the vent. Gill-opening wide. Ten abdominal and fourteen caudal vertebræ.

Inhabitants of tropical and temperate seas. Carnivorous. They appear first in cretaceous formations, where they are represented by *Platax* and some Caranx-like genera (*Vomer* and *Aipichthys* from the chalk of Comen in Istria). They are

Fig. 195.—Semiophorus velitans.

more numerous in various Tertiary formations, especially in the strata of Monte Bolca, where some still existing genera occur, as *Zanclus*, *Platax*, *Caranx* (*Carangopsis*), *Argyriosus* (*Vomer*), *Lichia*, *Trachynotus*. Of the extinct genera the following belong to this family :—*Pseudovomer* (*Licata*), *Amphistium*, *Archæus*, *Ductor*, *Plionemus* (?), and *Semiophorus*. *Equula* has been recently discovered in the Miocene marls of Licata in Sicily.

CARANX (including *Trachurus*).—Body more or less compressed, sometimes sub-cylindrical. Cleft of the mouth of moderate width. The first dorsal fin continuous, with about eight feeble spines, sometimes rudimentary; the soft dorsal and anal are succeeded by finlets in a few species. Two anal spines, somewhat remote

Fig. 196.—Plates of the lateral line of Caranx hippos.

from the fin. Scales very small. Lateral line with an anterior curved, and a posterior straight, portion, either entirely or posteriorly only covered by large plate-like scales, several of which are generally keeled, the keel ending in a spine. Dentition feeble. Air-bladder forked posteriorly.

The "Horse-mackerels" are found in abundance in almost all temperate and, especially, tropical seas. Many species wander to other parts of the coast, or to some distance from land, and have thus gradually extended their range over two or more oceanic areas; some are found in all tropical seas. The species described are very numerous, about ninety having been properly characterised and distinguished. Some attain to a length of three feet and more, and all are eatable. They feed on other fish and various marine animals.

Of the most noteworthy species the following may be mentioned:—*C. trachurus*, the common British Horse-mackerel, distinguished by having the lateral line in its whole length armed with large vertical plates; it is almost cosmopolitan within the temperate and tropical zones of the northern and southern hemispheres. *C. crumenophthalmus*, *C. carangus*, and *C. hippos*, three of the most common sea-fishes, equally abun-

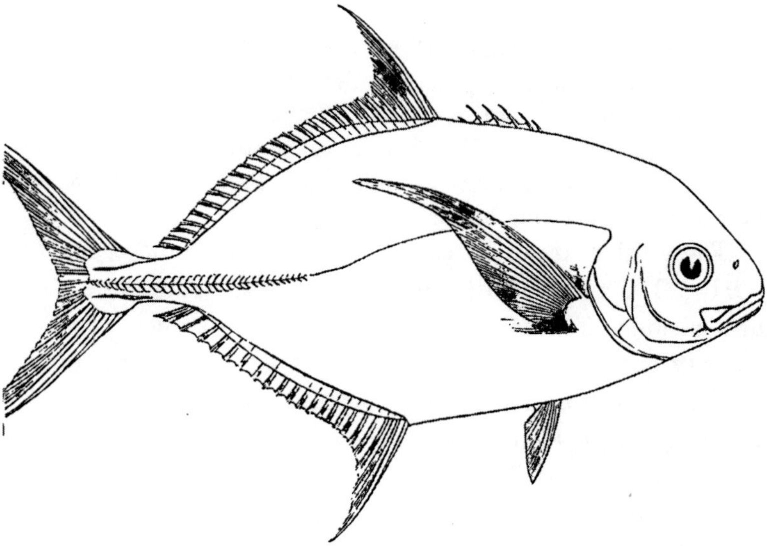

Fig. 197.—Caranx ferdau.

dant in the Atlantic and Indo-Pacific oceans; *C. ferdau*, from the Indo-Pacific, upwards of three feet in length. *C. armatus*, *ciliaris*, *gallus*, etc., which have an exceedingly short and compressed body, with rudimentary spinous dorsal fin, and with some of the rays of the dorsal and anal prolonged into filaments.

Argyriosus is closely allied to *Caranx*, especially to the last-named species, but the lateral line has no plates whatever; and the body is scaleless, chiefly of a bright silvery colour.

Two species from the tropical Atlantic.

Micropteryx.—Body much compressed, with prominent trenchant abdomen, covered with small scales; lateral line not shielded; præopercular margin entire. Cleft of the mouth rather small; præorbital of moderate width. The first dorsal continuous, with seven feeble spines. No detached finlets. Small teeth on the vomer and palatine bones.

Micropteryx chrysurus is a semi-pelagic fish, and very common in the tropical Atlantic, less so in the Indian Ocean.

SERIOLA.—Body oblong, slightly compressed, with rounded abdomen, covered with very small scales; lateral line not shielded; præopercular margin entire. Cleft of the mouth of moderate width, or rather wide. The first dorsal continuous, with feeble spines. No detached finlets. Villiform teeth in the jaws, on the vomer and palatine bones.

These fishes are often called "Yellow-tails," and occur in nearly all the temperate and tropical seas, sometimes at a great distance from land. Twelve species are known, and the majority have a wide geographical range. The larger grow to a length of from four to five feet, and are esteemed as food, especially at St. Helena, the Cape of Good Hope, in Japan, Australia, and New Zealand.

Seriolella and *Seriolichthys*, the latter from the Indo-Pacific, and distinguished by a finlet behind the dorsal and anal, are allied genera.

NAUCRATES.—Body oblong, sub-cylindrical, covered with small scales; a keel on each side of the tail. The spinous dorsal consists of a few short free spines; finlets none. Villiform teeth in the jaws, on the vomer and palatine bones.

The "Pilot-fish" (*N. ductor*) is a truly pelagic fish, known in all tropical and temperate seas. Its name is derived from its habit of keeping company with ships and large fish, especially Sharks. It is the *Pompilus* of the ancients, who describe it as pointing out the way to dubious or embarrassed sailors, and as announcing the vicinity of land by its sudden disappearance. It was therefore regarded as a sacred fish. The connection between the Shark and the Pilot-fish has received various interpretations, some observers having perhaps added more sentiment than is warranted by the actual facts. It was stated that the Shark never seized the Pilot-fish, that the latter was of great use to its big companion in conducting it and showing it the way to its food. Dr. Meyen in his "Reise um die Erde" states: "The pilot swims constantly in front of the Shark; we ourselves have seen three instances in which the Shark was led by the Pilot. When the Shark neared the ship the Pilot swam close to the snout, or near one of the pectoral fins of the animal. Sometimes he darted rapidly forwards or sidewards as if looking for something, and constantly went back again to the Shark. When we threw overboard a piece of bacon fastened on a great hook, the Shark was about twenty paces from the ship. With the quickness of lightning the Pilot came up, smelt at the dainty, and instantly swam back again to the Shark, swimming many times round his snout and splashing, as if to give him exact information as to the bacon. The Shark now began to put himself in motion, the Pilot showing him the way, and in a moment he was fast upon the hook.[1] Upon a later occasion we observed two Pilots in sedulous attendance on a Blue Shark, which we caught in the Chinese Sea. It seems probable that the Pilot feeds on the Sharks' excrements, keeps his company for that purpose, and directs his operations solely from this selfish view." We believe that Dr. Meyen's opinion, as expressed in his last words, is perfectly correct. The Pilot obtains a great part of his food directly from the Shark, in feeding on the parasitic crustaceans with which Sharks and other large fish are infested, and on the smaller pieces of flesh which are left unnoticed by the Shark when it tears its prey. The Pilot also, being a small fish, obtains greater security when in company of a Shark, which would keep at a distance all other fishes of prey that would be likely to prove dangerous to the Pilot. Therefore, in accompanying the Shark, the Pilot is led by the same instinct which makes it follow a ship.

[1] In this instance, one may entertain reasonable doubts as to the usefulness of the Pilot to the Shark.

With regard to the statement that the Pilot itself is never attacked by the Shark all observers agree as to its truth; but this may be accounted for in the same way as the impunity of the swallow from the hawk, the Pilot-fish being too nimble for the unwieldy Shark.

The Pilot-fish does not always leave the vessels on their approach to land. In summer, when the temperature of the sea-water is several degrees above the average, Pilots will follow ships to the south coast of England into the harbour, where they are generally speedily caught. Pilot-fish attain a length of 12 inches only. When very young their appearance differs so much from the mature fish that they have been described as a distinct genus, *Nauclerus*. This fry is exceedingly common in the open ocean, and constantly obtained in the tow-net; therefore the Pilot-fish retains its pelagic habits also during the spawning season, and some of the spawn found by voyagers floating on the surface is, without doubt, derived from this species.

CHORINEMUS.—Body compressed, oblong; covered with small scales, singularly shaped, lanceolate, and hidden in the skin. The first dorsal is formed by free spines in small numbers; the posterior rays of the second dorsal and anal are detached finlets. Small teeth in the jaws, on the vomer and palatine bones.

Twelve species are known from the Atlantic and Indo-Pacific; some enter brackish water, whilst others are more numerous at some distance from the shore. They attain to a length of from 2 to 4 feet. In the young, which have been described as *Porthmeus*, the spines and finlets are connected by membrane with the rest of the fin.

Lichia is an allied genus from the Mediterranean, tropical Atlantic, and the coast of Chili; five species.

TEMNODON.—Body oblong, compressed, covered with cycloid scales of moderate size. Cleft of the mouth rather wide. Jaws with a series of strong teeth; smaller ones on the vomer and the palatine bones. The first dorsal with eight feeble spines connected by membrane; finlets none. Lateral line not shielded. The second dorsal and anal covered with very small scales.

Temnodon saltator, sometimes called "Skip-jack," is spread over nearly all the tropical and sub-tropical seas; it frequents principally the coasts, but is also met with in the open sea. On the coasts of the United States it is well known by the name of "Blue-fish," being highly esteemed as food, and furnishing excellent sport. It is one of the most rapacious fishes, destroying an immense number of other shore-fishes, and killing many more than they can devour. It grows to a length of 5 feet, but the majority of those brought to market are not half that length.

TRACHYNOTUS.—Body more or less elevated, compressed, covered with very small scales. Cleft of the mouth rather small, with short convex snout. Opercles entire. The first dorsal composed of free spines in small number. No finlets. Teeth always small, and generally lost with age.

Ten species are known from the tropical Atlantic and Indo-Pacific; they rarely exceed a length of 20 inches. Some of the most common marine fishes belong to this genus, for instance *T. ovatus*, which ranges over the entire tropical zone.

PAMMELAS (*perciformis*) is allied to the preceding genus; from the coast of New York.

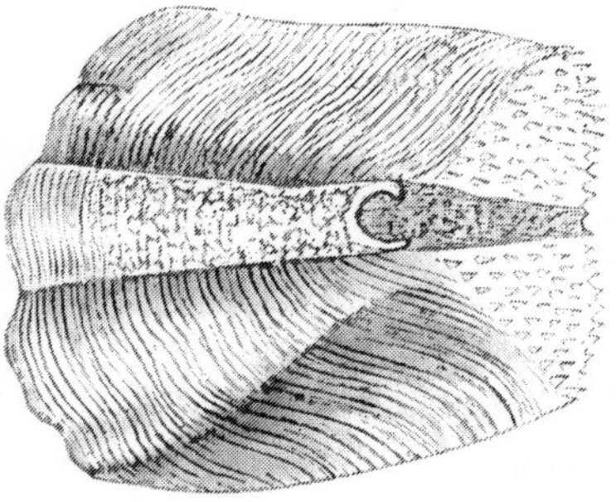

Fig. 198.—Magnified scale of Psettus argenteus.

PSETTUS.—Body much compressed and elevated; snout rather short. One dorsal, entirely covered with scales, with seven or eight spines; anal fin with three. Ventrals very small, rudimentary. Teeth villiform; no teeth on the palate. Scales small, ctenoid.

Only three species are known; one, *P. sebæ*, from the west coast of Africa, the two others from the Indo-Pacific. *P. argenteus* is a very common fish, attaining to a length of about 10 inches.

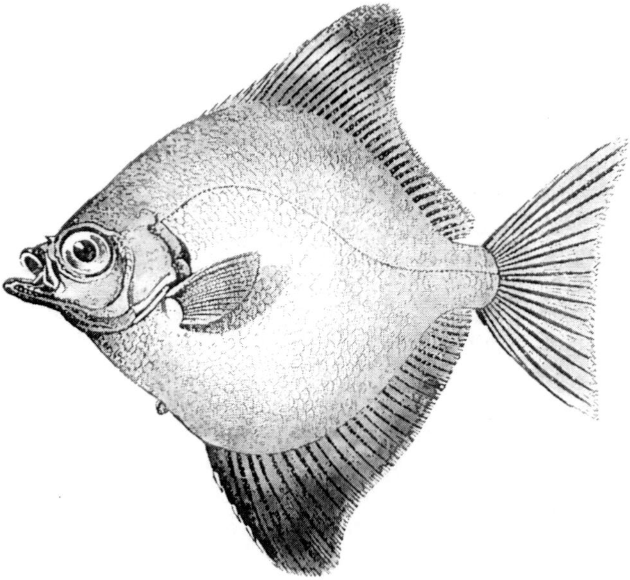

Fig. 199.—Psettus argenteus.

PLATAX.—Body much compressed and elevated; snout very short. One dorsal, with the spinous portion nearly entirely hidden, and formed by front three to seven spines; anal with three. Ventrals well developed, with one spine and five rays. Teeth setiform, with an outer series of rather larger teeth, notched at the top; palate toothless. Scales of moderate size or rather small.

These fishes are called "Sea-bats," from the extraordinary length of some portion of their dorsal and anal fins and of their ventrals. These long lobes are generally of a deep black colour. In mature and old individuals the fin-rays are much shorter than in the young, which have been described as distinct species. There are probably not more than seven species of "Sea-bats," if so many, and they all belong to the Indian Ocean and Western Pacific, where they are very common.

Zanclus.—Body much compressed and elevated. One dorsal, with seven spines, the third of which is very elongate. No teeth on the palate. Scales minute, velvety.

One species (*Z. cornutus*), which is extremely common in the Indo-Pacific. It is easily recognised by its snout, which is produced like that of *Chelmon*, and by the broad black bands crossing the yellow ground-colour. It attains to a length of eight inches, and undergoes during growth similar changes as *Acanthurus*.

Anomalops.—Body oblong, covered with small, rough scales. Snout very short, convex, with wide cleft of the mouth. Eyes very large; below the eye, in a cavity of the infraorbital ring, there is a glandular phosphorescent organ. Villiform teeth in the jaws and on the palatine bones, none on the vomer. First dorsal fin short, with a few feeble spines connected by membrane.

This genus, of which one species only is known (*A. palpebratus*), represents the family of Horse-Mackerels in the depths of the sea; but we do not know, at present, at what depth it lives. Only six specimens have been obtained hitherto from the vicinity of Amboyna, the Fidji, and Paumotu Islands; the largest was twelve inches long.

Capros.—Body compressed and elevated. Mouth very protractile. Scales rather small, spiny. First dorsal with nine spines, anal with three. Ventral fins well developed. Minute teeth in the jaws and on the vomer; none on the palatine bones.

The "Boar-fish" (*C. aper*) is common in the Mediterranean, and not rarely found on the south coast of England.

Allied are *Antigonia* and *Diretmus*, known from a few individuals obtained at Madeira and Barbadoes; they are probably fishes which but rarely come to the surface.

Equula.—Body more or less compressed, elevated or oblong, covered with small, deciduous, cycloid scales. Mouth very protractile. Minute teeth in the jaws; none on the palate. One dorsal. Formula of the fins: D. $\frac{8}{15\text{-}16}$, A. $\frac{3}{14}$, V. 1/5. The lower præopercular margin serrated.

Small species, abundant in the Indo-Pacific, disappearing

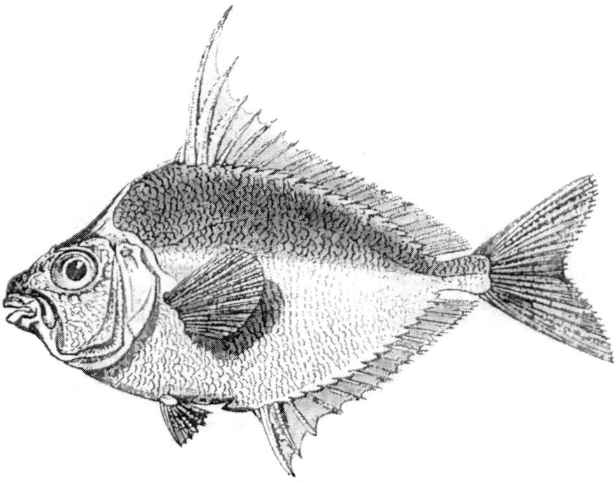

Fig. 200.—Equula edentula.

on the coasts of Japan and Australia. Some eighteen species have been described.

Gazza is very similar to *Equula*, but armed with canine teeth in the jaws.

Other genera referred to this family are *Lactarius* (*L. delicatulus*, common, and eaten on the East Indian coasts), *Seriolella*, *Paropsis*, and *Platystethus*.

THIRD FAMILY—CYTTIDÆ.

Body elevated, compressed, covered with small scales, or with bucklers, or naked; eye lateral. Teeth conical, small. No bony stay for the præoperculum. Dorsal fin composed of two distinct portions. Ventrals thoracic. No prominent papilla near the vent. Gill-opening wide. More than ten abdominal and more than fourteen caudal vertebræ.

The fishes of the "Dory" family are truly marine, and inhabit the temperate zone of the Northern and Southern Hemispheres. Some fossils from tertiary formations (one from Licata) belong to the genus *Zeus*.

ZEUS.—A series of bony plates runs along the base of the dorsal and anal fins; another series on the abdomen. Three or four anal spines.

"John Dorys" are found in the Mediterranean, on the eastern temperate shores of the Atlantic, on the coasts of Japan and Australia. Six species are known, all of which are highly esteemed for the table. The English name given to one of the European species (*Zeus faber*) seems to be partly a corruption of the Gascon "Jau," which signifies cock, "Dory" being derived from the French Dorée, so that the entire name means Gilt-Cock. Indeed, in some other localities of Southern Europe it bears the name of *Gallo*. The same species occurs also on the coasts of South Australia and New Zealand. The fishermen of Roman Catholic countries hold this fish in special respect, as they recognise in a black round spot on its side the mark left by the thumb of St. Peter when he took the piece of money from its mouth.

CYTTUS.—Body covered with very small scales; no osseous

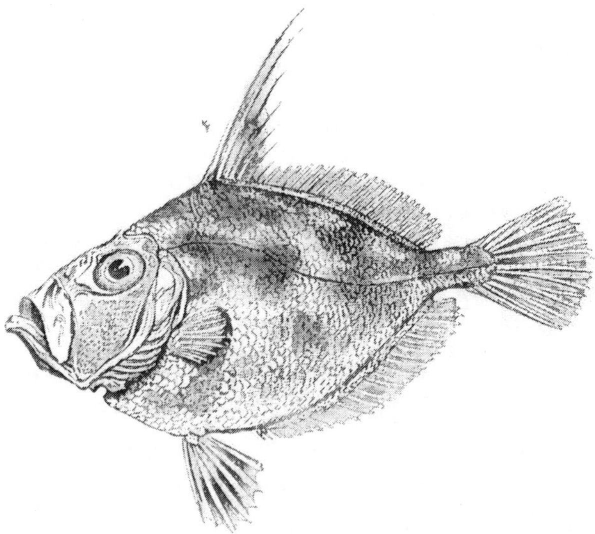

Fig. 201.—Cyttus australis.

bucklers on any part of the body. Two anal spines; ventral fins composed of one spine and six or eight rays.

Three species are known from Madeira, South Australia, and New Zealand.

Fourth Family—Stromateidæ.

Body more or less oblong and compressed, covered with very small scales; eye lateral. Dentition very feeble; œsophagus armed with numerous horny, barbed processes. No bony stay for the præoperculum. Dorsal fin single, long, without distinct spinous division. More than ten abdominal and more than fourteen caudal vertebræ.

This small family consists of strictly marine and partly pelagic species referred to two genera, *Stromateus* and *Centrolophus*. The former lacks ventral fins, at least in the adult stage, and is represented by about ten species in almost all the tropical and warmer seas. *Centrolophus*, hitherto known from two or three European species only (of which one occasionally reaches the south coast of England, where it is named "Black-fish"), has recently been discovered on the coast of Peru, and has probably a much wider range.

Fifth Family—Coryphænidæ.

Body compressed; eye lateral. Teeth small, conical, if present; œsophagus smooth. No bony stay for the præoperculum. Dorsal fin single, long, without distinct spinous division. More than ten abdominal and more than fourteen caudal vertebræ.

All the members of this family have pelagic habits. Representatives of it have been recognized in some fossil remains: thus *Goniognathus* from the Isle of Sheppey, and the living genus *Mene* (*Gastrocnemus*) at Monte Bolca.

Coryphæna.—Body compressed, rather elongate; adult specimens with a high crest on the top of the head; cleft of the mouth wide. A single dorsal extending from the occiput almost to the caudal, which is deeply forked; no distinct dorsal and anal spines. The ventrals are well developed, and can be received in a groove on the abdomen. Scales very small. Rasp-like teeth in the jaws, on the vomer and the palatine bones. Air-bladder absent.

Generally, though by misapplication of the name, called "Dolphins." About six species are known, each of which is

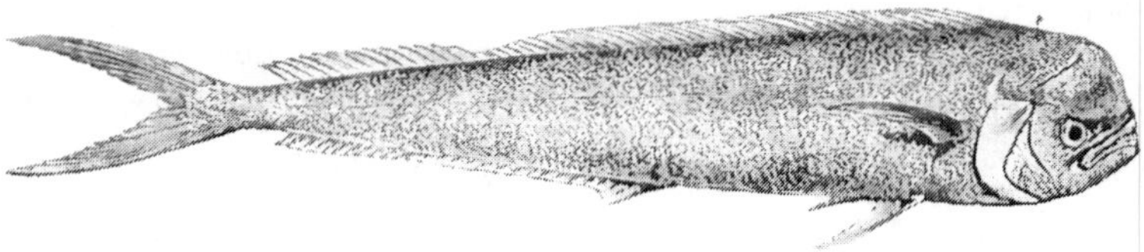

Fig. 202.—Dolphin from the Atlantic.

probably distributed over all the tropical and sub-tropical seas. Strictly pelagic in their habits, they are most powerful swimmers; they congregate in shoals, and pursue unceasingly the Flying-Fish, which try to escape their enemies by long flying leaps. They attain to a length of six feet, and are eagerly caught by sailors on account of their well-flavoured flesh. The beauty of their, unfortunately fugitive, colours has ever been a subject of admiration. As far as the colours are capable of description, those of the common species (*C. hippurus*), which is often seen in the Mediterranean, are silvery blue above, with markings of a deeper azure, and reflections of pure gold, the lower parts being lemon-yellow, marked with pale blue. The pectoral fins are partly lead colour, partly yellow; the anal is yellow, the iris of the eye golden. These iridescent colours change rapidly whilst the fish is dying, as in the Mackerel. The form of the body, and especially of the head, changes considerably with age. Very young specimens, from one to six inches in length, are abundant in the open sea, and frequently obtained in the tow-net. Their body is cylindrical, their head as broad as high, and the eye relatively very large, much longer than the snout. As the fish grows the body is more compressed, and finally a high crest is developed on the head, and the anterior part of the dorsal fin attains a height equal to that of the body.

Brama.—Body compressed, and more or less elevated, covered with rather small scales; cleft of the mouth very oblique, with the lower jaw longest. Dorsal and anal fins many-rayed, the former with three or four, the latter with two or three, spines; caudal deeply forked. Ventrals thoracic, with one spine and five rays. The jaws with an outer series of stronger teeth.

Pelagic fishes which, like the allied genus *Taractes*, range over almost all the tropical and temperate seas.

Lampris.—Body compressed and elevated, covered with very small deciduous scales; cleft of the mouth narrow. A single dorsal, without a spinous portion. Ventrals composed of numerous rays. Teeth none.

The "Sun-fish" (*L. luna*) is one of the most beautiful

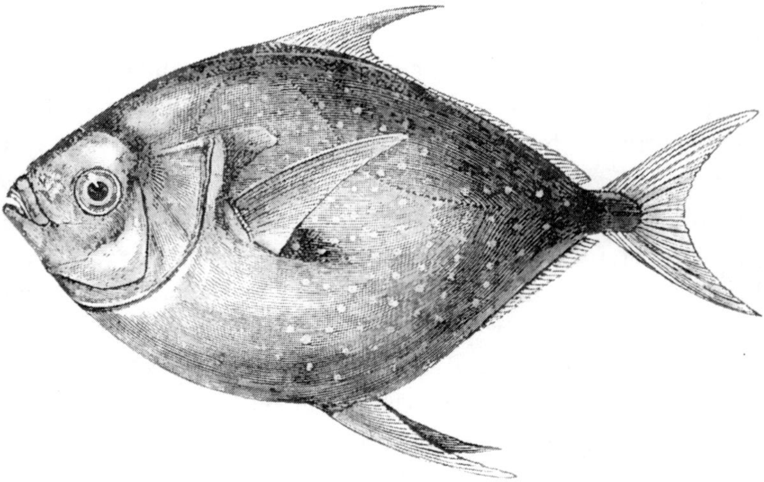

Fig. 203.—Lampris luna.

fishes of the Atlantic. It attains to the large size of four feet in length, is bluish on the back, with round silvery spots, which colour prevails on the lower parts; the fins are of a deep scarlet. It is said to be excellent eating. It is a pelagic fish, not rare about Madeira, but extending far northwards in the Atlantic; it seems to be rarer in the Mediterranean. All the specimens hitherto obtained were full-grown or nearly so. The skeleton exhibits several peculiarities, viz. an extraordinary development and dilatation of the humeral arch, and great strength of the numerous and closely-set ribs.

Other Coryphænoid genera are *Pteraclis, Schedophilus, Diana, Ausonia,* and *Mene;* all pelagic forms.

SIXTH FAMILY—NOMEIDÆ.

Body oblong, more or less compressed, covered with cycloid scales of moderate size; eye lateral. No bony stay for the præoperculum. Dorsal fin with a distinct spinous portion separated from the soft; sometimes finlets; caudal forked. More than ten abdominal, and more than fourteen caudal vertebræ.

Marine fishes; pelagic, at least when young.

GASTROCHISMA.—Cleft of the mouth wide. Finlets behind the

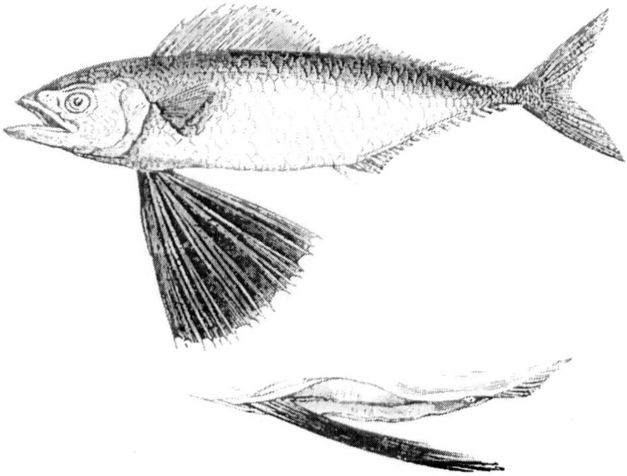

Fig. 204.—Gastrochisma melampus.

dorsal and anal fins. The ventral fins are exceedingly broad and long, and can be completely concealed in a fold of the abdomen.

G. melampus, from the coast of New Zealand; scarce.

NOMEUS.—Cleft of the mouth narrow. No finlets. The ventral fin is long and broad, attached to the abdomen by a membrane, and can be received in a fissure of the abdomen.

N. gronovii is a common pelagic fish in the Atlantic and Indian Oceans; of small size.

Other genera belonging to this family are *Psenes* and *Cubiceps*.

SEVENTH FAMILY—SCOMBRIDÆ.

Body oblong, scarcely compressed, naked or covered with small scales; eye lateral. Dentition well developed. No bony stay for the præoperculum. Two dorsal fins; generally finlets. Ventrals thoracic, with one spine and five rays. More than ten abdominal, and more than fourteen caudal vertebræ.

The fishes of the "Mackerel" family are pelagic forms, abundant in all the seas of the tropical and temperate zones. They are one of the four families of fishes which are the most useful to man, the others being the Gadoids, Clupeoids, and Salmonoids. They are fishes of prey, and unceasingly active, their power of endurance in swimming being equal to the rapidity of their motions. Their muscles receive a greater supply of blood-vessels and nerves than in other fishes, and are of a red colour, and more like those of birds or mammals. This energy of muscular action causes the temperature of their blood to be several degrees higher than in other fishes. They wander about in shoals, spawn in the open sea, but periodically approach the shore, probably in the pursuit of other fishes on which they feed.[1]

[1] Mackerel, like other marine fishes, birds, and mammals of prey, follow the shoals of young and adult Clupeoids in their periodical migrations; on the British coasts it is principally the fry of the Pilchard and Sprat which wanders from the open sea towards the coast, and guides the movements of the Mackerel.

Scombridæ are well represented in tertiary formations: in the Eocene schists of Glaris two extinct genera, *Palimphyes* and *Isurus*, have been discovered. In Eocene and Miocene formations *Scomber*, *Thynnus*, and *Cybium* are not uncommon.

SCOMBER.—The first dorsal continuous, with feeble spines; five or six finlets behind the dorsal and anal. Scales very small, and equally covering the whole body. Teeth small. Two short ridges on each side of the caudal fin.

Mackerels proper are found in almost all temperate and tropical seas, with the exception of the Atlantic shores of temperate South America, where they have not been found hitherto. In Europe, and probably also on the coast of England, three species occur: *S. scomber*, the common Mackerel, which lacks an air-bladder; *S. pneumatophorus*, a more southern species, with an air-bladder; and *S. colias*, like the former, but with a somewhat different coloration, and often called "Spanish" Mackerel. On the Cape of Good Hope, in Japan, on the coast of California, in South Australia, and New Zealand, Mackerels are abundant, which are either identical with, or very closely allied to, the European species. On the coasts of the United States the same species occur which tenant the western parts of the Atlantic. Altogether seven species are known.

THYNNUS.—The first dorsal continuous, with the spines rather

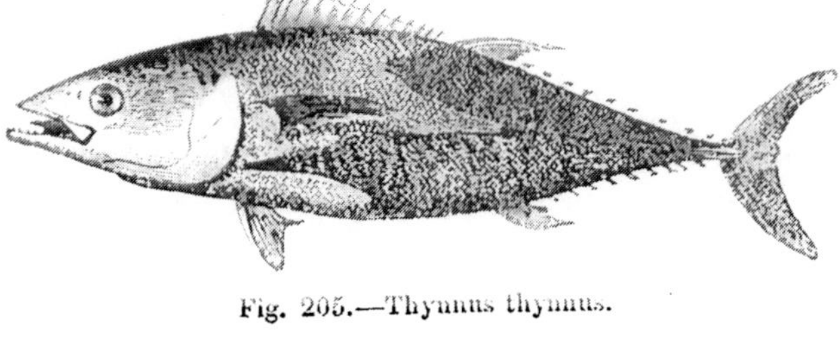

Fig. 205.—Thynnus thynnus.

feeble; from six to nine finlets behind the dorsal and anal.

Scales of the pectoral region crowded, forming a corslet. Teeth rather small. A longitudinal keel on each side of the tail.

The best-known species of this genus is the "Tunny" (*Thynnus thynnus*), abundant in the Mediterranean, and ranging to the south coast of England and to Tasmania. It is one of the largest fishes of the ocean, attaining to a length of 10 feet, and to a weight of more than 1000 lbs. The fishery of the Tunny is systematically carried on in the Mediterranean, and dates from the most remote antiquity. Its salted preparation was esteemed by the Romans under the name of *Saltamentum sardicum.* Its flesh is extensively eaten now, fresh as well as preserved.

Thynnus pelamys, or the "Bonito," is equally well known, and ranges over all the tropical and temperate seas; it eagerly pursues the Flying-fish, and affords welcome sport and food to the sailor. In its form it resembles the Tunny, but is more slender and rarely above three feet long.

Some of the other species are provided with very long pectoral fins, and generally called by sailors "Albacore." They are said to grow to a length of six feet; Bennett in his "Whaling Voyage," vol. ii. p. 278, makes the following observations on *Th. germo*, from the Pacific: "Ships when cruising slowly in the Pacific Ocean, are usually attended by myriads of this fish for many successive months. A few days' rapid sailing is, nevertheless, sufficient to get rid of them, however numerous they may be, for they seldom pay more than very transient visits to vessels making a quick passage. When the ship is sailing with a fresh breeze they swim pertinaciously by her side and take the hook greedily, but should she be lying motionless or becalmed they go off to some distance in search of prey, and cannot be prevailed upon to take the most tempting bait the sailor can devise. It is probably as a protection from their chief enemy, the Sword-fish, that they seek the society of a ship. I am not aware that the Shark is also their enemy; but they seemed to have an instinctive dread of this large fish, and when it approached the ship, would follow it in shoals, and annoy it in the same manner as the smaller birds may be seen to annoy those of a larger and predaceous kind, as the hawk or owl. They are very voracious and miscellaneous feeders. Flying-fish, Calmars, and small shoal-fish are their most natural food; though they do not refuse the animal offal from a ship. Amongst the other food contained in their maw, we have found small Ostracions, File-fish, Sucking-fish, Janthina shells, and pelagic crabs; in one instance a small Bonita, and in a second a Dolphin eight inches long, and a Paper-nautilus shell containing its sepia-tenant. It was often amusing to watch an Albacore pursuing a Flying-fish, and to mark the precision with which it swam beneath the feeble æronaut, keeping him steadily in view, and preparing to seize him at the moment of his descent. But this the Flying-fish would often elude by instantaneously renewing his leap, and not unfrequently escape by extreme agility."

PELAMYS.—The first dorsal continuous, with the spines rather feeble; from seven to nine finlets behind the dorsal and anal. Scales of the pectoral region forming a corslet. Teeth moderately strong. A longitudinal keel on each side of the tail.

Five species are known, of which *P. sarda* is common in the Atlantic and Mediterranean.

AUXIS.—Differing from the preceding two genera in having very small teeth in the jaws only, none on the palate.

Auxis rochei common in the Atlantic, Mediterranean, and Indian Ocean.

CYBIUM.—The first dorsal continuous, with the spines rather feeble; generally more than seven finlets behind the dorsal and anal. Scales rudimentary or absent. Teeth strong; a longitudinal keel on each side of the tail.

Twelve species from the tropical Atlantic and Indian

Ocean; frequenting more the coast-region than the open sea; attaining to a length of four or five feet.

ELACATE.—Body covered with very small scales; head depressed; cleft of the mouth moderately wide; no keel on the tail. The spinous dorsal is formed by eight small free spines; finlets none. Villiform teeth in the jaws, on the vomer and the palatine bones.

Elacate nigra, a coast fish common in the warmer parts of the Atlantic and the Indian Ocean.

ECHENEIS.—The spinous dorsal fin is modified into an adhesive disk, occupying the upper side of the head and neck.

This genus is closely allied to the preceding, from which it differs only by the transformation of the spinous dorsal fin into a sucking organ. The spines being composed of two halves, each half is bent down towards the right and the left, forming a support to a double series of transverse lamellæ, rough on their edges, the whole disk being of an oval shape and surrounded by a membranous fringe. Each pair of lamellæ is formed out of one spine, which, as usual, is supported at the base by an interneural spine. By means of this disk the "Sucking-fishes" or "Suckers" are enabled to attach themselves to any flat surface, a series of vacuums being created by the erection of the usually recumbent lamellæ. The adhesion is so strong that the fish can only be dislodged with difficulty, unless it is pushed forward by a sliding motion. The Suckers attach themselves to sharks, turtles, ships, or any other object which serves their purpose. They cannot be regarded as parasites, inasmuch as they obtain their food independently of their host. Being bad swimmers they allow themselves to be carried about by other animals or vessels endowed with a greater power of locomotion. They were as well known to the ancients as they are to the modern navigators. Aristotle and Aelian mention the Sucker under the name of φθεὶρ, or the *Louse*; "In the sea between Cyrene and Egypt there is a fish about the Dolphin (*Delphinus*), which they call the Louse; this becomes the fattest of all fishes, because it partakes of the plentiful supply of food captured by the Dolphin." Later writers, then, repeat a story, the source of which is unknown, viz. that the "Remora" is able to arrest vessels in their course, a story which has been handed down to our own time. It need not be stated that this is an invention, though it cannot be denied that the attachment of one of the larger species may retard the progress of a sailing vessel, especially when, as is sometimes the case, several individuals accompany the same ship. An account of a somewhat ingenuous way of catching sleeping turtles by means of a Sucking-fish held by a ring fastened round its tail, appears to have originated rather from an experiment than from regular practice.

Ten different species are known, of which *Echeneis remora* and *Echeneis naucrates* are the most common. The former is short and grows to a length of eight inches only, the latter is a slender fish, not rarely found three feet long. The bulkiest is *Echeneis scutata*, which attains to a length of two feet; individuals of that size weighing about eight lbs.

The number of pairs of lamellæ varies in the various species, from 12 to 27. The caudal fin of some of the species undergoes great changes with age. In young specimens the middle portion of the fin is produced into a long filiform lobe. This lobe becomes gradually shorter, and the fin shows a rounded margin in individuals of middle age. When the fish approaches the mature state, the upper and lower lobes are produced, and the fin becomes subcrescentic or forked.

[See Günther, "On the History of Echeneis." Ann. and Mag. Nat. Hist., 1860.]

EIGHTH FAMILY.—TRACHINIDÆ.

Body elongate, low, naked or covered with scales. Teeth small, conical. No bony stay for the præoperculum. One or two dorsal fins, the spinous portion being always shorter and much less developed than the soft; the anal similarly developed as the soft dorsal; no finlets. Ventrals with one spine and five rays. Gill-opening more or less wide. Ten or more than ten abdominal, and more than fourteen caudal vertebræ.

Carnivorous coast-fishes of small size, found in every quarter of the globe, but scarcely represented in the Arctic zone (*Trichodon*); on the other hand, they are rather numerous towards the Antarctic circle. All are bad swimmers, generally moving along the bottom in small depths. Only one genus (*Bathydraco*) is known from the deep-sea.

A genus which shows the principal characters of this family (*Callipteryx*), has been found in the tertiary deposits of Monte Bolca; it is scaleless. A second genus, *Trachinopsis*, has been recently described by Sauvage from the Upper Tertiary of Lorca in Spain; and a third (*Pseudoeleginus*) from the Miocene of Licata.

This family may be subdivided into five groups:—

1. In the URANOSCOPINA the eyes are on the upper surface of the head, directed upwards; the lateral line is continuous.

URANOSCOPUS.—Head large, broad, thick, partly covered with bony plates; cleft of the mouth vertical. Scales very small. Two dorsal fins, the first with from three to five spines; ventrals jugular; pectoral rays branched. Villiform teeth in the jaws, on the vomer and palatine bones; no canines. Generally a long filament below and before the tongue. Gill-cover armed.

The position of the eyes on the upper surface of the head, which these fishes have in common with many others, is well expressed by the name *Uranoscopus* (Stare-gazer). Their eyes are very small, and can be raised or depressed at the will of the fish. They are inactive fishes, generally lying hidden at the bottom between stones, watching for their prey. The delicate filament attached to the bottom of their mouth, and playing in front of it in the current of water which passes through the mouth, serves to lure small animals within reach of the fish. Eleven species are known from the Indo-Pacific and Atlantic, and one (*U. scaber*) from the Mediterranean; they attain rarely a length of twelve inches.

LEPTOSCOPUS.—Form of the head as in *Uranoscopus*, but entirely covered with a thin skin. Scales small, cycloid. One continuous dorsal; ventrals jugular; pectoral rays branched. Villiform teeth in both jaws, on the vomer and palatine bones; canines none. No oral filament. Gill-cover unarmed.

Fig. 206.—Leptoscopus macropygus.

Leptoscopus macropygus, not rare on the coast of New Zealand.

Other genera of Stare-gazers are *Agnus* from the Atlantic coasts of North America; *Anema* from the Indian Ocean and New Zealand; and *Kathetostoma* from Australia and New Zealand.

2. In the TRACHININA the eyes are more or less lateral; the lateral line is continuous; and the intermaxillary without a larger tooth on its posterior portion.

TRACHINUS.—Cleft of the mouth very oblique; eye lateral, but directed upwards. Scales very small, cycloid. Two dorsal fins, the first short, with six or seven spines; ventrals jugular; the lower pectoral rays simple. Villiform teeth in the jaws, on the vomer and palatine bones. Præorbital and præoperculum armed.

The "Weevers" are common fishes on the European coasts, and but too well known to all fishermen; singularly enough they do not extend across the Atlantic to the American coast, but reappear on the coast of Chili! Wounds by their dorsal and opercular spines are much dreaded, being extremely painful, and sometimes causing violent inflammation of the wounded part. No special poison-organ has been found in these fishes, but there is no doubt that the mucous secretion in the vicinity of the spines has poisonous properties. The dorsal spines as well as the opercular spine have a deep double groove in which the poisonous fluid is lodged, and by which it is inoculated in the punctured wound. On the British coasts two species occur, *T. draco*, the Greater Weever, attaining to a length of twelve inches, and *T. vipera*, the Lesser Weever, which grows only to half that size.

CHAMPSODON.—Body covered with minute granular scales; lateral lines two, with numerous vertical branches. Cleft of the mouth wide, oblique. Eye lateral, but directed upwards. Two dorsal fins; ventral fins jugular; pectoral rays branched. Teeth in the jaws in a single series, thin, long, of unequal size. Teeth on the vomer, none on the palate. Gill-openings exceedingly wide. Præoperculum with a spine at the angle and a fine serrature on the posterior margin.

Champsodon vorax is not uncommon at small depths off the Philippine Islands, Admiralty Islands, and in the Arafura Sea.

PERCIS.—Body cylindrical, with small ctenoid scales; cleft of the mouth slightly oblique; eye lateral, but directed upwards. Dorsal fins more or less continuous, the spinous with four or five short stiff spines; ventrals a little before the pectorals. Villiform teeth in the jaws, with the addition of canines; teeth on the vomer, none on the palatines. Opercles feebly armed.

Fifteen species; small, but prettily coloured shore-fishes of the Indo-Pacific.

SILLAGO.—Body covered with rather small, ctenoid scales.

Cleft of the mouth small, with the upper jaw rather longer; eye lateral, large. Two dorsals, the first with from nine to twelve spines; ventrals thoracic. Villiform teeth in the jaws, and on the vomer, none on the palatine bones. Operculum unarmed; præoperculum serrated. The bones of the head with wide muciferous channels.

Eight species; small, plain-coloured shore-fishes, common in the Indian Ocean to the coasts of Australia.

BOVICHTHYS.—Head broad and thick; cleft of the mouth horizontal, with the upper jaw rather longer; eye lateral, more or less directed upwards. Scales none. Two separate dorsal fins, the first with eight spines; ventrals jugular; the lower pectoral rays simple. Villiform teeth in the jaws, on the vomer and the palatine bones; no canines. Operculum with a strong spine; præorbital and præoperculum not armed.

Three species are known from the South Pacific.

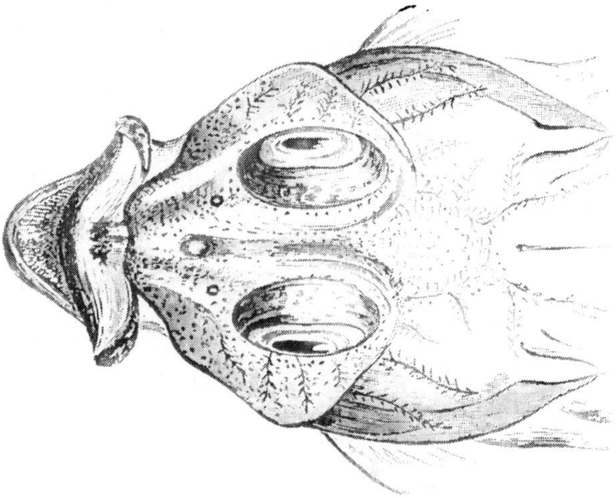

Fig. 207.—Head of Bovichthys variegatus, from New Zealand.

BATHYDRACO.—Body elongate, sub-cylindrical; head depressed, with the snout much elongate, spatulate; mouth wide, horizontal, with the lower jaw prominent; eyes very large, lateral, close together. Scales very small, imbedded in the skin. Lateral line wide, continuous. One dorsal fin; ventrals jugular; the lower pectoral rays branched. Teeth in the jaws in villiform bands; none on the vomer or the palatine bones. Opercles unarmed; ten branchiostegals; the gill-membranes free from the isthmus, and but slightly united in front. Air-bladder none.

A deep-sea fish, found at a depth of 1260 fathoms in the Antarctic Ocean (south of Heard Island).

CHÆNICHTHYS.—Head very large, with the snout spatulate, and with the cleft of the mouth very wide. Eye lateral. Scales none; lateral line sometimes with granulated scutes. Two dorsals, the first with seven spines; ventrals jugular. Jaws with rasp-like teeth; palate toothless.

Chænichthys rhinoceratus from Kerguelen's Land (see Fig. 108, p. 291); and *Ch. esox* from the Straits of Magelhaen.

Other genera belonging to this group are *Aphritis, Acanthaphritis, Eleginus, Chænichthys,* and *Chimarrhichthys* from the South Pacific and Antarctic zone; *Cottoperca* from the west coast of Patagonia; *Percophis* from the coast of Southern Brazil; and *Trichodon* from the coast of Kamtschatka.

3. In the PINGUIPEDINA the body is covered with small scales; the eye lateral; the lateral line continuous; and the intermaxillary is armed with a larger tooth on its posterior portion, as in many Labroids.

Two genera, *Pinguipes* and *Latilus,* from various parts of tropical and sub-tropical seas, belong to this group.

4. In the PSEUDOCHROMIDES, the lateral line is interrupted or not continued to the caudal fin; they have one continuous dorsal only.

These fishes are inhabitants of coral reefs or coasts: *Opisthognathus, Pseudochromis, Cichlops,* and *Pseudoplesiops.*

5. In the NOTOTHENIINA the lateral line is interrupted; and the dorsal fin consists of two separate portions.

They (with others) represent in the Antarctic zone the Cottoids of the Northern Hemisphere: they have the same habits as their northern analogues. In *Notothenia,* which on the southern extremity of South America, in New Zealand, Kerguelen's Land, etc., is represented by about twenty species, the body is covered with ctenoid scales, and the bones of the head are unarmed; whilst *Harpagifer*, a small species with a similar range as *Notothenia*, has the body naked, and the operculum and sub-operculum armed with long and strong spines.

Ninth Family—Malacanthidæ.

Body elongate, with very small scales; mouth with thick lips; a strong tooth posteriorly on the intermaxillary. Dorsal and anal fins very long, the former with a few simple rays anteriorly; ventrals thoracic, with one spine and five rays. Gill-opening wide, with the gill-membranes united below the throat. Ten abdominal and fourteen caudal vertebræ.

One genus only, *Malacanthus*, with three species from tropical seas.

Tenth Family—Batrachidæ.

Head broad and thick; body elongate, compressed behind; skin naked or with small scales. No bony stay for the præoperculum. Teeth conical, small or of moderate size. The spinous dorsal consists of two or three spines only; the soft and the anal long. Ventrals jugular, with two soft rays; pectorals not pediculated. Gill-opening a more or less vertical slit before the pectoral, rather narrow.

Carnivorous fishes, of small size, living on the bottom of the sea near the coast in the tropical zone, some species advancing into the warmer parts of the temperate zones.

Batrachus.—The spinous dorsal is formed by three stout spines. Gill-covers armed with spines. Circumference of the mouth and other parts of the head frequently provided with small skinny tentacles.

Some of the fishes of this genus possess a subcutaneous

spacious cavity behind the base of the pectoral fin, the inside of which is coated with a reticulated mucous membrane. It opens by a foramen in the upper part of the axil. This apparatus is the same which is found in many Siluroid fishes, and which has been noticed above, p. 192. There cannot be any doubt that it is a secretory organ, but whether the secretion has any poisonous properties, as in the Siluroids, or as in *Thalassophryne*, has not been determined. No instance of poisonous wounds having been inflicted by these fishes is on record. Twelve species are known, the distribution of which coincides with that of the family; one very fine species, *B. didactylus*, occurs in the Mediterranean.

THALASSOPHRYNE.—The spinous dorsal is formed by two spines only, each of which is hollow, like the opercular spine, and conveys the contents of a poison-bag situated at its base. Canine teeth none.

Two species are known from the Atlantic and Pacific coasts of Central America. The poison-apparatus is more perfectly developed than any other known at present in the class of fishes; it has been described above, p. 192. The

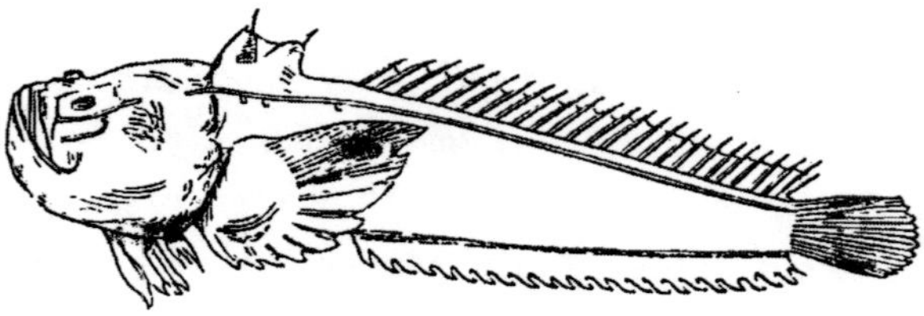

Fig. 208.—Thalassophryne reticulata.

species figured, *Th. reticulata*, is not uncommon at Panama, and attains to a length of fifteen inches.

PORICHTHYS.—Two small dorsal spines; a canine tooth on each side of the vomer.

Two species, from the Atlantic and Pacific sides of Central and South America.

ELEVENTH FAMILY—PSYCHROLUTIDÆ.

Body rather elongate, naked; head broad. Spinous dorsal separate or absent. Ventral fins close together, thoracic, composed of a few rays. Teeth small. Three gills and a half; pseudobranchiæ well developed; gill-openings of moderate width, the gill-membranes being attached to the isthmus.

Of this family only two representatives are known, viz. *Psychrolutes paradoxus*, from Vancouver's Islands, without first dorsal fin; and *Neophrynichthys latus*, from New Zealand, with two dorsal fins. Both are very scarce marine fishes.

TWELFTH FAMILY—PEDICULATI.

Head and anterior part of the body very large, without scales. No bony stay for the præoperculum. Teeth villiform or rasp-like. The spinous dorsal is advanced forwards, composed of a few more or less isolated spines, often transformed into tentacles; or entirely absent. Ventral fins jugular, with four or five soft rays, sometimes absent. The carpal bones are prolonged, forming a sort of arm, terminating in the pectoral. Gill-opening reduced to a small foramen, situated in or near the axil. Gills two and a half, or three, or three and a half; pseudobranchiæ generally absent.

This family contains a larger number of bizarre forms than any other; and there is, perhaps, none in which the singular organisation of the fish is more distinctly seen to be in consonance with its habits. Pediculates are found in all seas. The habits of all are equally sluggish and inactive; they are very bad swimmers; those found near the coasts lie on the bottom of the sea, holding on with their arm-like pectoral fins by sea-weed or stones, between which they are hidden; those of pelagic habits attach themselves to floating sea-weed or other objects, and are at the mercy of wind and current. A large proportion of the genera, therefore, have gradually found their way to the greatest depths of the ocean; retaining all the characteristics of their surface-ancestors, but assuming the modifications by which they are enabled to live in abyssal depths.

LOPHIUS.—Head exceedingly large, broad, depressed, with the eyes on its upper surface; cleft of the mouth very wide. Jaws and palate armed with rasp-like depressible teeth of unequal size. Body naked; bones of the head armed with numerous spines. The three anterior dorsal spines are isolated, situated on the head, and modified into long tentacles; the three following spines form a continuous fin; the soft dorsal and anal short. Gills three. Young individuals have the tentacles beset with lappets, and most of the fin-rays prolonged into filaments.

These fishes are well known under the names "Fishing-Frog," "Frog-fishes," "Anglers," or "Sea-devils." They are coast-fishes, living at very small depths. Four species are known: the British species (*L. piscatorius*) found all round the coasts of Europe and Western North America, and on the Cape of Good Hope; a second (Mediterranean) species, *L.*

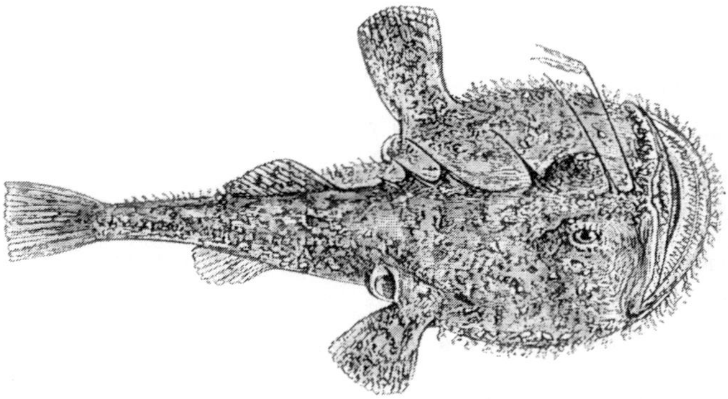

Fig. 209.—Lophius piscatorius.

budegassa; L. setigerus from China and Japan; and *L. naresii* from the Admiralty Islands.

The habits of all these species are identical. The wide

mouth extends all round the anterior circumference of the head, and both jaws are armed with bands of long pointed

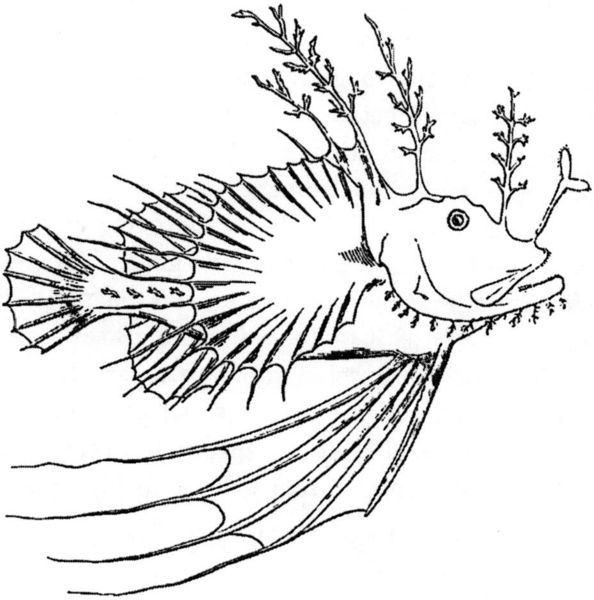

Fig. 210.—A young Fishing-Frog.

teeth, which are inclined inwards, and can be depressed so as to offer no impediment to an object gliding towards the stomach, but prevent its escape from the mouth. The pectoral and ventral fins are so articulated as to perform the functions of teeth, the fish being enabled to move, or rather to walk, on the bottom of the sea, where it generally hides itself in the sand, or amongst sea-weed. All round its head, and also along the body, the skin bears fringed appendages, resembling short fronds of sea-weed; a structure which, combined with the extraordinary faculty of assimilating the colours of the body to its surroundings, assists this fish greatly in concealing itself in places which it selects on account of the abundance of prey. To render the organisation of these creatures perfect in relation to their wants, they are provided with three long filaments inserted along the middle of the head, which are, in fact, the detached and modified three first spines of the anterior dorsal fin. The filaments most important in the economy of the fishing-frogs is the first, which is the longest, terminates in a lappet, and is movable in every direction. There is no doubt that the Fishing-frog, like many other fish provided with similar appendages, plays with this filament as with a bait, attracting fishes, which, when sufficiently near, are ingulfed by the simple act of the Fishing-frog opening its gape. Its stomach is distensible in an extraordinary degree, and not rarely fishes have been taken out of it quite as large and heavy as their destroyer. The British species grows to a length of more than five feet; specimens of three feet are common. Baird records that the spawn of the same species has been observed as a floating sheet of mucus, of from some 60 to 100 square feet.

CERATIAS.—Head and body much compressed and elevated; cleft of the mouth wide, subvertical. Eyes very small. Teeth in the jaws rasp-like, depressible; palate toothless. Skin covered with numerous prickles. The spinous dorsal is reduced to two long isolated spines, the first on the middle of the head, the second on the back. The soft dorsal and anal short; caudal very long. Ventrals none; pectorals very short. Two and a half gills. Skeleton soft and fibrous.

Ceratias holbölli, a deep-sea fish; only a few examples have been found near the coast of Greenland, and from the mid-Atlantic; the latter at a depth of 2400 fathoms. Deep black.

HIMANTOLOPHUS.—Head and body compressed and elevated; cleft of the mouth wide, oblique. Eyes very small. Teeth of the jaws rasp-like, depressible; palate toothless. Skin with scattered conical tubercles. The spinous dorsal is reduced to a single tentacle on the head. The soft dorsal, anal, caudal, and pectoral short. Ventrals none. Three and a half gills. Skeleton soft and fibrous.

This is another deep-sea form, hitherto found in very few examples in the Arctic and Mid-Atlantic Oceans. The single tentacle is beset with many long filaments at its extremity, thus answering the same purpose which is attained by a greater number of tentacles. Deep black.

Melanocetus.—Head and body compressed; head very large; cleft of the mouth exceedingly wide, vertical. Eyes very small.

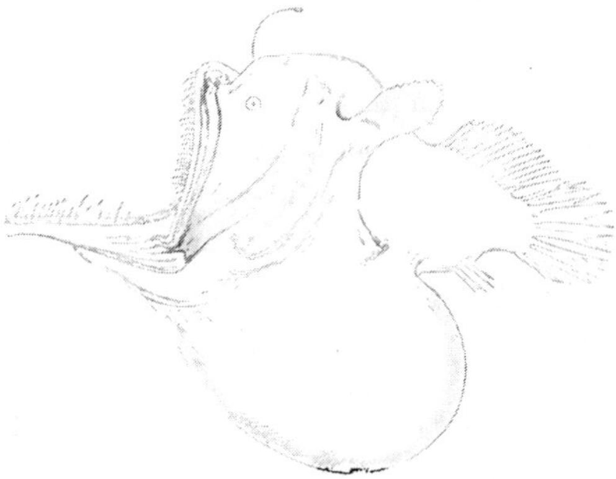

Fig. 211.—Melanocetus johnsonii.

Teeth of the jaws and vomer rasp-like, depressible. Skin smooth. The spinous dorsal is reduced to a single filament placed on the head. The soft dorsal and anal short. Ventrals none.

Two species are known from the Atlantic: *M. bispinossus* and *M. johnsonii*, obtained at depths of from 360 to 1800 fathoms. The specimen figured was not quite four inches long, and contained in its stomach, rolled up spirally into a ball, a Scopeline fish which measured 7½ inches in length and one inch in depth.

Oneirodes.—A deep-sea fish from the Arctic Ocean, differing from the preceding in possessing a second isolated dorsal ray on the back.

Antennarius.—Head very large, high, compressed; cleft of the mouth vertical or sub-vertical, of moderate width. Jaws and palate armed with rasp-like teeth. Eye small. Body naked or covered with minute spines; generally with tentacles. The spinous dorsal is reduced to three isolated spines, the anterior of which is modified into a tentacle, situated above the snout. The soft dorsal of moderate length; anal short. Ventrals present.

The fishes of this genus are pelagic, frequently met with in mid-ocean between the tropics, especially in parts of the sea with floating vegetation; not rarely individuals are found far from their native latitudes, carried by currents to the coasts of Norway and New Zealand. Their power of swimming is most imperfect. When near the coast they conceal themselves between corals, stones, or fucus, holding on to the ground by means of their arm-like pectoral fins. Their coloration is so similar to their surroundings that it is hardly possible to distinguish the fish from a stone or coral overgrown with vegetation. Their way of attracting and seizing their prey is evidently the same as in the other fishes of this family. The extraordinary range of some of the species which inhabit the Atlantic as well as the Indo-Pacific Oceans, is the consequence of their habit of attaching themselves to floating objects. Almost all the species are highly coloured, but the pattern of the various colours varies exceedingly. These fishes do not attain to any considerable size, and probably never exceed a length of ten inches. A great number of species have been distinguished by ichthyologists, but probably not more than twenty are known at present. The species figured on p. 295 (*A. caudomaculatus*) is common in the Red Sea, and probably occurs in other parts of the Indian Ocean.

Brachionichthys and *Saccarius* are allied genera from South Australia, Tasmania, and New Zealand.

CHAUNAX.—Head very large, depressed; cleft of the mouth wide, sub-vertical; eye small; rasp-like teeth in the jaws and palate. Skin covered with minute spines. The spinous dorsal is reduced to a small tentacle above the snout; the soft dorsal of moderate length; anal short; ventrals present.

A deep-sea fish (*Ch. pictus*), of uniform pink colour; hitherto found near Madeira and the Fidji Islands, at a depth of 215 fathoms.

MALTHE.—Anterior portion of the body very broad and depressed. The anterior part of the snout is produced into a more or less prominent process, beneath which there is a tentacle retractile into a cavity. Jaws and palate with villiform teeth. Skin with numerous conical protuberances. Soft dorsal fin and anal very short. Gill-opening superiorly in the axil; gills two and a half.

Although the rostral tentaele is situated at the lower side of the projeetion of the snout, it must be regarded as the homologue of a dorsal spine. In some of the preceding genera, *Oncirodes* and *Chaunax*, the first dorsal spine is so far advaneed on the snout as to eome into conneetion with the intermaxillary proeesses; and the position of the rostral tentaele in *Malthe* is only a still more advanced step towards the same speeial purpose for whieh the first dorsal spine is used in this family, viz. for the purpose of obtaining food. In *Malthe* it is obviously an organ of toueh. This genus belongs to the Ameriean shores of the Atlantie; *M. vespertilio* being a tropical, *M. cubifrons* a northern species.

HALIEUTÆA.—Head exceedingly large, depressed, nearly circular in its circumference. Cleft of the mouth wide, horizontal. Jaws with small rasp-like teeth; palate smooth. Forehead with a transverse bony bridge, beneath which is a tentacle (rostral spine) retractile into a cavity. Body and head covered with small stellate spines. Soft dorsal and anal very short. Gill-opening superiorly in the axil; gills two and a half.

A eoast-fish (*H. stellata*) from China and Japan. Frequently found dry in Chinese insect-boxes.

This genus appears to be represented in the Atlantic Ocean by *Halieutichthys* from Cuba, and by *Dibranchus*, dredged at a depth of 360 fathoms off the coast of West Africa; the latter genus possesses two gills only. Another genus, covered with large scattered tubercles, *Aegæonichthys*, has recently been described from New Zealand.

THIRTEENTH FAMILY—COTTIDÆ.

Form of the body oblong, sub-cylindrical. Cleft of the mouth lateral. Dentition feeble, generally in villiform bands. Some bones of the head are armed; and a bony stay connects the præopercular spine with the infraorbital ring. Two dorsal fins (rarely one), the spinous being less developed than the soft and than the anal. Ventrals thoracic, with five or less soft rays.

The fishes of this family are of small size, bad swimmers, and generally living on the bottom, near the coasts, of almost all the arctic, temperate, and tropical seas. Only a few live in fresh water. They prefer shallow to deep water; and there is only one instance known of a member of this family living at a great depth, viz. *Cottus bathybius* from the Japanese sea, which is stated to have been dredged in a depth of 565 fathoms. Fossil representatives are few in number: two or three species of *Trigla;* others, although having a general resemblance to the genus *Cottus*, were covered with ctenoid scales, and therefore are referred to a distinct genus, *Lepidocottus;* they are from tertiary formations.

COTTUS.—Head broad, depressed, rounded in front; body subcylindrical, compressed posteriorly. Scaleless; lateral line present. Pectoral rounded, with some or all the rays simple. Jaws and vomer with villiform teeth; palatine teeth none.

The "Bull-heads" or "Miller's Thumbs" are small fishes from the shores and fresh waters of the northern temperate zone. Some forty species are known; the greater number live in the northern half of the temperate zone. On the shore, as well as in rivers, they prefer rocky or stony to muddy ground, lying concealed between the stones, and watching for their prey, which consists of small crustaceans and other aquatic animals. The common British Miller's Thumb (*C. gobio*) is found in almost all suitable fresh waters of Northern and Central Europe, especially in small streams, and extends into Northern Asia. Other freshwater species abound in North America and Northern Asia. *Cottus scorpius* and *C. bubalis*, the common European marine species, range across the Atlantic to the American coasts. The male is said to construct a nest, for the reception of the spawn, of sea-weeds and stones, and to anxiously watch and defend his offspring. The spine at the angle of the præoperculum, which is simple in the majority of the freshwater species, is frequently armed with accessory processes, and antler-like, in marine.

CANTRIDERMICHTHYS differs from *Cottus* in having teeth on the palatine bones.

Eleven species are known, distributed like *Cottus*, but absent in Europe and North-western Asia.

ICELUS.—Head large, armed at the gill-covers and on the neck; body with a dorsal series of bony plates from the neck to the base of the caudal; lateral line with osseous tubercles; scattered scales on the sides and abdomen. Ventrals thoracic, with less than five rays. No pectoral filaments. Villiform teeth in the jaws, on the vomer and palatine bones.

Represents Cottus in the far north; *I. hamatus* is common in Spitzbergen and Greenland, and has been found in abundance in lat. 81° 44′.

PLATYCEPHALUS.—Head broad, much depressed, more or less armed with spines; body depressed behind the head, subcylindrical towards the tail, covered with ctenoid scales. Two dorsal fins; the first spine isolated from the others. Ventrals thoracic, but rather remote from the base of the pectorals. Villiform teeth in the jaws, on the vomer and palatine bones.

Fig. 212.—Platycephalus cirrhonasus, from Port Jackson.

About forty species are known, of which some attain a length of two feet. This genus represents in the tropical Indian Ocean the *Cotti* of the Arctic, and the *Notothenia* of the Antarctic zone. Like these, they live on the bottom in shallow water, hidden in the sand, the colours of which are assimilated by those of their body.

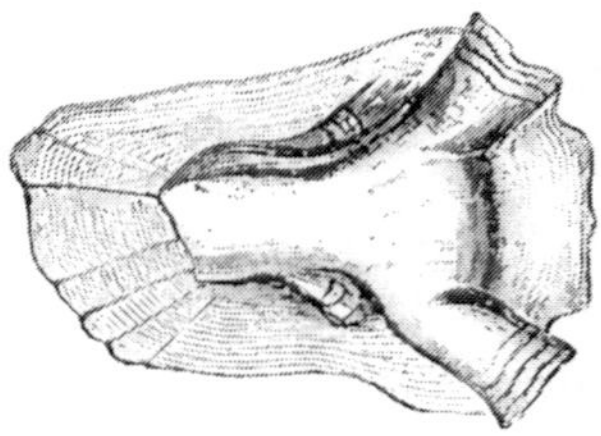

Fig. 213.—Scale from the lateral line of the same fish.

Therefore, they are very scarce near coral islands which are surrounded by great depths; whilst the number of species is rather considerable on many points of the shelving Australian coasts. Their long and strong ventral fins are of great use to them in locomotion. *P. insidiator* is one of the most common Indian and Australian fishes, and readily recognised by two oblique black bands on the upper and lower caudal lobes.

HOPLICHTHYS, similar to *Platycephalus*, but with the back and sides of the body covered with bony spiny plates. No separate dorsal spine.

One species, *H. langsdorffii*, is common on the coast of Japan, and frequently placed dry by the Chinese into their insect-boxes.

TRIGLA.—Head parallelopiped, with the upper surface and the sides entirely bony, the enlarged infraorbital covering the cheek.

Two dorsal fins. Three free pectoral rays. Villiform teeth. Air-bladder generally with lateral muscles, often divided into two lateral halves. The species may be referred to three groups :—

1. Palatine teeth none ; scales exceedingly small, except those of the lateral line : *Trigla.*
2. Palatine teeth none ; scales of moderate size : *Lepidotrigla.*
3. Palatine teeth present : *Prionotus.*

About forty species of "Gurnards" are known from tropical and temperate zones. They are too well known to need detailed description ; one of their principal characteristics is the three free finger-like pectoral appendages, which serve as

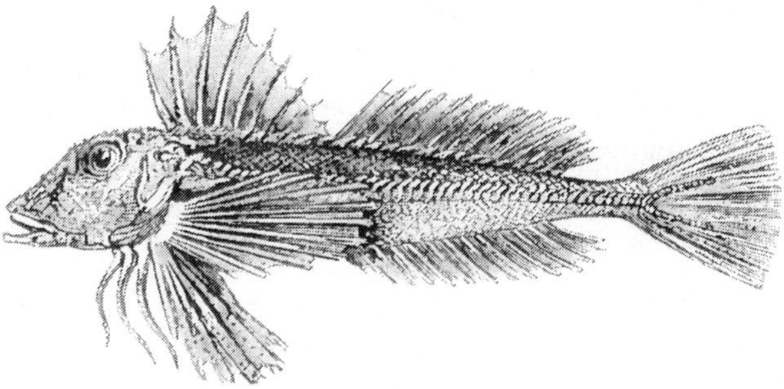

Fig. 214.—Trigla pleuracanthica.

organs of locomotion as well as touch, and which are supplied with strong nerves, as noticed above (pp. 108 and 120). The fins are frequently beautifully ornamented, especially the inner side of the long and broad pectorals, which is most exposed to the light when the fish is floating on the surface of the water, with pectorals spread out like wings. The grunting noise made by Gurnards when taken out of the water is caused by the escape of gas from the air-bladder through the open pneumatic duct. Gurnards are

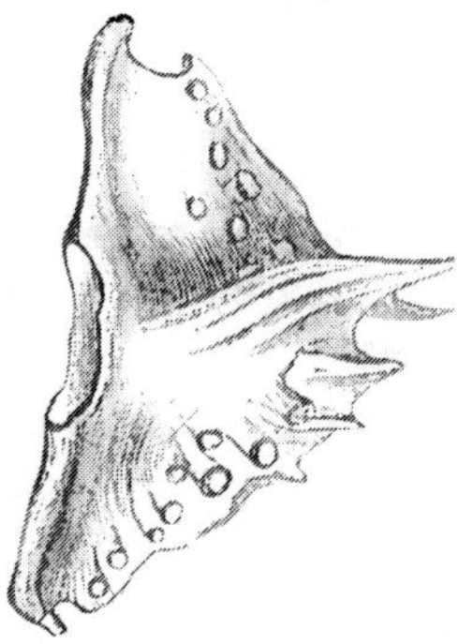

Fig. 215.—Scute of the lateral line of the same fish.

generally used as food; seven species occur on the British coast: the Red Gurnard (*T. pini*), the Streaked Gurnard (*T. lineata*), the Sapphirine Gurnard (*T. hirundo*), the Grey Gurnard (*T. gurnardus*), Bloch's Gurnard (*T. cuculus*), the Piper (*T. lyra*), and the Long-finned Gurnard (*T. obscura* or *T. lucerna*). Singularly, the European species cross the Atlantic but rarely, the American species belonging chiefly to the division *Prionotus*.

Several other genera belong to this family; for completeness' sake they are mentioned here, viz. *Bunocottus* from Cape Horn; *Rhamphocottus, Triglops* from Arctic North America; *Podabrus, Blepsias, Nautichthys, Scorpænichthys, Hemilepidotus, Artedius,* from the North Pacific; *Ptyonotus,* from Lake Ontario; *Polycaulus* from Indian Seas; *Bembras* from the Japanese Sea.

FOURTEENTH FAMILY—CATAPHRACTI.

Form of the body elongate, sub-cylindrical. Dentition feeble. Body completely cuirassed with osseous keeled scales or plates. A bony stay connects the angle of the præoperculum with the infraorbital ring. Ventrals thoracic.

Marine fishes, and partly pelagic. *Petalopteryx,* from the chalk of Mount Lebanon, is supposed to have a resemblance to *Dactylopterus.*

AGONUS.—Head and body angular, covered with bony plates. Two dorsal fins; no pectoral appendages. Small teeth in the jaws.

Small fishes, from the northern parts of the temperate zone and extending into the Arctic Ocean; the genus reappears in the Southern Hemisphere on the coast of Chile. Of the eleven species known, one (*A. cataphractus*) is not uncommon on the coast of Great Britain.

ASPIDOPHOROIDES, from Greenland, has a very similar form of the body, but possesses one short dorsal fin only.

SIPHAGONUS.—With the snout produced into a long tube like a Syngnathus; chin prominent, with a barbel.

From Behring's Strait and Japan.

PERISTETHUS.—Head parallelopiped, with the upper surface and the sides entirely bony; each præorbital prolonged into a long flat process, projecting beyond the snout. Body cuirassed with large bony plates. One continuous dorsal, or two dorsals, of which the second is the more developed. Two free pectoral appendages. Teeth none; lower jaw with barbels.

Singularly shaped fishes, of rather small size, from the Mediterranean, the warmer parts of the Atlantic, and the Indian Ocean; of the ten species known one species only has been found in the Pacific, near the Sandwich Islands. The European species is *P. cataphractum*. They are not common, and probably inhabit greater depths than the Gurnards, with which they have much in common as regards their habits.

DACTYLOPTERUS.—Head parallelopiped, with the upper surface and the sides entirely bony; scapula and angle of the præoperculum produced into long spines. Body with strongly keeled scales of moderate size; lateral line none. Two dorsal fins,

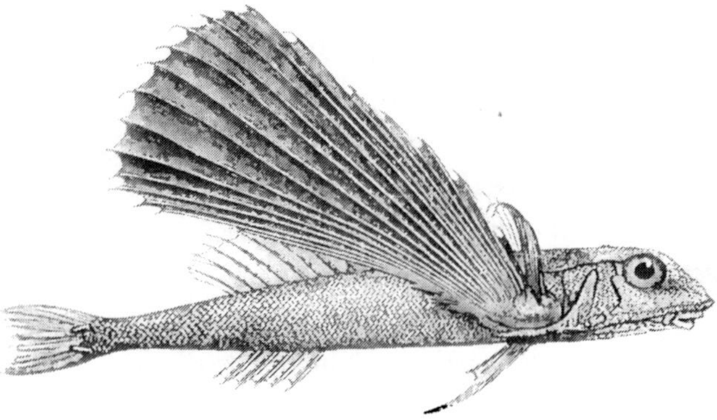

Fig. 216.—Dactylopterus volitans.

the second not much longer than the first; pectoral very long, an organ of flying, with the upper portion detached and shorter. Granular teeth in the jaws; none on the palate. Air-bladder divided into two lateral halves, each with a larger muscle.

Of "Flying Gurnards" three species only are known, which are very abundant in the Mediterranean, the tropical Atlantic, and Indo-Pacific. They, and the Flying Herrings (Exocoetus), are the only fishes which are enabled by their long pectoral fins to take flying leaps out of the water, and deserve the name of "Flying-Fishes." They are much heavier, and attain to a larger size, than the Exocoeti, specimens of eighteen inches in length not being scarce. When young, their pectorals are much shorter, and, consequently, they are unable to raise themselves out of the water (*Cephalacanthus*).

The vertebral column shows a singular coalescence of the anterior vertebræ, which form a simple tube, as in *Fistularia*.

We insert here as an appendix to this division the small family of *Pegasidæ*, the natural affinities of which are not yet clearly understood, but which resembles in some of its characters the *Cataphracti*.

Fifteenth Family—Pegasidæ.

Body entirely covered with bony plates, anchylosed on the trunk and movable on the tail. Barbels none. The margin of the upper jaw is formed by the intermaxillaries and their cutaneous prolongation, which extends downwards to the extremity of the maxillaries. Gill-cover formed by a large plate, homologous to the operculum, præoperculum, and suboperculum; interoperculum a long fine bone, hidden below the gill-plate. One rudimentary branchiostegal. The gill-plate is united with the isthmus by a narrow membrane; gill-openings narrow, in front of the base of the pectoral fin. Gills four, lamellated. Pseudobranchiæ and air-bladder absent. One short dorsal and anal fin, opposite to each other. Ventral fin present. Ovarian sacs closed.

One genus only is known, *Pegasus*. Its pectoral fins are broad, horizontal, long, composed of simple rays, some of which are sometimes spinous. Ventral fins one- or two- rayed. Upper part of the snout produced into a shorter or longer process. Mouth inferior, toothless. Suborbital ring well developed, forming a suture with the gill-cover. Vertebræ in small number, thin; no ribs. Four species are known, two of which are of a shorter, and the two others of a longer form. The former are *P. draconis*, common in the Indian Ocean, and *P. volans*, which is frequently stuck by the Chinese into the insect-boxes which they manufacture for sale. The

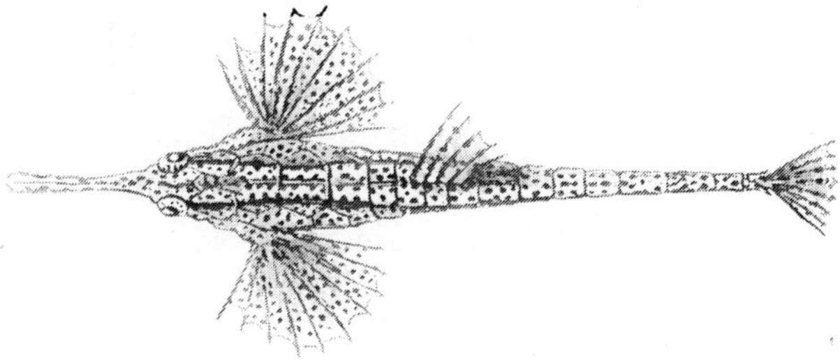

Fig. 217.—Pegasus natans.

two elongate species, *P. natans* and *P. lancifer*, are from the Chinese and Australian coasts. They are all very small fishes, probably living on sandy shoal places near the coast.

NINTH DIVISION—ACANTHOPTERYGII GOBIIFORMES.

The spinous dorsal, or spinous portion of the dorsal is always present, short, either composed of flexible spines, or much less developed than the soft; the soft dorsal and anal of equal extent. No bony stay for the angle of the præoperculum. Ventrals thoracic or jugular, if present, composed of one spine and five, rarely four, soft rays. A prominent anal papilla.

Shore-fishes, mostly exclusively marine, but some entering and living in fresh waters.

FIRST FAMILY—DISCOBOLI.

Body thick or oblong, naked or tubercular. Teeth small.

Ventral fins with one spine and five rays, all being rudimentary and forming the osseous support of a round disk, which is surrounded by a cutaneous fringe. Gill-openings narrow, the gill-membranes being attached to the isthmus.

Carnivorous fishes, living at the bottom of the shores of northern seas. By their ventral disk they are enabled to attach themselves very firmly to rocks.

CYCLOPTERUS.—Body thick, short, covered with a viscous, tubercular skin. Head large, snout short. Villiform teeth in the jaws, none on the palate. Skeleton soft, with but little earthy matter.

Three species of "Lump-suckers" are known from the northern temperate and the arctic zones. The common North European and North American species, *C. lumpus*, is

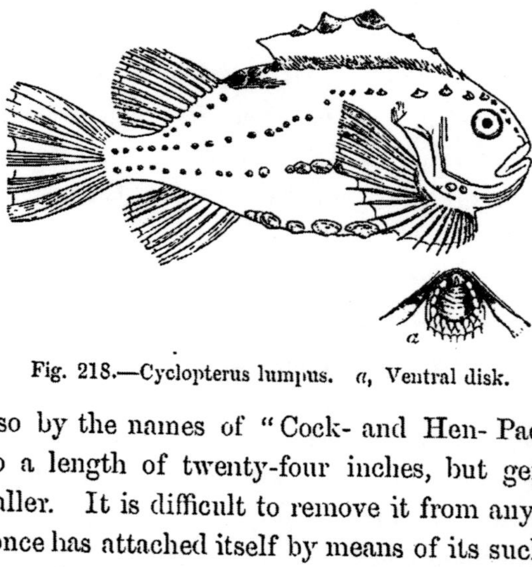

Fig. 218.—Cyclopterus lumpus. *a*, Ventral disk.

known also by the names of "Cock- and Hen- Paddle." It attains to a length of twenty-four inches, but generally is much smaller. It is difficult to remove it from any object to which it once has attached itself by means of its sucking-disk. Its skin is so thick as to more or less entirely conceal the first dorsal fin; it is covered with rough tubercles, the larger ones being arranged in four series along each side of the body. In young specimens these tubercles are absent. The arctic species, *C. spinosus*, has large conical plates on the head and body, each plate with a spine in the centre. Also of this species the young are naked, the plates making only gradu-

Fig. 219.—Young of Cyclopterus spinosus, from the Arctic Ocean, natural size.

ally their appearance, in the form of groups of tubercles. Their development is irregular, as young specimens of the same size may be entirely naked or tubercular. This species ranges beyond the 81° lat. N.

Liparis.—Body sub-cylindrical, enveloped in a more or less loose naked skin; head broad, obtuse. The infraorbital bone is styliform posteriorly, extending backwards to the margin of the præoperculum. One dorsal fin, with feeble flexible rays. Villiform teeth in the jaws, none on the palate.

Small fishes from the northern coasts of the temperate zone, ranging beyond the arctic circle. Eight species are known, of which two (*L. lineatus* and *L. montagui*) occur on the British coasts.

Second Family—Gobiidæ.

Body elongate, naked or scaly. Teeth generally small, sometimes with canines. The spinous dorsal fin, or portion of the dorsal fin, is the less developed, and composed of flexible spines; anal similarly developed as the soft dorsal. Sometimes the ventrals are united into a disk. Gill-opening more or less narrow, the gill-membranes being attached to the isthmus.

Small carnivorous littoral fishes, many of which have become acclimatised in fresh water. They are very abundant with regard to species as well as individuals, and found on or near the coasts of all temperate and tropical regions. Geologically they appear first in the chalk.

GOBIUS.—Body scaly. Two dorsal fins, the anterior generally with six flexible spines. Ventral fins united, forming a disk which is not attached to the abdomen. Gill-opening vertical, moderately wide.

The "Gobies" are distributed over all temperate and tropical coasts, and abundant, especially on the latter. Nearly three hundred species have been described. They live especially on rocky coasts, attaching themselves firmly with their ventrals to a rock in almost any position, and thus withstanding the force of the waves. Many of the species seem to delight in darting from place to place in the rush of water which breaks upon the shore. Others live in quiet brackish water, and not a few have become entirely acclimatised in fresh water, especially lakes. The males of some species construct nests for the eggs, which they jealously watch, and defend even for some time after the young are hatched. Several species are found on the British coast: *G. niger, paganellus, auratus, minutus, ruthensparri.* Fossil species of this genus have been found at Monte Bolca.

Fig. 220.—Gobius lentiginosus, from New Zealand.

A very small Goby, *Latrunculus pellucidus*, common in some localities of the British Islands and other parts of Europe, is distinguished by its transparent body, wide mouth, and uniserial dentition. According to R. Collett it offers some very remarkable peculiarities. It lives one year only, being the first instance of an *annual vertebrate.* It spawns in June and July, the eggs are hatched in August, and the fishes attain their full growth in the months from October to December. In this stage the sexes are quite alike, both having very small teeth and feeble jaws. In April the males lose the small teeth, which are replaced by very long and strong teeth, the jaws themselves becoming stronger. The teeth of the females remain unchanged. In July and August all the adults die off, and in September only the fry are to be found.

There are several other genera, closely allied to Gobius, as *Euctenogobius, Lophiogobius, Doliichthys, Apocryptes, Evorthodus, Gobiosoma and Gobiodon* (with scaleless body) *Triænophorichthys.*

SICYDIUM.—Body covered with ctenoid scales of rather small size. Cleft of the mouth nearly horizontal, with the upper jaw prominent; lips very thick; the lower lip generally with a series of minute horny teeth. A series of numerous small teeth in upper jaw, implanted in the gum, and generally movable; the lower jaw with a series of conical widely-set teeth. Two dorsal fins, the anterior with six flexible spines. Ventral fins united, and forming a short disk, more or less adherent to the abdomen.

Small freshwater fishes inhabiting the rivers and rivulets of the islands of the tropical Indo-Pacific. About twelve species are known; one occurs in the West Indies. *Lentipes* from the Sandwich Islands is allied to *Sicydium.*

PERIOPHTHALMUS.—Body covered with ctenoid scales of small or moderate size. Cleft of the mouth nearly horizontal, with the upper jaw somewhat longer. Eyes very close together, immediately below the upper profile, prominent, but retractile, with a well-developed outer eyelid. Teeth conical, vertical in both jaws. Two dorsal fins, the anterior with flexible spines; caudal fin with the lower margin oblique; base of the pectoral fin free, with strong muscles. Ventral fins more or less coalesced. Gill-openings narrow.

The fishes of this genus, and the closely-allied *Boleophthalmus,* are exceedingly common on the coasts of the tropical Indo-Pacific, especially on parts covered with mud or fucus. During ebb they leave the water and hunt for small crustaceans, and other small animals disporting themselves on the ground which is left uncovered by the receding water. With the aid of their strong pectoral and ventral fins and their tail, they hop freely over the ground, and escape danger by rapid leaps. The peculiar construction of their eyes, which

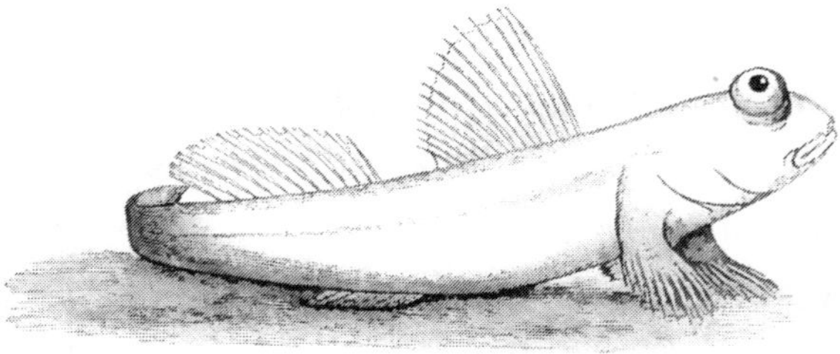

Fig. 221.—Periophthalmus koelreuteri.

are very movable, and can be thrust far out of their sockets, enables them to see in the air as well as in the water; when the eyes are retracted they are protected by a membranous eyelid. These fishes are absent in the eastern parts of the Pacific and on the American side of the Atlantic; but singularly enough one species reappears on the West African coast. About seven species are known (including *Boleophthalmus*), *P. koelreuteri* being one of the most common fishes of the Indian Ocean.

Eleotris.—Body scaly; eyes of moderate size, lateral, not prominent. Teeth small. Two dorsal fins, the anterior generally with six spines. Ventrals not united, though close together, with one spine and five rays.

About sixty species are known from the tropics, only a few extending into the temperate zone. As regards form, they repeat almost all the modifications observed among the Gobies, from which they differ only in having the ventral fins non-coalescent. On the whole they are somewhat larger than the Gobies, and rather freshwater than marine species, some of them being abundant in the rivulets of the islands of the Indo-Pacific and Atlantic. Others have even penetrated into the inland-waters of the African continent.

TRYPAUCHEN.—Body elongate, covered with minute scales; head compressed, with a deep cavity on each side, above the operculum. Teeth small, in a band. One dorsal, the spinous portion composed of six spines; dorsal and anal fins continuous with the caudal, ventral fins united.

Small fishes of singular aspect, from the East Indian coasts. Three species, of which *T. vagina* is common.

CALLIONYMUS.—Head and anterior part of the body depressed, the rest cylindrical, naked. Snout pointed, with the cleft of the mouth narrow, horizontal, and with the upper jaw very protractile. Eyes rather large, more or less directed upwards. Teeth very small, palate smooth. A strong spine at the angle of the præoperculum. Two dorsal fins, the anterior with three or four flexible spines; ventrals five-rayed, widely apart from each other. Gill-openings very narrow, generally reduced to a foramen on the upper side of the operculum.

The "Dragonets" are small, and generally beautifully coloured marine fishes, inhabitants of the coasts of the temperate zone of the Old World; the minority of species live in tropical parts of the Indo-Pacific; and these seem to descend to somewhat greater depths than the littoral species of the northern hemisphere. Secondary sexual characters are developed in almost all the species, the mature males having the fin-rays prolonged into filaments, and the fin-membranes brightly ornamented. On the British coast one species (*C. draco*) is very common, and locally called "Skulpin." About thirty species are known, many of which have the præopercular spine armed with processes or barbs. *Vulsus* is allied to *Callionymus*.

Other genera belonging to this family are—*Benthophilus* from the Caspian Sea; *Amblyopus, Orthostomus, Platyptera, Luciogobius, Oxymetopon,* and, perhaps, *Oxuderces*.

TENTH DIVISION—ACANTHOPTERYGII BLENNIIFORMES.

Body low, sub-cylindrical or compressed, elongate. Dorsal fin very long; the spinous portion of the dorsal, if distinct, is very long, as well developed as the soft, or much more; sometimes the entire fin is composed of spines only; anal more or less long; caudal fin subtruncated or rounded, if present. Ventral fins thoracic or jugular, if present.

FIRST FAMILY—CEPOLIDÆ.

Body very elongate, compressed, covered with very small cycloid scales; eyes rather large, lateral. Teeth of moderate size. No bony stay for the angle of the præoperculum. One very long dorsal fin, which, like the anal, is composed of soft rays. Ventrals thoracic, composed of one spine and five rays. Gill-opening wide. Caudal vertebræ exceedingly numerous.

The "Band-fishes" (*Cepola*) are small marine fishes, belonging principally to the fauna of the northern temperate zone; in the Indian Ocean the genus extends southwards to Pinang. The European species (*C. rubescens*) is found in isolated examples on the British coast, but is less scarce in some years than in others. These fishes are of a nearly uniform red colour.

SECOND FAMILY—TRICHONOTIDÆ.

Body elongate, sub-cylindrical, covered with cycloid scales of moderate size. Eyes directed upwards. Teeth in villiform bands. No bony stay for the angle of the præoperculum. One long dorsal fin, with simple articulated rays, and without a spinous portion; anal long. Ventrals jugular, with one spine and five rays. Gill-opening very wide. The number of caudal vertebræ much exceeding that of the abdominal.

Small marine fishes, belonging to two genera only, *Tricho-*

notus (*setigerus*) from Indian Seas, with some of the anterior dorsal rays prolonged into filaments; and *Hemerocoetes* (*acanthorhynchus*) from New Zealand, and sometimes found far out at sea on the surface.

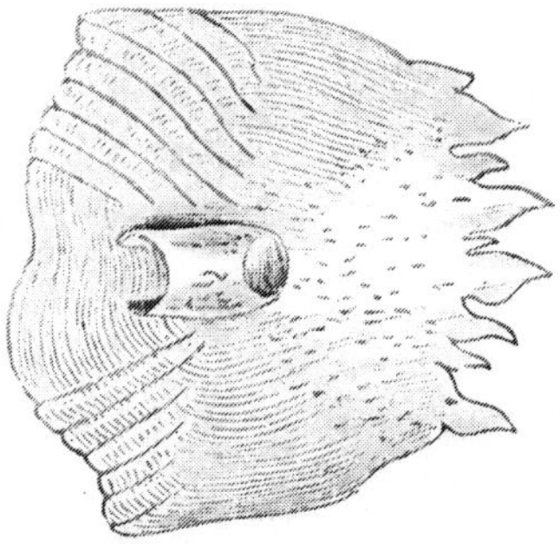

Fig. 222.—Scale from the lateral line of Hemerocœtes acanthorhynchus, with lacerated margin.

THIRD FAMILY—HETEROLEPIDOTIDÆ.

Body oblong, compressed, scaly; eyes lateral; cleft of the mouth lateral; dentition feeble. The angle of the præoperculum connected by a bony stay with the infraorbital ring. Dorsal long, with the spinous and soft portions equally developed; anal elongate. Ventrals thoracic, with one spine and five rays.

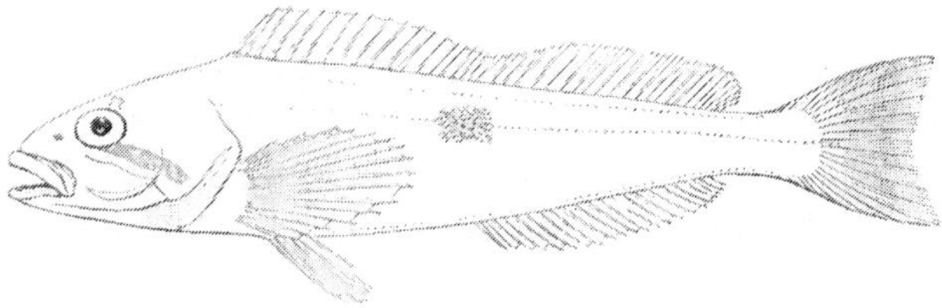

Fig. 223.—*Chirus hexagrammus*, from Japan.

Small shore-fishes, characteristic of the fauna of the Northern Pacific, some of the species occurring on the American as well as Asiatic side. They have been referred to several genera, as

CHIRUS, which is distinguished by the presence of several lateral lines;

OPHIODON, with one lateral line only, cycloid scales, and slightly armed præoperculum;

AGRAMMUS, with one lateral line only, ctenoid scales, and unarmed præoperculum; and

ZANIOLEPIS, with one lateral line and minute comb-like scales.

Fourth Family—Blenniidæ.

Body elongate, low, more or less cylindrical, naked or covered with scales, which generally are small. One, two, or three dorsal fins occupying nearly the whole length of the back, the spinous portion, if distinct, being as much developed as the soft, or more; sometimes the entire fin is composed of spines; anal fin long. Ventrals jugular, composed of a few rays, and sometimes rudimentary or entirely absent. Pseudobranchiæ generally present.

Littoral forms of great generic variety, occurring abundantly in all temperate and tropical seas. Some of the species have become acclimatised in fresh water, and many inhabit brackish water. With very few exceptions they are very small, some of the smallest fishes belonging to the family of "Blennies." One of the principal characteristics of the Blennies is the ventral fin, which is formed by less than five rays, and has a jugular position. The Blennies have this in common with many Gadoids, and it is sometimes difficult to decide to which of these two families a fish should be referred. In such doubtful cases the presence of the pseudobranchiæ (which are absent in Gadoids) may be of assistance.

In many Blennies the ventral fins have ceased to have any function, and become rudimentary, or are even entirely absent. In others the ventral fins, although reduced to cylindrical stylets, possess a distinct function, and are used as organs of locomotion, by the aid of which the fish moves rapidly over the bottom.

The fossil forms are scarcely known; *Pterygocephalus* from Monte Bolca appears to have been a Blennioid.

Anarrhichas.—Body elongate, with rudimentary scales; snout rather short; cleft of the mouth wide; strong conical teeth in the jaws, those on the sides with several pointed tubercles; a biserial band of large molar teeth on the palate. Dorsal fin long, with flexible spines; caudal separate. Ventrals none. Gill-openings wide.

The "Sea-wolf," or "Sea-cat" (*A. lupus*), is a gigantic

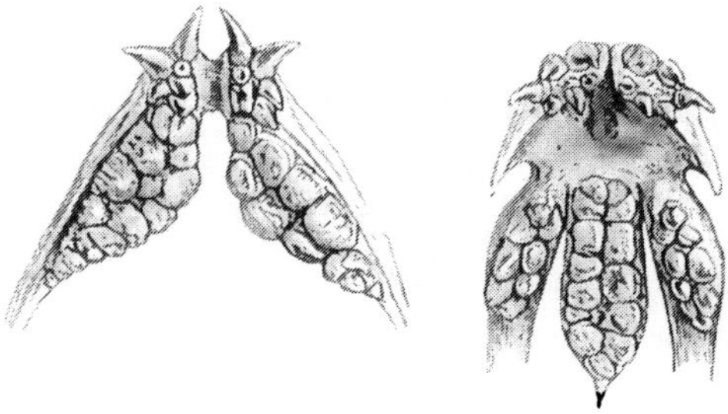

Fig. 224.—Teeth of the Wolf-fish, *Anarrhichas lupus.*

Blenny, attaining to a length of more than six feet. With its enormously strong tubercular teeth it is able to crush the hardest shells of Crustaceans or Mollusks, on which it feeds voraciously. It is an inhabitant of the northern seas, like two other allied species, all of which are esteemed as food by the inhabitants of Iceland and Greenland. Two other species of Sea-wolves occur in the corresponding latitudes of the North Pacific.

BLENNIUS.—Body moderately elongate, naked; snout short. A single dorsal, without detached portion; ventrals jugular, formed by a spine and two rays. Cleft of the mouth narrow; a single series of immovable teeth in the jaws; generally a curved tooth behind this series in both jaws, or in the lower only. A more or less developed tentacle above the orbit. Gill-opening wide.

About forty species of Blennius (in the restricted generic sense) are known from the northern temperate zone, the tropical Atlantic, Tasmania, and the Red Sea. But in the tropical Indian Ocean they are almost entirely absent, and replaced by other allied genera. Three species, found near the Sandwich Islands, are immigrants into the Pacific from the American Continent. They generally live on the coast, or attach themselves to floating objects, some species leading a pelagic life, hiding themselves in floating seaweed, in which they even propagate their species. All species readily accustom themselves to fresh water, and some (*B. vulgaris*) have become entirely acclimatised in inland lakes. British species are *B. gattorugine* (growing to a length of twelve inches), *B. ocellaris*, *B. galerita*, and *B. pholis*, the common "Shanny."

Chasmodes is a genus allied to *Blennius*, from the Atlantic coasts of temperate North America.

PETRÒSCIRTES.—Body moderately elongate, naked. Snout

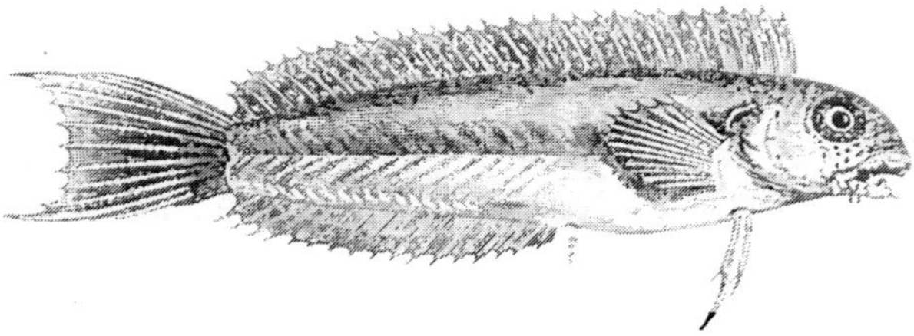

Fig. 225.—Petroscirtes bankieri, from Hong-Kong.

generally short. A single dorsal fin; ventrals composed of two or three rays. Cleft of the mouth narrow; a single series of immovable teeth in the jaws; a strong curved canine tooth behind this series, that of the lower jaw much stronger than that of the upper. Head sometimes with tentacles. Gill-opening reduced to a small fissure above the root of the pectoral.

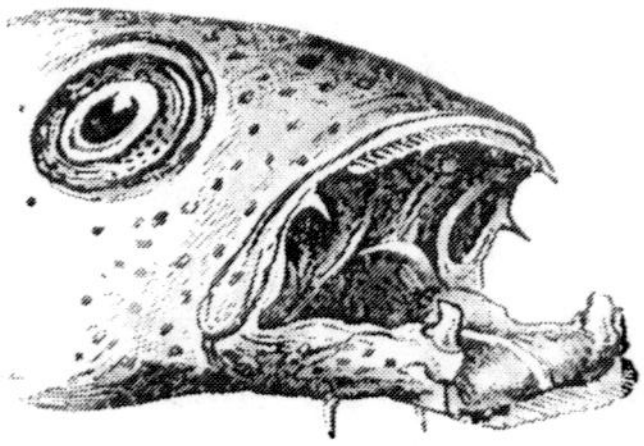

Fig. 226.—Dentition of the same, enlarged.

Thirty species, from the tropical Indo-Pacific, of small size.

SALARIAS.—Body moderately elongate, naked; snout short, with transverse cleft of the mouth; a series of numerous small teeth in the jaws, implanted in the gum and movable; generally a curved canine tooth on each side of the lower jaw, behind the series of small teeth. Dorsal fin continuous, sometimes divided into two portions by a more or less deep notch without a detached anterior part. Ventral fins with two or three rays. A tentacle above the orbit. Gill-openings wide.

Sixty species are known from the tropical zone, extending northwards to Madeira, southwards to Chile and Tasmania. In certain individuals of some of the species a longitudinal cutaneous crest is developed; all young individuals lack it, and in some other species it is invariably absent. Singularly enough this crest is not always a sexual character, as one might have supposed from analogy, but in some species at least it is developed in both sexes. Mature males, however, have generally higher dorsal fins and a more intense and variegated coloration than females and immature males, as is also the case in *Blennius*.

CLINUS—Body moderately elongate, covered with small scales; snout rather short; a narrow band or series of small teeth in the jaws and on the palate. Dorsal fin formed by numerous spines and a few soft rays, without a detached anterior portion; anal spines two. Ventrals with two or three rays. A tentacle above the orbit. Gill-opening wide.

Thirty species, from the coasts of tropical America and the southern temperate zone. Three other genera are closely allied to Clinus, viz. *Cristiceps* and *Cremnobates*, in which the three anterior dorsal spines are detached from the rest of the fin; and *Tripterygium*, with three distinct dorsal fins, of which the two anterior are spinous. The species of these genera are as numerous as those of *Clinus*, occurring in many parts of tropical seas, in the Mediterranean, and being especially well represented in South Australia and New Zealand.

STICHÆUS.—Body elongate, covered with very small scales; lateral line more or less distinct, sometimes several lateral lines. Snout short; very small teeth in the jaws, and generally on the palate. Dorsal fin long, formed by spines only. Ventrals with two or three rays. Caudal fin distinct. Gill-openings rather wide.

Small fishes, peculiar to the coasts near the arctic circle, ranging southwards to the coasts of Japan and Scandinavia. Ten species.

BLENNIOPS.—Body moderately elongate, covered with very small scales; lateral line none. Snout short; small teeth in the jaws, none on the palate. Dorsal fin long, formed by spines only. Ventrals with one spine and three rays. Caudal distinct. Gill-openings of moderate width, the gill-membranes coalescent across the isthmus.

A fine but not common kind of Blenny (*B. ascanii*), from the British and Scandinavian coasts.

CENTRONOTUS.—Body elongate, covered with very small scales; lateral line none. Snout short; very small teeth in the jaws. Dorsal fin long, formed by spines only. Ventrals none or rudimentary; caudal separate. Gill-openings of moderate width, gill-membranes coalescent.

Ten species are known from the northern coasts; southwards the genus extends to the coasts of France, New York, California, and Japan. *C. gunellus*, or the "Gunnel-fish" or "Butter-fish," is common on the British coasts. *Apodichthys* is allied to *Centronotus*, but the vertical fins are confluent; and a very large, excavated, pen-like spine lies hidden in a pouch in front of the anal fin. This spine is evidently connected in some way with the generative organs, as a furrow leads from the orifice of the oviduct to the groove of the spine. One species from the Pacific coast of North America. *Xiphidion* is another closely allied genus from the same locality.

CRYPTACANTHODES.—Body very elongate, naked, with a single lateral line. Head with the muciferous system well developed. Eye rather small. Conical teeth in the jaws, on the vomer and palatine bones. One dorsal formed by spines only; caudal connected with dorsal and anal. Ventrals none. Gill-opening of moderate width, with the gill-membranes joined to the isthmus.

One species (*C. maculatus*) from the Atlantic coasts of North America.

PATÆCUS.—Body oblong, elevated anteriorly; snout short, with sub-vertical anterior profile; minute teeth in the jaws and

Fig. 227.—Patæcus fronto.

on the vomer. Dorsal fin with the anterior spines strong and long, continuous with the caudal; ventrals none. Gill-openings wide.

Three species of this singular form are known from South and West Australia.

ZOARCES.—Body elongate, with the scales rudimentary; conical teeth in the jaws. Dorsal fin long, with a depression on the tail, which is formed by a series of spines much shorter than the rays. No other fin-spines. No separate caudal fin. Ventrals short, formed by three or four rays. Gill-openings wide.

Two species are known, one from the European, and the other from the North American side of the Atlantic. The former, *Z. viviparus*, is well known by the name of "Viviparous Blenny;" as is signified by this name it produces its young alive. These are so matured at the time of their birth that on their first exclusion they swim about with the utmost agility. No fewer than from two to three hundred young are sometimes produced by one female, and the abdomen of the mother is so distended before parturition that it is impossible to touch it without causing them to be extruded. Full grown individuals are about twelve inches long, but the

American species (*Z. anguillaris*) attains to a length of two or three feet.

Other genera of the family of Blennoids are:—*Blennophis, Nemophis, Plagiotremus, Neoclinus, Cebidichthys, Myxodes, Heterostichus, Dictyosoma, Lepidoblennius, Dactyloscopus, Gunellichthys, Urocentrus, Stichæopsis, Sticharium, Notograptus, Pholidichthys,* and *Pseudoblennius.*

Fifth Family—Acanthoclinidæ.

Body elongate, low, compressed, covered with small scales. One dorsal fin, occupying nearly the whole of the back, and chiefly composed of spines. Anal fin long, with numerous spines. Ventrals jugular, composed of a few rays only.

Of this family one fish only is known (*Acanthoclinus littoreus*), a small Blenny abundant on the coast of New Zealand.

Sixth Family—Mastacembelidæ.

Body elongate, eel-like, covered with very small scales. Mandible long, but little moveable. Dorsal fin very long, the anterior portion composed of numerous short isolated spines; anal fin with spines anteriorly. Ventrals none. The humeral arch is not suspended from the skull. Gill-openings reduced to a slit at the lower part of the side of the head.

Freshwater-fishes characteristic of and almost confined to the Indian region. The structure of the mouth and of the branchial apparatus, the separation of the humeral arch from the skull, the absence of ventral fins, the anatomy of the abdominal organs, affords ample proof that these fishes are Acanthopterygian eels. Their upper jaw terminates in a pointed moveable appendage, which is concave and transversely striated inferiorly in *Rhynchobdella,* and without transverse striæ in *Mastacembelus:* the only two genera of this family.

Thirteen species are known, of which *Rh. aculeata*, *M. pancalus* and *M. armatus* are extremely common, the latter attaining

Fig. 228.—Mastacembelus argus, from Siam.

to a length of two feet. Outlying species are *M. aleppensis* from Mesopotamia and Syria, and *M. cryptacanthus*, *M. marchei*, and *M. niger*, from West Africa.

ELEVENTH DIVISION—ACANTHOPTERYGII MUGILIFORMES.

Two dorsal fins more or less remote from each other; the anterior either short, like the posterior, or composed of feeble spines. Ventral fins with one spine and five rays, abdominal.

FIRST FAMILY—SPHYRÆNIDÆ.

Body elongate, sub-cylindrical, covered with small cycloid scales; lateral line continuous. Cleft of the mouth wide, armed with strong teeth. Eye lateral, of moderate size. Vertebræ twenty-four.

This family consists of one genus only, *Sphyræna*, generally called "Barracudas," large voracious fishes from the tropical and sub-tropical seas, which prefer the vicinity of the coast to the open sea. They attain to a length of eight feet, and a weight of forty pounds; individuals of this large size are dangerous to bathers. They are generally used as food, but sometimes (especially in the West Indies) their flesh assumes poisonous qualities, from having fed on smaller poisonous fishes. Seventeen species.

The Barracudas existed in the tertiary epoch, their remains being frequently found at Monte Bolca. Some other fossil genera have been associated with them, but as they are known from jaws and teeth or vertebræ only, their position in the system cannot be exactly determined; thus *Sphyrænodus* and *Hypsodon* from the chalk of Lewes, and the London clay of Sheppey. The American *Portheus* is allied to *Hypsodon*. Another remarkable genus from the chalk, *Saurocephalus*, has been also referred to this family.[1]

Second Family—Atherinidæ.

Body more or less elongate, sub-cylindrical, covered with scales of moderate size; lateral line indistinct. Cleft of the mouth of moderate width, with the dentition feeble. Eye lateral, large or of moderate size. Gill-openings wide. Vertebræ very numerous.

Small carnivorous fishes inhabiting the seas of the temperate and tropical zones; many enter fresh water, and some have been entirely acclimatised in it. This family seems to have been represented in the Monte Bolca formation by *Mesogaster*.

Atherina.—Teeth very small; scales cycloid. The first dorsal is short and entirely separated from the second. Snout obtuse, with the cleft of the mouth straight, oblique, extending to or beyond the anterior margin of the eye.

The Atherines are littoral fishes, living in large shoals, which habit has been retained by the species acclimatised in fresh water. They rarely exceed a length of six inches, but are nevertheless esteemed as food. From their general resemblance to the real Smelt they are often thus misnamed, but may always be readily recognised by their small first spinous dorsal fin. The young, for some time after they are hatched, cling together in dense masses, and in numbers almost incredible. The inhabitants of the Mediterranean

[1] The systematic affinities of these extinct genera are very obscure. Cope places them, with others (for instance *Protosphyræna*, which has a sword-like prolongation of the ethmoid), in a distinct family, *Saurodontidæ*: see ' Vertebrata of the Cretaceous Formations of the West," 1875.

coast of France call these newly hatched Atherines "Nonnat" (unborn). Some thirty species are known, of which *A. presbyter* and *A. boyeri* occur on the British coast.

ATHERINICHTHYS, distinguished from *Atherina* in having the snout more or less produced; and the cleft of the mouth generally does not extend to the orbit.

These Atherines are especially abundant on the coasts and in the fresh waters of Australia and South America. Of the twenty species known, several attain a length of eighteen inches and a weight of more than a pound. All are highly esteemed as food; but the most celebrated is the "Pesce Rey" of Chile (*A. laticlavia*).

TETRAGONURUS.—Body rather elongate, covered with strongly keeled and striated scales. The first dorsal fin is composed of numerous feeble spines, and continued on to the second. Lower jaw elevated, with convex dental margin, and armed with compressed, triangular, rather small teeth, in a single series.

This very remarkable fish is more frequently met with in the Mediterranean than in the Atlantic, but generally scarce. Nothing is known of its habits; when young it is one of the fishes which accompany Medusæ, and, therefore, it must be regarded as a pelagic form. Probably, at a later period of its life, it descends to greater depths, coming to the surface at night only. It grows to a length of eighteen inches.

THIRD FAMILY—MUGILIDÆ.

Body more or less oblong and compressed, covered with cycloid scales of moderate size; lateral line none. Cleft of the mouth narrow or of moderate width, without or with feeble teeth. Eye lateral, of moderate size. Gill-opening wide. The anterior dorsal fin composed of four stiff spines. Vertebræ twenty-four.

The "Grey Mullets" inhabit in numerous species and in great numbers the coasts of the temperate and tropical zones. They frequent brackish waters, in which they find an abundance of food which consists chiefly of the organic substances mixed with mud or sand; in order to prevent larger bodies from passing into the stomach, or substances from passing through the gill-openings, these fishes have the organs of the pharynx modified into a filtering apparatus. They take in a quantity of sand or mud, and, after having worked it for some time between the pharyngeal bones, they eject the roughest and indigestible portion of it. The upper pharyngeals have a rather irregular form; they are slightly arched, the convexity being directed towards the pharyngeal cavity, tapering anteriorly and broad posteriorly. They are coated with a thick soft membrane, which reaches far beyond the margin of the bone, at least on its interior posterior portion; this membrane is studded all over with minute horny cilia. The pharyngeal bone rests upon a large fatty mass, giving it a considerable degree of elasticity. There is a very large venous sinus between the anterior portion of the pharyngeal and the basal portion of the branchial arches. Another mass of fat, of elliptical form, occupies the middle of the roof of the pharynx, between the two pharyngeal bones. Each branchial arch is provided on each side, in its whole length, with a series of closely-set gill-rakers, which are laterally bent downwards, each series closely fitting into the series of the adjoining arch; they constitute together a sieve, admirably adapted to permit a transit for the water, retaining at the same time every other substance in the cavity of the pharynx.

The lower pharyngeal bones are elongate, crescent-shaped, and broader posteriorly than anteriorly. Their inner surface is concave, corresponding to the convexity of the upper pharyngeals, and provided with a single series of lamellæ, similar to those of the branchial arches, but reaching across the bone from one margin to the other.

The intestinal tract shows no less peculiarities. The lower portion of the œsophagus is provided with numerous long thread-like papillæ, and continued into the oblong-ovoid membranaceous cœcal portion of the stomach, the mucosa of which forms several longitudinal folds. The second portion of the stomach reminds one of the stomach of birds; it communicates laterally with the other portion, is globular, and surrounded by an exceedingly strong muscle. This muscle is not divided into two as in birds, but of great thickness in the whole circumference of the stomach, all the muscular fasciculi being circularly arranged. The internal cavity of this stomach is rather small, and coated with a tough epithelium, longitudinal folds running from the entrance opening to the pyloric, which is situated opposite to the other. A low circular valve forms a pylorus. There are five rather short pyloric appendages. The intestines make a great number of circumvolutions, and are seven feet long in a specimen thirteen inches in length.

Some seventy species of Grey Mullets are known, the majority of which attain to a weight of about four pounds, but

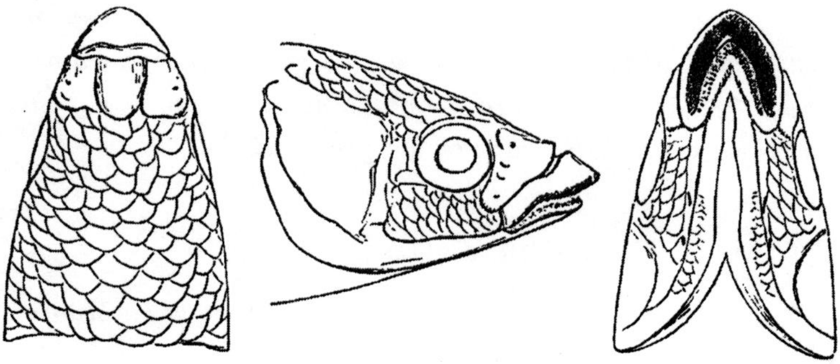

Fig. 229.—Mugil proboscideus.

there are many which grow to ten and twelve pounds. All are eaten, and some even esteemed, especially when taken out of fresh water. If attention were paid to their cultivation, great profits could be made by fry being transferred into suitable backwaters on the shore, in which they rapidly grow to a marketable size. Several species are more or less abundant on the British coasts, as *Mugil octo-radiatus* (Fig. 105, p. 254), *M. capito, M. auratus* (Fig. 106, p. 254), and *M. septentrionalis* (Fig. 107, p. 254), which, with the aid of the accompanying figures, and by counting the rays of the anal fin, may be readily distinguished—*M. octo-radiatus* having eight, and *M. capito* and *M. auratus* nine soft rays. A species inhabiting fresh waters of Central America (*M. proboscideus*) has the snout pointed and fleshy, thus approaching certain other fresh-water and littoral Mullets, which, on account of a modification of the structure of the mouth, have been formed into a distinct genus, *Agonostoma*. *Myxus* comprises Mullets with teeth more distinct than in the typical species.

This genus existed in the tertiary epoch, remains of a species having been found in the gypsum of Aix, in Provence.

Twelfth Division—Acanthopterygii Gastrosteiformes.

The spinous dorsal is composed of isolated spines if present; the ventrals are either thoracic or have an abdominal position in consequence of the prolongation of the pubic bones which are attached to the humeral arch. Mouth small, at the end of the snout which is generally more or less produced.

First Family—Gastrosteidæ.

Body elongate, compressed. Cleft of the mouth oblique; villiform teeth in the jaws. Opercular bones not armed; infraorbitals covering the cheek; parts of the skeleton forming incomplete external mails. Scales none, but generally large scutes along the side. Isolated spines in front of the soft dorsal fin. Ventral fins abdominal, joined to the pubic bone, composed of a spine and a small ray. Branchiostegals three.

Of "Sticklebacks" (*Gastrosteus*) about ten species are satisfactorily known, one of which (*G. spinachia*) lives in salt and brackish water, whilst the others inhabit principally fresh waters, although they all are able to exist in the sea. They are confined to the Temperate and Arctic zones of the northern hemisphere. The British freshwater species are the Three-spined Stickleback (*G. aculeatus*), which sometimes, especially in Central Europe, lacks scutes, sometimes has a series of scutes along the side of the body; the Four-spined Stickleback (*G. spinulosus*) and the Nine-spined Stickleback (*G. pungitius*). The commonest North American species is *G. noveboracensis*. The habits of all the freshwater species

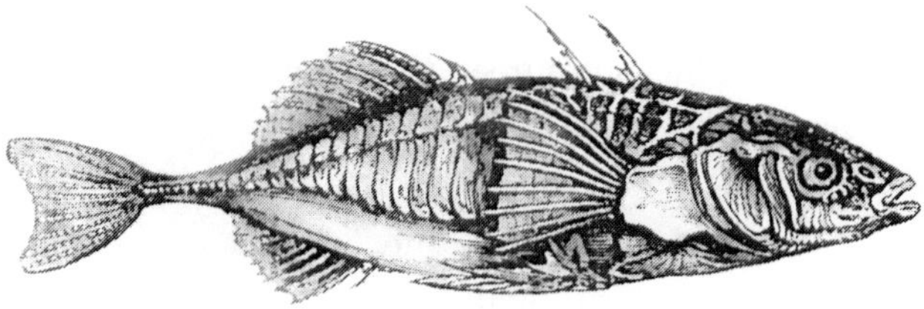

Fig. 230.—Gastrosteus noveboracensis.

are very similar. The common European species (*G. aculeatus*) is an active and greedy little fish, extremely destructive to the fry of other species, and consequently injurious in ponds where these are sought to be preserved. It is scarcely to be conceived what damage these little fishes do, and how greatly detrimental they are to the increase of all the fishes in general among which they live; for it is with the utmost industry, sagacity, and greediness that they seek out and destroy all the young fry that come their way. A small Stickleback, kept in an aquarium, devoured, in five hours' time, seventy-four young dace, which were about a quarter of an inch long, and of the thickness of a horse hair. Two days after it swallowed sixty-two; and would, probaby, have eaten as many every day could they have been procured. The Stickleback sometimes swarms in prodigious numbers. Pennant states that at Spalding, in Lincolnshire, there was once in seven years amazing shoals, which appear in the Welland, coming up the river in the form of a vast column. The quantity may, perhaps, be conceived from the fact that a man employed in collecting them, gained, for a considerable time, four shillings a-day by selling them at the rate of a halfpenny a bushel. Costa, who studied the manners of these small fishes, relates that, on the approach of spawning time, the male builds a nest of stalks of grass and other matters in a hollow of the bottom, a little above three inches wide and about six inches and a half deep, creeping over the materials on his belly, and cementing them with the mucus that exudes from his skin. The bottom of the nest is first laid, then the sides are raised, and lastly the top is covered over. A small hole is left on one side for an entrance. When the erection is complete, he seeks out a female, and conducting her, Costa says, with many caresses, to the nest, introduces her by the door into the chamber. In a few minutes she has laid two or three eggs, after which she bores a hole on the opposite side of the nest to that by which she entered, and makes her escape. The nest has now two doors, and the eggs are exposed to the cool stream of water, which, entering by one door flows out at the other. Next day the male goes again in quest of a female, and sometimes brings back the same, sometimes finds a new mate. This is repeated until the nest contains a considerable number of eggs, and each time the male rubs his side against the female and passes over the eggs. Next the male watches a whole month over his treasure, defending it stoutly against all invaders, and especially against his wives, who have a great desire to get at the eggs. When the young are hatched and able to do for themselves his cares cease.

The Sea-Stickleback (*G. spinachia*) is likewise a nest

builder, choosing for its operations especially the shallows of brackish water, which are covered with *Zostera*.

SECOND FAMILY—FISTULARIIDÆ.

Fishes of greatly elongated form; the anterior bones of the skull are much produced, and form a long tube, terminating in a narrow mouth. Teeth small; scales none, or small. The spinous dorsal fin is either formed by feeble isolated spines or entirely absent; the soft dorsal and anal of moderate length, ventral fins thoracic or abdominal, composed of five or six rays, without spine; if abdominal, they are separate from the pubic bones, which remain attached to the humeral arch. Branchiostegals five.

The "Flute-mouths" are also frequently called "Pipe-fishes," a name which they have in common with the Syngnathidæ. They are gigantic marine Sticklebacks, living near the shore, from which they are frequently driven into the open sea; some of the species, therefore, have a wide geographical range. Probably all enter brackish water. They are distributed over the whole of the tropical and sub-tropical parts of the Atlantic and Indo-Pacific. The species are few in number, but some of them are very common.

This family is well represented in Eocene formations; some of the remains belonging to the existing genera, *Fistularia*, *Aulostoma*, and *Auliscops*, the two former of which occur not rarely at Monte Bolca and in the schists of Glaris. Well-preserved remains of *Auliscops* have been found in the Marl-slates of the highlands of Padang in Sumatra. Extinct genera from Monte Bolca are *Urosphen*, the cylindrical body of which is terminated by a large cuneiform fin; and *Rhamphosus*, which has an immense spinous ray, denticulated behind, inserted on the nape.

FISTULARIA.—Body scaleless; caudal fin forked, with the two middle rays produced into a filament; no free dorsal spines.

Three species are known, common on the shores of the Tropical Atlantic (*F. tabaecaria*) and Indian Oceans (*F. serrata* and *F. depressa*); they attain to a length of from four to six feet.

The anterior portion of the vertebral column shows the same peculiarity as in *Dactylopterus*; it is a long compressed tube, composed of four elongate vertebræ, which are perfectly anchylosed; each of them has a pair of small foramina for blood-vessels. The neural spines and parapophyses of this tubiform portion are confluent into thin laminæ, the lateral of which are wing-like, and expanded in their anterior half.

AULOSTOMA.—Body covered with small scales. Caudal fin rhombic, without prolonged rays; a series of isolated feeble dorsal spines. Teeth rudimentary.

Two species from the Tropical Atlantic and Indian Oceans.

AULISCOPS.—Body naked. Ventrals thoracic. Numerous spines in front of the dorsal fin.

One species (*A. spinescens*) from the Pacific coast of North America. *Aulorhynchus* from the same sea, and *Aulichthys* from Japan, are allied genera.

THIRTEENTH DIVISION—ACANTHOPTERYGII CENTRISCIFORMES.

Two dorsal fins; the spinous short, the soft and the anal of moderate extent. Ventral fins truly abdominal, imperfectly developed.

This division consists of one family, *Centriscidæ*, with two genera. The fishes belonging to it are very small, marine, and, in consequence of their limited power of swimming, often driven out into the open sea. They have the same structure of the mouth and snout as the Fistulariidæ, but combine with it peculiarities of the shape of body, of the structure of the vertical fins, and of the relations between endo- and exo-skeleton, which render them altogether a singular and interesting type. *Amphisile* has been found in a fossil state at Monte Bolca.

CENTRISCUS.—Body oblong or elevated, compressed, covered with small rough scales; lateral line none; some bony strips on the side of the back, and on the margin of the thorax and abdomen; the former, in one species, are confluent and form a shield. Teeth none. Two dorsal fins, the first with one of the spines very strong. Ventral fins small, abdominal, composed of five soft rays. Four branchiostegals.

Of the four species the most generally known is *C. scolopax*, the "Trumpet-fish" or "Bellows-fish," which rarely occurs on the south coast of England, is more common farther

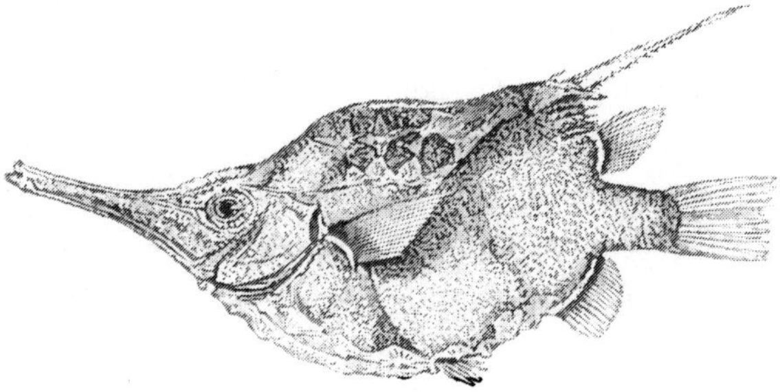

Fig. 231.—Centriscus humerosus.

south, and reäppears in Tasmania. The allied *C. gracilis* is one of the fishes common to the Mediterranean and Japanese Seas. The species figured, *C. humerosus*, occurs on the coast of South Australia, and is very scarce.

AMPHISILE.—Body elongate, strongly compressed, provided with a dorsal cuirass, which is formed by portions of the skeleton; the longitudinal axis of the tail is not in the same line with that of the trunk. Scales none. Teeth none. Two dorsal fins situated on the hindmost part of the back; ventral fins rudimentary, abdominal. Three or four branchiostegals.

The three species known of this genus are found in the tropical Indo-Pacific. Their body is so thin that it has the appearance of being artificially compressed between two sheets of paper; it is semi-transparent, especially in the region of the air-bladder. The structure of the vertebral column is extremely singular and unique among Acanthopterygians. The abdominal portion is more than four times as long as the caudal; nevertheless it is composed of only six vertebræ, whilst the latter consists of fourteen. The abdominal vertebræ are extremely slender, the third alone being nearly as long as the whole caudal portion; they have a slight ridge superiorly and inferiorly, and on each side; the whole portion lying in the uppermost concavity of the dorsal cuirass. The caudal vertebræ are extremely short, and the strength of their neural and hæmal spines is in proportion to their size. The dorsal cuirass is not a dermal production, but formed by modified parts of the endoskeleton; its composition, the number and condition of its single parts, and, finally, the first dorsal spine, which in *A. punctulata* is so singularly attached to it, favour this opinion. The plates, which occupy the vertebral line, would correspond to the neural spines, and the lateral plates on which the ribs are suspended to the parapophyses. *Amphisile* may be considered as a Chelonian form among fishes.

FOURTEENTH DIVISION—ACANTHOPTERYGII GOBIESOCIFORMES.

No spinous dorsal; the soft and the anal short or of moderate length, situated on the tail; ventral fins subjugular, with an adhesive apparatus between them. Body naked.

These fishes are well characterised by their single dorsal fin, and by their adhesive ventral apparatus, which has only an external similarity to the organ observed in *Cyclopterus* and *Liparis;* its structure is typically different from it.

Whilst in those genera the ventral fins occupy the centre of the disk forming its base, these fins are here widely apart from each other, as in *Callionymus*, forming only a portion of the periphery of the disk, which is completed by a cartilaginous expansion of the coracoid bones. The following description of its structure is taken from *Sicyases sanguineus*, but it is essentially the same in all the genera.

The whole disk is exceedingly large, subcircular, longer than broad, its length being one-third of the whole length of the fish. The central portion is formed merely by skin, which is separated from the pelvic or pubic bones by several layers of muscles. The peripheric portion is divided into an anterior and posterior part by a deep notch behind the ventrals. The anterior peripheric portion is formed by the four ventral rays, the membrane between them, and a broad fringe which extends anteriorly from one ventral to the other; this fringe is a fold of the skin, containing on each side the rudimentary ventral spine, but no cartilage. The posterior peripheric portion is suspended on each side from the coracoid, the upper bone of which is exceedingly broad, becoming a free movable plate behind the pectoral. A broad cartilage is firmly attached to it. The lower bone of the coracoid is of a triangular form, and supports a very broad fold of the skin, extending from one side to the other, and containing a cartilage which runs through the whole of that fold. Five processes of the cartilage are continued into the soft striated margin in which the disk terminates posteriorly. The surface of the disk is coated with thick epidermis, like the sole of the foot of higher animals. The epidermis is divided into many polygonal plates; there are no such plates between the roots of the ventral fins.

Not less unique is the structure of the bones which have some relation to this external adhesive apparatus. As exemplified by *Chorismochismus dentex* the coracoid is well developed, and, as usual, composed of two pieces, the upper of which is not suspended from the humerus, but fixed by a ligament to the hinder margin of the carpal bones. It is a broad lamella, dilated posteriorly into the cartilage, which is externally visible; the lower piece is narrower, and fixed to the extremity of the pubic bone of its side. The pubic bones are united by suture, and form together a heart-shaped disk, the point of which is produced backwards. The anterior portion of the disk is concave, with a bony longitudinal bridge and a feeble transverse ridge. The disk is fixed to the humeral bones by the convex portions of its anterior margin, whilst the convex portions of the lateral margins serve as base for the ventral fins. The latter are composed of one spine, which is transformed into a broad, thin, and curved plate, hidden below the skin, and apparently of four rays; but on closer examination we find that the hidden ray has a longitudinal groove anteriorly, in which another thinner ray lies concealed. This ray is quite free, and not joined to the pubic bone.

The fishes belonging to the single family of this division, *Gobiesocidæ*, are strictly marine but littoral fishes. They are scattered over the temperate zones of both hemispheres, and more numerous than between the Tropics. All are of small or very small size.

The adhesive disk consists of an anterior and posterior division. In some of the genera the posterior division has no

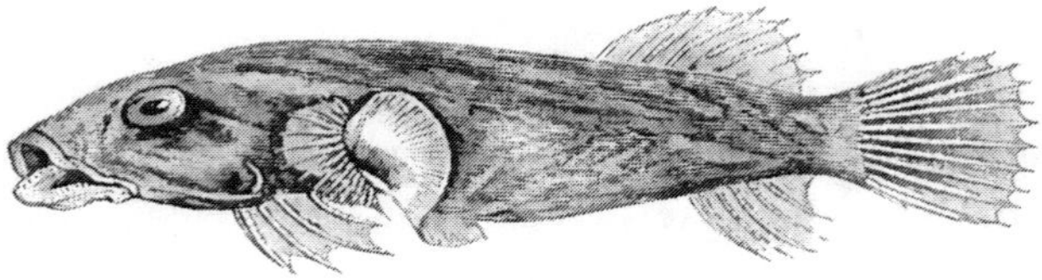

Fig. 232.—Gobiesox cephalus.

free anterior margin, the teeth being either all conical, as in *Chorisochismus* (Cape of Good Hope) and *Cotylis* (Red Sea and Indian Ocean); or incisor-like in both jaws, as in *Sicyases* (coast of Chili and West Indies); or incisor-like at least in the lower jaw, as in *Gobicsox* (West Indies and Pacific coasts of South America). In other genera the posterior portion of the adhesive disk has a free anterior margin. Only one of these genera has incisor-like teeth, viz. *Diplocrepis* from New

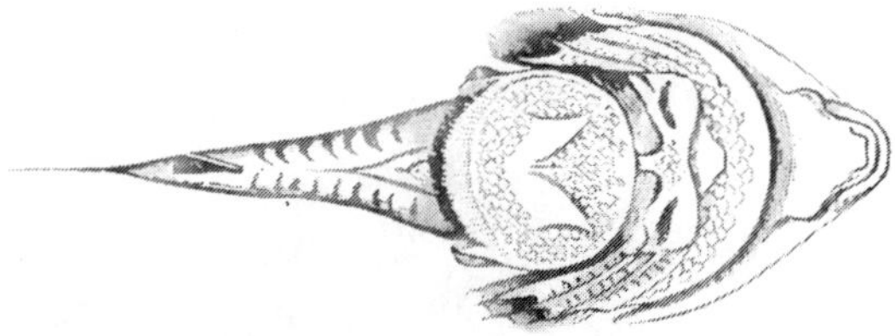

Fig. 233.—Diplocrepis puniceus.

Zealand. In the remaining genera, *Crepidogaster* (from Tasmania and South Australia), *Trachelochismus* (from New Zealand and the Fiji Islands), *Lepadogaster*, and *Leptopterygius*, the teeth are very small and fine. The two last genera are European, and *Lepadogaster* at least is common on the Southern British coasts. The three species known as British —*L. gouanii*, *L. candollii*, and *L. bimaculatus*—are prettily coloured, but subject to great variation.

Fifteenth Division—Acanthopterygii Channiformes.

Body elongate, covered with scales of moderate size; no spine in any of the fins; dorsal and anal long. No superbranchial organ, only a bony prominence on the anterior surface of the hyomandibular.

These fishes belong to the single family *Ophiocephalidæ*, Freshwater-fishes characteristic of the Indian region, which, however, have found their way into Africa, where they are represented by one or two species. Thirty-one species are known altogether, most of which are extremely abundant; some attain to a length of more than two feet. Like other tropical freshwater fishes, they are able to survive droughts, living in semi-fluid mud, or lying in a torpid state below the hard-baked crusts of the bottom of a tank from which every every drop of water has disappeared. Respiration is probably entirely suspended during the state of torpidity, but whilst the mud is still soft enough to allow them to come to the surface, they rise at intervals to take in a quantity of air, by means of which their blood is oxygenised. This habit has been observed in some species to continue also to the period

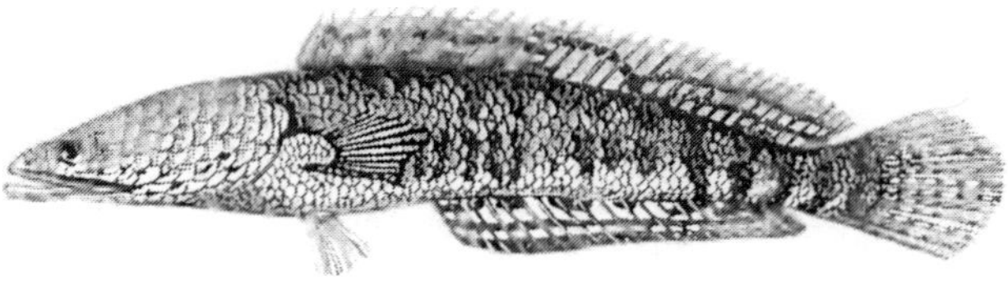

Fig. 234.—Ophiocephalus striatus, India.

of the year in which the fish lives in normal water, and individuals which are kept in a basin and prevented from coming to the surface and renewing the air for respiratory purposes, are suffocated. The particular manner in which the accessory branchial cavity participates in respiratory functions is not known. It is a simple cavity, without an accessory branchial organ, the opening of which is partly closed by a fold of the mucous membrane.

SIXTEENTH DIVISION—ACANTHOPTERYGII LABYRINTHIBRANCHII.

Body compressed, oblong or elevated, with scales of moderate size. A superbranchial organ in a cavity accessory to that of the gills.

FIRST FAMILY—LABYRINTHICI.

Dorsal and anal spines present, but in variable numbers; ventrals thoracic. Lateral line absent, or more or less distinctly interrupted. Gill-opening rather narrow, the gill-membranes of both sides coalescent below the isthmus, and scaly ; gills four ; pseudobranchiæ rudimentary or absent.

Freshwater-fishes of the Cyprinoid division of the Equatorial zone. They possess the faculty of being able to live for some time out of the water, or in thick or hardened mud, in a still greater degree than the fishes of the preceding family. In the accessory branchial cavity there is lodged a laminated organ which evidently has the function of assisting in the oxygenisation of the blood. In *Anabas* it is formed by several exceedingly thin bony laminæ, similar in form to the auricle, and concentrically situated one above the other, the innermost being the largest. The degree in which these laminæ are developed is dependent on age. In specimens from one inch and a half to two inches and a half long there are only two such laminæ, a third being indicated by a small protuberance at the central base of the second or outer laminæ. In specimens of from three to four inches in length the third lamina is developed, covering one-half of the second. The edges of all the laminæ are straight, not valanced. In specimens of from four to five inches a fourth lamina makes its appearance in the basal centre of the third lamina. The other laminæ continue to grow in their circumference, and their edges now become undulated and slightly frilled. Cuvier and Valenciennes have examined still larger specimens. The figure given by them and reproduced here was taken from a specimen six or seven inches long, and shows the superbranchial organ composed of six laminæ.

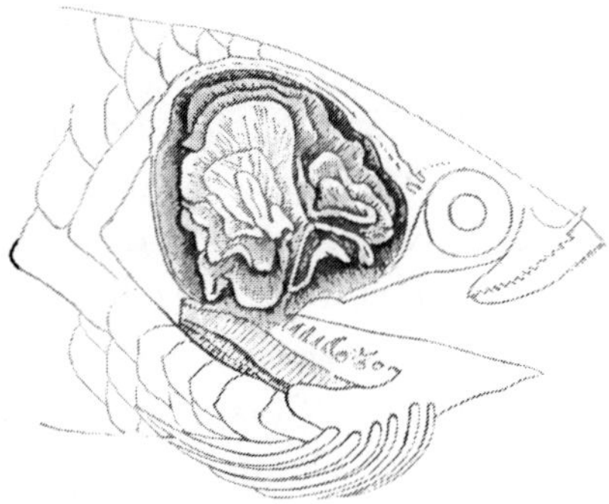

Fig. 235.—Superbranchial organ of Anabas.

The air-bladder of the majority of these fishes is very large, extending far into the tail, and, therefore, divided behind by the hæmal spines into two lateral portions.

The Labyrinthici are generally of small size; they are capable of being domesticated, and some of them deserve particular attention on account of the dazzling beauty of their colours or the flavour of their flesh.

ANABAS.—Body compressed, oblong; præorbital and orbitals serrated. Small teeth in the jaws and on the vomer; none on the palatines. Dorsal and anal spines numerous. Lateral line interrupted.

The "Climbing Perch" (*A. scandens*) is generally distributed over the Indian Region, and well known from its faculty of moving for some distance over land, and even up inclined surfaces. In 1797 Daldorf, in a memoir communicated to the Linnean Society of London, mentions that in 1791 he had himself taken an Anabas in the act of ascending a palm tree which grew near a pond. The fish had reached the height of five feet above the water, and was going still higher. In the effort to do this it held on to the bark of the tree by the preopercular spines, bent its tail, and stuck in the spines of the anal; then released its head, and, raising it, took a new hold with the preoperculum higher up. The fish is named in the Malayan language the "Tree Climber." It rarely attains a length of seven inches.

Spirobranchus from the Cape, and *Ctenopoma* from Tropical Africa, represent *Anabas* in that continent.

POLYACANTHUS.—Body compressed, oblong; operculum without spines or serrature; cleft of the mouth small, more or less oblique, not extending beyond the vertical from the orbit, and little protractile. Small fixed teeth in the jaws, none on the palate. Dorsal and anal spines numerous; the soft dorsal and anal, the caudal, and the ventral, more or less elongate in mature specimens. Caudal rounded. Lateral line interrupted or absent.

This genus is represented chiefly in the East Indian Archipelago; seven species are known; some of them have been domesticated on account of the beauty of their colours, and several varieties have been produced. One of them is to be mentioned, as, under the name of "Paradise-fish," it has been introduced into the aquaria of Europe, where it readily breeds. It was known already to Lacépède, and has been mentioned since his time in all ichthyological works as *Macropus viridi-auratus*. In adult males some of the rays, and especially the caudal lobes, are much prolonged.

Osphromenus.—Body compressed, more or less elevated; operculum without spine or serrature. Small fixed teeth in the jaws, none on the palate. Dorsal spines in small or moderate number; anal spines in moderate or great number; ventral fins with the outer ray very long, filiform. Lateral line not interrupted or absent.

To this genus belongs the celebrated "Gourami" (*Osphromenus olfax*), reputed to be one of the best flavoured Fresh-

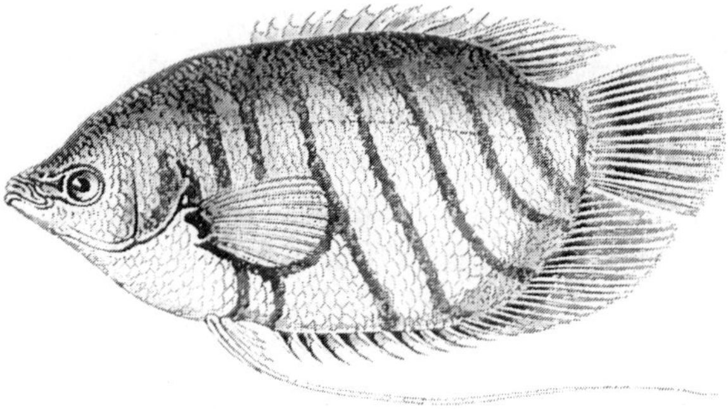

Fig. 236.—Osphromenus olfax.

water-fishes in the East-Indian Archipelago. Its original home is Java, Sumatra, Borneo, and several other islands; but thence it has been transported to, and acclimatised in, Penang, Malacca, Mauritius, and even Cayenne. Being an almost omnivorous fish and tenacious of life, it seems to recommend itself particularly for acclimatisation in other tropical countries, and specimens kept in captivity become as tame as carps. It attains the size of a large turbot. A second, but much smaller, species of this genus, *O. trichopterus*, is frequently kept in vessels on account of the exquisite beauty of its varying iridescent metallic tints; like other fishes of this family it is very pugnacious.

Trichogaster, a very common Bengalese fish, differs from *Osphromenus* in having the ventral fins reduced to a single long filament.

BETTA.—Body compressed, oblong; operculum without spine or serrature. Small fixed teeth in the jaws, none on the palate. Dorsal fin short, on the middle of the back, without any pungent spine; anal fin long. Ventral fin with five soft rays, the outer one being produced. Lateral line interrupted or absent.

A species of this genus (*B. pugnax*) is, on account of its pugnacious habits, reared by the Siamese. Cantor gives the following account:—"When the fish is in a state of quiet, its dull colours present nothing remarkable; but if two be brought together, or if one sees its own image in a looking-glass, the little creature becomes suddenly excited, the raised fins and the whole body shine with metallic colours of dazzling beauty, while the projected gill membrane, waving like a black frill round the throat, adds something of grotesqueness to the general appearance. In this state it makes repeated darts at its real or reflected antagonist. But both, when taken out of each other's sight, instantly become quiet. This description was drawn up in 1840, at Singapore, by a gentleman who had been presented with several by the King of Siam. They were kept in glasses of water, fed with larvæ of mosquitoes, and had thus lived for many months. The Siamese are as infatuated with the combats of these fish as the Malays are with their cock-fights; and stake on the issue considerable sums, and sometimes their own persons and families. The license to exhibit fish-fights is farmed, and brings a considerable annual revenue to the King of Siam. The species abounds in the rivulets at the foot of the hills of Penang. The inhabitants name it 'Pla-kat,' or the 'Fighting-fish;' but the kind kept especially for fighting is an artificial variety cultivated for the purpose."

MICRACANTHUS.—This genus represents the three last-named genera in Africa, where it has been recently discovered in tributaries of the river Ogooué. It seems to differ from the Indian genera chiefly by its more elongate body, the structure of the fins being scarcely different (D. $\frac{3}{7}$, A. $\frac{4}{23}$, V. $\frac{1}{4}$).

SECOND FAMILY—LUCIOCEPHALIDÆ.

Body elongate, covered with scales of moderate size. Lateral line present. Teeth small. Gill-opening wide; pseudobranchiæ none. The superbranchial organ is formed by two branchial arches, which are dilated into a membrane. One short dorsal fin; dorsal and anal spines none; ventrals composed of one spine and five rays. Air-bladder none.

A small Freshwater-fish (*Luciocephalus pulcher*), from the East-Indian Archipelago.

SEVENTEENTH DIVISION—ACANTHOPTERYGII LOPHOTIFORMES.

Body riband shaped, with the vent near its extremity; a short anal behind the vent; dorsal fin as long as the body.

Only one species is known of this division or family, *Lophotes cepedianus*. It is most probably a deep-sea fish, but does not descend to so great a depth as the *Trachypteridæ*, its bony and soft parts being well coherent. It is a scarce fish, hitherto found in the Mediterranean, off Madeira, and in the Sea of Japan; its length is known to exceed five feet. The head is elevated into a very high crest, and the dorsal fin commences with an exceedingly strong and long spine on the head. Silvery, with rose-coloured fins.

EIGHTEENTH DIVISION—ACANTHOPTERYGII TÆNIIFORMES.

Body riband shaped; dorsal fin as long as the body; anal absent; caudal rudimentary, or not in the longitudinal axis of the fish.

The "Ribbon-fishes" are true deep-sea fishes, met with in all parts of the oceans, generally found when floating dead on the surface, or thrown ashore by the waves. Their body is like a band, specimens of from fifteen to twenty feet long being from ten to twelve inches deep, and about an inch or two broad at their thickest part. The eye is large and lateral; the mouth small, armed with very feeble teeth; the head deep and short. A high dorsal fin runs along the whole length of the back, and is supported by extremely numerous

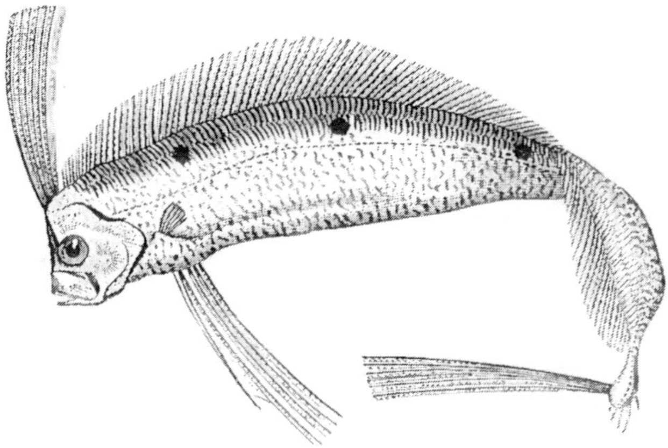

Fig. 237.—Trachypterus tænia.

rays; its foremost portion, on the head, is detached from the rest of the fin, and composed of very elongate flexible spines. The anal fin is absent. The caudal fin (if preserved, which is rarely the case, in adult specimens) has an extra-axial position, being directed upwards like a fan. The ventrals are thoracic, either composed of several rays or reduced to a single long filament. The coloration is generally silvery, with rosy fins.

When these fishes reach the surface of the water the expansion of the gases within their body has so loosened all parts of their muscular and bony system, that they can be lifted out of the water with difficulty only, and nearly always portions of the body and fins are broken and lost. The bones contain very little bony matter, are very porous, thin and light. At what depths Ribbon-fishes live is not known; probably the depths vary for different species; but although none have been yet obtained by means of the deep-sea dredge, they must be abundant at the bottom of all oceans, as dead fishes or fragments of them are frequently obtained. Some writers have supposed from the great length and narrow shape of these fishes that they have been mistaken for "Sea-serpents;" but as these monsters of the sea are always represented by those who have had the good fortune of meeting with them as remarkably active, it is not likely that harmless Ribbon-fishes, which are either dying or dead, have been the objects described as "Sea-serpents."

Young Ribbon-fishes (from two to four inches) are not

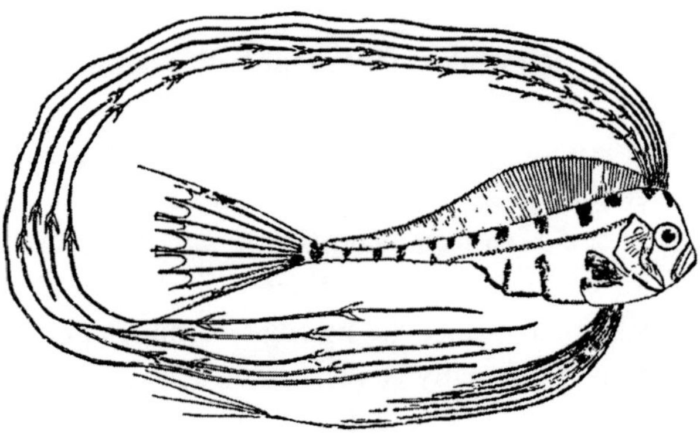

Fig. 238.—Young Trachypterus.

rarely met with near the surface; they possess the most

extraordinary development of fin rays observed in the whole class of fishes, some of them being several times longer than the body, and provided with lappet-like dilatations. There is no doubt that fishes with such delicate appendages are bred and live in depths where the water is absolutely quiet, as a sojourn in the disturbed water of the surface would deprive them at once of organs which must be of some utility for their preservation.

Ribbon-fishes are divided into three genera :—

TRACHYPTERUS.—In which the ventral rays are well developed, and composed of several more or less branched rays. Specimens of this genus have been taken in the Mediterranean, Atlantic, at Mauritius, and in the Eastern Pacific. The "Deal-fish" (*T. arcticus*) is often met with in the North Atlantic, and specimens are generally found after the equinoctial gales on the coasts of the Orkneys and North Britain.

STYLOPHORUS.—Without ventrals, and with the tail terminating in an exceedingly long cord-like appendage. Known from one specimen only, found at the beginning of this century between Cuba and Martinique. It is eleven inches long, and preserved in the Museum of the Royal College of Surgeons in London.

Regalecus.—Each ventral fin is reduced to a long filament, dilated at the extremity; caudal fin rudimentary or absent. These are the largest of all Ribbon-fishes, specimens being on record the length of which exceeded twenty feet. They have been taken in the Mediterranean, North and South Atlantic, Indian Ocean, and on the coast of New Zealand. They are frequently called "Kings of the herrings," from the erroneous notion that they accompany the shoals of herrings; or "Oar-fishes," from their two ventral fins, which have a dilatation at their extremity not unlike the blade of an oar. One or more species (*R. banksii*) are sometimes found on the British coasts, but they are very scarce, not more than sixteen captures having been recorded between the years 1759 and 1878.

NINETEENTH DIVISION—ACANTHOPTERYGII NOTACANTHIFORMES.

Dorsal fin short, composed of short, isolated spines, without a soft portion. Anal fin very long, anteriorly with many spines; ventrals abdominal, with more than five soft and several unarticulated rays.

Notacanthus is the most aberrant type of Acanthopterygians. Of the characteristics of this order the development of spines in the vertical fins is the only one preserved in the fishes of this genus. Their body is elongate, covered with very small scales; the snout protrudes beyond the mouth. Eyes lateral, of moderate size; dentition feeble. Five species are known from the Arctic Ocean, Mediterranean, Atlantic, and Southern Pacific. They inhabit considerable depths, probably from 100 to 400 fathoms, and during the "Challenger" expedition specimens have been obtained from an alleged depth of 1875 fathoms.

SECOND ORDER:

ACANTHOPTERYGII PHARYNGOGNATHI.

Part of the rays of the dorsal, anal, and ventral fins are non-articulated spines. The lower pharyngeals coalesced. Air-bladder without pneumatic duct.

Fig. 239.—Coalescent Pharyngeals of Scarus cretensis. *a*, upper; *b*, lower pharyngeals.

First Family—Pomacentridæ.

Body short, compressed, covered with ctenoid scales. Dentition feeble; palate smooth. The lateral line does not extend to the caudal fin, or is interrupted. One dorsal fin, with the spinous portion as well developed as the soft, or more. Two, sometimes three, anal spines; the soft anal similar to the soft dorsal. Ventral fins thoracic, with one spine and five soft rays. Gills three and a half; pseudobranchiæ and air-bladder present. Vertebræ, twelve abdominal and fourteen caudal.

The fishes of this family are marine; they resemble the Chætodonts with regard to their mode of life, living chiefly in the neighbourhood of coral formations. Like them they are beautifully coloured, the same patterns being sometimes reproduced in members of both families, proving that the

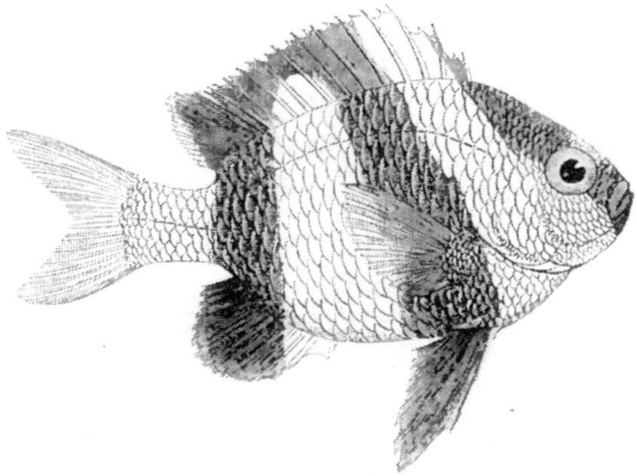

Fig. 240.—Dascyllus aruanus. Natural size, from the Indo-Pacific.

development and distribution of colours is due to the agencies of climate, of the surroundings and of the habits of animals. The geographical range of the *Pomacentridæ* is co-extensive with that of the Chætodonts, the species being most numerous in the Indo-Pacific and Tropical Atlantic, a few extending northwards to the Mediterranean and Japan, southwards to the coasts of South Australia. They feed chiefly on small marine animals, and such as have compressed teeth appear to feed on the small Zoophytes covering the banks, round which these "Coral-fishes" abound. In a fossil state this family is known from a single genus only, *Odonteus*, from Monte Bolca, allied to *Heliastes*. The recent genera belonging to this family are:—*Amphiprion, Premnas, Dascyllus, Lepidozygus, Pomacentrus, Glyphidodon, Parma*, and *Heliastes*. About 120 species are known.

SECOND FAMILY—LABRIDÆ.

Body oblong or elongate, covered with cycloid scales. The lateral line extends to the caudal, or is interrupted. One dorsal fin, with the spinous portion as well developed as, or more than, the soft. The soft anal similar to the soft dorsal. Ventral fins thoracic, with one spine and five soft rays. Palate without teeth. Branchiostegals five or six; gills three and a half; pseudobranchiæ and air-bladder present. Pyloric appendages none; stomach without cæcal sac.

The "Wrasses" are a large family of littoral fishes, very abundant in the temperate and tropical zones, but becoming scarcer towards the Arctic and Antarctic circles, where they disappear entirely. Many of them are readily recognised by their thick lips, which are sometimes internally folded, a peculiarity which has given to them the German term of "Lip-fishes." They feed chiefly on mollusks and crustaceans, their dentition being admirably adapted for crushing hard substances. Many species have a strong curved tooth at the posterior extremity of the intermaxillary, for the purpose of pressing a shell against

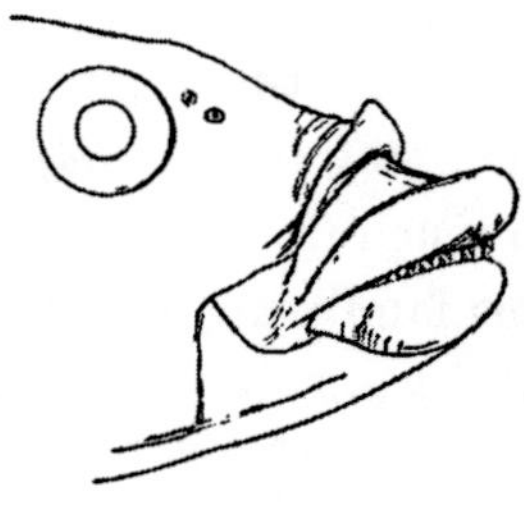

Fig. 241.—Lips of a Wrasse, *Labrus festivus*.

the lateral and front teeth by which it is crushed. Other Wrasses feed on corals, others on zoophytes; a few are herbivorous. In all Wrasses the upper pharyngeal bones seem to be jointed to the basi-occipital; but whilst in *Labrus* the basi-occipital is raised on each side into a large flattish condyle, fitting into a concavity of the upper pharyngeals, in *Scarus* the mode of articulation is reversed, the basi-occipital having a pair of long grooves, in which the oblong condyles of the upper pharyngeals slide forwards and backwards. Beautiful colours prevail in this family, permanent pigmentary colours as well as passing iridescent reflections of the scales. Some species remain very small, others grow to a weight of fifty pounds. The larger kinds especially are prized as food, the smaller less so.

Remains of Labridæ, recognised by their united pharyngeals, which bear molar-like teeth, are not scarce in tertiary formations of France, Germany, Italy, and England. Such remains from Monte Bolca and the Swiss Molasse have been referred to the genus *Labrus*. Others, *Nummopalatus* and *Phyllodus*, are allied, but cannot be assigned, to one of the recent genera; the latter genus is first represented in cretaceous formations of Germany. Another genus, *Taurinichthys*, from the Miocene of France, represents the *Odacina* of the living fauna. *Egertonia*, from the Isle of Sheppey, differs so much from all recent Labroid genera that its pertinence to this family appears doubtful.

[See *J. Cocchi*, Monografia dei Pharyngodopilidæ, 1866; and *E. Sauvage*, Sur le genre Nummopalatus, in Bull. Soc. Geol. France, 1875.]

Labrus.—Body compressed, oblong, covered with scales of moderate size, in more than forty transverse series; snout more or less pointed; imbricate scales on the cheeks and opercles; none or only a few on the interoperculum. Teeth in the jaws conical, in a single series. Dorsal spines numerous, thirteen or twenty-one, none of which are prolonged; anal spines three. Lateral line not interrupted.

Young "Wrasses" differ from mature specimens in having the præoperculum serrated. The headquarters of this genus are the Mediterranean, whence it ranges, gradually diminishing towards the north, along all the shores of Europe. Nine species are known; British are the "Ballan Wrasse" (*L. maculatus*), and the "Striped or Red Wrasse" or "Cook" (*L. mixtus*). The two sexes of the latter species are very differently coloured; the male being generally ornamented with blue streaks, or a blackish band along the body, whilst the female has two or three large black blotches across the back of the tail.

CRENILABRUS are Labrus with serrated præoperculum; the number of their dorsal spines varies from thirteen to eighteen, and the scales are arranged in less than forty transverse series.

The range of this genus is coextensive with *Labrus*. *C. melops*, the "Gold-sinny," or "Cork-wing," is common on the British coasts.

TAUTOGA.—Body compressed, oblong, covered with small scales; scales on the cheek rudimentary, opercles naked. Teeth in the jaws conical, in double series; no posterior canine tooth. Dorsal spines seventeen, anal spines three. Lateral line not interrupted.

The "Tautog," or "Black-fish," is common on the Atlantic coasts of temperate North America, and much esteemed as food.

CTENOLABRUS.—Body oblong, covered with scales of moderate size; imbricate scales on the cheeks and opercles. Teeth in the jaws in a band, with an outer series of stronger conical teeth; no posterior canine tooth. Dorsal spines from sixteen to eighteen; anal spines three. Lateral line not interrupted.

Four species, from the Mediterranean and the temperate parts of the North Atlantic, *Ct. rupestris* being common on the British, and *Ct. burgall* on the North American coasts.

ACANTHOLABRUS.—A Wrasse with five or six anal spines, and with the teeth in a band.

From the Mediterranean and British coasts (*A. palloni*).

CENTROLABRUS.—Wrasses with four or five anal spines, and with the teeth in a single series.

Two species are known from Madeira and the Canary Islands, and one from northern Europe and Greenland. The latter is scarce on the British coasts, but bears a distinct name on the south coast, where it is called "Rock-cook."

LACHNOLAEMUS from the West Indies, and MALACOPTERUS from Juan Fernandez, are Labroids, closely allied to the preceding North Atlantic genera.

COSSYPHUS.—Body compressed, oblong, with scales of moderate size; snout more or less pointed; imbricate scales on the cheeks and opercles; basal portion of the vertical fins scaly. Lateral line not interrupted. Teeth in the jaws in a single series; four canine teeth in each jaw anteriorly; a posterior canine tooth. Formula of the fins: D. $\frac{12}{9-11}$, A. $\frac{3}{12}$.

Twenty species are known from the tropical zone and coasts adjoining it; some, like *C. gouldii* from Tasmania, attain a length of three or four feet.

CHILINUS.—Body compressed, oblong, covered with large scales; lateral line interrupted; cheeks with two series of scales; præoperculum entire; teeth in a single series, two canines in each jaw; no posterior canine tooth; lower jaw not produced backwards. Dorsal spines subequal in length; formula of the fins: D. $\frac{9-10}{10-9}$, A. $\frac{3}{8}$.

Common in the tropical Indo-Pacific, whence more than twenty species are known. Hybrids between the different species of this genus are not uncommon.

EPIBULUS.—Closely allied to the preceding genus, but with a very protractile mouth, the ascending branches of the intermaxillaries, the mandibles, and the tympanic being much prolonged.

This fish (*E. insidiator*) is said to seize marine animals by suddenly thrusting out its mouth and engulphing those that come within the reach of the elongated tube. It attains a length of twelve inches, is common in the tropical Indo-Pacific, and varies much in coloration.

ANAMPSES.—Distinguished by its singular dentition, the two front teeth of each jaw being prominent, directed forwards, compressed, with cutting edge. D. $\frac{9}{12}$, A. $\frac{3}{12}$.

Beautifully coloured fishes from the tropical Indo-Pacific. Ten species.

PLATYGLOSSUS.—Scales in thirty or less transverse series; lateral line not interrupted. A posterior canine tooth. Dorsal spines nine.

Small beautifully coloured Coral-fishes, abundant in the equatorial zone and the coasts adjoining it. Some eighty species are known (inclusive of the allied genera *Stethojulis*, *Leptojulis*, and *Pseudojulis*).

NOVACULA.—Body strongly compressed, oblong, covered with scales of moderate size; head compressed, elevated, obtuse, with the supero-anterior profile more or less parabolic; head nearly entirely naked. Lateral line interrupted. No posterior canine tooth. D. $\frac{9}{12}$, A. $\frac{3}{12}$; the two anterior dorsal spines sometimes remote or separate from the others.

Twenty-six species are known from the tropical zone, and the warmer parts of the temperate zones. They are readily recognised by their compressed, knife-shaped body, and peculiar physiognomy; they scarcely exceed a length of twelve inches.

JULIS.—Scales of moderate size; lateral line not interrupted. Head entirely naked. Snout of moderate extent, not produced; no posterior canine tooth. Dorsal spines ten.

Co-extensive with *Platyglossus* in their geographical distribution, and of like beautiful coloration and similar habits. Some of the most common fishes of the Indo-Pacific, as *J. lunaris*, *trilobata*, and *dorsalis*, belong to this genus.

CORIS.—Scales small, in fifty or more transverse series; lateral line not interrupted. Head entirely naked. Dorsal spines nine.

Twenty-three species, distributed like *Platyglossus;* two reach the south coast of England, *Coris julis* and *C. giofredi*, said to be male and female of the same species. Some belong to the most gorgeously coloured kinds of the whole class of fishes.

Genera allied to the preceding Labroids are—*Chocrops, Xiphochilus, Semicossyphus, Trochocopus, Decodon, Pteragogus, Clepticus, Labrichthys, Labroides, Duymæria, Cirrhilabrus, Doratonotus, Pseudochilinus, Hemigymnus, Gomphosus, Cheilio*, and *Cymolutes.*

PSEUDODAX.—Scales of moderate size; lateral line continuous; cheeks and opercles scaly. Each jaw armed with two pairs of broad incisors, and with a cutting lateral edge; teeth of the lower pharyngeal confluent, pavement-like. Dorsal spines eleven.

One species (*P. moluccensis*) from the East Indian Archipelago.

SCARUS.—Jaws forming a sharp beak, the teeth being soldered together. The lower jaw projecting beyond the upper. A single series of scales on the cheek; dorsal spines stiff, pungent; the upper lip double in its whole circuit. The dentigerous plate of the lower pharyngeal is broader than long.

The fishes of this genus, and the three succeeding, are known by the name of "Parrot-wrasses." Of *Scarus* one species (*S. cretensis*) occurs in the Mediterranean, and nine others in the tropical Atlantic. The first was held in high repute by the ancients, and Aristotle has several passages respecting its rumination. It was most plentiful and of the best quality in the Carpathian Sea, between Crete and Asia Minor, but was not unknown even in early times on the Italian coasts, though Columella says that it seldom passed beyond Sicily in his day. But in the reign of Claudius, according to Pliny, Optatus Elipentius brought it from the Troad, and introduced it into the sea between Ostium and Campagna. For five years all that were caught in the nets were thrown into the sea again, and from that time it was an abundant fish in that locality. In the time of Pliny it was considered to be the first of fishes (*Nunc Scaro datur principatus*); and the expense incurred by Elipentius was justified, in the opinion of the Roman gourmands, by the extreme delicacy of the fish. It was a fish, said the poets, whose very excrements the gods themselves were unwilling to reject. Its flesh was tender, agreeable, sweet, easy of digestion, and quickly assimilated; yet if it happened to have eaten an Aplysia, it produced violent diarrhœa. In short, there is no fish of which so much has been said by ancient writers. In the present day the Scarus of the Archipelago is considered to be a fish of exquisite flavour; and the Greeks still name it *Scaro,* and eat it with a sauce made of its liver and intestines. It feeds on fucus; and Valenciennes thinks that the necessity for masticating its vegetable diet thoroughly, and the working of it with that intent backwards and forwards in the mouth, may have given rise to the notion of its being a ruminant; and it is certain that its aliment is very finely divided when it reaches the stomach.

SCARICHTHYS.—Differing from *Scarus* only in having flexible dorsal spines.

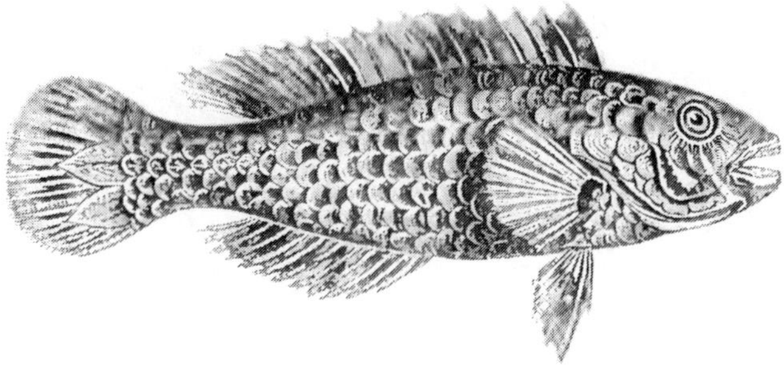

Fig. 242.—Scarichthys auritus.

Two species from the Indo-Pacific.

Callyodon.—Differing from *Scarichthys* in having the upper lip double posteriorly only.

Nine species from the tropical zone.

Pseudoscarus.—Jaws forming a strong beak, the teeth being soldered together. The upper jaw projecting beyond the lower. Two or more series of scales on the cheek. The dentigerous plate of the lower pharyngeal longer than broad.

This tropical genus contains by far the greatest number of Scaroid Wrasses, some seventy species being known, and a still greater number of names being introduced into the various Ichthyological works. They are beautifully coloured, but the colours change with age, and vary in an extraordinary degree in the same species. They rapidly fade after death, so that it is almost impossible to recognise in preserved specimens the species described from living individuals. Many attain to a rather large size, upwards of three feet in length. The majority are eaten, but some acquire poisonous properties from their food, which consists either of corals or of fucus.

Odax.—The edge of each jaw is sharp, without distinct teeth. The dentigerous plate of the lower pharyngeal triangular, much broader than long. Cheeks and opercles scaly; scales of the body small or rather small; lateral line continuous. Snout conical. Dorsal spines rather numerous, flexible.

Six species from the coasts of Australia and New Zea-

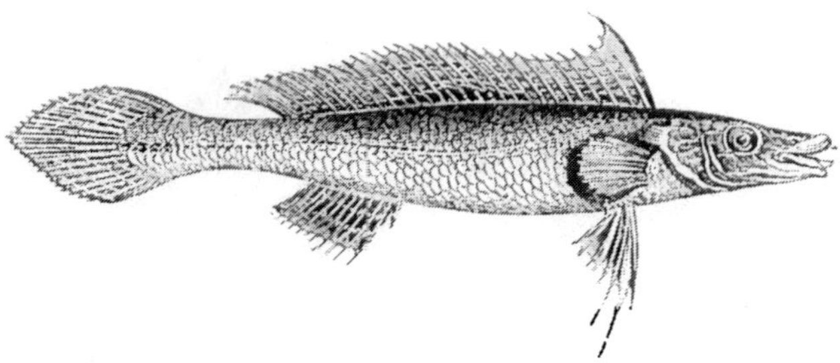

Fig. 243.—Odax radiatus.

land. Small. The species figured (*O. radiatus*) is from Western Australia.

CORIDODAX.—Jaws as in *Odax*, head naked. Scales of the body small; lateral line continuous. Snout of moderate extent. Dorsal spines numerous, flexible.

The "Butter-fish," or "Kelp-fish" of the colonists of New Zealand (*C. pullus*), is prized as food, and attains to a weight of four or five pounds. It feeds on zoophytes, scraping them from the surface of the kelp, with its curiously formed teeth. Its bones are green, like those of *Belone*.

OLISTHEROPS, from King George's Sound, has scales of moderate size, but agrees otherwise with *Coridodax*.

SIPHONOGNATHUS.—Head and body very elongate, snout long, as in *Fistularia*; upper jaw terminating in a long, pointed, skinny appendage; opercles and cheeks scaly; scales of moderate size; lateral line continuous. Dorsal spines numerous, flexible. Jaws as in *Odax*; the dentigerous plate of the lower pharyngeal very narrow.

S. argyrophanes, from King George's Sound, is the most aberrant type of Wrasses, whose principal characters are retained, but united with a form of the body which resembles that of a Pipe-fish.

THIRD FAMILY—EMBIOTOCIDÆ.

Body compressed, elevated or oblong, covered with cycloid scales; lateral line continuous. One dorsal fin, with a spinous portion, and with a scaly sheath along the base, which is separated by a groove from the other scales; anal with three spines and numerous rays; ventral fins thoracic, with one spine and five rays. Small teeth in the jaws, none on the palate. Pseudobranchiæ present. Stomach siphonal, pyloric appendages none. Viviparous.

Marine Fishes characteristic of the fauna of the temperate North Pacific, the majority living on the American side, and only a few on the Asiatic. All are viviparous (see Fig. 70, p. 159). Agassiz describes the development of the embryoes as a normal ovarian gestation, the sac containing the young not being the oviduct but the ovarian sheath, which fulfils the functions of the ovary. This organ presents two modes of arrangement: in one there is a series of triangular membranous flaps communicating with each other, between which the young are arranged, mostly longitudinally, the head of one to the tail of another, but sometimes with the bodies curved, to the number of eighteen or twenty; in the other, the cavity is divided by three membranes converging to a point, into four compartments, not communicating with each other except towards the genital opening, the young being arranged in the same longitudinal manner. The proportionate size of the young is very remarkable. In a female specimen 10½ inches long, and 4½ inches high, the young were nearly 3 inches long and 1 inch high. Seventeen species are known, the majority of which belong to *Ditrema*, and one to *Hystcrocarpus*. They do not attain to a large size, varying from three-quarters to three pounds in weight.

FOURTH FAMILY—CHROMIDES.

Body elevated, oblong or elongate, scaly, the scales being generally ctenoid. Lateral line interrupted or nearly so. One dorsal fin, with a spinous portion; three or more anal spines; the soft anal similar to the soft dorsal. Ventral fins thoracic, with one spine and five rays. Teeth in the jaws small, palate smooth. Pseudobranchiæ none. Stomach cœcal; pyloric appendages none.

Freshwater-fishes of rather small size from the tropical parts of Africa and America; one genus from Western India. The species with lobate teeth, and with many circumvolutions of the intestines, are herbivorous, the other carnivorous.

ETROPLUS.—Body compressed, elevated, covered with ctenoid scales of moderate size. Lateral line indistinct. Dorsal and anal spines numerous. Teeth compressed, lobate, in one or two series. Anterior prominences of the branchial arches not numerous, short, conical, hard. Dorsal fin not scaly.

Two species from Ceylon and Southern India.

CHROMIS.—Body compressed, oblong, covered with cycloid scales of moderate size. Dorsal spines numerous, anal spines three. Teeth compressed, more or less lobate, in one series. Anterior prominences of the branchial arches short, thin, lamelliform, non-serrated. Dorsal fin not scaly.

Some twenty species are known from the fresh waters of Africa and Palestine; the most celebrated is the "Bulti," or "Bolty," of the Nile, one of the few well-flavoured fishes of

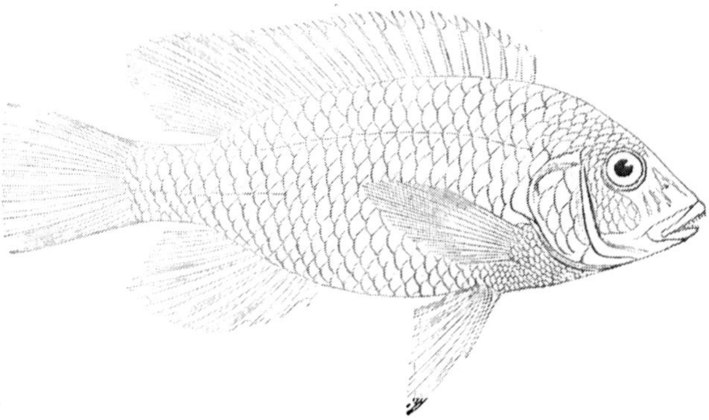

Fig. 244.—*Chromis andreæ*, from the Lake of Galilee.

that river; it grows to the length of twenty inches. Two or three species of this genus occur in the Jordan and Lake of Galilee.

HEMICHROMIS, differing from *Chromis* in having conical teeth in one or two series.

Ten species, the range of which is coextensive with that of *Chromis*. One species, *H. sacra*, is abundant in the Lake of Galilee.

PARETROPLUS, differing from *Hemichromis* in having nine anal spines.

One species from Madagascar.

ACARA.—Body compressed, oblong, covered with ctenoid scales of moderate size. Dorsal spines numerous, anal spines three or four; base of the soft dorsal nearly uncovered by scales. Teeth in a band, small, conical. Anterior prominences of the first branchial arch very short tubercles.

Some twenty species are known from the fresh waters of Tropical America, *A. bimaculata* being one of the most common fishes of that region. All are very small.

HEROS.—Differing from *Acara* in having more than four anal spines.

Some fifty species are known from the fresh waters of Tropical America, especially Central America, where almost every large lake or river is tenanted by one or more peculiar

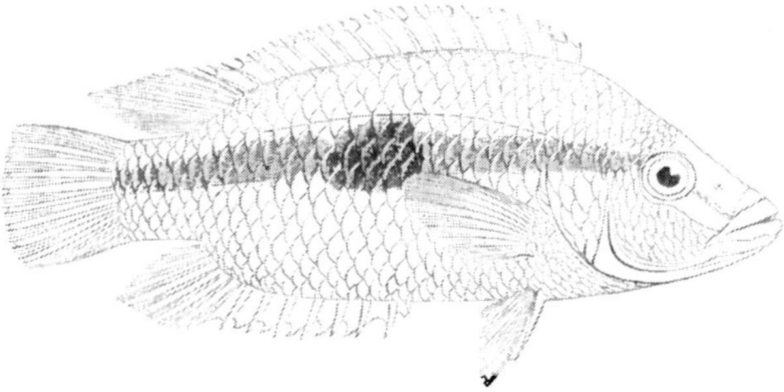

Fig. 245.—*Heros salvini*, from Central America.

species. They are of rather small size, rarely exceeding a length of twelve inches.

Genera allied to *Heros*, and likewise from Tropical America, are *Neetroplus, Mesonauta, Petenia, Uaru*, and *Hygrogonus*.

CICHLA.—Form of the body perch-like. Scales small; the spinous and soft portions of the dorsal fin of nearly equal extent, and separated by a notch; anal spines three. Each jaw with a broad band of villiform teeth. The outer branchial arch with lanceolate crenulated prominences along its concave side. Dorsal and anal fins scaly.

Four species from Brazil, Guyana, and Peru.

CRENICICHLA.—Body low, sub-cylindrical; scales small or rather small. The spinous portion of the dorsal is much more developed than the soft, both being continuous, and not separated by a notch; anal spines three. Præopercular margin serrated. Each jaw with a band of conical teeth. The outer branchial arch with short tubercles. Dorsal and anal fins naked.

Ten species from Brazil and Guyana.

The following genera complete the list of South American Chromides: *Chætobranchus, Mesops, Satanoperca, Geophagus, Symphysodon,* and *Pterophyllum*.

THIRD ORDER—ANACANTHINI.

Vertical and ventral fins without spinous rays. The ventral fins, if present, are jugular or thoracic. Air-bladder, if present, without pneumatic duct.

These characters are common to all the members of this order, with the exception of a freshwater-fish from Tasmania and South Australia (*Gadopsis*), which has the anterior portion of the dorsal and anal fins formed of spines.

FIRST DIVISION—ANACANTHINI GADOIDEI.

Head and body symmetrically formed.

FIRST FAMILY—LYCODIDÆ.

Vertical fins confluent. Ventral fin, if present, small, attached to the humeral arch, jugular. Gill-opening narrow, the gill-membrane being attached to the isthmus.

Marine littoral fishes of small size, resembling Blennies, chiefly represented in high latitudes, but a few living within the tropical zone.

LYCODES.—Body elongate, covered with minute scales imbedded in the skin, or naked; lateral line more or less indistinct. Eye of moderate size. Ventral small, short, rudimentary, jugular, composed of several rays. Upper jaw overlapping the lower. Conical teeth in the jaws, on the vomer, and on the palatine bones. Barbel none. Five or six branchiostegals; gill-opening narrow, the gill-membranes being attached to the isthmus. Pseudobranchiæ present. Air-bladder none. Pyloric appendages two, or rudimentary, or entirely absent. No prominent anal papilla.

Fig. 246.—Lycodes mucosus, from Northumberland Sound.

Nine species are known from the Arctic Ocean, four from the southern extremity of the American continent.

GYMNELIS.—Body elongate, naked. Eye of moderate size or rather small. Ventrals none. Vent situated at some distance backwards from the head. Small conical teeth in the jaws, on the vomer and palatine bones. Jaws equal anteriorly. Barbel none. Six branchiostegals; gill-opening narrow, the gill membranes being attached to the isthmus. Pseudobranchiæ present; air-bladder none. Pyloric appendages two; no prominent anal papilla.

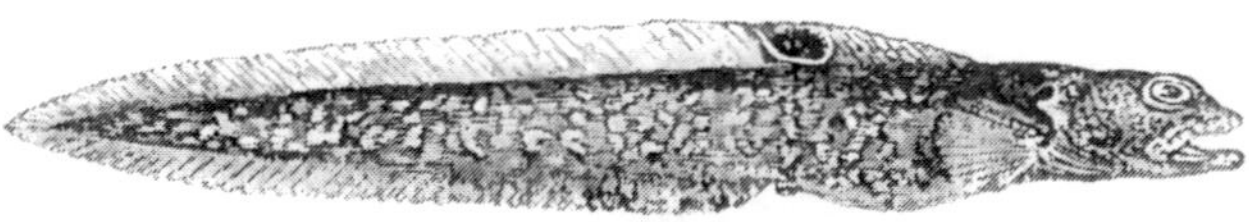

Fig. 247.—Gymnelis viridis.

One species (*G. viridis*) from Greenland, the other (*G. pictus*) from the Straits of Magelhæn.

The other genera belonging to this family are *Uronectes* from Baffin's Bay, *Microdesmus* from Panama, *Blennodesmus* from the coast of North-Eastern Australia, and *Maynea* from the Straits of Magelhæn.

SECOND FAMILY—GADIDÆ.

Body more or less elongate, covered with small smooth scales. One, two, or three dorsal fins, occupying nearly the whole of the back; rays of the posterior dorsal well developed; one or two anal fins. Caudal free from dorsal and anal, or, if they are united, the dorsal with a separate anterior portion. Ventrals jugular, composed of several rays, or, if they are reduced to a filament, the dorsal is divided into two. Gill-opening wide; the gill-membranes generally not attached to the isthmus. Pseudobranchiæ none, or glandular, rudimentary. An air-bladder and pyloric appendages generally present.

The family of "Cod-fishes" consists partly of littoral and surface species (and they form the majority), partly of deep-sea forms. The former are almost entirely confined to the temperate zones, extending beyond the Arctic Circle; the latter have, as deep-sea fishes generally, a much wider range, and hitherto have been found chiefly at considerable depths of rather low latitudes. Only two or three species inhabit fresh waters. They form one of the most important articles of food and subsistence to the fishermen in Europe and North America, and to whole tribes bordering upon the Arctic Ocean.

Fossil remains are scarce. *Nemopteryx* and *Palæogadus* have been described from the schists of Glaris, a formation believed to have been the bottom of a very deep sea. In the clay of Sheppey species occur allied to *Gadus, Merluccius,* and *Phycis;* others, not readily determinable, have been found at Licata in Sicily (Miocene).

GADUS.—Body moderately elongate, covered with small scales. A separate caudal, three dorsal, and two anal fins; ventrals narrow, composed of six or more rays. Teeth in the upper jaw in a narrow band; vomerine teeth; none on the palatines.

Arctic and temperate zones of the Northern Hemisphere. Eighteen species are known, of which the following are the most important:—

Gadus morrhua, the common "Cod-fish"—in German called "Kabeljau" when fresh and old, "Dorsch" when young and fresh, "Stock-fish" when dried, "Labberdan" when salted—measures from two to four feet, and attains to a weight of one hundred pounds. On the British coasts and in the German Ocean it is generally of a greenish or brownish-olive colour, with numerous yellowish or brown spots. Farther northwards darker-coloured specimens, frequently without any spots, predominate; and on the Greenland, Iceland, and North Scandinavian coasts the Cod have often a large irregular black blotch on the side. The Cod-fish occurs between 50° and 75° lat. N., in great profusion, to a depth of 120 fathoms, but is not found nearer the Equator than 40° lat. Close to the coast it is met with singly all the year round, but towards the spawning-time it approaches the shore in numbers, which happens in January in England and not before May on the American coasts. The English resorted to the cod-fisheries of Iceland before the year 1415, but since the sixteenth century most vessels go to the banks of Newfoundland, and almost all the preserved Cod consumed during Lent in the various continental countries is imported from across the Atlantic. At one time the Newfoundland cod-fishery rivalled in importance the whale-fishery and the fur trade of North America. Cod-liver oil is prepared from the liver on the Norwegian coast, but also other species of this genus contribute to this most important drug.

Gadus tomcodus abundantly occurs on the American coasts; it remains within smaller dimensions than the common Cod-fish. *Gadus æglefinus*, the "Haddock" ("Schell-fisch" of the Germans, "Hadot" of the French), is distinguished by a black lateral line and a blackish spot above the pectoral fin. It attains to a length of three feet in the higher latitudes, but remains smaller on more southern coasts; like the Cod it extends across the Atlantic. The largest specimens are taken on the British coast in winter, because at that time they leave the deep water to spawn on the coast. *Gadus merlangus*, the "Whiting," with a black spot in the axil of the pectoral fin. *Gadus luscus*, the "Bib," "Pout," or "Whiting-pout," with cross-bands during life, and with a black axillary spot, rarely exceeding a weight of five pounds. *Gadus fabricii*, a small species, but occurring in incredible numbers on the shores near the Arctic circle, and ranging to 80° lat. N. *Gadus pollachius*, the "Pollack," without a barbel at the chin, and with the lower jaw projecting beyond the upper. *Gadus virens*, the "Coal-fish," valuable on account of its size and abundance, and therefore preserved for export like the Cod.

The fishes of the genus *Gadus* are bathymetrically succeeded by several genera, as *Gadiculus*, *Mora*, and *Strinsia*; however these do not descend to sufficiently great depths to be included into the deep-sea Fauna; the two following are true deep-sea fishes.

HALARGYREUS.—Body elongate, covered with small scales. Two dorsal and two anal fins; ventrals composed of several rays. Jaws with a band of minute villiform teeth; vomer and palatines toothless. No barbel.

The single species known, *H. johnsonii*, proves to be a deep-sea fish by its organisation as well as geographical distribution. Originally known from a single specimen, which was obtained at Madeira, it has since been found off the coast of New Zealand. There is no doubt that it will be discovered also in intermediate seas.

MELANONUS.—Head and body rather compressed, covered with cycloid scales of moderate size, and terminating in a long tapering tail, without caudal. Eye of moderate size. Villiform teeth in the jaws, on the vomer and palatine bones. Barbel none. A short anterior dorsal, the second extending to the end of the tail, and the anal being of similar length. Ventrals composed of several rays. Bones soft and flexible.

This is one of the discoveries made during the expedition of the "Challenger." The single specimen obtained is of a deep-black colour, and was dredged up at a depth of 1975 fathoms in the Antarctic Ocean.

MERLUCCIUS.—Body elongate, covered with very small scales. A separate caudal; two dorsal fins and one anal; ventrals well developed, composed of seven rays. Teeth in the jaws and on the vomer rather strong, in double or triple series. No barbel.

Two species are known of this genus, widely separated in their distribution. The European species, *M. vulgaris*, the "Hake," is found on both sides of the Atlantic, and grows to a length of four feet. It is caught in great numbers, and preserved as "Stock-fish." The second species *M. gayi*, is common in the Straits of Magellhæn and on the coast of Chili, less so in New Zealand.

The vertebral column of this genus shows a singular modification of the apophyses. The neural spines of all the abdominal vertebræ are extremely strong, dilated, wedged into one another. The parapophyses of the third to sixth vertebræ are slender, styliform, whilst those of all the following abdominal vertebræ are very long and broad, convex on the upper and concave on the lower surface; the two or three anterior pairs are, as it were, inflated. The whole forms a strong roof for the air-bladder, reminding us of a similar structure in *Kurtus*.

PSEUDOPHYCIS.—Body of moderate length, covered with rather small scales. A separate caudal, two dorsals, and one anal; ventral fins very narrow and styliform, but composed of several rays. Jaws with a band of small teeth; vomer and palatines toothless. Chin with a barbel.

Two species, of which *Ps. bachus* is common on the coast of New Zealand.

Allied genera are *Lotella, Physiculus, Uraleptus,* and *Læmonema,* from moderate depths, obtained chiefly off Madeira and the Southern Temperate Zone.

Fig. 248.—Pseudophycis bachus.

PHYCIS.—Body of moderate length, covered with small scales. Fins more or less enveloped in loose skin. A separate caudal; two dorsal fins and one anal; the anterior dorsal composed of from eight to ten rays; ventrals reduced to a single long ray, bifid at its end. Small teeth in the jaws and on the vomer; palatine bones toothless. Chin with a barbel.

Six species from the temperate parts of the North Atlantic and the Mediterranean, one, *Ph. blennioides,* is occasionally found on the British coast.

HALOPORPHYRUS.—Body elongate, covered with small scales. A separate caudal, two dorsal fins, and one anal; the first dorsal with four rays; ventrals narrow, composed of six rays. Jaws and vomer with villiform teeth; palatine bones toothless. Chin with a barbel.

A small genus of deep-sea fishes, of which three species are known. They offer a striking instance of the extraordinary distribution of deep-sea fishes; *H. lepidion* occurs in from 100 to 600 fathoms in the Mediterranean and the neighbouring parts of the Atlantic, off the coast of Japan, and various parts of the South Atlantic; *H. australis* in from 55 to 70 fathoms in the Straits of Magelhæn; and finally *H. rostratus* in from 600 to 1375 fathoms, midway between the Cape of Good Hope and Kerguelen's Land, and in the South Atlantic.

LOTA.—Body elongate, covered with very small scales. A separate caudal, two dorsal fins, and one anal; ventrals narrow, composed of six rays. Villiform teeth in the jaws and on the vomer; none on the palatines. The first dorsal with from ten to thirteen well-developed rays. Chin with a barbel.

The "Burbot," or "Eel-pout" (*L. vulgaris*, Fig. 8, p. 43), is a Freshwater-fish which never enters salt water. It is locally distributed in Central and Northern Europe and North America; it is one of the best Freshwater-fishes, and exceeds a length of three feet.

MOLVA.—Differs from *Lota* in having several large teeth in the lower jaw and on the vomer.

The "Ling" (*M. vulgaris*) is a very valuable species, common on the northern coasts of Europe, Iceland, and Greenland; and generally found from three to four feet long. The larger number of the specimens caught are cured and dried.

MOTELLA.—Body elongate, covered with minute scales. A separate caudal. Two dorsal fins, the anterior of which is reduced to a narrow rayed fringe, more or less concealed in a longitudinal groove; the first ray is prolonged. One anal fin. Ventrals composed of from five to seven rays. A band of teeth in the jaws and on the vomer.

Eight species of "Rocklings" are known from the coasts of Europe, Iceland, Greenland, Japan, the Cape of Good Hope, and New Zealand. They are of small size, and chiefly distinguished by the number of their barbels. British are the Five-bearded Rockling (*M. mustela*), the Three-bearded Rocklings (*M. tricirrhata*, *macrophthalma*, and *maculata*), and the Four-bearded Rockling (*M. cimbria*). *M. macrophthalma* comes from a depth of from 80 to 180 fathoms. The young are known as "Mackerel Midge" (*Couchia*), and sometimes met with in large numbers at some distance from the coast.

RANICEPS.—Head large, broad, and depressed; body of moderate length, covered with minute scales. A separate caudal.

Two dorsal fins, the anterior of which is very short, rudimentary. One anal fin. Ventrals composed of six rays. Card-like teeth in the jaws and on the vomer.

The "Trifurcated Hake," *R. trifurcus*, not uncommon on the coasts of Northern Europe.

BREGMACEROS.—Body fusiform, compressed posteriorly, covered with cycloid scales of moderate size. Two dorsal fins; the anterior reduced to a single long ray on the occiput; the second and the anal much depressed in the middle; ventrals very long, composed of five rays. Teeth small.

A dwarf Gadoid, the only one found at the surface between

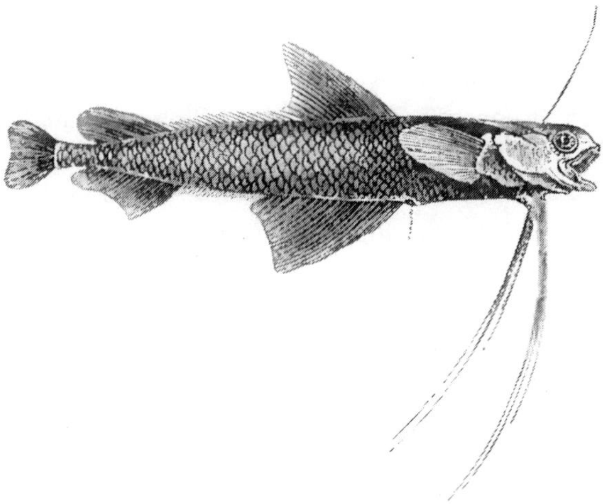

Fig. 249.—Bregmaceros macclellandii.

the Tropics. *B. macclellandii* scarcely exceeds three inches in length, is not uncommon in the Indian Ocean, and has found its way to New Zealand; specimens have been picked up in mid-ocean.

MURÆNOLEPIS.—Body covered with lanceolate epidermoid productions, intersecting each other at right angles like those of a Freshwater-eel. Vertical fins confluent, no caudal being discernible; an anterior dorsal fin is represented by a single filamentous ray; ventral fins narrow, composed of several rays. A barbel. Jaws with a band of villiform teeth; palate toothless.

One species (*M. marmoratus*) from Kerguelen's Land.

CHIASMODUS.—Body naked; stomach and abdomen distensible. Two dorsal fins and one anal; a separate caudal; ventral fins rather narrow, with several rays. Upper and lower jaws with two series of large pointed teeth, some of the anterior being very large and movable; teeth on the palatine bones, but none on the vomer. Chin without barbel.

This Gadoid (*Ch. niger*, Fig. 111, p, 311), inhabits great depths in the Atlantic (to 1500 fathoms). The specimen figured was taken with a large Scopeloid in its stomach.

BROSMIUS.—Body moderately elongate, covered with very small scales. A separate caudal, one dorsal, and one anal; ventrals narrow, composed of five rays. Vomerine and palatine teeth. A barbel.

The "Torsk" (*B. brosme*) is confined to the northern parts of the temperate zone, and probably extends to the arctic circle.

THIRD FAMILY—OPHIDIIDÆ.

Body more or less elongate, naked, or scaly. Vertical fins generally united; no separate anterior dorsal or anal; dorsal occupying the greater portion of the back. Ventral fins rudimentary or absent, jugular. Gill-openings wide, the gill-membranes not attached to the isthmus.

Marine fishes (with the exception of *Lucifuga*), partly littoral, partly bathybial. They may be divided into five groups.

I. *Ventral fins present, attached to the humeral arch:* BROTULINA.

BROTULA.—Body elongate, covered with minute scales. Eye of moderate size. Each ventral reduced to a single filament, sometimes bifid at its extremity. Teeth villiform; snout with barbels. One pyloric appendage.

Five species of small size from the Tropical Atlantic and Indian Ocean.

Lucifuga are *Brotula* organised for a subterranean life. The eye is absent, or quite rudimentary, and covered by the skin; the barbels of Brotula are replaced by numerous minute

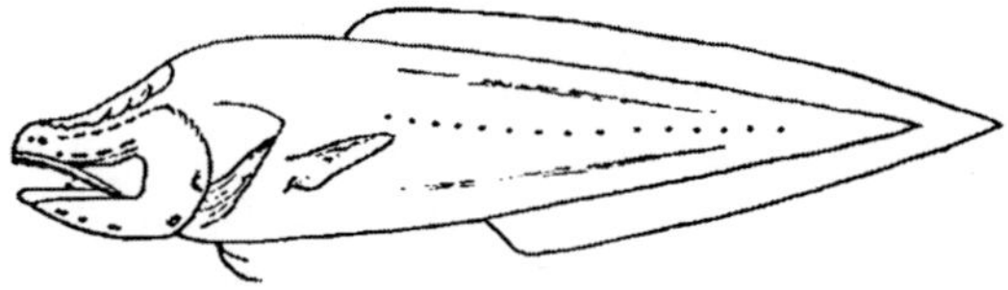

Fig. 250.—Lucifuga dentata, from caves in Cuba.

ciliæ or tubercles. They inhabit the subterranean waters of caves in Cuba, and never come to the light.

BATHYNECTES.—Body produced into a long tapering tail, without caudal. Mouth very wide, villiform teeth in the jaws, on the vomer and palatine bones. Barbel none. Ventral fins reduced to simple or bifid filaments, placed close together, and near to the humeral symphysis. Gill-membranes not united; gill-laminæ remarkably short. Bones of the head soft and cavernous; operculum with a very feeble spine above.

Deep-sea fishes, inhabiting depths varying from 1000 to 2500 fathoms. Three species are known, the largest specimen obtained being seventeen inches long.

ACANTHONUS.—Head large and thick, armed in front and on the opercles with strong spines; trunk very short, the vent being below the pectoral; tail thin, strongly compressed, tapering,

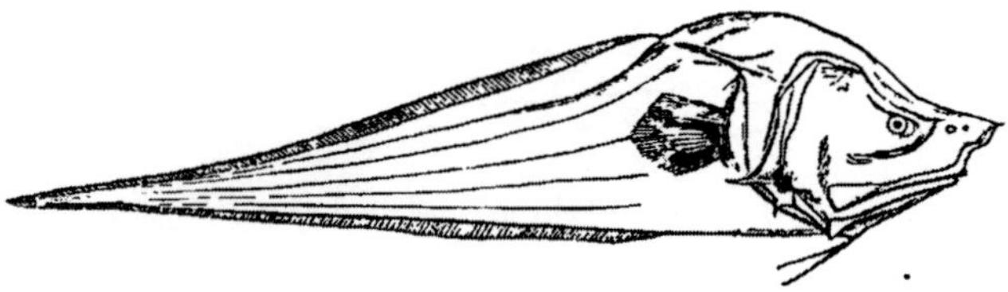

Fig. 251.—Acanthonus armatus.

without caudal. Eye small. Mouth very wide; villiform teeth in the jaws, on the vomer and palatine bones. Barbel none.

Ventrals reduced to simple filaments placed close together on the humeral symphysis. Scales extremely small. Bones of the head soft.

Only two specimens, thirteen inches long, of this remarkable deep-sea form have been obtained, at a depth of 1075 fathoms, in the Indian Ocean.

TYPHLONUS.—Head large, compressed, with most of the bones in a cartilaginous condition; the superficial bones with large muciferous cavities, not armed. Snout a thick protuberance projecting beyond the mouth, which is rather small and inferior. Trunk very short, the vent being below the pectoral; tail thin, strongly compressed, tapering, without separate caudal. Eye externally not visible. Villiform teeth in the jaws, on the vomer and palatine bones. Barbel none. Scales thin, deciduous, small.

Also of this deep-sea fish two specimens only are known, 10 inches long, from a depth of 2200 fathoms in the Western Pacific.

APHYONUS.—Head, body, and tapering tail strongly compressed, enveloped in a thin, scaleless, loose skin. Vent far behind the pectoral. Snout swollen, projecting beyond the wide mouth. No teeth in the upper jaw, small ones in the lower. No externally visible eye. Barbel none. Head covered with a system of wide muciferous channels, the dermal bones being almost membranaceous, whilst the others are in a semi-cartilaginous condition. Notochord persistent, but with a superficial indication of vertebral segments.

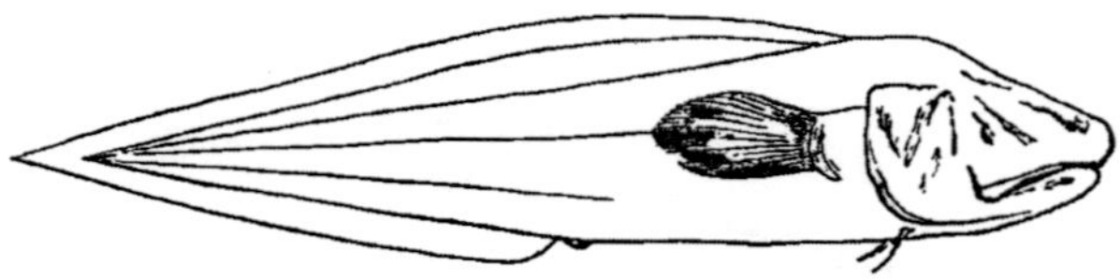

Fig. 252.—Aphyonus gelatinosus.

One specimen only of this most remarkable form is known; it is $5\frac{1}{2}$ inches long, and was obtained at a depth of 1400 fathoms south of New Guinea.

Of the remaining genera belonging to this group, *Brotulophis*, *Halidesmus*, *Dinematichthys*, and *Bythites* are surface forms; *Sirembo* and *Pteridium* inhabit moderate depths; *Rhinonus* is a deep-sea fish.

II. *Ventral fins replaced by a pair of bifid filaments (barbels) inserted below the glosso-hyal:* OPHIDIINA.

OPHIDIUM.—Body elongate, compressed, covered with very small scales. Eye of moderate size. All the teeth small.

Small fishes from the Atlantic and Pacific. Seven species are known, differing from one another in the structure of the air-bladder (see p. 145).

GENYPTERUS is a larger form of *Ophidium*, in which the outer series of teeth in the jaws and the single palatine series contains strong teeth.

Three species from the Cape of Good Hope, South Australia, New Zealand, and Chili are known. They grow to a length of five feet, and have an excellent flesh, like cod, well adapted for curing. At the Cape they are known by the name of "Klipvisch," and in New Zealand as "Ling" or "Cloudy Bay Cod."

III. *No ventral fins whatever; vent at the throat:* FIERASFERINA.

These fishes (*Fierasfer* and *Encheliophis*) are of very small size and eel-like in shape; the ten species known are found in the Mediterranean, Atlantic, and Indo-Pacific. As far as is known they live parasitically in cavities of other marine animals, accompany Medusæ, and more especially penetrate into the respiratory cavities of Star-fishes and Holothurians. Not rarely they attempt other animals less suited for their habits, as, for instance, Bivalves; and cases are known in which they have been found imprisoned below the mantle of the Mollusk, or covered over with a layer of the pearly substance secreted by it. They are perfectly harmless to their host, and merely seek for themselves a safe habitation, feeding on the animalcules which enter with the water the cavity inhabited by them.

IV. *No ventral fins whatever; vent remote from the head; gill-openings very wide, the gill-membranes not being united:* AMMODYTINA.

The "Sand-eels" or "Launces" (*Ammodytes*) are extremely common on sandy shores of Europe and North America. They live in large shoals, rising as with one accord to the surface, or diving to the bottom, where they bury themselves with incredible rapidity in the sand. They are much sought after for bait by fishermen, who discover their presence on the surface by watching the action of Porpoises which feed on them. These Cetaceans, when they meet with a shoal, know how to keep it on the surface by diving below and swimming round it, thus destroying large numbers of them. The most common species on the British coast is the Lesser Sand-eel (*A. tobianus*); the Greater Sand-eel (*A. lanceolatus*), which attains to a length of eighteen inches; *A. siculus*, from the Mediterranean, scarcer in British seas. Two species live on the American coasts, *A. americanus* and *A. dubius;* one in California, *A. personatus. Bleekeria* from Madras is the second genus of this group.

V. *No ventral fins whatever; vent remote from the head;*

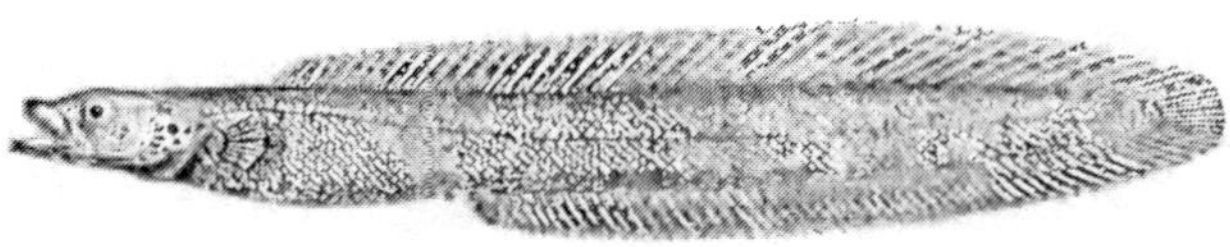

Fig. 253.—Congrogadus subducens.

gill-openings of moderate width, the gill-membranes being united below the throat, not attached to the isthmus: CONGROGADINA.

Only two fishes belong to this group—*Congrogadus* from the Australian coasts, and *Haliophis* from the Red Sea.

Fourth Family—Macruridæ.

Body terminating in a long, compressed, tapering tail,

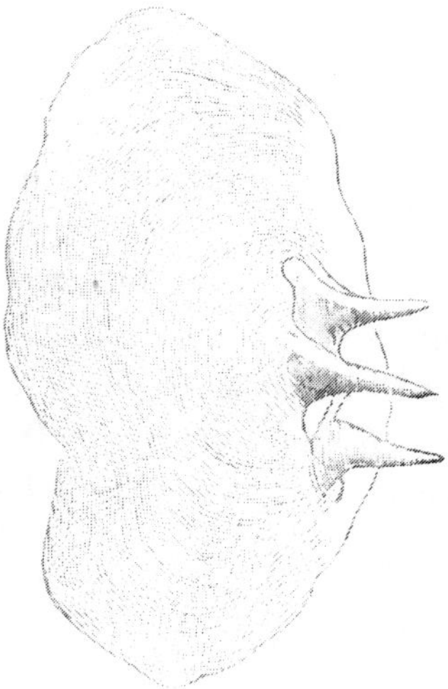

Fig. 254.—Scale of Macrurus trachyrhynchus.

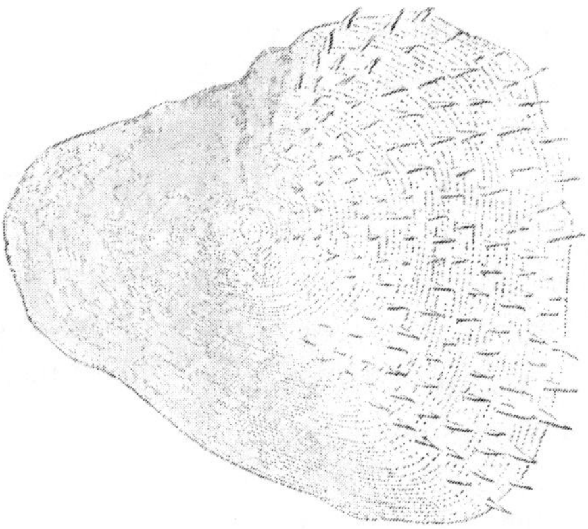

Fig. 255.—Scale of Macrurus cœlorhynchus.

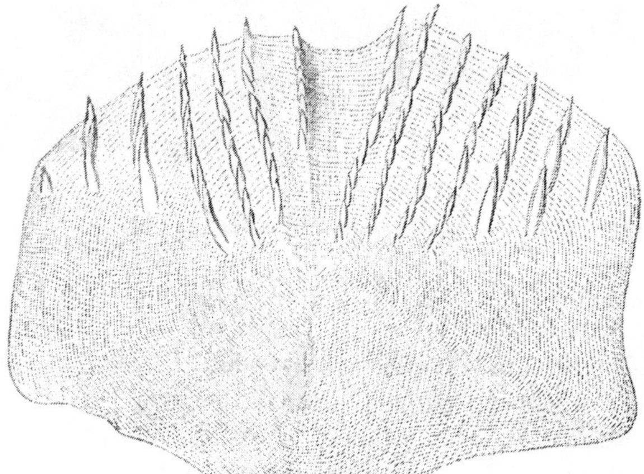

Fig. 256.—Scale from the lateral line of Macrurus australis.

covered with spiny, keeled, or striated scales. One short ante-

rior dorsal; the second very long, continued to the end of the tail, and composed of very feeble rays; anal of an extent similar to that of the second dorsal; no caudal. Ventral fins thoracic or jugular, composed of several rays.

This family, known a few years ago from a limited number of examples, representing a few species only, proves to be one which is distributed over all oceans, occurring in considerable variety and great abundance at depths of from 120 to 2600 fathoms. They are, in fact, Deep-sea Gadoids, much resembling each other in the general shape of their body, but differing in the form of the snout and in the structure of their scales. About forty species are known, of which many attain a length of three feet. They have been referred to the following genera:—

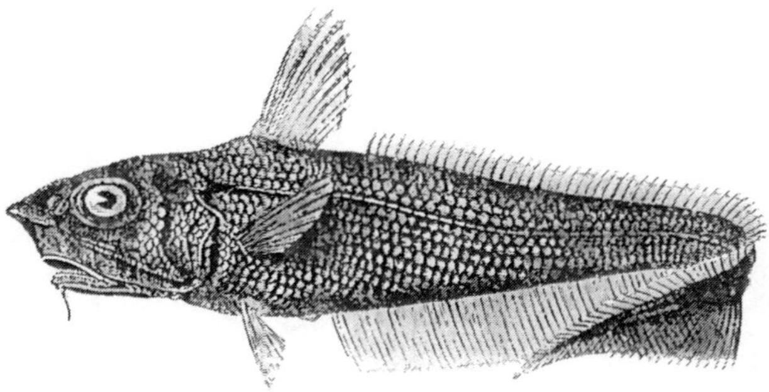

Fig. 257.—Macrurus australis.

MACRURUS.—Scales of moderate size; snout produced, conical; mouth inferior.

CORYPHÆNOIDES.—Scales of moderate size; snout obtuse, obliquely truncated; cleft of the mouth lateral.

MACRURONUS.—Scales of moderate size, spiny; snout pointed; mouth anterior and lateral, with the lower jaw projecting.

MALACOCEPHALUS.—Scales very small, ctenoid; snout short, obtuse, obliquely truncated.

BATHYGADUS.—Scales small, cycloid; snout not projecting beyond the mouth; mouth wide, anterior, and lateral.

Atelcopus from Japan and *Xenocephalus* from New Ireland are genera belonging to the Gadoid Anacanths, but are very imperfectly known.

Second Division—Anacanthini Pleuronectoidei.

Head and part of the body unsymmetrically formed.

This division consists of one family only:

Pleuronectidæ.

The fishes of this family are called "Flat-fishes," from their strongly compressed, high, and flat body; in consequence of the absence of an air-bladder, and of the structure of their paired fins, they are unable to maintain their body in a vertical position, resting and moving on one side of the body only. The side turned towards the bottom is sometimes the left, sometimes the right, colourless, and termed the "blind" side; that turned upwards and towards the light is variously, and in some tropical species even vividly, coloured. Both eyes are on the coloured side, on which side also the muscles are more strongly developed. The dorsal and anal fins are exceedingly long, without division. All the Flat-fishes undergo remarkable changes with age, which, however, are very imperfectly known and not yet fully understood, from the difficulty of referring larval forms to their respective parents. The larvæ are, singularly enough, much more frequently met in the open ocean than near the coast; they are transparent, like *Leptocephali;* perfectly symmetrical, with an eye on each side of the head, and swim in a vertical position like other fishes. The manner in which one eye is transferred from the blind to the coloured side is subject to discussion. Whilst some naturalists believe that the eye turning round its axis pushes its way through the yielding bones from the blind to the upper side, others hold that, as soon as the body of the fish commences to rest on one side only, the eye of that side, in its tendency to turn towards the light, carries the surrounding parts of the head with it; in fact, the whole of the forepart of the head is twisted towards the coloured side, which is a process of but little difficulty as long as the framework of the head is still cartilaginous.

Flat-fishes when adult live always on the bottom, and swim with an undulating motion of their body. Sometimes they rise to the surface; they prefer sandy bottom, and do not descend to any considerable depth. They occur in all seas, except in the highest latitudes and on rocky, precipitous coasts, becoming most numerous towards the equator; those of the largest size occur in the temperate zone. Some enter fresh water freely, and others have become entirely acclimatised in ponds and rivers. All are carnivorous.

Flat-fishes were not abundant in the tertiary epoch; the only representative known is a species of *Rhombus* from Monte Bolca.

The size and abundance of Flat-fishes, and the flavour of the flesh of the majority of the species, render this family one of the most useful to man; and especially on the coasts of the northern temperate zone, their capture is one of the most important sources of profit to the fishermen.

PSETTODES.—Mouth very wide, the maxillary being more than one-half of that of the head. Each jaw armed with two series of long, slender, curved, distant teeth, the front teeth of the inner series of the lower jaw being the longest, and received in a groove before the vomer; vomerine and palatine teeth. The dorsal fin commences on the nape of the neck.

This genus fitly heads the list of Flat-fishes, having retained more of symmetrical structure than the other members of the family, and, therefore, their eyes are as often found on the right as on the left side. It seems to swim, not unfrequently, in a vertical position. Only one species is known, *Ps. erumei*, common in the Indian Ocean.

HIPPOGLOSSUS.—Eyes on the right side; mouth wide, the length of the maxillary being one-third of that of the head. Teeth in the upper jaw in a double series; the anterior of the upper jaw and the lateral of the lower strong. The dorsal fin commences above the eye.

The "Holibut" (*H. vulgaris*) is the largest of all Flat-fishes, attaining to a length of five and six feet, and a weight of several hundredweights. It is found along the northern coasts of Europe, on the coasts of Kamtschatka and California, particularly frequenting banks situated at some distance from the coast, and at a depth of 50 to 120 fathoms.

Other genera, with nearly symmetrical mouth, in which the dorsal fin commences above the eye, are *Hippoglossoides* (the "Rough Dab") and *Tephritis*.

RHOMBUS.—Eyes on the left side. Mouth wide, the length of the maxillary being more than one-third of that of the head. Each jaw with a band of villiform teeth, without canines; vomerine teeth, none on the palatines. The dorsal fin commences on the snout. Scales none or small.

Seven species from the North Atlantic and Mediterranean, of which the most noteworthy are the "Turbot," *Rh. maximus*, one of the most valued food-fishes, and growing to a length of three feet; the "Turbot of the Black Sea," *Rh. mæoticus*, the body of which is covered with bony, conical tubercles, which are as large as the eye; the "Brill," *Rh. lævis*, represented on the North American coasts by *Rh. aquosus;* the "Whiff," or "Mary-sole," or "Sail-fluke," *Rh. megastoma;* "Bloch's Top-knot," *Rh. punctatus* (described by Yarrell as *Rh. hirtus*, and often confounded with the following species).

PHRYNORHOMBUS, differing from *Rhombus* in lacking vomerine teeth. The scales are very small and spiny.

The "Top-knot" (*Ph. unimaculatus*) occurs occasionally

on the south coast of England, and is more common in the Mediterranean; it is a small species.

ARNOGLOSSUS.—Mouth wide, the length of the maxillary being more or not much less than one-third of that of the head. Teeth minute, in a single series in both jaws; vomerine or palatine teeth none. The dorsal fin commences on the snout. Scales of moderate size, deciduous; lateral line with a strong curve above the pectoral. Eyes on the left side.

Seven species from European and Indian Seas. The "Scald-fish" (*A. laterna*) is common in the Mediterranean, and extends to the south coast of England; it is a small species.

PSEUDORHOMBUS.—Mouth wide, the length of the maxillary being more than one-third of that of the head. Teeth in both jaws in a single series, of unequal size; vomerine or palatine teeth none. The dorsal fin commences on the snout. Scales small; lateral line with a strong curve anteriorly. Eyes on the left side. Interorbital space not concave.

A tropical genus with a few outlying species, represented chiefly in the Indo-Pacific, and also in the Atlantic. Seventeen species.

RHOMBOIDICHTHYS.—Mouth of moderate width or small. Teeth minute, in a single or double series; vomerine or palatine teeth none. Eyes separated by a concave more or less broad space. The dorsal fin commences on the snout. Scales ciliated; lateral line with a strong curve anteriorly. Eyes on the left side.

A tropical genus, but also represented in the Mediterranean and on the coast of Japan. Sixteen species, the majority of which are prettily coloured and ornamented with ocellated spots; in some species the adult males have some of the fin-rays prolonged into filaments.

Other genera with nearly symmetrical mouth, in which the dorsal fin commences before the eye, on the snout, are *Citharus*, *Anticitharus*, *Brachypleura*, *Samaris*, *Psettichthys*, *Citharichthys*, *Hemirhombus*, *Paralichthys*, *Liopsetta*, *Lophonectes*, *Lepidopsetta*, and *Thysanopsetta*.

PLEURONECTES.—Cleft of the mouth narrow, with the dentition much more developed on the blind side than on the coloured. Teeth in a single or in a double series, of moderate size; palatine and vomerine teeth none. The dorsal fin commences above the eye. Scales very small or entirely absent. Eyes generally on the right side.

This genus is characteristic of the littoral fauna of the northern temperate zone, a few species ranging to the Arctic circle. Twenty-three species are known, of which the following are the most noteworthy: *P. platessa*, the "Plaice," ranging from the coast of France to Iceland; *P. glacialis*, from the Arctic coasts of North America; *P. americanus*, the transatlantic representative of the Plaice; *P. limanda*, the common "Dab;" *P. microcephalus*, the "Smear-dab;" *P. cynoglossus*, the "Craig-fluke;" *P. flesus*, the "Flounder."

RHOMBOSOLEA.—Eyes on the right side, the lower in advance of the upper. Mouth narrower on the right side than on the left; teeth on the blind side only, villiform; palatine and vomerine teeth none. The dorsal fin commences on the foremost part of the snout. Only one ventral which is continuous with the anal. Scales very small, cycloid; lateral line straight.

This genus represents *Pleuronectes* in the Southern Hemisphere, but consists of three species only, which occur on the coasts of New Zealand, and are valued as food-fishes.

Other genera, with narrow unsymmetrical mouth, in which the upper eye is not in advance of the lower, and which have pectoral fins, are *Parophrys, Psammodiscus, Ammotretis, Peltorhamphus, Nematops, Laeops,* and *Poecilopsetta.*

SOLEA.—Eyes on the right side, the upper being more or less in advance of the lower. Cleft of the mouth narrow, twisted round to the left side. Villiform teeth on the blind side only; vomerine or palatine teeth none. The dorsal fin commences on the snout, and is not confluent with the caudal. Scales very small, ctenoid; lateral line straight.

"Soles" are numerously represented in all suitable localities within the temperate and tropical zones, with the exception of the southern parts of the southern temperate zone, in which they are absent. Some enter or live in fresh water. Nearly forty species are known. British are *S. vulgaris*, the common "Sole;" *S. aurantiaca*, the "Lemon-sole," which is rather a southern species, and inhabits, on the south coast of England, deeper water than the common Sole; *S. variegata*, the "Banded Sole," with very small pectoral fins; and *S. minuta*, the "Dwarf-Sole."—Allied to *Solea* are *Pardachirus* and *Liachirus* from the Indian coasts.

SYNAPTURA.—Eyes on the right side, the upper in advance of the lower. Cleft of the mouth narrow, twisted round to the left side; minute teeth on the left side only. Vertical fins confluent. Scales small, ctenoid; lateral line straight.

Twenty species; with the exception of two from the Mediterranean and coast of Portugal, all belong to the fauna of the Indian Ocean.—Closely allied is *Aesopia*.

GYMNACHIRUS.—Mouth very small, toothless. Scales none, lateral line straight. Eyes on the right side. The dorsal fin commences on the snout; caudal free. Pectorals rudimentary or entirely absent.

Two species from the Tropical Atlantic.

CYNOGLOSSUS.—Eyes on the left side; pectorals none; vertical fins confluent. Scales ctenoid; lateral line on the left side double or triple; upper part of the snout produced backwards into a hook; mouth unsymmetrical, rather narrow. Teeth minute, on the right side only.

Abundant in the Indian seas, and especially on the flat sandy shores of China. About thirty-five species are known, which rarely exceed a length of eighteen inches. They are easily recognised by their long narrow shape (which has been compared to a dog's tongue) and the peculiar form of their snout.

To complete the list of Pleuronectoid genera, the following

have to be mentioned: *Soleotalpa* and *Apionichthys*, Soles with rudimentary eyes; *Ammopleurops*, *Aphoristia*, and *Plagusia*, which are closely allied to *Cynoglossus*, the latter genus having the lips provided with tentacles.

FOURTH ORDER—PHYSOSTOMI.

All the fin-rays articulated, only the first of the dorsal and pectoral fins is sometimes ossified. Ventral fins, if present, abdominal, without spine. Air-bladder, if present, with a pneumatic duct (except in Scombresocidæ).

FIRST FAMILY—SILURIDÆ.

Skin naked or with osseous scutes, but without scales. Barbels always present; maxillary bone rudimentary, almost always forming a support to a maxillary barbel. Margin of the upper jaw formed by the intermaxillaries only. Suboperculum absent. Air-bladder generally present, communicating with the organ of hearing by means of the auditory ossicles. Adipose fin present or absent.

A large family, represented by numerous genera, which exhibit a great variety of form and structure of the fins; they inhabit the fresh waters of all the temperate and tropical regions; a few enter the sea but keep near the coast. The first appearance of Siluroids is indicated by some fossil remains in tertiary deposits of the highlands of Padang in Sumatra, where *Pseudeutropius* and *Bagarius*, types well represented in the living Indian fauna, have been found. Also in North America spines referable to Cat-fishes have been found in tertiary formations.

The skeleton of the typical Siluroids shows many peculiarities. The cranial cavity is not membranous on the sides, but closed as in the Cyprinidæ, by the orbitosphenoids and the ethmoid that unite with the pre-frontals, carrying forward the cranial cavity to the nasal bone, without leaving a membranous septum between the orbits. But the supraoccipital is greatly developed, and in many the post-temporal is united by suture to the sides of the cranium. In numerous members of the family the skull is enlarged posteriorly, by dermal ossifications, to form a kind of helmet which spreads over the nape; the lateral angles of this production are formed by the suprascapulæ, augmented and fixed by suture, and the median part is the extension of the supraoccipital, which is generally very large, is connected anteriorly with the frontal, and passing backwards between the post-frontals, the parietals, the mastoids, and the suprascapulæ, goes past them all on to the nape. The mastoids interpose between the post-frontals and the parietals, so as to come in contact with the supraoccipital, and the parietals but little developed are pressed to the back part of the cranium, and in some instances wholly disappear.

The suprascapula most frequently unites to the mastoid by an immovable suture, which includes the parietal when that bone is present, and extends even to the supraoccipital. It gives out besides two processes, one of them resting on the exoccipital and basioccipital, or wedging itself between them, and the other going to the first vertebra; sometimes a plate from the exoccipital supports the same vertebra. This vertebra, though it presents a pretty continuous centrum beneath, is in reality composed of three or four coalescent vertebræ, as we ascertain by its diapophyses, by the circular elevations of the neural canal, and by the holes for the exit of the pairs of spinal nerves. There is great variety in the development of the various processes of the bones we have mentioned, and there is no less in the magnitude and connections of the first three interneurals.

In general in the species which have a strong dorsal

spine the second and third interneurals unite to form a single plate, the "buckler;" the great spine is articulated to the third interneural, and there is only the vestige of a spine on the second interneural in form of a small oval bone, forked below, whose function is to act as a bolt or fulcrum to the great spine when the fish wishes to use it as an offensive weapon. The great spine itself is joined by a ring to a second spine, which belongs to the third interneural. This articulation by ring exists in Lophius and a few other fishes not of this family.

The first interneural does not carry a ray, and it varies much in the species whose helmet is continuous with the buckler, as in many of the Bagri and Pimelodi. In these cases the supraoccipital, extending backwards, conceals the first interneural, passing over it to touch with its point the buckler formed by the second and third interneurals. In other instances, as in Synodontis and Auchenipterus, the supra-occipital and second interneural, forking and expanding, inclose and join themselves to the first interneural, but leave a larger or smaller space in the middle of the nuchal armour which they contribute to form. When the point of the supra-occipital does not reach quite to the second interneural, the first interneural remains free from connection, and occasionally shows as a narrow plate interposed between the other two; in such a case the helmet is not continuous with the buckler.

The neural spines of the coalescent centra, which form the apparently single first vertebra, concur also in sustaining the nuchal plate-armour and the first great dorsal spine. They carry the interneurals, are joined to them by suture, and one of them is often inclined towards the occiput to assist in sustaining the head; in fact, this part of the skeleton is constructed to give firm mutual support.

The shoulder-girdle of the Siluroids is also formed to give

resistance to the strong weapon with which it is frequently armed. The post-temporal, as we have said above, is often united by suture to the cranium, and it obtains support below by one or two processes that are fixed on the basioccipitals and on the diapophysis of the first vertebra.

In most osseous fishes the clavicle completes the lower key of the scapular arch in joining its fellow by suture or synchondrosis without the intervention of the coracoid; but in the Siluroids the coracoid descends to take part in this joint, and sometimes even to occupy the half of the suture, which is not unfrequently constructed of very deep interlocking serratures. The solidity of this base of the pectoral spine is further augmented by the intimate union of the coracoid and scapula, which often extends to junction by suture, or even to coalescence; and these bones, moreover, give off two bony arches—the first a slender one, arising from the salient edge of the coracoid near the pectoral fin, and going to the interior face of the scapular that is applied to the interior surface of the ascending branch of the clavicle; the second and broader supplementary arch is often perforated by a large hole; it also emanates from the same salient edge of the radius, but proceeds in opposite direction to the inferior edge of the clavicle, a little before the insertion of the pectoral spine. The two arches give attachments to the muscles that move this spine; in the Synodontes and many Bagri the upper arch remains in a cartilaginous or ligamentous condition, while in Malapterurus it is the lower arch that does not ossify, but both are fully formed in the Siluri and many other Siluroids more closely allied to that typical genus. The post-clavicle is also wanting in the Siluroids. The pterygoid and entopterygoid are reduced to a single bone, the symplectic is wholly wanting, and the palatine is merely a slender cylindrical bone. The sub-operculum is likewise constantly absent in all the Siluroids.

The great number of different generic types has necessitated a further division of this family into eight subdivisions:

I. SILURIDÆ HOMALOPTERÆ.—*The dorsal and anal fins are very long, nearly equal in extent to the corresponding parts of the vertebral column.*

a. CLARIINA.

CLARIAS.—Dorsal fin extending from the neck to the caudal, without adipose division. Cleft of the mouth transverse, anterior, of moderate width; barbels eight; one pair of nasal, one of maxillary, and two pairs of mandibulary barbels. Eyes small. Head depressed; its upper and lateral parts are osseous, or covered with only a very thin skin. A dendritic accessory branchial organ is attached to the convex side of the second and fourth branchial arches, and received in a cavity behind the gill-cavity proper. Ventrals six-rayed; only the pectoral has a pungent spine. Body eel-like.

Twenty species from Africa, the East Indies, and the intermediate parts of Asia; some attain to a length of six feet. They inhabit muddy and marshy waters; the physiological function of the accessory branchial organ is not known. Its skeleton is formed by a soft cartilaginous substance covered by mucous membrane, in which the vessels are imbedded. The vessels arise from branchial arteries, and return the blood into branchial veins. The vernacular name of the Nilotic species is "Carmoot."

HETEROBRANCHUS differs from *Clarias* only in the structure of the dorsal fin, the posterior portion of which is adipose.

The geographical range of this genus is not quite co-extensive with that of *Clarias*, inasmuch as it is limited to Africa and the East-Indian Archipelago. Six species.

b. PLOTOSINA.

PLOTOSUS.—A short dorsal fin in front, with a pungent spine; a second long dorsal coalesces with the caudal and anal. Vomerine teeth molar-like. Barbels eight or ten; one immediately before the posterior nostril, which is remote from the

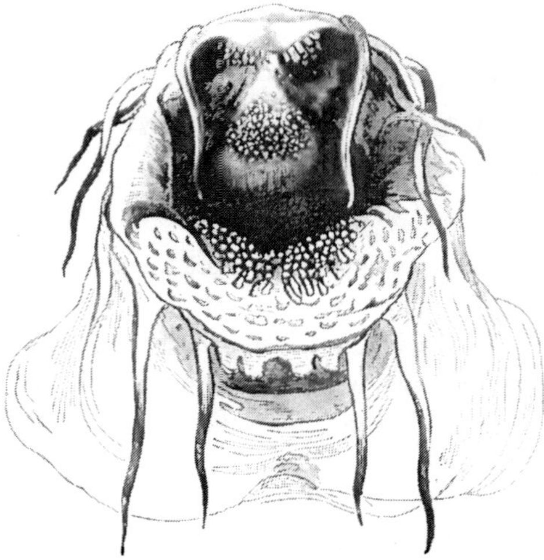

Fig. 258.—Mouth of Cnidoglanis megastoma, Australia.

anterior, the latter being quite in front of the snout. Cleft of the mouth transverse. Eyes small. The gill-membranes are not confluent with the skin of the isthmus. Ventral fins many-rayed. Head depressed; body elongate.

Three species are known from brackish waters of the Indian Ocean freely entering the sea. *Plotosus anguillaris* is distinguished by two white longitudinal bands, and is one of the most generally distributed and common Indian fishes.—

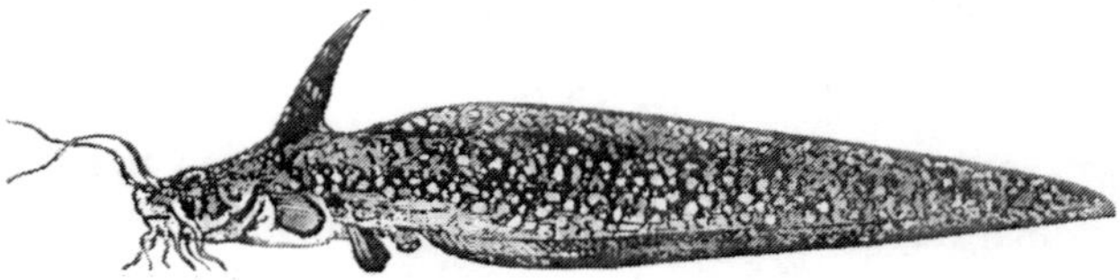

Fig. 259.—Cnidoglanis microcephalus.

Copidoglanis and *Cnidoglanis* are two very closely allied forms, chiefly from rivers and brackish waters of Australia. None of these Siluroids attain to a considerable size. *Chaca*, from the East Indies, belongs likewise to this sub-family.

II. SILURIDÆ HETEROPTERÆ.—*The rayed dorsal fin is very little developed, and, if it is present, it belongs to the abdominal portion of the vertebral column; the adipose fin is exceedingly small or absent. The extent of the anal is not much inferior to that of the caudal vertebral column. The gill-membranes overlap the isthmus, remaining more or less separate:* SILURINA.

SACCOBRANCHUS.—Adipose fin none; dorsal very short, without pungent spine, placed above the ventrals. Cleft of the mouth transverse, anterior, of moderate width; barbels eight. Eyes rather small. The upper and lateral parts of the head osseous or covered with a very thin skin. Gill-cavity with an accessory posterior sac, extending backwards between the muscles along each side of the abdominal and caudal portions of the vertebral column. Ventrals six-rayed.

Small fishes from East Indian rivers; four species are known. The lung-like extension of the branchial cavity receives water, and is surrounded by contractile transverse muscular fibres by which the water is expelled at intervals. The vessels of the sac take their origin in the last branchial artery, and pass into the aorta.

SILURUS.—No adipose fin; one very short dorsal, without pungent spine. Barbels four or six, one to each maxillary, and one or two to each mandible. Nostrils remote from each other. Head and body covered with soft skin. The eye is situated above the level of the angle of the mouth. The dorsal fin is anterior to the ventrals which are composed of more than eight rays. Caudal rounded.

This genus, of which five species are known, inhabits the temperate parts of Europe and Asia. The species which has given the name to the whole family, is the "Wels" of the Germans, *Silurus glanis.* It is found in the fresh waters east of the Rhine, and is, besides the Sturgeons, the largest of European Freshwater-fishes, and the only species of this family which occurs in Europe. Barbels six. It attains to a weight of 300 or 400 lbs., and the flesh, especially of smaller specimens, is firm, flaky, and well flavoured. Aris-

Fig. 260.—The "Wels," Siluris glanis.

totle described it under the name of *Glanis.* Its former occurrence in Scotland has justly been denied. In China it is represented by a similar species, *S. asotus*, which, however, has four barbels only.

This sub-family is well represented by various other genera in the fresh waters of the African as well as Indian region. African genera are *Schilbe* and *Eutropius;* East Indian: *Silurichthys, Wallago, Belodontichthys, Eutropiichthys, Cryptopterus, Callichrous, Hemisilurus, Siluranodon, Ailia, Schilbichthys, Lais, Pseudeutropius, Pangasius, Helicophagus,* and *Silondia.*

III. Siluridæ Anomalopteræ.—*Dorsal and adipose fins very short, the former belonging to the caudal vertebral column; anal very long. Ventrals in front of the dorsal. Gill-membranes entirely separate, overlapping the isthmus:* (Hypophthalmina.

Hypophthalmus.—Dorsal fin with seven rays, the first of which is slightly spinous. The lower jaw is rather the longer. Barbels six, those of the mandible long. No teeth; intermaxillaries very feeble. Head covered with skin. Eye of moderate size, situated behind and below the angle of the mouth. Ventrals small, six-rayed.

Four species from tropical America. The second genus of this sub-family is *Helogenes* from the Essequibo.

IV. SILURIDÆ PROTEROPTERÆ.—*The rayed dorsal fin is always present, short, with not more than twelve short rays, and belongs to the abdominal portion of the vertebral column, being placed in advance of the ventrals. The adipose fin is always present and well developed, although frequently short. The extent of the anal is much inferior to that of the caudal vertebral column. The gill-membranes are not confluent with the skin of the isthmus, their posterior margin always remaining free even if they are united with each other. Whenever the nasal barbel is present it belongs to the posterior nostril.*

a. BAGRINA.

BAGRUS.—Adipose fin long; a short dorsal with a pungent spine and nine or ten soft rays; anal fin short, with less than twenty rays. Barbels eight. The anterior and posterior nostrils are remote from each other, the posterior being provided with a barbel. Teeth on the palate in a continuous band. Eyes with a free orbital margin. Caudal forked; ventrals six-rayed.

This genus consists of two species only, common in the Nile, viz. the "Bayad," *B. bayad*, and *B. docmac*. Both grow to a large size, exceeding a length of five feet, and are eaten. *Chrysichthys* and *Clarotes* are two other Siluroid genera from African rivers, closely allied to Bagrus. Similar Siluroids are common in the East Indies, and have been referred to the following genera: *Macrones, Pseudobagrus, Liocassis, Bagroides, Bagrichthys, Rita, Acrochordonichthys, Akysis.*

b. AMIURINA.

AMIURUS.—Adipose fin of moderate length; a short dorsal with a pungent spine and six soft rays; anal fin of moderate length. Barbels eight. The anterior and posterior nostrils are remote from each other, the posterior being provided with a barbel. Palate edentulous. Head covered with skin above. Ventrals eight-rayed.

The "Cat-fishes" of North America, of which about a dozen different species are known. One species occurs in China. Allied, but smaller forms are *Hopladclus and Noturus,* likewise from North America.

c. PIMELODINA.

PLATYSTOMA.—Adipose fin of moderate length; a short dorsal fin with a pungent spine and six or seven soft rays; anal fin rather short. Snout very long, spatulate, with the upper jaw more or less projecting; the upper surface of the head not covered by the skin. Barbels six; the anterior and posterior nostrils remote from each other, none with a barbel. Palate toothed. Caudal forked; ventrals six-rayed, inserted behind the dorsal.

Twelve species from South America, some attaining a length of six feet, the majority being ornamented with deep-black spots or bands. Allied genera from South America, likewise distinguished by a long spatulate snout, are *Sorubim, Hcmisorubim,* and *Platystomatichthys,* whilst *Phractoccphalus, Piramutana, Platynematichthys, Piratinga, Bagropsis,* and *Sciadcs,* have a snout of ordinary length. The barbels of some are of extraordinary length, and not rarely dilated and bandlike.

PIMELODUS.—Adipose fin well developed; dorsal fin short, with a more or less pungent spine and six rays; anal fin short. Barbels six, cylindrical or slightly compressed, none of them belonging to either of the nostrils, which are remote from each other. Palate edentulous. Ventrals six-rayed, inserted behind the dorsal.

Of all South American genera this is represented by the greatest number of species, more than forty being well characterised; they differ chiefly with regard to the length of the adipose fin and barbels, and the strength of the dorsal spine. Singularly, two species (*P. platychir* and *P. balayi*), are found in West Africa. The majority are of but moderate size and plain coloration.—Allied South American genera (also without teeth on the palate), are *Pirinampus, Conorhynchus, Notoglanis, Callophysus, Lophiosilurus.*

AUCHENOGLANIS.—Adipose fin rather long, dorsal short, with a pungent spine and seven rays; anal short. Snout produced, pointed, with narrow mouth. Barbels six, none of which belongs to either of the nostrils, which are remote from each other. The teeth of each jaw form a pair of small elliptic patches which are longer than broad; palate edentulous. Eyes of moderate size. Ventrals six-rayed.

One species, *Au. biscutatus*, from the Nile, Senegal, and other West African rivers.

d. ARIINA.

ARIUS.—Adipose fin of moderate length or short; a short dorsal fin with a pungent spine and seven soft rays; anal fin rather short. Head osseous above; barbels six, four at the mandible, none at either of the nostrils which are close together. Eyes with a free orbital margin. Caudal fin forked; ventrals six-rayed, behind the dorsal.

Of all Siluroid genera this has the greatest number of species (about seventy), and the widest distribution, being represented in almost all tropical countries which are drained by large rivers; some of the species prefer brackish to fresh water, and a few enter the sea, but keep near to the coast. Some of the species are of small size, whilst others exceed a length of five feet. The extent of the armature of the neck and the dentition vary much in the different species, and affords two of the principal characters by which the species are separated.—The following genera are allied to *Arius*, *Galeichthys* from South Africa; *Genidens* and *Paradiplomystax* from Brazil; *Diplomystax* from Chile; *Aelurichthys* from Central and South America; *Hemipimelodus, Ketengus, Osteo-*

geniosus, and *Batrachocephalus* from the East Indies; and *Atopochilus* from West Africa.

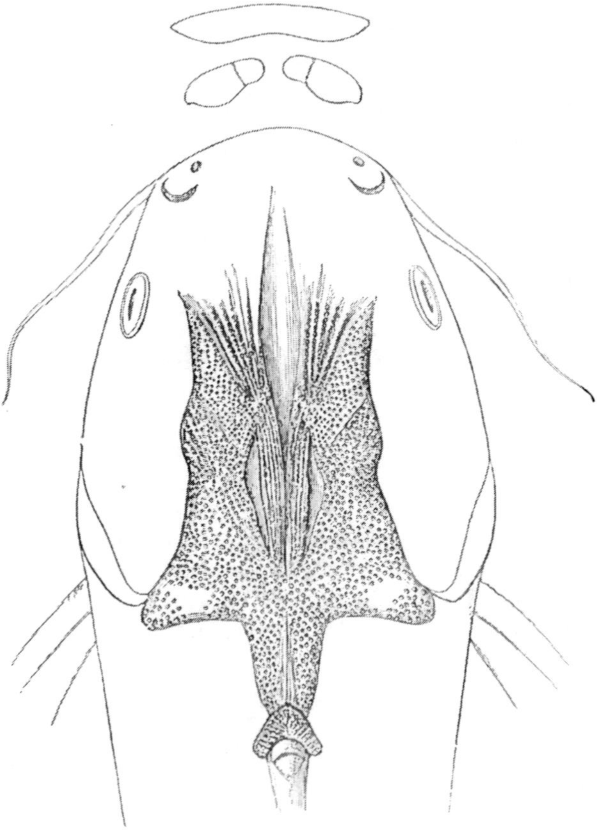

Fig. 261.—Arius australis, from Queensland.

c. BAGARIINA.

BAGARIUS.—Adipose fin rather short; a short dorsal with one spine and six rays; anal fin of moderate length. Barbels eight, of which one pair stands between the anterior and posterior nostrils which are close together. Head naked above. Caudal fin deeply forked; ventrals rays six. Thorax without longitudinal plaits of the skin.

A large Siluroid (*B. bagarius*) from rivers of India and Java; exceeding a length of six feet.

EUGLYPTOSTERNUM.—Adipose fin of moderate length; a short dorsal with a pungent spine and six rays; anal fin short. Barbels eight, of which one pair is placed between the anterior and

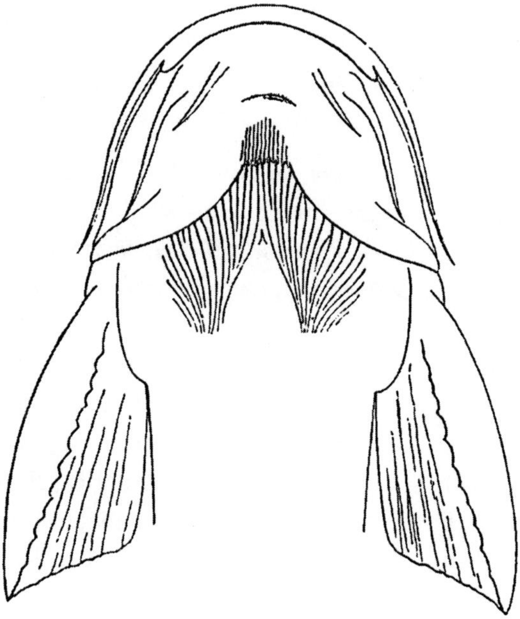

Fig. 262.—Euglyptosternum coum, thoracic adhesive apparatus.

posterior nostrils which are close together. Teeth on the palate villiform, in two separate patches. Eyes small, below the skin. Caudal forked; ventral rays six. Pectorals horizontal, with a thoracic adhesive apparatus between, which is formed by longitudinal plaits of the skin.

This fish (*Eu. coum*) inhabits the river Coic in Syria, and is about twelve inches long. The plaited structure on the thorax probably increases the capability of the fish of maintaining its position in the rapid current of the stream, a function which appears to be chiefly performed by the horizontally expanded pectoral fins. A similar structure is found in

Glyptosternum, a genus represented by eight species in mountain streams of the East Indies, and differing from the Syrian species in lacking the teeth on the palate.

V. SILURIDÆ STENOBRANCHIÆ.—*The rayed dorsal fin is short, if present, belonging to the abdominal portion of the vertebral column, the ventrals being inserted behind it (except in Rhinoglanis). The gill-membranes are confluent with the skin of the isthmus.*

a. DORADINA.

Some of the genera have no bony shields along the lateral line, and a small adipose fin or none whatever; all of these are South American—*Ageniosus, Tetranematichthys, Euanemus, Auchenipterus, Glanidium, Centromochlus, Trachelyopterus, Cetopsis,* and *Astrophysus.*

Others have a series of bony scutes along the middle of the side; they form the genus *Doras* with two closely allied forms, *Oxydoras* and *Rhinodoras.* Some twenty-five species are known, all from rivers of tropical America, flowing into the Atlantic. These fishes have excited attention by their habit of travelling, during the dry season, from a piece of water about to dry up, in quest of a pond of greater capacity. These journeys are occasionally of such a length that the fish spends whole nights on the way, and the bands of scaly travellers are sometimes so large that the Indians who happen to meet them, fill many baskets of the prey thus placed in their hands. The Indians supposed that the fish carry a supply of water with them, but they have no special organs, and can only do so by closing the gill-openings, or by retaining a little water between the plates of their bodies, as Hancock supposes. The same naturalist adds that they make regular nests, in which they cover up their eggs with care and defend them, male and female uniting in this parental duty until the eggs are hatched. The nest is constructed at the beginning of the rainy season, of leaves, and is sometimes placed in a hole scooped out in the beach.

Finally, in the last genus, the lateral scutes are likewise absent, viz. in

SYNODONTIS.—The adipose fin is of moderate length or rather long; the dorsal fin has a very strong spine and seven soft rays. The teeth in the lower jaw are movable, long, very thin at the base, and with a slightly-dilated brown apex. Mouth small. Barbels six, more or less fringed with a membrane or with filaments. Neck with broad dermal bones.

Synodontis is characteristic of the fauna of tropical Africa, where it is represented by fifteen species. Several

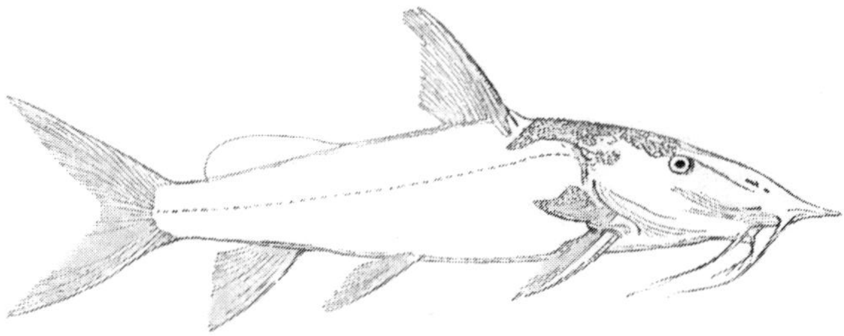

Fig. 263.—Synodontis xiphias.

occur in the Nile, and are known by the vernacular name "Schal." Some attain a length of two feet. The species figured is from West Africa, and characterised by its long upper jaw.

b. RHINOGLANINA.

RHINOGLANIS.—Two dorsal fins, both composed of rays, the first with a strong spine; anal rather short. Barbels six; anterior and posterior nostrils close together, the posterior very large, open. Neck with broad dermal bones. Ventrals with seven rays, inserted below the posterior rays of the first dorsal fin.

This Siluroid is known from a single example only one and a half inches long, obtained at Gondokoro on the Upper Nile. *Callomystax* represents this type in the Ganges and Indus.

c. Malapterurina.

Malapterurus.—One dorsal fin only, which is adipose and situated before the caudal; anal of moderate length or short; caudal rounded; ventrals six-rayed, inserted somewhat behind the middle of the body; pectorals without pungent spine. Barbels six: one to each maxillary and two on each side of the mandible. The anterior and posterior nostrils are remote from each other. No teeth on the palate. The entire head and body covered with soft skin. Eyes small. Gill-opening very narrow, reduced to a slit before the pectoral.

The "Electric Cat- or Sheath-fishes" are not uncommon in the fresh waters of tropical Africa; three species have been described, of which *M. electricus* occurs in the Nile; they

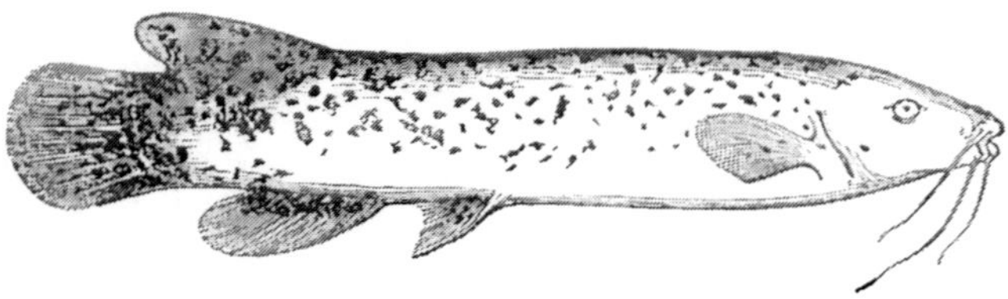

Fig. 264.—Malapterurus electricus.

grow to a length of about four feet. Although the first dorsal fin is absent, its position (if it had been developed) is indicated by a rudimentary interneural spine, which rests in the cleft of the neural process of the first vertebra. The electric organ extends over the whole body, but is thickest on the abdomen; it lies between two aponeurotic membranes, below the skin, and consists of rhomboidal cells which contain a rather firm gelatinous substance. The electric nerve takes its origin from the spinal chord, does not enter into connection with ganglia, and consists of a single enormously-strong primitive fibre, which distributes its branches in the electric organ.

VI. Siluridæ Proteropodes.—*The rayed dorsal fin is always present and rather short; the ventrals are inserted below (very rarely in front of) the dorsal. The gill-membranes are confluent with the skin of the isthmus, the gill-opening being reduced to a short slit. Pectorals and ventrals horizontal. Vent before, or not much behind, the middle of the length of the body.*

a. Hypostomatina.

Stygogenes.—Adipose fin short; dorsal and anal short; the outer fin-rays somewhat thickened and rough; palate toothless; cleft of the mouth of moderate width, with a maxillary barbel on each side; a short broad flap on each side between the nostrils, which are close together. Lower lip very broad, pendent. Eyes small, covered with transparent skin. Head covered with soft skin. Ventrals six-rayed.

These small Siluroids, which are called "Preñadillas" by the natives, together with the allied *Arges, Brontes,* and *Astroplebus,* have received some notoriety through Humboldt's accounts, who adopted the popular belief that they live in subterranean waters within the bowels of the active volcanoes of the Andes, and are ejected with streams of mud and water during eruptions. Humboldt himself considers it very singular that they are not cooked and destroyed whilst they are vomited forth from craters or other openings. The explanation of their appearance during volcanic eruptions is, that they abound in the numerous lakes and torrents of the Andes, that they are killed by the sulphuretted gases escaping during an eruption, and swept down by the torrents of water issuing from the volcano.

Callichthys.—Adipose fin short, supported anteriorly by a short movable spine; dorsal with a feeble spine and seven or eight rays; anal short. Teeth minute or entirely absent; cleft of the mouth rather narrow, with a pair of maxillary barbels on each side, which are united at the base. Eyes small. Head covered with osseous plates; body wholly protected by two series of large imbricate shields on each side. Ventrals six-rayed.

Twelve species of this genus are known; they are small, and similarly distributed as *Doras,* with which they have

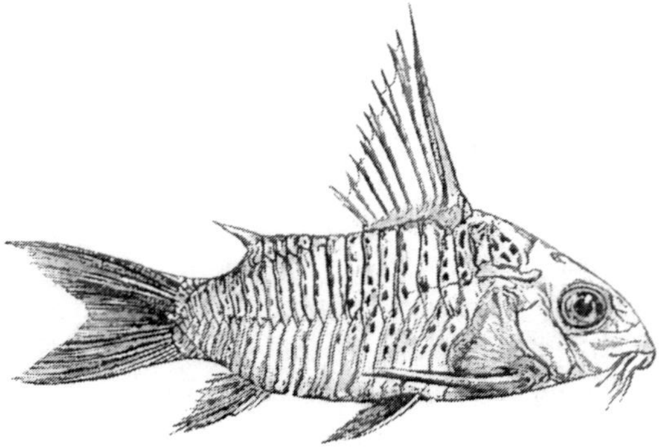

Fig. 265.—Callichthys armatus, from the Upper Amazon. Natural size.

much in common as regards their mode of life. They likewise are able to travel over land, and construct nests for their ova.

CHÆTOSTOMUS.—A short adipose fin, supported anteriorly by a short, compressed, curved spine; dorsal fin of moderate length, with from eight to ten rays, the first of which is simple; anal fin short; ventral six-rayed; pectoral with a strong spine. Head and body completely cuirassed, the lower parts being sometimes naked; body rather short, with four or five longitudinal series of large imbricate scutes on each side; tail not depressed. Snout produced, obtuse in front; mouth inferior, transverse, with a single series of generally very fine bent teeth in both jaws. Interoperculum very movable and armed with erectile spines.

This genus, with the allied *Plecostomus, Liposarcus, Pterygoplichthys, Rhinelepis, Acanthicus,* and *Xenomystus,* is well represented in the fresh waters of South America, whence about sixty species are known. The majority do not exceed a length of twelve inches, but some attain to more than double that size. In some of the species the male is provided with long bristles round the margin of the snout and inter-operculum.

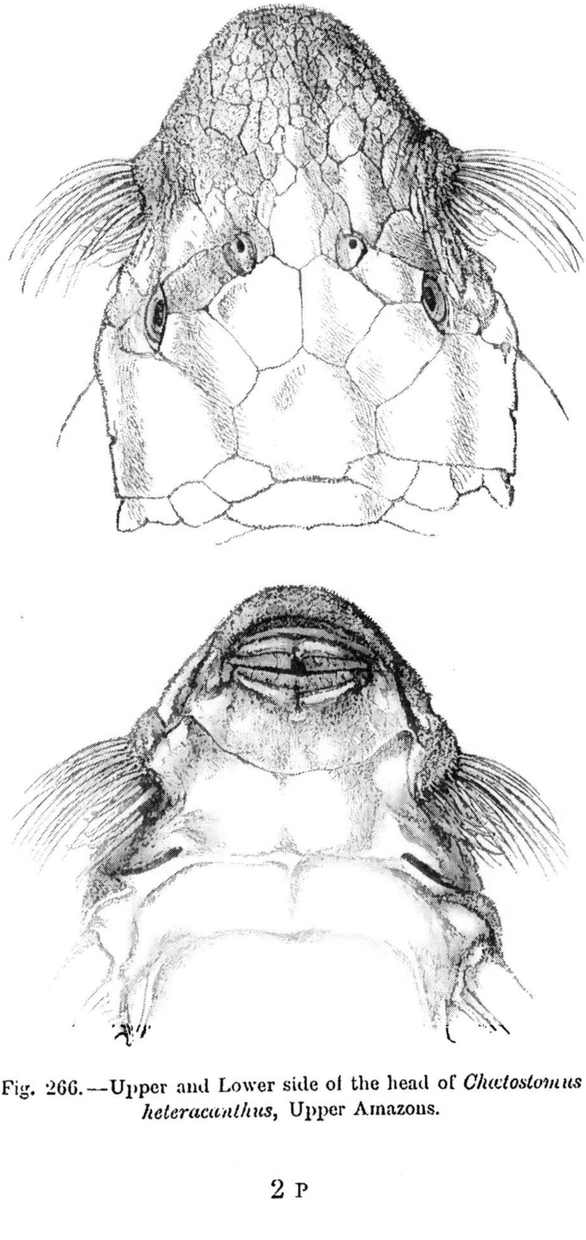

Fig. 266.—Upper and Lower side of the head of *Chaetostomus heteracanthus*, Upper Amazons.

Hypoptopoma.—Differing from *Chætostomus* in the peculiar formation of the head, which is depressed, spatulate, the eyes

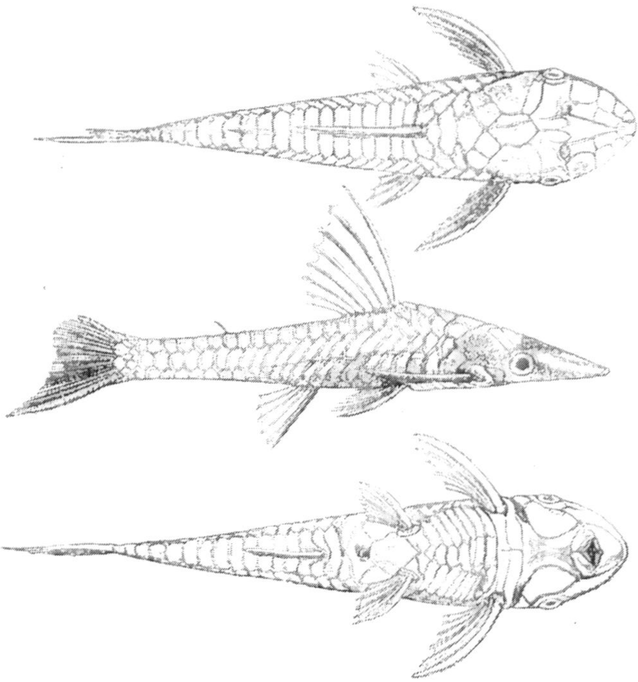

Fig. 267.—*Hypoptopoma thoracatum*, Upper Amazons. Natural size.

being on the lateral edge of the head. The movable gill-covers are reduced to two bones, neither of which is armed, viz.—the operculum small and placed as in *Chætostomus*, and a second, larger one, separated from the eye by the narrow sub-orbital ring, and placed at the lower side of the head.

LORICARIA.—One short dorsal fin ; anal short ; the outer ray of each fin thickened, but flexible. Head depressed, with the snout more or less produced and spatulate. Mouth situated at the lower side of the snout, remote from its extremity, transverse, surrounded by broad labial folds which are sometimes fringed ; a short barbel at each corner of the mouth. Teeth in the jaws small, bent, with a dilated, notched apex, in a single series, sometimes absent. Head and body cuirassed; tail depressed, long; eye rather small or of moderate size.

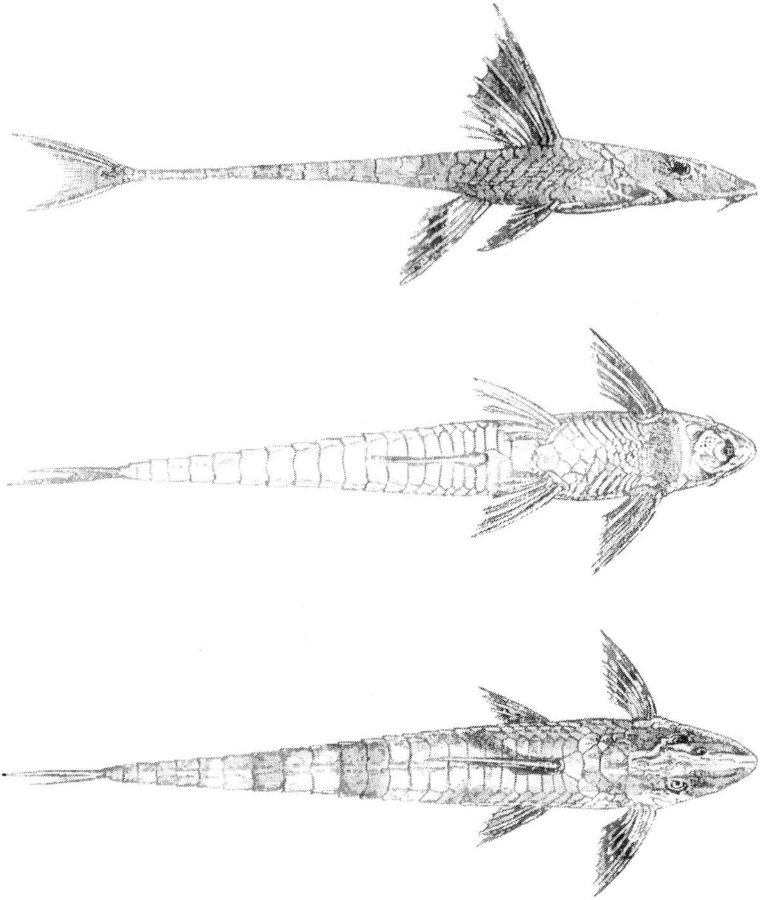

Fig. 268.—*Loricaria lanceolata*, Upper Amazons. Natural size.

Small fishes from rivers of tropical America; about twenty-six species are known. The male of some species has a bearded or bristly snout.

ACESTRA differs from *Loricaria* in having the snout much prolonged.

SISOR.—Head depressed, spatulate; trunk depressed; tail long and thin. One short dorsal fin; anal short; ventrals seven-rayed. Head partially osseous, rough; a series of bony plates along the median line of the back; lateral line rough. Eyes very small. Mouth inferior, small, transverse, with barbels; teeth none.

A single species, *S. rhabdophorus*, from rivers of northern Bengal. Allied to this genus is *Erethistes* from Assam.

PSEUDECHENEIS.—Adipose fin of moderate length; a short dorsal with one spine and six rays; anal fin rather short. Barbels eight. Mouth small, inferior. Head depressed, covered with soft skin above; eyes small, superior. Caudal fin forked; pectorals horizontal, with a thoracic adhesive apparatus between, formed by transverse plaits of the skin. Ventrals six-rayed.

A very small species, inhabiting the mountain-streams of Khassya; by means of the adhesive apparatus it is enabled to hold on to stones, thus preventing the current from sweeping it away. *Exostoma* is a similar small Siluroid from Indian mountain-streams, but without the thoracic apparatus; probably its mouth performs the same function.

b. ASPREDININA.

ASPREDO.—Adipose fin none; dorsal short, without pungent spine; anal very long, but not united with the caudal. Head broad, much depressed; tail very long and slender. Barbels not less than six, one of which is attached to each intermaxillary; none at the nostrils. Eyes very small. Head covered with soft skin; the anterior and posterior nostrils are remote from each other. Ventrals six-rayed.

Six species are known from Guyana; the largest grows to a length of about eighteen inches. The remarkable mode of taking care of their ova has been noticed above (p. 161, Fig. 72). *Bunocephalus*, *Bunocephalichthys*, and *Harttia*, from tropical America, are other genera of this sub-family which remain to be mentioned.

VII. SILURIDÆ OPISTHOPTERÆ.—*The rayed dorsal fin is always present, short, and placed above or behind the middle of the length of the body, above or behind the ventrals which, however, are sometimes absent; anal short. Nostrils remote from each other; if a nasal barbel is present, it belongs to the anterior nostril. Lower lip not reverted. The gill-membranes are not confluent with the skin of the isthmus:* NEMATOGENYINA and TRICHOMYCTERINA.

The genera *Heptapterus, Nematogenys, Trichomycterus, Eremophilus,* and *Pariodon,* belong to this sub-family. They are small South American Siluroids, the majority of which inhabit waters at high altitudes, up to 14,000 feet above the level of the sea. In the Andes they replace the Loaches of the Northern Hemisphere, which they resemble in appearance and habits, and even in coloration, offering a striking example of the fact that similar forms of animals are produced under similar external physical conditions.

VIII. SILURIDÆ BRANCHICOLÆ.—*The rayed dorsal fin is present, short, and placed behind the ventrals; anal short. Vent far behind the middle of the length of the body. Gill-membranes confluent with the skin of the isthmus.*

Stegophilus and *Vandellia,* two genera from South America, comprising the smallest and least developed Siluroids. Their body is narrow, cylindrical, and elongate; a small barbel at each maxillary; the operculum and inter-operculum are armed with short stiff spines. The natives of Brazil accuse these fishes of entering and ascending the urethra of persons while bathing, causing inflammation and sometimes death. This requires confirmation, but there is no doubt that they live parasitically in the gill-cavity of larger fishes (*Platystoma*), but probably they enter these cavities only for places of safety, without drawing any nourishment from their host.

SECOND FAMILY—SCOPELIDÆ.

Body naked or scaly. Margin of the upper jaw formed by the intermaxillary only; opercular apparatus sometimes incompletely developed. Barbels none. Gill-opening very wide; pseudobranchiæ well developed. Air-bladder none. Adipose fin present. The eggs are enclosed in the sacs of the ovary, and excluded by oviducts. Pyloric appendages few in number or absent. Intestinal tract very short.

Exclusively marine, the majority being either pelagic or deep-sea forms. Of fossil remains the following have been referred to this family:—*Osmeroides,* from Mount Lebanon, which others believe to be a marine salmonoid; *Hemisaurida,* from Comen, allied to *Saurus; Parascopelus* and *Anapterus,* from the miocene of Licata, the latter genus allied to *Paralepis.*

SAURUS (inclus. *Saurida*).—Body sub-cylindrical, rather elongate, covered with scales of moderate size; head oblong; cleft of the mouth very wide; intermaxillary very long, styliform, tapering; maxillary thin, long, closely adherent to the intermaxillary. Teeth card-like, some being elongate, slender; all can be laid downwards and inwards. Teeth on the tongue, and palatine bones. Eye of moderate size. Pectorals short; ventrals eight- or nine-rayed, inserted in advance of the dorsal, not far behind the pectorals. Dorsal fin nearly in the middle of the length of the body, with thirteen or less rays; adipose fin small; anal short or of moderate length; caudal forked.

Fifteen species of small size, from the shores of the tropical and sub-tropical zones. The species figured on p. 42, Fig. 5, occurs on the north-west coast of Australia and in Japan.

BATHYSAURUS.—Shape of the body similar to that of *Saurus,* sub-cylindrical, elongate, covered with small scales. Head depressed, with the snout produced, flat above. Cleft of the mouth very wide, with the lower jaw projecting; intermaxillary very long, styliform, tapering, not movable. Teeth in the jaws, in broad bands, not covered by lips, curved, unequal in size and barbed at the end. A series of similar teeth runs along the whole length of each side of the palate. Eye of moderate size, lateral. Pectoral of moderate length. Ventral eight-rayed, inserted immediately behind the pectoral. Dorsal fin in the middle of the length of the body, with about eighteen rays. Adipose fin absent or present. Anal of moderate length. Caudal emarginate.

Deep-sea fishes, obtained in the Pacific at depths varying from 1100 to 2400 fathoms. The largest example is twenty inches long. Two species.

Bathypterois.—Shape of the body like that of an *Aulopus*. Head of moderate size, depressed in front, with the snout projecting, the large mandible very prominent beyond the upper jaw. Cleft of the mouth wide; maxillary developed, very movable, much dilated behind. Teeth in narrow villiform bands in the jaws. On each side of the broad vomer a small patch of similar teeth; none on the palatines or on the tongue. Eye very small. Scales cycloid, adherent, of moderate size. Rays of the pectoral fin much elongated, some of the upper being separate from the rest, and forming a distinct division. Ventrals abdominal, with the outer rays prolonged, eight-rayed. Dorsal fin inserted in the middle of the body, above or immediately behind the root of the ventral, of moderate length. Adipose fin present or absent. Anal short. Caudal forked.

This very singular form is one of the discoveries of the "Challenger;" it is widely distributed over the seas of the Southern Hemisphere, in depths varying from 520 to 2600 fathoms. The elongate pectoral rays are most probably organs of touch. Four species were discovered, the largest specimen being thirteen inches long.

Harpodon.—Body elongate, covered with very thin, diaphanous, deciduous scales. Head thick, with very short snout; its bones are very soft, and the superficial ones are modified into wide muciferous cavities; the lateral canal of the body is also very wide, and a pair of pores corresponds to each scale of the lateral line, one being above, the other below the scale. Cleft of the mouth very wide; intermaxillary very long, styliform, tapering; maxillary absent. Teeth card-like, recurved, unequal in size; the largest are in the lower jaw, and provided with a single barb at the posterior margin of the point. Eye small. Ventral fins long, nine-rayed, inserted below the anterior dorsal rays; dorsal fin in the middle of the length of the body; adipose fin small; anal of moderate length; caudal fin three-lobed, the lateral line being continued along the central lobe. Centre of the vertebræ open in the middle.

Two species only are known of this singular genus; both are evidently inhabitants of considerable depths, and periodically come nearer to the surface. One (*H. nehereus*) is well known in the East Indies, being of excellent flavour. When newly taken its body is brilliantly phosphorescent. When salted and dry it is known under the names of "Bombay-ducks" or "Bummaloh," and exported in large quantities from Bombay and the coast of Malabar. The second species (*H. microchir*) exceeds the other in length, and has been found in the sea off Japan.

Scopelus.—Body oblong, more or less compressed, covered with large scales. Series of phosphorescent spots run along the lower side of the body, and a similar glandular substance sometimes occupies the front of the snout and the back of the tail.

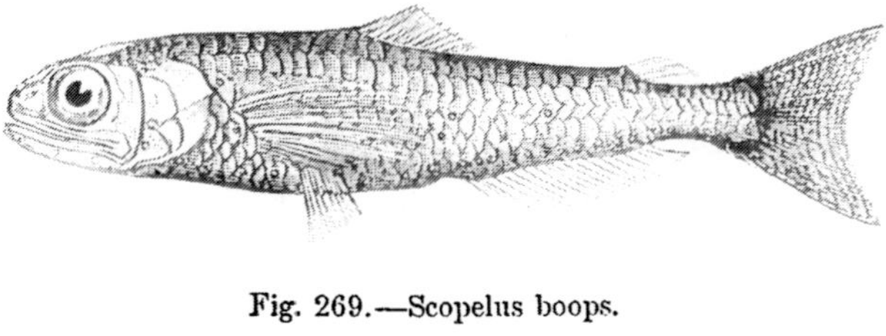

Fig. 269.—Scopelus boops.

Cleft of the mouth very wide. Intermaxillary very long, styliform, tapering; maxillary well developed. Teeth villiform. Eye large. Ventrals eight-rayed, inserted immediately in front of or below the anterior dorsal rays. Dorsal fin nearly in the middle of the length of the body; adipose fin small; anal generally long; caudal forked. Branchiostegals from eight to ten.

The fishes of this genus are small, of truly pelagic habits, and distributed over all the temperate and tropical seas; they are so numerous that the surface-net, when used during a night of moderate weather, scarcely ever fails to enclose some specimens. They come to the surface at night only; during the day and in very rough weather they descend to depths where they are safe from sunlight or the agitation of the water. Some species never rise to the surface; indeed, Scopeli have been brought up in the dredge from almost any depth to 2500 fathoms. Thirty species are known. *Gymnoscopelus* differs from *Scopelus* in lacking scales.

IPNOPS.—Body elongate, sub-cylindrical, covered with large, thin, deciduous scales, and without phosphorescent organs. Head depressed, with broad, long, spatulate snout, the whole upper surface of which is occupied by a most peculiar organ of vision (or luminosity), longitudinally divided in two symmetrical halves. Bones of the head well ossified. Mouth wide, with the lower jaw projecting; maxillary dilated behind. Both jaws with narrow bands of villiform teeth; palate toothless. Pectoral and ventral fins well developed, and, owing to the shortness of the trunk, close together. Dorsal fin at a short distance behind the vent; adipose fin none; anal fin moderately long; caudal subtruncated. Pseudobranchiæ none.

This singular genus, one of the "Challenger" discoveries, is known from four examples, obtained at depths varying between 1600 and 2150 fathoms, off the coast of Brazil, near Tristan d'Acunha and north of Celebes. All belong to one species, *I. murrayi.* The eye seems to have lost its function of vision and assumed that of producing light. The specimens are from 4 to 5½ inches long.

PARALEPIS.—Head and body elongate, compressed, covered with deciduous scales. Cleft of the mouth very wide; maxillary developed, closely adherent to the intermaxillary. Teeth in a single series, unequal in size. Eye large. Ventrals small, inserted opposite or nearly opposite the dorsal. Dorsal fin short, on the hinder part of the body; adipose fin small; anal elongate, occupying the end of the tail; caudal emarginate.

Three species; small pelagic fishes from the Mediterranean and Atlantic.—*Sudis,* from the Mediterranean, has a dentition slightly different from that of *Paralepis.*

PLAGYODUS.—Body elongate, compressed, scaleless; snout much produced, with very wide cleft of the mouth. Intermaxillary very long and slender; maxillary thin, immovable. Teeth in the jaws and of the palate very unequal in size, the

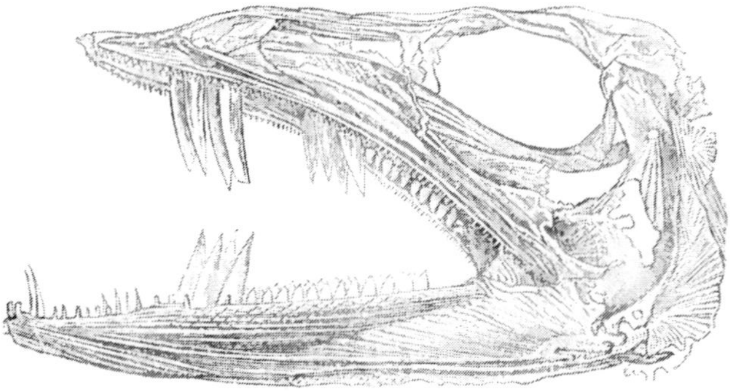

Fig. 270.—Plagyodus ferox.

majority pointed and sharp, some very large and lanceolate. Eye large. Pectoral and ventral fins well developed; the rayed dorsal fin occupies the whole length of the back from the occiput to opposite the anal fin; adipose and anal fins of moderate size. Caudal forked. Branchiostegals six or seven.

This is one of the largest and most formidable deep-sea fishes. One species only is well known, *P. ferox,* from Madeira and the sea off Tasmania; other species have been noticed from Cuba and the North Pacific, but it is not evident in what respects they differ specifically from *P. ferox.* This fish grows to a length of six feet, and from the stomach of one example have been taken several Octopods, Crustaceans, Ascidians, a young Brama, twelve young Boar-fishes, a Horse-mackerel, and one young of its own species. The stomach is coecal; the commencement of the intestine has extremely thick walls, its inner surface being cellular, like the lung of a reptile; a pyloric appendage is absent. All the bones are extremely thin, light, and flexible, containing very little earthy matter; singular is the development of a system of abdominal ribs, symmetrically arranged on both sides, and extending the whole length of the abdomen. Perfect specimens are rarely obtained on account of the want of coherence of the muscular and osseous parts, caused by the diminution of pressure when the fish reaches the surface of the water. The exact depth at which *Plagyodus* lives is not known; probably it never rises above a depth of 300 fathoms.

The other less important genera belonging to this family are *Aulopus*, *Chlorophthalmus*, *Scopelosaurus*, *Odontostomus*, and *Nannobrachium*.

THIRD FAMILY—CYPRINIDÆ.

Body generally covered with scales; head naked. Margin

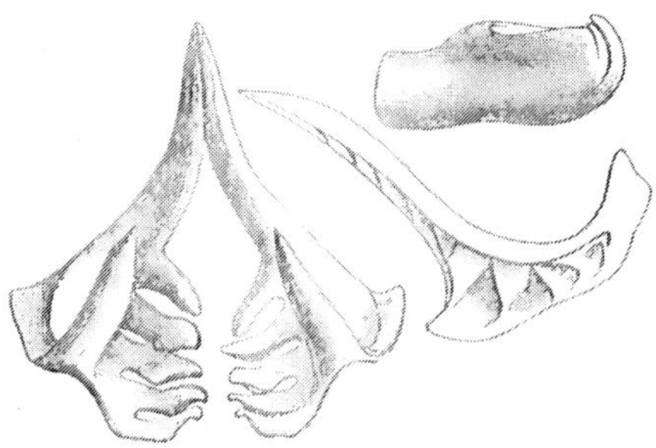

Fig. 271.—Pharyngeal bones and teeth of the Bream, Abramis brama.

of the upper jaw formed by the intermaxillaries. Belly rounded, or, if trenchant, without ossifications. No adipose fin. Stomach without blind sac. Pyloric appendages none. Mouth toothless; lower pharyngeal bones well developed, falciform, subparallel to the branchial arches, provided with teeth, which are arranged in one, two, or three series. Air-bladder large, divided into an anterior and posterior portion by a constriction, or into a right or left portion, enclosed in an osseous capsule. Ovarian sacs closed.

The family of "Carps" is the one most numerously represented in the fresh waters of the Old World and of North America. Also numerous fossil remains are found in tertiary freshwater-formations, as in the limestones of Oeningen and Steinheim, in the lignites of Bonn, Stöchen, Bilin, and Ménat, in the marl slates and carbonaceous shales of Licata in Sicily, and of Padang in Sumatra, in corresponding deposits of Idaho in North America. The majority can be referred to existing genera: *Barbus, Thynnichthys, Gobio, Leuciscus, Tinca, Amblypharyngodon, Rhodeus, Cobitis, Acanthopsis,* only a few showing characters different from those of living genera: *Cyclurus, Hexapsephus, Mylocyprinus* (tertiary of North America).

Most Carps feed on vegetable and animal substances; a few only are exclusive vegetable feeders. There is much less diversity of form and habits in this family than in the Siluroids; however, the genera are sufficiently numerous to demand a further subdivision of the family into groups.

I. CATOSTOMINA.—*Pharyngeal teeth in a single series, exceedingly numerous and closely set. Dorsal fin elongate, opposite to the ventrals; anal short, or of moderate length. Barbels none.*

These fishes are abundant in the lakes and rivers of North America, more than thirty species having been described, and many more named, by American ichthyologists. Two species are known from North-Eastern Asia. They are generally called "Suckers," but their vernacular nomenclature is very arbitrary and confused. Some of the species which inhabit the large rivers and lakes grow to a length of three feet and a weight of fifteen pounds. The following genera may be distinguished:—*Catostomus*, "Suckers," "Red-horses," "Stone-rollers," "White Mullets;" *Moxostoma; Sclerognathus*, "Buffaloes," "Black Horses;" and *Carpiodes*, "Spear-fish," "Sail-fish."

II. CYPRININA.—*Anal fin very short, with not more than five or six, exceptionally seven, branched rays. Dorsal fin opposite ventrals. Abdomen not compressed. Lateral line running along the median line of the tail. Mouth frequently with barbels, never more than four in number. Pharyngeal teeth generally in a triple series in the Old World genera; in a double or single series in the North American forms, which are small and feebly developed. Air-bladder present, without osseous covering.*

CYPRINUS.—Scales large. Dorsal fin long, with a more or less strong serrated osseous ray; anal short. Snout rounded, obtuse, mouth anterior, rather narrow. Pharyngeal teeth, 3. 1. 1.–1. 1. 3, molar-like. Barbels four.

Fig. 272.—The Carp, Cyprinus carpio.

The "Carp" (*C. carpio*, "Karpfen," "La carpe,") is originally a native of the East, and abounds in a wild state in China, where it has been domesticated for many centuries; thence it was transported to Germany and Sweden, and the year 1614 is assigned as the date of its first introduction into England. It delights in tranquil waters, preferring such as have a muddy bottom, and the surface partially shaded with plants. Its food consists of the larvæ of aquatic insects, minute testacea, worms, and the tender blades and shoots of plants. The leaves of lettuce, and other succulent plants of a similar kind, are said to be particularly agreeable to them, and to fatten them sooner than any other food. Although the Carp eats with great voracity when its supply of aliment is abundant, it can subsist for an astonishing length of time without nourishment. In the winter, when the Carps assemble in great numbers, and bury themselves among the mud and the roots of plants, they often remain for many months without eating. They can also be preserved alive for a considerable length of time out of the water, especially if care be taken to moisten them occasionally as they become dry. Advantage is often taken of this circumstance to transport them alive, by packing them among damp herbage or damp linen; and the operation is said to be unattended with any risk to the animal, especially if the precaution be taken to put a piece of bread in its mouth steeped in brandy!

The fecundity of these fishes is very great, and their numbers consequently would soon become excessive but for the many enemies by which their spawn is destroyed. No fewer than 700,000 eggs have been found in the ovaries of a single Carp, and that, too, by no means an individual of the largest size. Their growth is very rapid, more so perhaps than that of any other Freshwater fish, and the size which they sometimes attain is very considerable. In certain lakes in Germany individuals are occasionally taken weighing thirty or forty pounds; and Pallas relates that they occur in the Volga five feet in length, and even of greater weight than the examples just alluded to. The largest of which we have any account is that mentioned by Bloch, taken near Frankfort-on-the-Oder, which weighed seventy pounds, and measured nearly nine feet in length,—a statement the accuracy of which is very much open to doubt.

Like other domesticated animals the Carp is subject to variation; some individuals, especially when they have been bred under unfavourable circumstances, have a lean and low body; others are shorter and higher. Some have lost every trace of scales, and are called "Leather-carps;" others retain them along the lateral line and on the back only ("Spiegelkarpfen" of the Germans). Finally, in some are the fins much prolonged, as in certain varieties of the Gold-fish. Cross-breeds between the Carp and the Crucian Carp are of common occurrence. The Carp is much more esteemed as food in inland countries than in countries where the more delicate kinds of sea fishes can be obtained.

Carassius differs from *Cyprinus* in lacking barbels; its pharyngeal teeth are compressed, in a single series, 4–4.

Two well-known species belong to this genus. The "Crucian Carp" (*C. carassius*, "Karausche") is generally distributed over Central and Northern Europe, and extends into Italy and Siberia. It inhabits stagnant waters only, and is so tenacious of life that it will survive a lengthened sojourn in the smallest pools, where, however, it remains stunted; whilst in favourable localities it attains to a length of twelve inches. It is much subject to variation of form; very lean examples are commonly called "Prussian Carps." Its usefulness consists in keeping ponds clean from a superabundance of vegetable growth, and in serving as food for other more esteemed fishes. The second species is the "Goldfish," *Carassius auratus*. It is of very common occurrence in a wild state in China and the warmer parts of Japan, being entirely similar in colour to the Crucian Carp. In a domesticated state it loses the black or brown chromatophors, and becomes of a golden-yellow colour; perfect Albinos are comparatively scarcer. Many varieties and monstrosities have been produced during the long period of its domestication; the variety most highly priced at present being the so-called "Telescope-fish," of which a figure is annexed. The Gold-fish

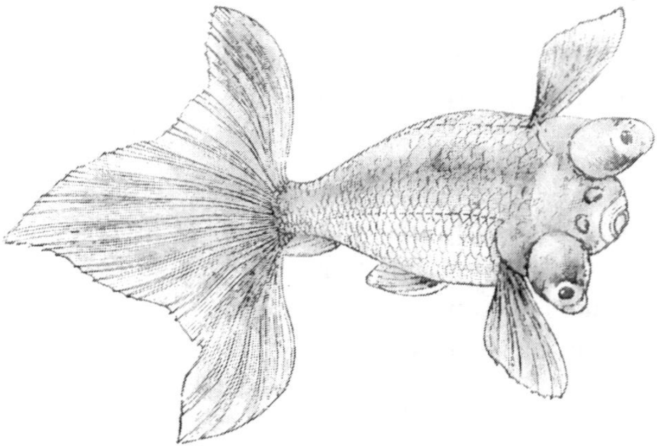

Fig. 273.—Cyprinus auratus, var.

is said to have been first brought to England in the year 1691, and is now distributed over nearly all the civilised parts of the world.

CATLA.—Scales of moderate size. Dorsal fin without osseous ray, with more than nine branched rays, commencing nearly opposite to the ventrals. Snout broad, with the integuments very thin; there is no upper lip, the lower with a free continuous posterior margin. Symphysis of the mandibulary bones loose, with prominent tubercles. Mouth anterior. Barbels none. Gill-rakers very long, fine, and closely set. Pharyngeal teeth, 5. 3. 2.–2. 3. 5.

The "Catla" (*C. buchanani*), one of the largest Carps of the Ganges, growing to a length of more than three feet, and esteemed as food.

LABEO.—Scales of moderate or small size. Dorsal fin without osseous ray, with more than nine branched rays, commencing somewhat in advance of the ventrals. Snout obtusely rounded, the skin of the maxillary region being more or less thickened, forming a projection beyond the mouth. Mouth transverse, inferior, with the lips thickened, each or one of them being provided with an inner transverse fold, which is covered with a deciduous horny substance forming a sharp edge, which, however, does not rest upon the bone as base, but is soft and movable. Barbels very small, two or four; the maxillary barbels more or less hidden in a groove behind the angle of the mouth. Anal scales not enlarged. Pharyngeal teeth uncinate, 5. 4. 2.-2. 4. 5. Snout generally more or less covered with hollow tubercles.

About thirteen species are known from rivers of tropical Africa and the East Indies.

DISCOGNATHUS.—Scales of moderate size. Dorsal fin without osseous ray, with not more than nine branched rays, commencing somewhat in advance of the ventrals. Snout obtusely rounded, more or less depressed, projecting beyond the mouth, more or less tubercular. Mouth inferior, transverse, crescent-shaped; lips broad, continuous, with an inner sharp edge of the jaws, covered with horny substance on the lower jaw; upper lip more or less distinctly fringed; lower lip modified into a suctorial disk, with free anterior and posterior margins. Barbels two or four; if two, the upper are absent. Anal scales not enlarged. Pectoral fins horizontal. Pharyngeal teeth, 5. 4. 2.–2. 4. 5.

A small fish (*D. lamta*), extremely abundant in almost all the mountain streams from Abyssinia and Syria to Assam.

CAPOËTA.—Scales small, of moderate or large size. Dorsal fin with or without a strong osseous ray, with not more than nine branched rays. Snout rounded, with the mouth transverse and at its lower side; each mandible angularly bent inwards in front, the anterior mandibular edge being nearly straight, sharpish, and covered with a horny brown layer. No lower labial fold. Barbels two (rarely four), or entirely absent. Anal scales not conspicuously enlarged. Pharyngeal teeth compressed, truncated, 5 or 4. 3. 2.–2. 3. 4 or 5.

Characteristic of the fauna of Western Asia; one species from Abyssinia. Of the fifteen species known *C. damascina* deserves to be specially mentioned, being abundant in the Jordan and other rivers of Syria and Asia Minor.

BARBUS.—Scales of small, moderate, or large size. Dorsal fin generally with the (third) longest simple ray ossified, enlarged, and frequently serrated; never, or only exceptionally, with more than nine branched rays, commencing opposite or nearly opposite to the root of the ventral fin. Eyes without adipose eyelid. Anal fin frequently very high. Mouth arched, without inner folds, inferior or anterior; lips without horny covering. Barbels short, four, two, or none. Anal scales not enlarged. Pharyngeal teeth 5. 4 or 3. 3 or 2.–2 or 3. 3 or 4. 5. Snout but rarely with tubercles or pore-like grooves.

No other genus of Cyprinoids is composed of so many species as the genus of "Barbels," about 200 being known from the tropical and temperate parts of the Old World; it is not represented in the New World. Although the species differ much from each other in the form of the body, number of barbels, size of the scales, strength of the first dorsal ray or spine, etc., the transition between the extreme forms is so perfect that no further generic subdivision should be attempted. Some attain a length of six feet, whilst others never exceed a length of two inches. The most noteworthy are the large Barbels of the Tigris (*B. subquincunciatus, B. esocinus, B. scheich, B. sharpeyi*); the common Barbel of Central Europe and Great Britain (*B. vulgaris*); the "Bynni" of the Nile (*B. bynni*); *B. canis* from the Jordan; the "Mahaseer" of the mountain streams of India (*B. mosal*), probably the largest of all species, the scales of which are sometimes as large as the palm of a hand. The small, large-scaled species are especially numerous in the East Indies and the fresh waters of Tropical Africa.

THYNNICHTHYS.—Scales small. Dorsal fin without an osseous ray, with not more than nine branched rays, commencing nearly opposite the ventrals. Head large, strongly compressed; eye without well-developed adipose membrane, in the middle of the depth of the head. Snout with the integuments very thin; there is no upper lip, and the lower jaw has a thin labial fold on the sides only. Mouth anterior and lateral; barbels none. Gill-rakers none; laminæ branchiales long, half as long as the postorbital portion of the head; pseudobranchiæ none. Pharyngeal teeth lamelliform, with flat oblong crown, 5. 3 or 4. 2–2. 4 or 3. 5, the teeth of the three series being wedged into one another.

Three species from the East Indies.

OREINUS.—Scales very small. Dorsal fin with a strong osseous serrated ray, opposite to the ventrals. Snout rounded, with the mouth transverse, and at its lower side; mandibles broad, short, and flat, loosely joined together; margin of the jaw covered with a thick horny layer; a broad fringe-like lower lip, with free posterior margin. Barbels four. Vent and anal fin in a sheath, covered with enlarged tiled scales. Pharyngeal teeth pointed, more or less hooked, 5. 3. 2–2. 3. 5.

Three species from mountain streams of the Himalayas.

SCHIZOTHORAX.—Hill-barbels, with the same singular sheath on each side of the vent, as in the preceding genus; but they differ in having the mouth normally formed, with mandibles of the usual length and width.

Seventeen species are known from fresh waters of the Himalayas, and north of them. Other genera from the same region, and with the anal sheath, are *Ptychobarbus*, *Gymnocypris*, *Schizopygopsis*, and *Diptychus*.

GOBIO.—Scales of moderate size; lateral line present. Dorsal fin short, without spine. Mouth inferior; mandible not projecting beyond the upper jaw when the mouth is open; both jaws with simple lips; a small but very distinct barbel at the angle of the mouth, quite at the extremity of the maxillary. Gill-rakers very short; pseudobranchiæ. Pharyngeal teeth, 5. 3 or 2.–2 or 3. 5, hooked at the end.

The "Gudgeons" are small fishes of clear fresh waters of Europe; they are, like the barbels, animal feeders. In Eastern Asia they are represented by two closely allied genera, *Ladislavia* and *Pseudogobio*.

CERATICHTHYS.—Scales of moderate or small size; lateral line present. Dorsal fin short, without spine, not or but slightly in advance of the ventrals. Mouth subinferior; the lower jaw does not project beyond the upper when the mouth is open; intermaxillaries protractile from below the maxillaries; both jaws with thickish lips; a small barbel at the angle of the mouth, quite at the extremity of the maxillary. Gill-rakers very short and few in number: pseudobranchiæ. Pharyngeal teeth 4–4. hooked at the end (sometimes 4, 1–1. 4).

About ten species are known from North America; they are small, and called "Chub" in the United States. *C. biguttatus* is, perhaps, the most widely-diffused Freshwater-fish in the United States, and common everywhere. Breeding males have generally a red spot on each side of the head.

Other similar genera from the fresh waters of North America, and generally called "Minnows," are *Pimephales* (the "Black Head"), *Hyborhynchus*, *Hybognathus*, *Campostoma* (the "Stone-lugger"), *Ericymba*, *Cochlognathus*, *Exoglossum* (the "Stone Toter" or "Cut-lips"), and *Rhinichthys* (the "Long-nosed Dace").

The remaining Old World genera belonging to the group *Cyprinina* are *Cirrhina*, *Dangila*, *Osteochilus*, *Barynotus*, *Tylognathus*, *Abrostomus*, *Crossochilus*, *Epalzeorhynchus*, *Barbichthys*, *Amblyrhynchichthys*, *Albulichthys*, *Aulopyge*, *Bungia*, and *Pseudorasbora*.

III. ROHTEICHTHYINA.—*Anal fin very short, with not more*

than six branched rays. Dorsal fin behind ventrals. Abdomen compressed. Lateral line running along the median line of the tail. Mouth without barbels. Pharyngeal teeth in a triple series.

One genus and species only, *Rohteichthys microlepis*, from Borneo and Sumatra.

IV. Leptobarbina.—*Anal fin very short, with not more than six branched rays. Dorsal fin opposite to ventrals. Abdomen not compressed. Lateral line running in the lower half of the tail. Barbels present, not more than four in number. Pharyngeal teeth in a triple series.*

One genus and species only, *Leptobarbus hoevenii*, from Borneo and Sumatra.

V. Rasborina.—*Anal fin very short, with not more than six branched rays. Dorsal fin inserted behind the origin of the ventrals. Abdomen not compressed. Lateral line running along the lower half of the tail, if complete. Mouth sometimes with barbels, which are never more than four in number. Pharyngeal teeth in a triple or single series. Air-bladder present, without osseous covering.*

Rasbora.—Scales large, or of moderate size, there being generally four and a half longitudinal series of scales between the origin of the dorsal fin and the lateral line, and one between the lateral line and the ventral. Lateral line curved downwards. Dorsal fin with seven or eight branched rays, not extending to above the anal, which is seven-rayed. Mouth of moderate width, extending to the front margin of the orbit, with the lower jaw slightly prominent, and provided with three prominences in front, fitting into grooves of the upper jaw; barbels none, in one species two. Gill-rakers short, lanceolate. Pseudobranchiæ. Pharyngeal teeth in three series, uncinate.

Thirteen species of small size from the East Indian Continent and Archipelago, and from rivers on the east coast of Africa.

AMBLYPHARYNGODON.—Scales small; lateral line incomplete. Dorsal fin without an osseous ray, with not more than nine branched rays, commencing a little behind the origin of the ventrals. Head of moderate size, strongly compressed; eye without adipose membrane; snout with the integuments very thin; there is no upper lip, and the lower jaw has a short labial fold on the sides only. Mouth anterior, somewhat directed upwards, with the lower jaw prominent. Barbels none. Gill-rakers extremely short; pseudobranchiæ. Pharyngeal teeth molar-like, with their crowns concave, 3. 2. 1.–1. 2. 3. Intestinal tract narrow, with numerous convolutions.

Three species of small size from the Continent of India.

To the same group belong *Luciosoma*, *Nuria*, and *Aphyocypris*, from the same geographical region.

VI. SEMIPLOTINA.—*Anal fin short, with seven branched rays, not extending forwards to below the dorsal. Dorsal fin elongate, with an osseous ray. Lateral line running along the middle of the tail. Mouth sometimes with barbels.*

Two genera: *Cyprinion*, from Syria and Persia, and *Semiplotus* from Assam.

VII. XENOCYPRIDINA.—*Anal fin rather short, with seven or more branched rays, not extending forwards to below the dorsal fin. Dorsal short, with an osseous ray. Lateral line running along the middle of the tail. Mouth sometimes with barbels. Pharyngeal teeth in a triple or double series.*

Three genera: *Xenocypris* and *Paracanthobrama* from China; and *Mystacoleucus* from Sumatra.

VIII. LEUCISCINA.—*Anal fin short or of moderate length, with from eight to eleven branched rays, not extending forwards to below the dorsal. Dorsal fin short, without osseous ray. Lateral line, if complete, running along, or nearly in, the middle of the tail. Mouth generally without barbels. Pharyngeal teeth in a single or double series.*

LEUCISCUS.—Body covered with imbricate scales. Dorsal

fin commencing opposite, rarely behind, the ventrals. Anal fin generally with from nine to eleven, rarely with eight (in small species only), and still more rarely with fourteen rays. Mouth without structural peculiarities; lower jaw not trenchant; barbels none. Pseudobranchiæ. Pharyngeal teeth conical or compressed, in a single or double series. Intestinal tract short, with only a few convolutions.

The numerous species of this genus are comprised under the name of "White-fish;" they are equally abundant in the northern temperate zone of both hemispheres, about forty species being known from the Old World, and about fifty from the New. The most noteworthy species of the former Fauna are the "Roach" (*L. rutilus*, see Fig. 21, p. 50), common all over Europe north of the Alps; the "Chub" (*L. cephalus*), extending into Northern Italy and Asia Minor; the "Dace" (*L. leuciscus*), a companion of the Roach; the "Id" or "Nerfling" (*L. idus*), from the central and northern parts of Continental Europe, domesticated in some localities of Germany, in this condition assuming the golden hue of semi-albinism, like a Gold-fish, and then called the "Orfe;" the "Rudd," or "Red-eye" (*L. erythrophthalmus*), distributed all over Europe and Asia Minor, and distinguished by its scarlet lower fins; the "Minnow" (*L. phoxinus*), abundant everywhere in Europe, and growing to a length of seven inches in favourable localities. The North American species are much less perfectly known; the smaller ones are termed "Minnows," the larger "Shiner" or "Dace." The most common are *L. cornutus* (Red-fin, Red Dace); *L. neogæus*, a minnow resembling the European species, but with incomplete lateral line; *L. hudsonius*, the "Spawn-eater" or "Smelt."

TINCA.—Scales small, deeply embedded in the thick skin; lateral line complete. Dorsal fin short, its origin being opposite the ventral fin; anal short; caudal subtruncated. Mouth anterior; jaws with the lips moderately developed; a barbel at the angle of the mouth. Gill-rakers short, lanceolate; pseudobranchiæ rudimentary. Pharyngeal teeth 4 or 5.–5, cuneiform, slightly hooked at the end.

Fig. 274.—The Tench (Tinca tinca).

Only one species of "Tench" is known (*T. tinca*), found all over Europe in stagnant waters with soft bottom. The "Golden Tench" is only a variety of colour, an incipient albinism like the Gold-fish and Id. Like most other Carps of this group it passes the winter in a state of torpidity, during which it ceases to feed. It is extremely prolific, 297,000 ova having been counted in one female; its spawn is of a greenish colour.

LEUCOSOMUS.—Scales of moderate or small size; lateral line present. Dorsal fin commencing opposite, or nearly opposite, to the ventral. Anal fin short. Mouth anterior or sub-anterior; intermaxillaries protractile. A very small barbel at the extremity of the maxillary. Lower jaw with rounded margin, and with the labial folds well developed laterally. Gill-rakers short; pseudobranchiæ. Pharyngeal teeth in a double series.

A North American genus, to which belong some of the most common species of the United States. *L. pulchellus* (the "Fall-fish," "Dace," or "Roach"), one of the largest White-fishes of the Eastern States, attaining to a length of 18 inches, and abundant in the rapids of the larger rivers. *L. corporalis* (the "Chub"), common everywhere from New England to the Missouri region.

CHONDROSTOMA.—Scales of moderate size or small. Lateral.

line terminating in the median line of the depth of the tail. Dorsal fin with not more than nine branched rays, inserted above the root of the ventrals. Anal fin rather elongate, with ten or more rays. Mouth inferior, transverse, lower jaw with a cutting edge, covered with a brown horny layer. Barbels none. Gill-rakers short, fine; pseudobranchiæ. Pharyngeal teeth 5 or 6 or 7.–7 or 6 or 5, knife-shaped, not denticulated. Peritoneum black.

Seven species from the Continent of Europe and Western Asia.

Other Old World genera belonging to the *Leuciscina* are *Mylolcucus*, *Ctenopharyngodon*, and *Paraphoxinus*; from North America: *Mylopharodon*, *Meda*, *Orthodon*, and *Acrochilus*.

IX. Rhodeina.—*Anal fin of moderate length, with from nine to twelve branched rays, extending forwards to below the dorsal. Dorsal fin short, or of moderate length. Lateral line, if complete, running along or nearly in the middle of the tail. Mouth with very small, or without any barbels. Pharyngeal teeth in a single series.*

Very small roach-like fishes inhabiting chiefly Eastern Asia and Japan, one species (*Rh. amarus*) advancing into Central Europe. The thirteen species known have been distributed among four genera, *Achilognathus*, *Acanthorhodeus*, *Rhodeus*, and *Pseudoperilampus*. In the females a long external urogenital tube is developed annually during the spawning season. The European species is known in Germany by the name of "Bitterling."

X. Danionina.—*Anal fin of moderate length or elongate, with not less, and generally more, than eight branched rays. Lateral line running along the lower half of the tail. Mouth with small, or without any, barbels. Abdomen not trenchant. Pharyngeal teeth in a triple or double series.*

Small fishes from the East Indian Continent, Ceylon, the East Asiatic Islands, and a few from East African Rivers,

The genera belonging to this group are *Danio*, *Pteropsarion*, *Aspidoparia*, *Barilius*, *Bola*, *Scharca*, *Opsariichthys*, *Squaliobarbus*, and *Ochetobius*: altogether about forty species.

XI. Hypophthalmichthyina.—*Anal fin elongate. Lateral line running nearly along the median line of the tail. Mouth without barbels. Abdomen not trenchant. No dorsal spine. Pharyngeal teeth in a single series.*

One genus (*Hypophthalmichthys*) with two species from China.

XII. Abramidina.—*Anal fin elongate. Abdomen, or part of the abdomen, compressed.*

Abramis.—Body much compressed, elevated, or oblong. Scales of moderate size. Lateral line present, running in the lower half of the tail. Dorsal fin short, with spine, opposite to the space between ventrals and anals. Lower jaw generally shorter, and rarely longer than the upper. Both jaws with simple lips, the lower labial fold being interrupted at the symphysis of the mandible. Upper jaw protractile. Gill-rakers rather short; pseudobranchiæ. The attachment of the branchial membrane to the isthmus takes place at some distance behind the vertical from the orbit. Pharyngeal teeth in one or two

Fig. 275.—The Bream (Abramis brama).

series, with a notch near the extremity. Belly behind the ventrals compressed into an edge, the scales not extending across it.

The "Breams" are represented in the temperate parts of

both northern hemispheres; in Europe there occur the "Common Bream, *A. brama;* the "Zope," *A. ballerus; A. sapa;* the "Zärthe," *A. vimba; A. elongatus;* the "White Bream," *A. blicca; A. bipunctatus.* Of these *A. brama* and *A. blicca* are British; the former being one of the most common fishes, and sometimes attaining to a length of two feet. Crosses between these two species, and even with other Cyprinoids, are not rare. Of the American species *A. americanus* ("Shiner," "Bream") is common and widely distributed; like the European Bream it lives chiefly in stagnant waters or streams with a slow current.

Aspius.—Body oblong; scales of moderate size; lateral line complete, terminating nearly in the middle of the depth of the tail. Dorsal fin short, without spine, opposite to the space between the ventrals and anal; anal fin elongate, with thirteen or more rays. Lower jaw more or less conspicuously projecting beyond the upper. Lips thin, simple, the lower labial fold being at the symphysis; upper jaw but little protractile. Gill-rakers short and widely set; pseudobranchiæ. The attachment of the branchial membrane to the isthmus takes place below the hind margin of the orbit. Pharyngeal teeth hooked, 5. 3.–3 or 2. 5 or 4 Belly behind the ventrals compressed, the scales covering the edge.

Four species from Eastern Europe to China.

Alburnus.—Body more or less elongate; scales of moderate size; lateral line present, running below the median line of the tail. Dorsal fin short, without spine, opposite to the space between ventrals and anal; anal fin elongate, with more than thirteen rays. Lower jaw more or less conspicuously projecting beyond the upper. Lips thin, simple, the lower labial fold being interrupted at the symphysis of the mandible. Upper jaw protractile. Gill-rakers slender, lanceolate, closely set; pseudobranchiæ. The attachment of the branchial membrane to the isthmus takes place below the hind margin of the orbit. Pharyngeal teeth in two series, hooked. Belly behind the ventrals compressed into an edge, the scales not extending across it.

"Bleak" are numerous in Europe and Western Asia, fifteen species being known. The common Bleak (*A. alburnus*) is found north of the Alps only, and represented by another species (*A. alburnellus*, "Alborella," or "Avola") in Italy.

Of the other genera referred to this group, *Leucaspius* and *Pelecus* belong to the European Fauna; *Pelotrophus* is East African; all the others occur in the East Indies or the temperate parts of Asia, viz. *Rasborichthys*, *Elopichthys*, *Acanthobrama* (Western Asia), *Osteobrama*, *Chanodichthys*, *Hemiculter*, *Smiliogaster*, *Toxabramis*, *Culter*, *Eustira*, *Chela*, *Pseudolabuca*, and *Cachius*.

XIII. Homalopterina.—*Dorsal and anal fins short, the former opposite to ventrals. Pectoral and ventral fins horizontal, the former with the outer rays simple. Barbels six or none. Air-bladder absent. Pharyngeal teeth in a single series, from ten to sixteen in number.*

Inhabitants of hill-streams in the East Indies; they are of small size and abundant where they occur. Thirteen species are known belonging to the genera *Homaloptera*, *Gastromyzon*, *Crossostoma*, and *Psilorhynchus*.

XIV. Cobitidina.—*Mouth surrounded by six or more barbels. Dorsal fin short or of moderate length; anal fin short. Scales small, rudimentary, or entirely absent. Pharyngeal teeth in a single series, in moderate number. Air-bladder partly or entirely enclosed in a bony capsule. Pseudobranchiæ none:* Loaches.

Misgurnus.—Body elongate, compressed. No sub-orbital spine. Ten or twelve barbels, four belonging to the mandible. Dorsal fin opposite to the ventrals; caudal rounded.

Four species from Europe and Asia. *M. fossilis* is the largest of European Loaches, growing to a length of ten inches; it occurs in stagnant waters of eastern and southern

Germany and northern Asia. In China and Japan it is replaced by an equally large species, *M. anguillicaudatus.*

NEMACHILUS.—No erectile sub-orbital spine. Six barbels, none at the mandible. Dorsal fin opposite to the ventrals.

The greater number of Loaches belong to this genus; about fifty species are known from Europe and temperate Asia; such species as extend into tropical parts inhabit streams of high altitudes. Loaches are partial to fast-running streams with stony bottom, and exclusively animal feeders. In spite of their small size they are esteemed as food where they occur in sufficient abundance. The British species, *N. barbatulus,* is found all over Europe except Denmark and Scandinavia.

COBITIS.—Body more or less compressed, elongate; back not arched. A small, erectile, bifid sub-orbital spine below the eye. Six barbels only on the upper jaw. Dorsal fin inserted opposite to ventrals. Caudal rounded or truncate.

Only three species are known, of which *C. tænia* occurs in Europe. It is scarce and very local in Great Britain.

BOTIA.—Body compressed, oblong; back more or less arched. Eyes with a free circular eyelid; an erectile bifid sub-orbital spine. Six barbels on the upper jaw, sometimes two others at the mandibulary symphysis. Dorsal fin commencing in advance of the root of the ventrals; caudal fin forked. Air-bladder consisting of two divisions: the anterior enclosed in a partly osseous capsule, the posterior free, floating in the abdominal cavity.

Fig. 276.—Botia rostrata. From Bengal.

This genus is more tropical than any of the preceding, and the majority of the species (eight in number) are finely coloured. The more elevated form of their body, and the imperfect ossification of the capsules of the air-bladder, the divisions of which are not side by side, but placed in the longitudinal axis of the body, indicate likewise that this genus is more adapted for still waters of the plains than for the currents of hill-streams.

Other genera from tropical India are *Lepidocephalichthys*, *Acanthopsis*, *Oreonectes* (hills near Hong-Kong), *Paramisgurnus* (Yan-tse-Kiang), *Lepidocephalus*, *Acanthophthalmus*, and *Apua*.

FOURTH FAMILY—KNERIIDÆ.

Body scaly, head naked. Margin of the upper jaw formed by the intermaxillaries. Dorsal and anal fins short, the former belonging to the abdominal portion of the vertebral column. Teeth none, either in the mouth or pharynx. Barbels none. Stomach siphonal; no pyloric appendages. Pseudobranchiæ none. Branchiostegals three; air-bladder long, not divided. Ovaries closed.

Small loach-like fishes from fresh waters of tropical Africa; two species only, *Kneria angolensis* and *K. spekii*, are known.

FIFTH FAMILY—CHARACINIDÆ.

Body covered with scales, head naked; barbels none. Margin of the upper jaw formed by the intermaxillaries in the middle and by the maxillaries laterally. Generally a small adipose fin behind the dorsal. Pyloric appendages more or less numerous; air-bladder transversely divided into two portions, and communicating with the organ of hearing by means of the auditory ossicles. Pseudobranchiæ none.

The fishes of this family are confined to the fresh waters of Africa, and especially of tropical America, where they replace the Cyprinoids, with which family, however, they have but little in common as far as their structural characteristics are concerned. Their co-existence in Africa with

Cyprinoids proves only that that continent is nearer to the original centre, from which the distribution of Cyprinoids commenced than tropical America. The family includes herbivorous as well as strictly carnivorous forms; some are toothless, whilst others possess a most formidable dentition. The family contains so many diversified forms as to render a subdivision into groups necessary. They have not yet been obtained in fossiliferous strata.

I. ERYTHRININA.—*Adipose fin absent.*

The sixteen species of this group belong to the fauna of tropical America, and are referred to the genera *Macrodon, Erythrinus, Lebiasina, Nannostomus, Pyrrhulina, and Corynopoma.*

II. CURIMATINA.—*A short dorsal and an adipose fin; dentition imperfect. Tropical America.*

CURIMATUS.—Dorsal fin placed nearly in the middle of the body; anal rather short or of moderate length; ventrals below the dorsal. Body oblong or elevated, with the belly rounded or flattened before the ventrals. Cleft of the mouth transverse, lips none, margins of the jaws trenchant. No teeth whatever. Intestinal tract very long and narrow.

About twenty species are known, of rather small size.

The other genera of this group have teeth, but they are either rudimentary or absent in some part of the jaws: *Prochilodus, Cænotropus, Hemiodus, Saccodon, Parodon.*

III. CITHARININA.—*A rather long dorsal and an adipose fin; minute labial teeth. Tropical Africa.*

One genus only, *Citharinus*, with two species, is known. Common in the Nile, attaining to a length of three feet.

IV. ANASTOMATINA.—*A short dorsal and an adipose fin; teeth in both jaws well developed; the gill-membranes grown to the isthmus; nasal openings remote from each other. Tropical America.*

Leporinus.—Dorsal fin placed nearly in the middle of the length of the body; anal short; ventrals below the dorsal. Body oblong, covered with scales of moderate size; belly rounded. Cleft of the mouth small, with the lips well developed; teeth in the intermaxillary and mandible, few in number, flattened, with the apex more or less truncated, and not serrated; the middle pair of teeth is the longest in both jaws; palate toothless.

This genus is generally distributed in the rivers east of the Andes; about twenty species are known, some of which, like *L. frederici*, *L. megalepis*, are very common. They are well marked by black bands or spots, and rarely grow to a length of two feet, being generally much smaller.—The other genera belonging to this group are *Anastomus* and *Rhytiodus*.

V. Nannocharacina.—*A short dorsal and an adipose fin; teeth in both jaws well developed; notched incisors. The gill-membranes are grown to the isthmus. Nostrils close together.*

One genus, *Nannocharax*, with two species only, from the Nile and Gaboon; very small.

VI. Tetragonopterina.—*A short dorsal and an adipose fin; the teeth in both jaws well developed, compressed, notched, or denticulated; the gill-membranes free from the isthmus, and the nasal openings close together. South America and Tropical Africa.*

Alestes.—The dorsal fin is placed in the middle of the length of the body, above or behind the ventrals; anal fin rather long. Body oblong, covered with scales of moderate or large size; belly rounded. Cleft of the mouth rather small. Maxillary teeth none; intermaxillary teeth in two series; those of the front series more or less compressed, more or less distinctly tricuspid; the teeth of the hinder series are broad, molar-like, each armed with several pointed tubercles. Teeth in the lower jaw in two series; those in the front series laterally compressed, broader behind than in front; the hinder series is composed of two conical teeth. All the teeth are strong, few in number.

Fourteen species from Tropical Africa; several inhabit the Nile, of which the "Raches" (*A. dentex* and *A. kotschyi*) are the most common.

TETRAGONOPTERUS.—The dorsal fin is placed in the middle of the length of the body, above or immediately behind the ventrals; anal fin long. Body oblong or elevated, covered with scales of moderate size; belly rounded. Cleft of the mouth of moderate width. Anterior teeth strong, lateral teeth small. Intermaxillary and mandibulary teeth sub-equal in size, with a compressed and notched crown, the former in a double, the latter in a single series; maxillary with a few teeth near its articulation, rarely with the entire edge denticulated.

Of all the genera of this family *Tetragonopterus* is represented by the greatest number of species; about fifty are known. Some of them seem to have a very wide range, whilst others are merely local. All are of small size, rarely exceeding a length of eight inches.

CHIRODON.—Dorsal fin placed in the middle of the length of the body, behind the ventrals; anal long or of moderate length. Body oblong, covered with scales of moderate size; lateral line not continued to the tail. Belly rounded before the ventrals. Cleft of the mouth narrow, maxillary short. A single series of small serrated teeth in the intermaxillary and mandibulary; maxillary teeth none.

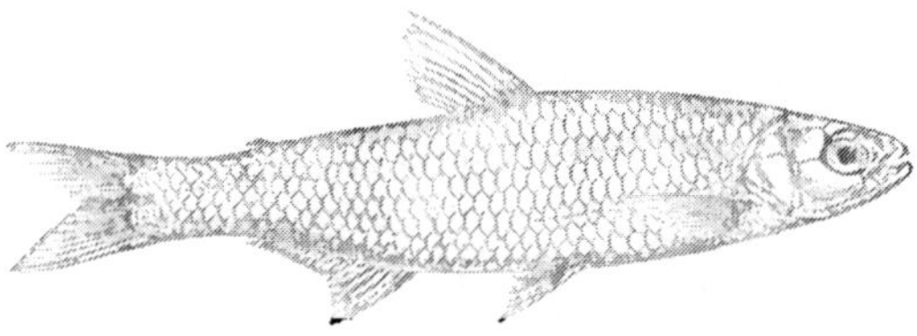

Fig. 277.—Chirodon alburnus.

Three species of small size from various parts of South America; the species figured is represented of the natural size, and comes from the Upper Amazons.

MEGALOBRYCON.—Dorsal fin placed in middle of the length of the body, immediately behind the ventrals. Anal long. Abdomen rounded in front of, and somewhat compressed behind, the ventrals. Cleft of the mouth of moderate width. Teeth notched, in a triple series in the intermaxillary, and in a single in the maxillary and mandible; no other teeth behind the mandibulary teeth or on the palate. Scales of moderate size, with the free portion striated.

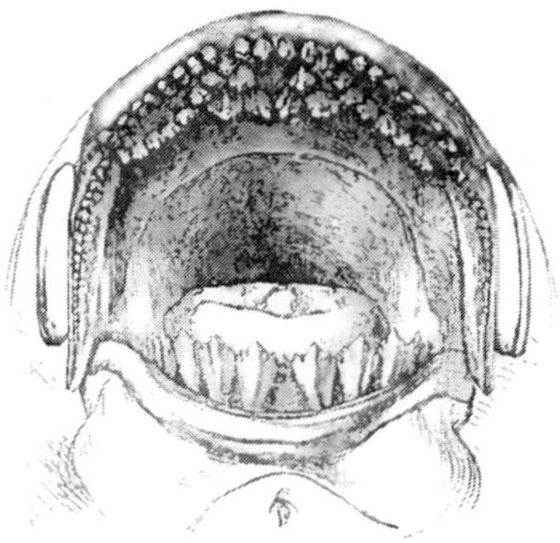

Fig. 278.—Megalobrycon cephalus.

One species from the Upper Amazons (*M. cephalus*). Specimens more than one foot long have been obtained.

GASTROPELECUS.—Dorsal fin placed behind the middle of the length of the body, above the anal; anal long; pectoral long; ventrals very small or rudimentary. Body strongly compressed, with the thoracic region dilated into a sub-semicircular disk. Scales of moderate size. Lateral line descending obliquely backwards towards the origin of the anal fin. The lower profile compressed into an acute ridge. Cleft of the mouth of moderate width; teeth compressed, tricuspid, in one or two series in the intermaxillary, and in a single in the mandible; maxillary with a few minute conical teeth; palate toothless.

Three specimens of this singular form are known from Brazil and the Guyanas; they are of very small size.

The majority of the other genera belonging to this group are South American, viz. *Piabucina*, *Scissor*, *Pseudochalceus*, *Aphyocharax*, *Chalceus*, *Brycon*, *Chalcinopsis*, *Bryconops*, *Creagrutus*, *Chalcinus*, *Piabuca*, *Paragoniates*, and *Agoniates*; two only are African, viz. *Nannæthiops*, which represents the South American *Tetragonopterus*, and *Brycanæthiops*, which is allied to Brycon.

VII. HYDROCYONINA.—*A short dorsal and an adipose fin;*

teeth in both jaws well developed and conical; gill membranes free from the isthmus; nasal openings close together. South America and Tropical Africa. Fishes of prey.

HYDROCYON.—The dorsal fin is in the middle of the length of the body, above the ventrals; anal of moderate length. Body oblong, compressed, covered with scales of moderate size; belly rounded. Cleft of the mouth wide, without lips; the intermaxillaries and mandibles are armed with strong pointed teeth, widely set and few in number; they are received in notches of the opposite jaw, and visible externally when the mouth is closed. Palate toothless. Cheeks covered with the enlarged suborbital bones. Orbit with an anterior and posterior adipose eyelid. Intestinal tract short.

Four species from Tropical Africa; two occur in the Nile, *H. forskalii* being abundant, and well known by the names "Kelb el bahr" and "Kelb el moyeh." Their formidable dentition renders them most destructive to other fishes; they grow to a length of four feet.

CYNODON.—Dorsal fin placed behind, or nearly in, the middle of the length of the body, behind the ventrals; anal long. Head and body compressed, oblong, the latter covered with very small scales; belly compressed, keeled. Teeth in the intermaxillary, maxillary, and mandible in a single series, conical, widely set, of unequal size; a pair of very large canine teeth anteriorly in the lower jaw, received in two grooves on the palate; palate with patches of minute granulated teeth. The outer branchial arch without gill-rakers, but with very short tubercles.

Four species from Brazil and the Guyanas; they are as formidable fishes of prey as the preceding, and grow to the same size.

With the exception of *Sarcodaces*, all the remaining genera of this group belong to the fauna of Tropical America, viz. *Anacyrtus, Hystricodon, Salminus, Oligosarcus, Xiphorhamphus*, and *Xiphostoma.*

VIII. DISTICHODONTINA.—*Dorsal fin rather elongate; adipose fin present. Gill-membranes attached to the isthmus; belly rounded. Tropical Africa.*

The species, ten in number, belong to one genus only (*Distichodus*), well known on the Nile under the name of "Nefaseh." They grow to a considerable size, being sometimes four feet long and one and a half foot deep. They are used as food.

IX. ICHTHYBORINA.—*An adipose fin; number of dorsal rays increased* (12-17); *gill-membranes free from the isthmus. Belly rounded; canine teeth. Tropical Africa.*

Two genera only: *Ichthyborus* from the Nile, and *Phago* from West Africa. Small fishes of very rare occurrence.

X. CRENUCHINA.—*Dorsal fin rather elongate; an adipose fin. Gill-membranes free from the isthmus, with the belly rounded, and without canine teeth.*

This small group is represented in the Essequibo by a single species, *Crenuchus spilurus*, and by another in West Africa, *Xenocharax spilurus.*

XI. SERRASALMONINA.—*Dorsal fin rather elongate; an adipose fin. Gill-membranes free from the isthmus; belly serrated. Tropical America.*

Although the fishes of this family do not attain any considerable size, the largest scarcely exceeding two feet in length, their voracity, fearlessness, and number renders them a perfect pest in many rivers of tropical America. In all, the teeth are strong, short, sharp, sometimes lobed incisors, arranged in one or more series; by means of them they cut off a mouthful of flesh as with a pair of scissors; and any animal falling into the water where these fishes abound is immediately attacked and cut in pieces in an incredibly short time. They assail persons entering the water, inflicting dangerous wounds before the victims are able to make their escape.

In some localities it is scarcely possible to catch fishes with the hook and line, as the fish hooked is immediately attacked by the "Caribe" (as these fishes are called), and torn to pieces before it can be withdrawn from the water. The Caribes themselves are rarely hooked, as they snap the hook or cut the line. The smell of blood is said to attract at once thousands of these fishes to a spot. They are most abundant in the Brazils and Guyanas; some forty species are known, and referred to the genera *Mylesinus, Serrasalmo, Myletes,* and *Catoprion.*

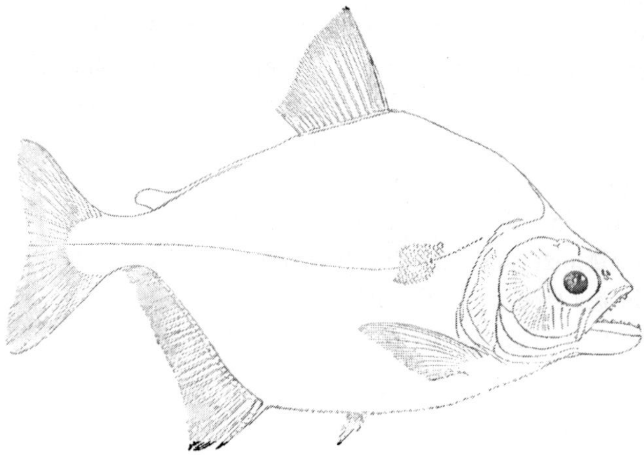

Fig. 279.—Serrasalmo scapularis, from the Essequibo.

SIXTH FAMILY—CYPRINODONTIDÆ.

Head and body covered with scales; barbels none. Margin of the upper jaw formed by the intermaxillaries only. Teeth in both jaws; upper and lower pharyngeals with cardiform teeth. Adipose fin none; dorsal fin situated on the hinder half of the body. Stomach without blind sac; pyloric appendages none. Pseudobranchiæ none; air-bladder simple, without ossicula auditus.

Small fishes, inhabiting fresh, brackish, and salt water of Southern Europe, Africa, Asia, and America. The majority are viviparous; and to facilitate copulation the anal fin of the adult male of many species is modified into a copulatory organ, which is probably (partially at least) introduced into the vulva of the female; but it is uncertain whether it serves to conduct the semen, or merely to give the male a firmer hold of the female during the act. Also secondary sexual differences are developed in the Cyprinodonts; the males are always the smaller, sometimes several times smaller than the females, quite diminutive; and they are perhaps the smallest fishes in existence. The fins generally are more developed in the males, and the coloration is frequently different also. Some species are carnivorous; others live on the organic substances mixed with mud. Fossil remains have been found in tertiary strata, all apparently referable to the existing genus *Cyprinodon;* they occur near Aix in Provence, in the marl of Gesso, St. Angelo, in the Brown coal near Bonn, near Frankfort, and in the freshwater-chalk of Oeningen. In the latter locality a *Poecilia* occurs likewise.

The genera can be divided into two groups:

I. CYPRINODONTIDÆ CARNIVORÆ.—*The bones of each ramus of the mandible are firmly united; intestinal tract short, or but little convoluted. Carnivorous or insectivorous.*

CYPRINODON.—Cleft of the mouth small, developed laterally and horizontally. Snout short. Teeth of moderate size, incisor-like, notched, in a single series. Scales rather large. Origin of the anal fin behind that of the dorsal in both sexes, both fins being larger in the male than in the female. Anal not modified into a sexual organ.

Seven species occur in the Mediterranean region, all of which seem able to live in briny springs or pools, the water of which contains a much greater percentage of salts than sea-water, as the brine-springs near the Dead Sea or in the Sahara. They are as little affected by the high temperature of some of these springs (91°), for instance of that at Sidi

Ohkbar in the Sahara. Like other fishes living in limited localities or concealing themselves in mud, Cyprinodonts lose sometimes their ventral fins; such specimens have been described as *Tellia.* The species of the New World are less known than those of the Old, but not less numerous.

Allied to Cyprinodon are *Fitzroyia* from Monte Video, and *Characodon* from Central America.

HAPLOCHILUS.—Snout flat, both jaws being much depressed, and armed with a narrow band of villiform teeth. Body oblong, depressed anteriorly, compressed posteriorly. Dorsal fin short, commencing behind the origin of the anal, which is more or less elongate.

Twenty species from the East Indies, tropical Africa, and temperate and tropical America.

FUNDULUS.—Cleft of the mouth of moderate width, developed laterally and horizontally. Snout of moderate length. Teeth in a narrow band, those of the outer series being largest, conical. Scales of moderate size. Dorsal fin commencing before or opposite the origin of the anal. Sexes not differentiated.

"Killifish," abundant in the New World, where about twenty species have been found; *F. heteroclitus, majalis, diaphanus,* being common on the Atlantic coast of the United States; from the Old World two species only are known, viz. *F. hispanicus* from Spain, and *F. orthonotus* from the east coast of Africa. Allied to *Fundulus* are the South American *Limnurgus, Lucania, Rivulus,* and *Cynolebias.*

ORESTIAS.—Ventral fins none. Cleft of the mouth of moderate width, directed upwards, with the lower jaw prominent, and with the upper protractile. Both jaws with a narrow band of small conical teeth. Scales rather small or of moderate size, those on the head and upper part of the trunk frequently enlarged, plate-like, and granulated. Dorsal and anal fins moderately developed, opposite to each other. Sexes not differentiated by modification of the anal fin. The gill-membranes of both sides are united for a short distance, and not attached to the isthmus.

Inhabitants of Lake Titicaca and other elevated sheets of water on the Cordilleras of Peru and Bolivia, between the 14th and 19th degrees of latitude, at an elevation of 13,000 and 14,000 feet above the level of the sea. Singularly, the fishes of this outlying genus attain to a greater size than any other members of this family, being about eight inches long and comparatively bulky. They are considered a delicacy. Six species.

JENYNSIA.—Cleft of the mouth small, developed laterally and horizontally; snout not produced. Both jaws with a series of tricuspid teeth of moderate size. Scales of moderate size. The origin of the anal fin is, in both sexes, behind that of the dorsal, although the anal of the male is modified into an intromittent organ, in which scarcely any of the rays remain distinct.

One species from Maldonado.

GAMBUSIA.—Cleft of the mouth developed laterally and horizontally. Snout not produced, with the lower jaw more or

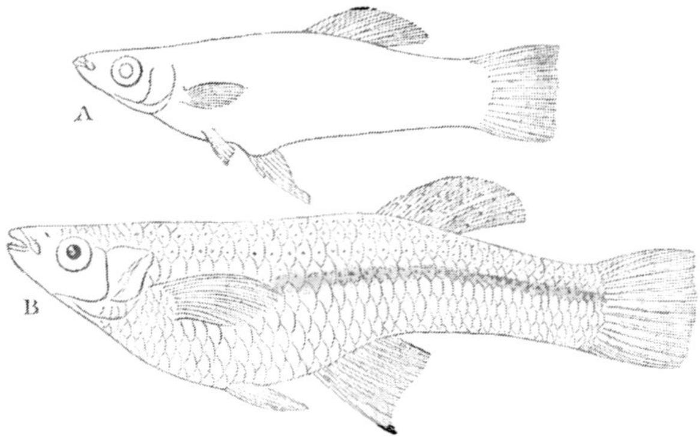

Fig. 280.—Gambusia punctata, from Cuba. A. Male; B. Female.

less prominent. Both jaws with a band of teeth, those of the outer series being strongest and conical. Scales rather large. Origin of the anal fin more or less in advance of that of the dorsal. Anal fin of the male modified into an intromittent organ and much advanced.

Eight species from the West Indies and the southern parts of South America.—Allied genera are the Central American *Pseudoxiphophorus* and *Belonesox*.

ANABLEPS.—Head broad and depressed, with the supraorbital part very much raised. Body elongate, depressed anteriorly and compressed posteriorly. Cleft of the mouth horizontal, of moderate width, the mandible being short; upper jaw protractile. Both jaws armed with a band of villiform teeth, those of the outer series being largest and somewhat movable. The integuments of the eye are divided into an upper and lower portion by a dark-coloured transverse band of the conjunctiva; also the pupil is completely divided into two by a pair of lobes projecting from each side of the iris. Scales rather small or of moderate size. Dorsal and anal fins short, the former behind the latter. The anal fin of the male is modified into a thick and long scaly conical organ with an orifice at its extremity.

Three species from tropical America. They are the longest Cyprinodonts, attaining to the length of nearly twelve inches. Their peculiar habit of swimming with part of the head out of the water has been noticed above (p. 113).

II. CYPRINODONTIDÆ LIMNOPHAGÆ.—*The bones of each ramus of the mandible are but loosely joined; intestinal tract with numerous circumvolutions. Sexes differentiated. Mud-eating. Tropical America.*

POECILIA.—Cleft of the mouth small, transverse; mandible very short. Both jaws with a narrow band of minute teeth. Scales rather large. Origin of the anal fin generally nearly opposite to that of the dorsal fin in the female, but in the male it is modified into an intromittent organ and much advanced. Dorsal fin short, with not more than eleven rays.

Sixteen species.

MOLLIENESIA.—Differing from *Poecilia* in having a larger dorsal fin, with twelve or more rays.

Five species. The males are most beautifully coloured, and their dorsal fin is much enlarged. In one species (*M.*

hellerii), besides, the lower caudal rays of the mature male are prolonged into a long, sword-shaped, generally black and yellow appendage.

Two other genera belong to this group: *Platypoecilus* and *Girardinus.*

SEVENTH FAMILY—HETEROPYGII.

Head naked; body covered with very small scales; barbels none. Margin of the upper jaw formed by the intermaxillaries. Villiform teeth in the jaws and on the palate. Adipose fin none. Dorsal fin belonging to the caudal portion of the vertebral column, opposite to the anal. Ventral fins rudimentary or absent. Vent situated before the pectorals. Stomach coecal; pyloric appendages present. Pseudobranchiæ none; air-bladder deeply notched anteriorly.

To this small family, which is closely allied to the Cyprinodonts and Umbridæ, belongs the famous Blind Fish of the Mammoth Cave in Kentucky, *Amblyopsis spelæus.* It is destitute of external eyes, and the body is colourless; although the eyes, with the optic nerve, are quite rudimentary, the optic lobes are as much developed as in fishes with perfect eyes. The loss of vision is compensated by the acuteness of its sense of hearing, as well as by a great number of tactile papillæ, arranged on transverse ridges on the head, and provided with nervous filaments coming from the fifth pair. The ovary is single, and the fish is viviparous, like the Cyprinodonts. It seems to occur in all the subterranean rivers that flow through the great limestone region underlying the carboniferous rocks in the central portion of the United States. As in *Cyprinodon*, so in this genus, specimens occur without ventral fins; they have been called *Typhlichthys.* The largest size to which *Amblyopsis* grows is five inches.

Chologaster is closely allied, but provided with small

external eyes; its body is coloured, but it is destitute of ventrals. It was found once in a rice field in South Carolina.

[See Tellkampf, Müll. Arch. 1844, p. 381; Packard and Putnam, "The Mammoth Cave and its Inhabitants." Salem. 1872. 8°.]

Eighth Family—Umbridæ.

Head and body covered with scales; barbels none. Margin of the upper jaw formed by the intermaxillaries mesially, and by the maxillaries laterally. Adipose fin none; the dorsal fin belongs partly to the abdominal portion of the vertebral column. Stomach siphonal; pyloric appendages none; pseudobranchiæ glandular, hidden; air-bladder simple.

Two small species only are known: *Umbra krameri* from Austria and Hungary, and *Umbra limi*, locally distributed in the United States; called "Hunds-fish" in Germany, "Dog-fish" or "Mud-fish" in America.

Ninth Family—Scombresocidæ.

Body covered with scales; a series of keeled scales along each side of the belly. Margin of the upper jaw formed by the intermaxillaries mesially, and by the maxillaries laterally. Lower pharyngeals united into a single bone. Dorsal fin opposite the anal, belonging to the caudal portion of the vertebral column. Adipose fin none. Air-bladder generally present, simple, sometimes cellular, without pneumatic duct. Pseudobranchiæ hidden, glandular. Stomach not distinct from the intestine, which is quite straight, without appendages.

The fishes of this family are chiefly marine, some living in the open ocean, whilst others have become acclimatised in fresh water; many of the latter are viviparous, all the marine forms being oviparous. They are found in all the temperate and tropical zones. Carnivorous.

This family is represented in the strata of Monte Bolca by rare remains of a fish named *Holosteus*, allied to *Belone* or

Scombresox, and by a species of *Belone* in the miocene of Licata.

BELONE.—Both jaws are prolonged into a long slender beak. All the dorsal and anal rays connected by membrane.

The long upper jaw of the "Gar-pike" is formed by the intermaxillaries, which are united by a longitudinal suture. Both jaws are beset with asperities, and with a series of longer, conical-pointed, widely-set teeth. Skimming along the surface of the water, the Gar-pike seize with these long jaws small fish as a bird would seize them with its beak; but their gullet is narrow, so that they can swallow small fish only. They swim with an undulating motion of the body; although they are in constant activity, their progress through the water is much slower than that of the Mackerels, the shoals of which sometimes appear simultaneously with them on our coasts. Young specimens are frequently met in the open ocean; when very young their jaws are not prolonged, and during growth the lower jaw is much in advance of the upper, so that these young fishes resemble a *Hemirhamphus*. About fifty species are known from tropical and temperate seas, *Belone belone* being a common fish on the British coast. Its bones, like those of all its congeners, are green; and therefore the fish, although good eating, is disliked by many persons. Some species attain a length of five feet.

SCOMBRESOX.—Both jaws are prolonged into a long slender beak. A number of detached finlets behind the dorsal and anal fins.

The "Saury" or "Skipper" resemble the Gar-pike, but the teeth in the jaws are minute; they seem to feed chiefly on soft pelagic animals. In their habits they are still more pelagic; and the young, in which the beak is still undeveloped, are met with everywhere in the open ocean, in the

Atlantic as well as in the Pacific. The European species, *Sc. saurus*, is not rare on the British coast; four other species have been described, closely allied to *Sc. saurus*.

HEMIRHAMPHUS.—The lower jaw only is prolonged into a long slender beak.

In the young both jaws are short; the upper is never prolonged, the intermaxillaries forming a triangular, more or less convex, plate. The "Half-beaks" are common between and near the tropics; some forty species are known, none of which attain to the same length as the Gar-pike, scarcely ever exceeding a length of two feet. Some of the tropical species live in fresh water only; they are of small size and viviparous.

ARRHAMPHUS.—Mouth formed as in *Hemirhamphus*, except that the lower jaw is not produced into a beak. Pectoral fins of moderate length.

One species (*A. sclerolepis*) from the coast of Queensland (not New Zealand); it may be regarded as a *Hemirhamphus*, with retarded development of the lower jaw.

EXOCOETUS.—Jaws short, intermaxillaries and maxillaries separate. Teeth minute, rudimental, and sometimes absent. Body moderately oblong, covered with rather large scales. Pectorals very long, an organ of flying.

Forty-four different kinds of "Flying-fishes" are known

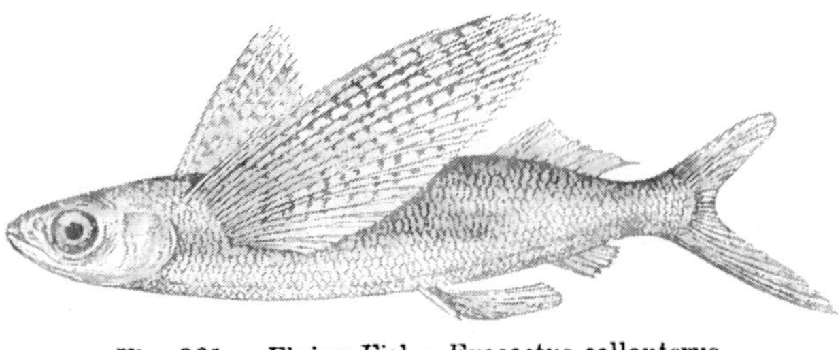

Fig. 281.—Flying Fish; Exocoetus callopterus.

from tropical and sub-tropical seas; some have a very wide range, whilst others seem to remain within one particular part of the ocean; thus, the species figured, *E. callopterus,* has been hitherto found on the Pacific side of the isthmus of Panama only. Their usual length is about 10 or 12 inches, but specimens of 18 inches have been caught. They always live in shoals, and their numbers at certain times and localities are immense; thus, at Barbadoes many boats engage in their capture, as they are excellent eating. The pectorals are in the various species of unequal length; in some they extend to the anal fin only; in others (and these are the best fliers) to the caudal. A few have curious, barbel-like appendages at the lower jaw, which may disappear with age or be persistent throughout life. The literature on the subject of Flying-fishes is very extensive, and great diversity of opinion exists among observers as regards the mode and power of their flight; but the most reliable agree that the fishes do not leave the water for the purpose of catching insects (!), and that they are unable to move their fins in the manner of a bat or bird, or to change voluntarily the direction of their flight, or to fly beyond a very limited distance. The most recent enquiries are those of K. Möbius ("Die Bewegungen der Fliegenden Fische durch die Luft," Leip. 1878, 8vo), the chief results of which may be summed up thus: Flying-fish are more frequently observed in rough weather and in a disturbed sea than during calm; they dart out of the water when pursued by their enemies, or frightened by an approaching vessel, but frequently also without any apparent cause, as is also observed in many other fishes; and they rise without regard to the direction of the wind or waves. The fins are kept quietly distended, without any motion, except an occasional vibration caused by the air whenever the surface of the wing is parallel with the current of the wind. Their flight is rapid, but gradually decreasing in velocity, greatly exceeding that of a ship going 10 miles an hour, and a distance of 500 feet. Generally, it is longer when the fishes fly against than with or at an angle to the wind. Any vertical or horizontal deviation from a straight line is not caused at the will of the fish, but by currents of the air; thus they retain a horizontally straight course when flying with or against the wind, but are carried towards the right or left whenever the direction of the wind is at an angle with that of their flight. However, it sometimes happens that the fish during its flight immerses its caudal fin in the water, and by a stroke of its tail turns towards the right or left. In a calm the line of their flight is always also vertically straight, or rather parabolic, like the course of a projectile, but it may become undulated in a rough sea, when they are flying against the course of the waves; they then frequently overtop each wave, being carried over it by the pressure of the disturbed air. Flying-fishes often fall on board of vessels, but this never happens during a calm, or from the lee side, but during a breeze only, and from the weather side. In daytime they avoid a ship, flying away from it; but during the night, when they are unable to see, they frequently fly against the weather-board, where they are caught by the current of air, and carried upwards to a height of 20 feet above the surface of the water, while, under ordinary circumstances, they keep close to it. All these observations point clearly to the fact that any deflection from a straight course is due to external circumstances, and not to voluntary action on the part of the fish.

TENTH FAMILY—ESOCIDÆ.

Body covered with scales; barbels none. Margin of the upper jaw formed by the intermaxillaries mesially, and by the maxillaries laterally. Adipose fin none; the dorsal fin belongs to the caudal portion of the vertebral column. Stomach without blind sac; pyloric appendages none. Pseudobranchiæ glandular, hidden; air-bladder simple; gill-opening very wide.

This family includes one genus only, *Esox*, the "Pikes," inhabitants of the fresh waters of the temperate parts of Europe, Asia, and America. The European species, *E. lucius*, inhabits

Fig. 282.—The Pike. (Esox lucius.)

all three continents, but the North American waters harbour five, or perhaps more, other species, of which the "Muskellunge," or "Maskinonge" (*E. estor*) of the Great Lakes attains to the same large size as the common Pike. The other species are generally called "Pickerell" in the United States.

Fossil Pike, belonging to the existing genus, have been found in the freshwater-chalk of Oeningen, and in the diluvial marl of Silesia. Remains of the common Pike occur in abundance in quaternary deposits.

Eleventh Family—Galaxiidæ.

Body naked, barbels none. Margin of the upper jaw chiefly formed by the intermaxillaries, which are short, and continued by a thick lip, behind which are the maxillaries. Belly rounded; adipose fin none; dorsal opposite to anal. Pyloric appendages in small number. Air-bladder large, simple; pseudobranchiæ none. The ova fall into the cavity of the abdomen before exclusion.

Small freshwater fishes of the southern hemisphere, belonging to two genera, *Galaxias* and *Neochanna*. Of the former genus five species are found in New Zealand, where this type is most developed, three in New South Wales, two

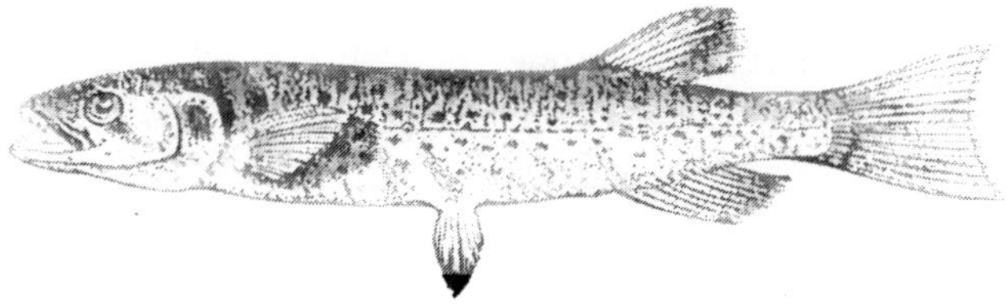

Fig. 283.—Galaxias truttaceus, from Tasmania.

in Tasmania, and four in the southern extremity of South America. Their native name in New Zealand is "Kokopu," and they were dignified with the name of "Trout" by the settlers before the introduction of true Salmonidæ. They rarely exceed a length of eight inches. *Neochanna* is a degraded form of *Galaxias*, from which it differs by the absence of ventral fins. This fish has hitherto been found only in burrows, which it excavates in clay or consolidated mud, at a distance from water.

Twelfth Family—Mormyridæ.

Body and tail scaly; head scaleless; barbels none. The margin of the upper jaw is formed in the middle by the intermaxillaries, which coalesce into a single bone, and laterally by the maxillaries. Sub- and inter-operculum present, the latter very small. On each side of the single parietal bone a cavity leading into the interior of the skull, and covered with a thin bony lamella. All the fins are well developed, in Mormyrus; *or caudal, anal, and ventral fins are absent, in* Gymnarchus. *No adipose fin. Pseudobranchiæ none; gill-openings reduced to a short slit. Air-bladder simple. Two cœca pylorica behind the stomach.*

This family is characteristic of the freshwater fauna of tropical Africa. Of *Mormyrus* (including *Hyperopisus* and

Mormyrops), fifty-one species are known, of which eleven occur in the Nile. Some attain a length of three or four feet, others remain small. Their flesh is said to have an excellent flavour. The species figured (and probably other allied

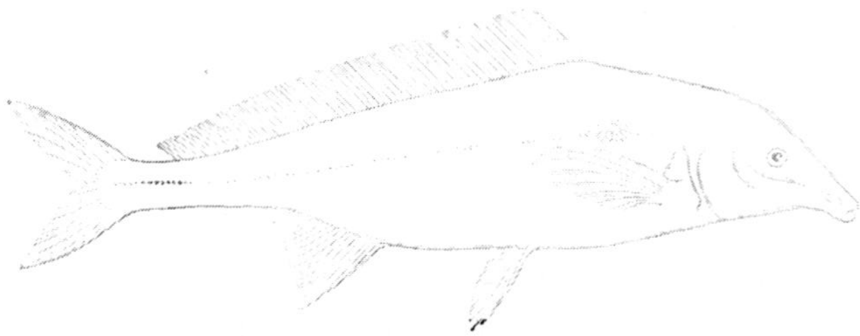

Fig. 284.—Mormyrus oxyrhynchus.

species) was an object of veneration to the ancient Egyptians, and, therefore, frequently occurs in their emblematic inscriptions. They abstained from eating it because it was one of three different kinds of fishes accused of having devoured a member of the body of Osiris, which, therefore, Isis was unable to recover when she collected the rest of the scattered members of her husband.

The *Mormyri* possess a singular organ on each side of the tail, without electric functions, but evidently representing a transitional condition from muscular substance to an electric organ. It is an oblong capsule divided into numerous compartments by vertical transverse septa, and containing a gelatinous substance. The *Mormyri* differ much with regard to the extent of the dorsal and anal fins, the former sometimes occupying the greater portion of the length of the back, sometimes being much shorter and limited to the tail. In some the snout is short and obtuse, in others long and decurved, with or without appendage.

Of *Gymnarchus* one species only is known, *G. niloticus*, which occurs in the Nile and West African rivers, and attains a length of six feet. The form of its body is eel-like, and each jaw is armed with a series of incisor-like teeth.

Like *Mormyrus, Gymnarchus* posseses a pseudo-electric organ, thickest on the tail, tapering in front, and extending nearly to the head. It consists of four membranaceous tubes intimately connected with the surrounding muscles, and containing prismatic bodies arranged in the manner of a paternoster. The air-bladder of Gymnarchus is cellular, very extensible, and communicates with the dorsal side of the œsophagus by a duct possessing a sphincter.

[See *Erdl*, Münchner Gelehrte Anzeigen, 1846, xxiii., and *Hyrtl*, Denkschr. Akad. Wiss. Wien. 1856. xii.]

THIRTEENTH FAMILY—STERNOPTYCHIDÆ.

Body naked, or with very thin deciduous scales; barbels none. Margin of the upper jaw formed by the maxillary and intermaxillary, both of which are toothed; opercular apparatus not completely developed. Gill-opening very wide; pseudobranchiæ present or absent; air-bladder simple, if present. Adipose fin present, but generally rudimentary. Series of phosphorescent bodies along the lower parts. The eggs are enclosed in the sacs of the ovarium, and excluded by oviducts.

Pelagic and Deep-sea fishes of small size.

STERNOPTYX.—Trunk much elevated and compressed, with the trunk of the tail very short. Body covered with a silvery pigment, without regular scales; series of phosphorescent spots run along the lower side of the head, body, and tail. Cleft of the mouth wide, vertical, with the lower jaw prominent. Jaws armed with small teeth. Eyes rather large, and although lateral, directed upwards and placed close together. Ventral fins very small. A series of imbricate scutes runs along the abdomen, forming a kind of serrature. The dorsal fin is short, and occupies about the middle of the length of the fish; it is preceded by the first commencement of the formation of a spinous dorsal, several neural spines being prolonged beyond the dorsal muscle forming a triangular osseous plate. Adipose fin rudimentary; anal short; caudal forked.

These small fishes are now and then picked up in the Mediterranean and Atlantic. According to the dredging-records of the "Challenger," they and the allied genera *Argyropclecus* and *Polyipnus* would descend to depths of respectively 1100 and 2500 fathoms; but the form of their body and their whole organisation render this statement very improbable; they most likely live at a small depth during the day-time, coming to the surface at night, like many *Scopelus.*

Coccia and *Maurolicus* are two other genera allied to the preceding.

CHAULIODUS.—Body elongate, compressed, covered with exceedingly thin and deciduous scales; series of luminous (phosphorescent) spots run along the lower side of the head, body, and tail. Head much compressed and elevated, with the bones thin, but ossified, and with the opercular portion very narrow, the interoperculum being rudimentary. Cleft of the mouth ex-

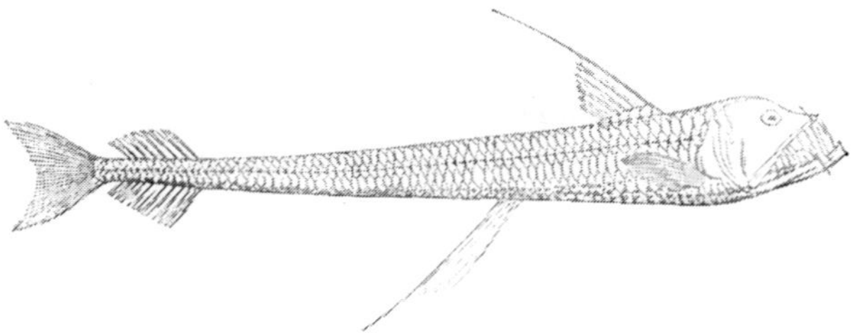

Fig. 285.—Chauliodus sloanii.

ceedingly wide, the intermaxillary forming one half of the upper jaw. Each intermaxillary with four long canine teeth; edge of the maxillary finely denticulated; mandible with pointed, widely set teeth, the anterior of which are exceedingly long; none of the large teeth are received within the mouth. Palatine with a single series of small pointed teeth; no teeth on the tongue. Eye of moderate size. Pectoral and ventral fins well developed. Dorsal fin anteriorly on the trunk, before the ventrals; adipose fin small, sometimes fimbriated; anal short, rather close to the caudal, which is forked. Gill-opening very wide, the outer branchial arch extending forward to behind the symphysis of the lower jaw; it has no gill-rakers. Branchiostegals numerous.

This genus, of which one species only (*Ch. sloanii*) is known, is generally distributed over the great depths of the oceans, and does not appear to be scarce; it attains to a length of 12 inches, and must be one of the most formidable fishes of prey of the deep-sea.

Allied genera are *Gonostoma*, *Photichthys*, and *Diplophos*, all of which have the teeth of much smaller size.

Fourteenth Family—Stomiatidæ.

Skin naked, or with exceedingly delicate scales; a hyoid barbel. Margin of the upper jaw formed by the intermaxillary and maxillary which are both toothed; opercular apparatus but little developed. Gill-opening very wide; pseudobranchiæ none. The eggs are enclosed in the sacs of the ovarium, and excluded by oviducts.

Deep-sea fishes, descending to the greatest depths, characterised by their barbel and their formidable dentition.

Fig. 286.—*Astronesthes niger*. The white spots in front of the eye are phosphorescent organs.

Some have two dorsal fins, the posterior of which is adipose; they belong to the genus *Astronesthes*, are the smallest of the family, and frequently met with in the Atlantic.

The others—viz. *Stomias*, *Echiostoma*, *Malacosteus*, and *Bathyophis*, lack the adipose fin, the rayed dorsal being opposite to the anal. Of these the one longest known is

Stomias.—Body elongate, compressed, covered with exceedingly fine and deciduous scales, which are scarcely imbricate, lying in subhexagonal impressions; vent situated at no great distance from the caudal fin. Head compressed, with the snout very short, and with the cleft of the mouth very wide. Teeth pointed, unequal in size, those of the intermaxillaries and of the mandible being the longest; maxillary finely denticulated; vomer with a pair of fangs; palatine bones and tongue with smaller pointed teeth. Eye of moderate size. Opercular portion of the head narrow. A fleshy barbel in the centre of the hyoid region. Dorsal opposite the anal, close to the caudal; pectoral and ventral fins feeble, the latter inserted behind the middle of the length of the body. Series of phosphorescent dots run along the lower side of the head, body, and tail. Gill-opening very wide. Pyloric appendages none.

Three species are known; beside specimens which were found floating on the surface, others have been dredged from depths varying between 450 and 1800 fathoms.

FIFTEENTH FAMILY—SALMONIDÆ.

Body generally covered with scales; head naked; barbels none. Margin of the upper jaw formed by the intermaxillaries mesially, and by the maxillaries laterally. Belly rounded. A small adipose fin behind the dorsal. Pyloric appendages generally numerous, rarely absent. Air-bladder large, simple; pseudobranchiæ present. The ova fall into the cavity of the abdomen before exclusion.

Inhabitants of the sea and freshwater; the majority of the marine genera are deep-sea forms. The freshwater forms are peculiar to the temperate and arctic zones of the Northern Hemisphere, one occurring in New Zealand; many freshwater species periodically or occasionally descending to the sea. One of the most valuable families of the class of fishes. No fossils of the freshwater forms are known; but of the marine genera, *Osmerus* occurs in the greensand of Ibbenbusen, and in the schists of Glaris and Licata; a species of

Mallotus, indistinguishable from the living *M. villosus*, occurs abundantly in nodules of clay of unknown geological age in Greenland. Other genera, as *Osmeroides*, *Acrognathus*, and *Aulolepis*, from the chalk of Lewes, belong to the same fauna as species of Beryx, and were probably deep-sea Salmonoids.

Salmo.—Body covered with small scales. Cleft of the mouth wide, the maxillary extending to below or beyond the eye. Dentition well developed; conical teeth in the jaw bones, on the vomer and palatines, and on the tongue, none on the pterygoid bones. Anal short, with less than fourteen rays. Pyloric appendages numerous; ova large. Young specimens with dark cross-bands (Parr-marks).

We know of no other group of fishes which offers so many difficulties to the ichthyologist with regard to the distinction of the species as well as to certain points in their life-history, as this genus, although this may be partly due to the unusual attention which has been given to their study, and which has revealed an almost greater amount of unexplained facts than of satisfactory solutions of the questions raised. The almost infinite variations of these fishes are dependent on age, sex and sexual development, food, and the properties of the water. Some of the species interbreed, and the hybrids mix again with one of the parent species, thus producing an off-spring more or less similar to the pure breed. The *coloration* is, first of all, subject to variation; and consequently this character but rarely assists in distinguishing a species, there being not one which would show in all stages of development the same kind of coloration. The young of all the species are *barred*; and this is so constantly the case that it may be used as a generic or even as a family character, not being peculiar to *Salmo* alone, but also to *Thymallus* and probably to *Coregonus*. The number of bars is not quite constant, but the migratory Trout have two (and even three) more than the River-Trout. In some waters River-trout remain small, and frequently retain the Parr-marks all their lifetime; at certain seasons a new coat of scales overlays the Parr-marks, rendering them invisible for a time; but they reappear in time, or are distinct as soon as the scales are removed. When the Salmones have passed this "Parr" state, the coloration becomes much diversified. The males, especially during and immediately after the spawning time, are more intensely coloured and variegated than the females; specimens which have not attained to maturity retaining a brighter silvery colour, and being more similar to the female fish. Food appears to have less influence on the coloration of the outer parts than on that of the flesh; thus the more variegated specimens are frequently out of condition, whilst well-fed individuals with pinkish flesh are of a more uniform though bright coloration. Chemistry has not supplied us yet with an analysis of the substance which gives the pink colour to the flesh of many Salmonoids; but there is little doubt that it is identical with, and produced by, the red pigments of many salt- and fresh-water Crustaceans, which form a favourite food of these fishes. The water has a marked influence on the colours; Trout with intense ocellated spots are generally found in clear rapid rivers, and in small open Alpine pools; in the large lakes with pebbly bottom the fish are bright silvery, and the ocellated spots are mixed with or replaced by X-shaped black spots; in pools or parts of lakes with muddy or peaty bottom, the trout are of a darker colour generally, and when enclosed in caves or holes, they may assume an almost uniform blackish coloration.

The change of scales (that is, the rapid reproduction of the worn part of the scales) coincides in the migratory species with their sojourn in the sea. The renovated scales give them a bright silvery appearance, most of the spots disappearing or being overlaid and hidden by the silvery scales. Now, some of the species, like *S. fario*, inhabit all the different waters indicated, even brackish water, and, in consequence, we find a great variation of colour in one and the same species; others are more restricted in their habitat, like *S. salar*, *S. ferox*, etc., and, therefore, their coloration may be more precisely defined.

With regard to *size* the various species do not present an equal amount of variation. Size appears to depend on the abundance of food and the extent of the water. Thus, the Salmon and the different kinds of great Lake-trout do not appear to vary considerably in size, because they find the same conditions in all the localities inhabited by them. A widely spread species, however, like *S. fario,* when it inhabits a small mountain pool with scanty food, may never exceed a weight of eight ounces, whilst in a large lake or river, where it finds an abundance and variety of food, it attains to a weight of fourteen or sixteen pounds. Such large River-trout are frequently named and described as Salmon-trout, Bull-trout, etc. Further, in Salmones, as in the majority of fishes and tailed Batracheans, there is an innate diversity of growth in individuals hatched from the same spawn. Some grow rapidly and normally, others more slowly, and some remain dwarfed and stationary at a certain stage of development.

The *proportions of the various parts of the body* to one another vary exceedingly in one and the same species. Beside the usual changes from the young to the sexually mature form observed in all fishes, the snout undergoes an extraordinary amount of alteration of shape. In the mature male the intermaxillaries and the mandible are produced in various degrees, and the latter is frequently more or less bent upwards. Hence the males have the snout much more pointed and produced, and the entire head longer, than the females; with the intermaxillary bone the teeth, with which it is armed, are also enlarged, sometimes to four times the size of those of the females. And if this development of the front part of the head happens to be going on while the individual is able to obtain only a scanty supply of food, the usual proportions of the head and trunk are so altered that the species is very difficult to recognise. Barren male fish approach the females in the proportions of the head and body, but hybrid fishes do not differ in this respect from their parents. The abundance or scarcity of food, and the disposition or indisposition of the Salmonoids to feed, are other causes affecting the growth or fulness of the various parts of the body. In well-fed fishes the head is proportionally not only smaller but also shorter, and *vice versa.*

The *fins* vary to a certain degree. The variation in the number of the rays is inconsiderable and of no value for specific distinction. The caudal fin undergoes considerable changes of form with age, and dependently upon the sexual development. Young specimens of all species have this fin more or less deeply excised, so that the young of a species which has the caudal emarginate throughout life, is distinguished by a deeper incision of the fin, from the young of a species which has it truncate in the adult state. As the individuals of a species do not all attain to maturity at the same age and at the same size, and as mature individuals generally have the caudal less deeply excised than immature ones of the same age and size, it is evident that the variations in the form of the caudal are considerable and numerous, and that it is a very misleading character if due regard be not paid to the age and sexual development of the fish. Further, species inhabiting rapid streams as well as still waters show considerable variations in the form and length of all the fins; those individuals which live in rapid streams, being in almost constant motion and wearing off the delicate extremities of the fins, have the fin-rays comparatively shorter and stouter, and the fins of a more rounded form, particularly at the corners, than individuals inhabiting ponds or lakes. Moreover, one and the same individual may pass a part of its life in a lake, and enter a river at certain periods, thus changing the form of its fins almost periodically.

Finally, to complete our enumeration of these variable characters, we must mention that in old males, during and after the spawning-season, the *skin* on the back becomes thickened and spongy, so that the scales are quite invisible, being imbedded in the skin.

After this cursory review of variable characters we pass on to those which are more constant, not subject to ready modification by external circumstances; and which, therefore, ought to be noticed in every description of a species of Salmo.

1. *The form of the præoperculum of the adult fish.*—The præoperculum is composed of a vertical (posterior) and horizontal (lower) part (limb), both meeting at a more or less rounded angle. The development of the lower limb is a very constant character; in some species (as in the Salmon) it is long, in others (*S. ferox*, *S. brachypoma*) exceedingly short. The adjoining woodcuts will readily show this difference.

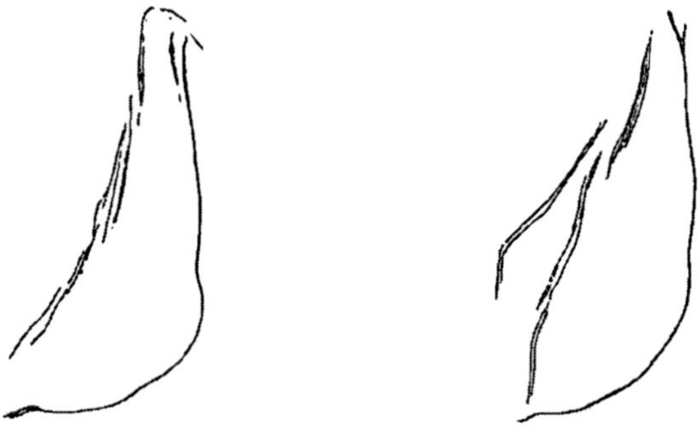

A B

Fig. 287.—Præoperculum of A. Salmo salar; B. Salmo brachypoma.

In young specimens of all Salmonoids the præoperculum has a very short lower limb; but whilst in some species it lengthens with age, its development in a horizontal direction is arrested in others.

2. *The width and strength of the maxillary of the adult fish.*—To show this character in two distinct species, we have given woodcuts of the maxillaries of females (12 inches long) of *S. fario* and *S. levenensis* of the same size.

Fig. 288.—Maxillary of A. Salmo fario; B. Salmo levenensis.

In young specimens of all Salmonoids the maxillary is comparatively shorter and broader, somewhat resembling that of *Coregonus;* yet this bone offers a valuable character for the determination of the young of some species; for instance, in a young *S. cambricus* it extends scarcely to below the centre of the eye, whilst in *S. fario* of the same size it reaches to, or even beyond, this point.

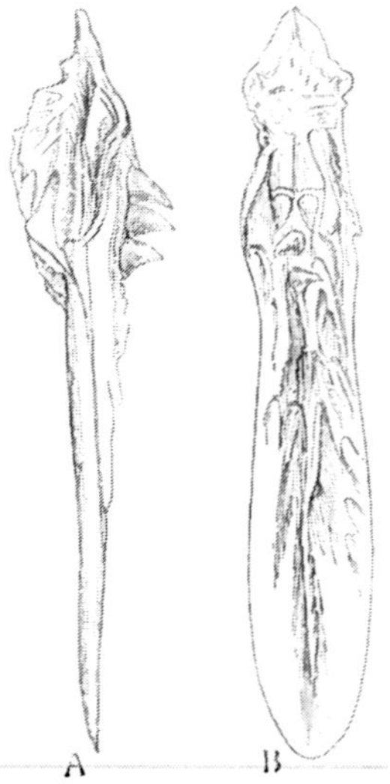

Fig. 289.—Vomerine teeth of Salmo salar (Salmon). A. Side view. B. Lower view.

3. *The size of the teeth, those of the intermaxillaries excepted.*

4. *The arrangement, and the permanence or deciduousness of the vomerine teeth.*—In some species the vomer is normally armed with a double or single series throughout life, although, of course, some of the teeth are frequently accidentally lost; in others, these teeth disappear gradually with age, the hinder ones first, so that finally the anterior only remain. In order to ascertain the arrangement of the teeth, it is necessary to remove the gengiva. Frequently the teeth stand in a distinctly double or single series, or they are placed alternately; but frequent irregularities occur which render

the character vague, or even unsafe, so that some zoologists have rejected it entirely as unreliable. However, when a

Fig. 290.—Vomerine teeth of Salmo fario, lower view.

Fig. 291.—Vomerine teeth of a Charr, side view.

greater number of individuals really belonging to the same species are examined, a pretty safe conclusion may be arrived at as regards the arrangement of the teeth.

5. *The form of the caudal fin* in specimens of a given size, age, and sexual development.

6. *A great development of the pectoral fins*, when constant in individuals from the same locality.

7. *The size of the scales*, as indicated by the number of transverse rows above the lateral line: one of the most constant characters.

8. *The number of vertebræ.*—Considering the great number of vertebræ in Salmonoids the constancy of this character is truly surprising. An excess or a diminution of the normal number by two, is of rare occurrence, and generally to be explained by the fact that one vertebra has been abnormally divided into two, two such vertebræ being considerably smaller than the others; or, on the other hand, that two have merged into one centrum, which is then unusually large, and provided with two neural spines. We have seen one case only, in which three vertebræ were united. The number of vertebræ can be easily ascertained in specimens destined for preservation in spirits, by an incision made along one side of the fish, a little above the lateral line.

9. *The number of pyloric appendages.*—There can be no doubt that this character may materially assist in fixing a species. We shall see that in some species it varies from 30 to 50; but in others, as in the Salmon and Charr, it has been

found very constant (see Fig. 56, p. 131). If unexpected variations occur, their cause may be found in a partial confluence of the cæca, as we have observed that specimens of *S. levenensis* (a species normally with from 70 to 90 cæca), had those appendages of unusual width when the normal number was diminished.

We have mentioned above that many points in the *life-history* of the Salmonoids still remain very obscure :—

1. Johnson, a correspondent of Willughby ("Hist. Pisc.," p. 194), had already expressed his belief that the different Salmonoids interbreed; and this view has since been shared by many who have observed these fishes in nature. Hybrids between the Sewin (*S. cambricus*) and the River Trout (*S. fario*) were numerous in the Rhymney and other rivers of South Wales, before Salmonoids were almost exterminated by the pollutions allowed to pass into those streams, and so variable in their characters that the passage from one species to the other could be demonstrated in an almost unbroken series, which might induce some naturalists to regard both species as identical. Abundant evidence of a similar character has accumulated, showing the frequent occurrence of hybrids between *S. fario* and *S. trutta;* hybrids between *S. fario* and species of Charr have been abundantly bred by continental pisciculturists. In some rivers the conditions appear to be more favourable to hybridism than in others, in which hybrids are of comparatively rare occurrence. Hybrids between the Salmon and some other species are very scarce everywhere. The hybrids are sexually as much developed as the pure breed, but nothing whatever is known of their further propagation and progeny.

2. Siebold has shown that some individuals of every species are not sexually developed, and that such individuals differ also externally from those normally developed. How-

ever, he appears to have gone too far when he stated that this state of sterility extends over the whole existence of such individuals, and that, therefore, the external peculiarities also remain permanent throughout life. According to Widegren this sterility is merely a temporary immaturity, and a part of the individuals arrive at a full sexual development at a later or much later period than others. To this we may add that many Salmonoids cease to propagate their species after a certain age, and that all so called overgrown individuals (that is, specimens much exceeding the usual size of the species) are barren. Externally they retain the normal specific characters.

The Salmon offers a most remarkable instance of irregularity as regards the age at which the individuals arrive at maturity. Shaw has demonstrated, in the most conclusive manner, that those small Salmonoids, which are generally called Parr, are the offspring of the Salmon, and that many males, from 7 to 8 inches long, have their sexual organs fully developed, and that their milt has all the impregnating properties of the seminal fluid of a much older and larger fish. That this Parr is not a distinct species—as has been again maintained by Couch—is further proved by the circumstance that these sexually mature Parr are absolutely identical in their zoological characters with the immature Parr, which are undoubtedly young Salmon, and that no Parr has ever been found with mature ova. But whether these Parr produce normal Salmon, impregnating the ova of female salmon, or mingle with the River-trout, or whether they continue to grow and propagate their species as fully developed Salmon, are questions which remain to be answered. We may only add that, as far as we know, barren old Salmon are extremely scarce.

3. The question whether any of the migratory species can be retained by artificial means in fresh water, and finally accommodate themselves to a permanent sojourn therein,

must be negatived for the present. Several instances of successful experiments made for this purpose have been brought forward; but all these accounts are open to serious doubts, inasmuch as they do not afford us sufficient proof that the young fish introduced into ponds were really young migratory Salmonoids, or that the full-grown specimens were identical with those introduced, and not hybrids or non-migratory Trout of a somewhat altered appearance in consequence of the change of their locality. We have seen the experiment tried at two places in South Wales, and in both cases the Salmon and the pure Sewin died when not allowed to return to the sea. On the other hand, hybrid fishes from the Sewin and the Trout survived the experiment, and continued to grow in a pond perfectly shut up from communication with the sea. In that locality neither those hybrids nor the trout spawn.

4. Although the majority of the mature individuals of a migratory species ascend a river at a certain fixed time before the commencement of spawning, others enter the fresh-water at a much earlier period, either singly or in small troops; and many appear to return to the sea before they reascend at the time of the regular immigration. It is not improbable that one and the same individual may change the salt- or fresh-water several times in the year. However, this is the case in certain rivers only, for instance, in those falling into the Moray Firth; in others one immigration only is known to occur. The cause of the irregular ascents previous to the autumnal ascents is unknown. A part, at least, of the hybrid fishes retain the migratory instinct; but it is not known whether sterile individuals accompany the others in their migrations.

5. It is said that the migratory species invariably return to the river in which they are bred. Experiments have shown that this is normally the case; but a small proportion

appear to stray so far away from their native place as to be unable to find their way back. Almost every year Salmon and Sea-trout in the Grilse-state make their appearance at the mouth of the Thames (where the migrating Salmonoids have become extinct for many years), ready to reascend and to restock this river as soon as its poisoned water shall be sufficiently purified to allow them a passage.

6. There has been much dispute about the time required for the growth of Salmonoids. The numerous and apparently contradictory observations tend to show that there is a great amount of variation even among individuals of the same origin living under the same circumstances, some of them growing much more quickly than others, and being ready to descend to the sea twelve months before their brethren. The cause of this irregularity is not explained. On the other hand, when we consider the fibrous condition of the Salmonoid skeleton, which is much less solid, and more wanting in calcareous substance, than that of the majority of Teleosteous fishes, we shall be quite prepared to adopt the truth of the observation that the young Salmonoids return to the fresh water, after a few months sojourn in the sea, and after having feasted on nourishing Crustaceans, Sand-eels, etc., with their former weight in ounces increased to pounds.

7. Liability to variation in form indicates that an animal can adapt itself to a variety of circumstances; therefore, such species as show the greatest pliability in this respect, are those which most recommend themselves for domestication and acclimatisation within certain climatic limits. Thus, the River-trout or Sea-trout were very proper subjects for those eminently successful attempts to establish them in similar latitudes of the Southern Hemisphere, whilst the attempt of transferring them into the low hill-streams of India ended (as could be foreseen) in a total failure. Those two species must

now be considered to be fully acclimatised in Tasmania and New Zealand, and with but little protection may be expected to hold their own in the freshwaters of those colonies. Whether the acclimatisation of the Salmon will be in the end equally and permanently successful, remains to be seen. The true *S. salar* is not subject to variation, and is very sensitive to any change of external conditions, and to every kind of interference with its economy. The fourth species, with which attempts of acclimatisation in Southern Australia have been made, is a migratory Salmon from the Sacramento river in California. This experiment is still in progress, and believed to be promising of success. It will be a most curious problem to ascertain, how much the original characters and habits of those species will be affected by their transference to so distant a part of the globe. At present it would be too hazardous to offer an opinion on this point, especially as it is a fact that numerous cross-breeds have been introduced into, and reared in, Tasmania, which must more or less interfere with the characters of the pure breeds.

It is apparent, from the foregoing remarks, that the distinction of the various species of Salmonidæ is a matter of considerable difficulty, and that there is scope for great diversity of opinion. At any rate it is only by a close, long-continued study, and constant comparison of specimens of various ages and from various localities, that one is enabled to find a guide through the labyrinth of confusing variations. However, it is a significant fact that the very same characters by which we are enabled to distinguish European species occur again, though in an exaggerated form, in American Salmonoids (which everybody will admit to be of distinct species), and therefore our faith in them necessarily becomes strengthened. In accordance with acknowledged principles in zoology, forms which differ from their congeners by a combination of two or more of constant characters, are to be distinguished under distinct specific names. Most likely they have been derived, at a not very remote period, from common ancestors, but the question of their specific distinctness is no more affected by this consideration than the question whether *Salmo* and *Coregonus* are distinct genera. Whenever the zoologist observes two forms distinguished by peculiarities of organisation, such as cannot be conceived to be the effects of an external or internal cause, disappearing with the disappearance of that cause, and which forms have been propagated and are being propagated uniformly through all the generations within the limits of our observations, and are yet most probably to be propagated during the existence of mankind, he is obliged to describe these forms as distinct, and they will commonly be called species.

The species of the genus *Salmo* are inhabitants of the temperate and arctic zones of the Northern Hemisphere; the species are most abundant in the northern parts of the temperate zone, becoming scarcer beyond the Arctic circle, and in the warmer parts towards the south. The southernmost points in which Salmones are found, are, on the American continent, the rivers falling into the head of the Californian Gulf, and in the Old World the mountain rivers of the Atlas and Hindu Kush. The Salmones from those localities are migratory Trout in the New World, non-migratory and small in the Old. Those species which range to the highest latitudes (lat. 82°) belong to the division of Charr, a group which generally are more intolerant of a moderate temperature, than real Trout. The genus is subdivided into

a. Salmones—Salmon and Trout—with teeth on the body of the vomer as well as its head (see Figs. 289 and 290).

b. Salvelini—Charr—with teeth on the head of the vomer only (see Fig. 291).

Of the host of species (the majority of which is unfortunately very insufficiently characterised) we enumerate the following:—[1]

a. SALMONES.

1. *S. salar* (Salmon; Lachs or Salm; Saumon) (Fig. 6, p. 43). The Salmon can generally be readily recognised, but there are instances in which the identification of specimens is doubtful, and in which the following characters (besides others) will be of great assistance. The tail is covered with relatively large scales, there being constantly eleven, or sometimes twelve in a transverse series running from behind the adipose fin forwards to the lateral line, whilst there are from thirteen to fifteen in the different kinds of Sea-trout and River-trout. The number of pyloric appendages (see Fig. 56, p. 131) is great, generally between 60 and 70, more rarely falling to 53 or rising to 77. The body of the vomer is armed with a single series of small teeth, which at an early age are gradually lost from behind towards the front, so that half-grown and old individuals have only a few (1-4) left. The Salmon inhabits temperate Europe southwards to 43° N. lat., and is not found in any of the rivers falling into the Mediterranean. In the New World its southern boundary is 41° N. lat.

2. *S. trutta* (Sea-trout, Salmon-trout).[2]—Especially numerous in North Britain.

3. *S. cambricus* (Sewin).—Wales, South of England, Ireland, Norway, and Denmark.

4. *S. fario* (Common River-trout).

5. *S. macrostigma* (Algeria).

6. *S. lemanus* (Lake of Geneva).

7. *S. brachypoma.*—A migratory species from the rivers Forth, Tweed, and Ouse.

8. *S. gallivensis* (Galway Sea-trout).

---

[1] For specific characters and detailed descriptions we refer to Günther, "Catal. of Fishes," vol. vi.

[2] The names "Bull-trout" and "Peal" are not attributable to definite species. We have examined specimens of *S. salar*, *S. trutta*, and *S. cambricus* and *S. fario*, to which the name "Bull-trout" had been given; and that of "Peal" is given indiscriminately to Salmon-grilse and to *S. cambricus*.

9. *S. orcadensis.*—A non-migratory trout from Lough Stennis, in the Orkney Islands.

10. *S. ferox.*—The great Lake-trout of North Britain, Wales, and Ireland.

11. *S. stomachicus* (the Gillaroo of Ireland).

12. *S. nigripinnis* from mountain-pools of Wales.

13. *S. levenensis* (Lochleven Trout).

14. *S. oxi* from the rivers of the Hindu Kush.

15. *S. purpuratus* from the Pacific coast of Asia and North America.

16. *S. macrostoma.*—Japan.

17. *S. namaycush.*—The great Lake-trout of North America.

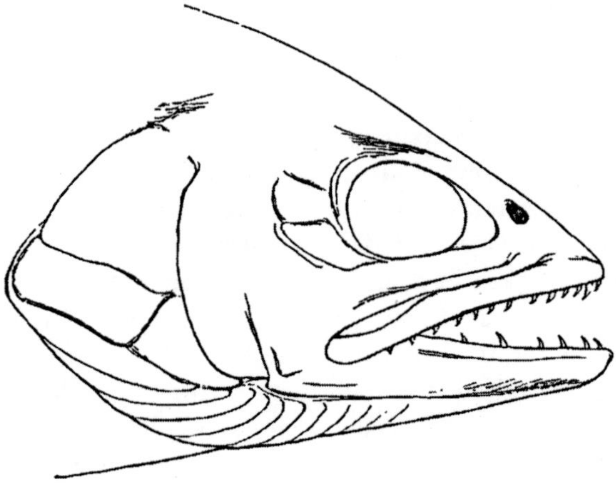

Fig. 292.—Salmo brachypoma.

b. Salvelini: Charr.

1. *S. umbla.*—The "Ombre chevalier" of the Swiss lakes.

2. *S. salvelinus.*—The "Sælbling" of the Alpine lakes of Bavaria and Austria.

3. *S. alpinus.*—The common Northern Charr, growing to a length of four feet, and migratory.

4. *S. killinensis.*—The Loch Killin Charr, Inverness-shire.

5. *S. willughbii.*—The Loch Windermere Charr.

6. *S. perisii.*—The "Torgoch" of Wales.

7. *S. grayi.*—The "Freshwater Herring" of Lough Melvin, Ireland.

8. *S. colii.*—Charr of Loughs Eske and Dan.

9. *S. hucho.*—The "Huchen" of the Danube, growing to the size of the Salmon.

10. *S. alipes* from lakes in Boothia Felix and Greenland.

11. *S. arcturus.*—The most northern species from 82° lat.

12. *S. fontinalis.*—The common "Brook - trout" of the United States.

13. *S. oquassa.*—A lake species from the State of Maine.

Oncorhynchus differs from *Salmo* only in the increased number of anal rays, which are more than fourteen. All the species are migratory, ascending American and Asiatic rivers flowing into the Pacific. The Californian Salmon (*O. quinnat?*) belongs to this genus.

Other allied genera are *Brachymystax* and *Luciotrutta.*

PLECOGLOSSUS.—Body covered with very small scales. Cleft of the mouth wide; maxillary long. Dentition feeble; intermaxillaries with a few small, conical, pointed teeth; the teeth of the maxillaries and mandibles are broad, truncated, lamellated and serrated, movable, seated in a fold of the skin. The mandibles terminate each in a small knob, and are not jointed at the symphysis. The mucous membrane in the interior of the mouth —between the terminal halves of the mandibles—forms a peculiar organ, being raised into folds, with a pair of pouches in front and a single one behind. Tongue very small, with minute teeth, its apical part being toothless; palate apparently without teeth.

A small aberrant form of Freshwater-Salmonoids abundantly found in Japan and the Island of Formosa.

OSMERUS.—Body covered with scales of moderate size. Cleft of the mouth wide; maxillary long, extending to, or nearly to, the hind margin of the orbit. Dentition strong; intermaxillary and maxillary teeth small, much smaller than those of the mandible. Vomer with a transverse series of teeth, several of which are large, fang-like; a series of conical teeth along the palatine and pterygoid bones. Tongue with very strong fang-like teeth anteriorly, and with several longitudinal series of smaller ones posteriorly. Pectoral fins moderately developed. Pyloric appendages very short, in small number; ova small.

The "Smelt" (*O. eperlanus*) is common on many places of the coasts of Northern Europe and America. In the sea it grows to a length of eight inches; but, singularly, it frequently migrates from the sea into rivers and lakes, where its growth is very much retarded. That this habit is one of very old date, is evident from the fact that this small freshwater form occurs, and is fully acclimatised in lakes which have now no open communication with the sea. And still more singularly, this same habit, with the same result, has been observed in the Smelt of New Zealand (*Retropinna richardsonii*). The Smelt is considered a delicacy in Europe, as well as in America, where the same species occurs. Two other allied genera, *Hypomesus* and *Thaleichthys*, are found on the Pacific coast of North America, the latter being caught in immense numbers, and known by the name "Eulachon" and "Oulachan;" it is so fat, that it is equally used as food and as candle.

Mallotus.—Body covered with minute scales, which are somewhat larger along the lateral line and along each side of the belly; in mature males these scales become elongate, lanceolate, densely tiled, with free projecting points, forming villous bands. Cleft of the mouth wide; maxillary very thin, lamelliform, extending to below the middle of the eye. Lower jaw the longer, partly received between the maxillaries. Dentition very feeble; the teeth forming single series; only the teeth on the tongue are somewhat larger and disposed in an elliptical patch. Pectoral fins large, horizontal, with broad base. Pyloric appendages very short, in small number; ova small.

The "Capelin" (*M. villosus*) is found on the Arctic coasts of America and of Kamtschatka. It is caught in immense numbers by the natives, who consume it fresh, or dry it for use in the winter. Its length does not exceed nine inches.

Coregonus.—Body covered with scales of moderate size. Cleft of the mouth small; maxillary broad, short or of moderate length, not extending behind the orbit. Teeth, if present, extremely minute and deciduous. Dorsal fin of moderate length; caudal deeply forked. Ova small.

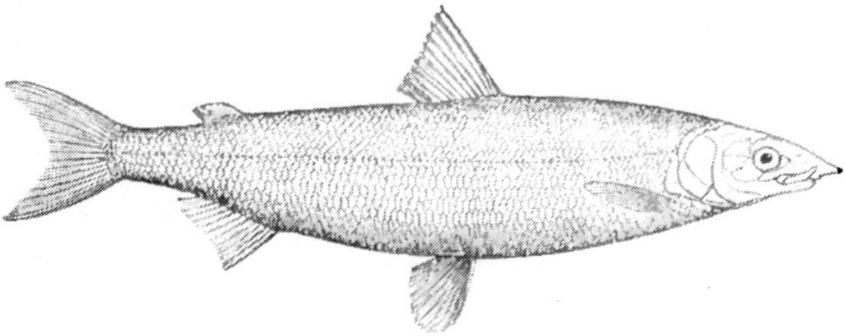

Fig. 293.—Coregonus oxyrhynchus.

The majority of the species, of which more than forty are known, are lacustrine species; and comparatively few are

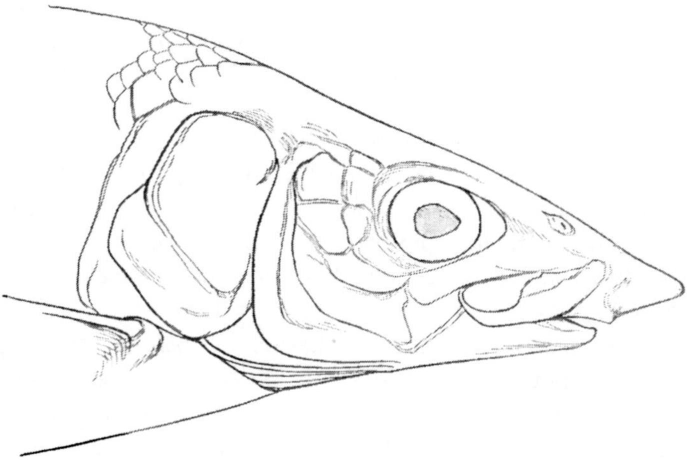

Fig. 294.—Head of Coregonus oxyrhynchus.

subject to periodical migrations to the sea, like *Salmo*. They are confined to the northern parts of temperate Europe, Asia, and North America. Their distribution is local, but sometimes three and more species are found in the same lake. They abound in every lake and river of the northern parts of North America, and are known by the name of "White-fish."

They are of vital importance to some tribes of the native population. The European *C. oxyrhynchus* is as much a marine as a freshwater species. In the British Islands several small species occur, viz. *C. clupeoides*, the "Gwyniad,"

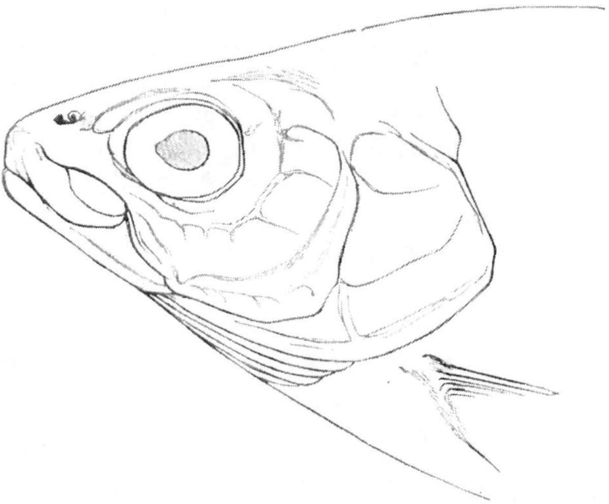

Fig. 295.—Coregonus clupeoides.

"Schelly," or "Powen" from the great lakes; *C. vandesius*, the "Vendace" of Lochmaben; and *C. pollan*, the "Pollan" of the Irish lakes. The latter is brought in quantities to Belfast market during the season, that is, at the time when it rises from the depths of Lough Neagh to deposit its spawn near the shore. Thomson says that in September 1834 some 17,000 were taken there at three or four draughts of the net. Some of the species of the continent of Europe and America attain to a much larger size than the British species, viz. to a length of two feet.

THYMALLUS.—Principally distinguished from *Coregonus* by its long many-rayed dorsal fin.

"Graylings"—five species, inhabiting clear streams of the north of Europe, Asia, and North America. The best known are the "Poisson bleu" of the Canadian voyageurs (*Th. signifer*), and the European Grayling (*T. vulgaris*).

SALANX.—Body elongate, compressed, naked or covered with small, exceedingly fine, deciduous scales. Head elongate and much depressed, terminating in a long, flat, pointed snout. Eye small. Cleft of the mouth wide; jaws and palatine bones with conical teeth, some of the intermaxillaries and mandibles being enlarged; no teeth on the vomer; tongue with a single series of curved teeth. Dorsal fin placed far behind the ventrals, but in front of the anal; anal long; adipose fin small; caudal forked. Pseudobranchiæ well developed; air-bladder none. The entire alimentary canal straight, without bend; pyloric appendages none. Ova small.

This small, transparent, or whitish fish (*S. chinensis*) is well known at Canton and other places of the coast of China as "Whitebait," and considered a delicacy. It is evidently a fish which lives at a considerable depth in the sea, and approaches the coast only at certain seasons.

Finally, this family is represented in the deep sea by three genera, *Argentina*, *Microstoma*, and *Bathylagus*, of which the two former live at moderate depths, and have been known for a long time, whilst the last was discovered during the "Challenger" expedition in the Atlantic and Antarctic Oceans at depths of 1950 and 2040 fathoms. As *Argentina* is sometimes found in the North Atlantic, and even near the British coasts, we give its principal characters.

ARGENTINA.—Scales rather large; cleft of the mouth small; intermaxillaries and maxillaries very short, not extending to below the orbit. Eye large. Jaws without teeth; an arched series of minute teeth across the head of the vomer and on the fore part of the palatines; tongue armed with a series of small curved teeth on each side. Dorsal fin short, in advance of the ventrals; caudal deeply forked. Pseudobranchiæ well developed. Pyloric appendages in moderate numbers. Ova small.

Four species are known, of which *A. silus* and *A. hebridica* have been found occasionally on the North British, and, more frequently, on the Norwegian coast. The other species are from the Mediterranean. Attaining to a length of 18 inches.

Sixteenth Family—Percopsidæ.

Body covered with ctenoid scales; head naked. Margin of the upper jaw formed by the intermaxillaries only; opercular apparatus complete. Barbels none. Gill-openings wide. Adipose fin present.

One genus and species only (*Percopsis guttatus*); interesting as having the general characters of Salmonoids, but the mouth and scales of a Percoid. Freshwaters of the northern United States.

Seventeenth Family—Haplochitonidæ.

Body naked or scaly (cycloid). Margin of the upper jaw formed by the intermaxillary; opercular apparatus complete. Barbels none. Gill-opening wide; pseudobranchiæ. Air-bladder simple. Adipose fin present. Ovaries laminated; the eggs fall into the cavity of the abdomen, there being no oviduct. Pyloric appendages none.

Freshwater-fishes which represent the Salmonoids in the

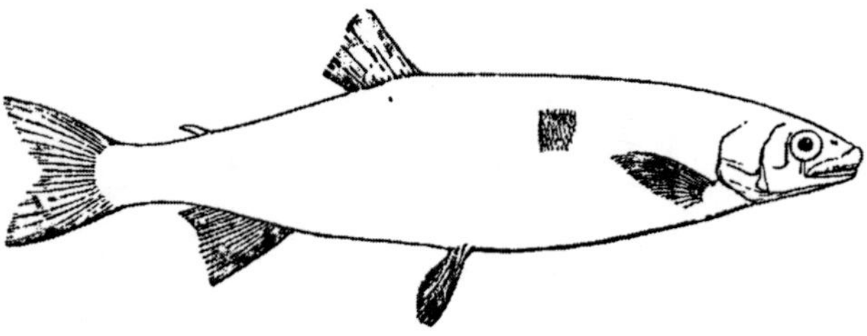

Fig. 296.—Prototroctes oxyrhynchus, New Zealand.

southern hemisphere. Two genera only are known. *Haplochiton* (Fig. 104, p. 250) abundant in lakes and the streams falling into the Straits of Magelhæn and in the rivers of Chile and the Falkland Islands. It has the general appearance of a Trout, but is naked. *Prototroctes*, with the habit of a *Coregonus*, scaly, and provided with minute teeth; one species (*P. maræna*) is common in South Australia, the other (*P. oxyrhynchus*) in New Zealand. The settlers in these colonies call them Grayling; the Maori name of the second species is "Upokororo."

EIGHTEENTH FAMILY—GONORHYNCHIDÆ.

Head and body entirely covered with spiny scales; mouth with barbels. Margin of the upper jaw formed by the intermaxillary, which, although short, is continued downwards as a thick lip, situated in front of the maxillary. Adipose fin none; the dorsal fin is opposite to the ventrals, and short, like the anal. Stomach simple, without blind sac; pyloric appendages in small number. Pseudobranchiæ; air-bladder absent. Gill-openings narrow.

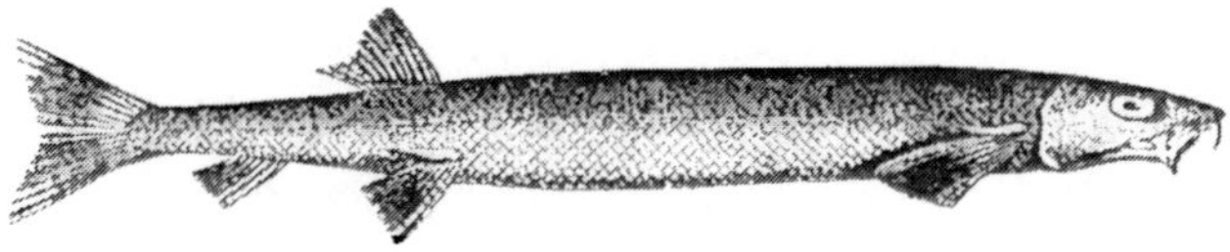

Fig. 297.—Gonorhynchus greyi.

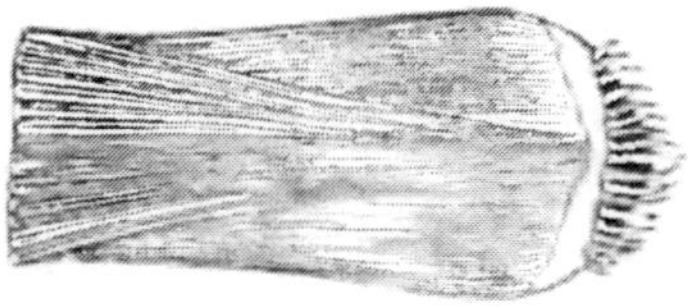

Fig. 298.—Scale of Gonorhynchus greyi.

One genus and species only (*Gonorhynchus greyi*) is known; it is a semi-pelagic fish, not very rare off the Cape of Good Hope, and in the Australian and Japanese seas. From 12 to 18 inches long. The colonists in New Zealand name it "Sand-eel," as it frequents bays with sandy bottom. It is eaten.

NINETEENTH FAMILY—HYODONTIDÆ.

Body covered with cycloid scales; head naked; barbels none. Margin of the upper jaw formed by the intermaxillaries mesially, and by the maxillaries laterally, the latter being articulated to the end of the former. Opercular apparatus complete. Adipose fin none; the dorsal fin belongs to the caudal portion of the vertebral column. Stomach horseshoe-shaped, without blind sac; intestine short; one pyloric appendage. Pseudobranchiæ none; air-bladder simple. Gill-openings wide. The ova fall into the abdominal cavity before exclusion.

One genus and species only (*Hyodon tergisus*) is known, generally called "Moon-eye." It is abundant in the western streams and great lakes of North America. From 12 to 18 inches long.

TWENTIETH FAMILY—PANTODONTIDÆ.

Body covered with large cycloid scales; sides of the head osseous. Margin of the upper jaw formed by the single intermaxillary mesially, and by the maxillaries laterally. The dorsal fin belongs to the caudal portion of the vertebral column, is short, opposite and similar to the anal. Gill-openings wide; gill-covers consisting of a præoperculum and operculum only. Branchiostegals numerous. Pseudobranchiæ none; air-bladder simple. Stomach without cœcal sac; one pyloric appendage. Sexual organs with a duct.

A small freshwater-fish (*Pantodon buchholzi*), singularly alike to a Cyprinodont, from the west coast of Africa.

TWENTY-FIRST FAMILY—OSTEOGLOSSIDÆ.

Body covered with large hard scales, composed of pieces like mosaic. Head scaleless; its integuments nearly entirely replaced by bone; lateral line composed of wide openings of the mucus-duct. Margin of the upper jaw formed by the intermaxillaries mesially, and by the maxillaries laterally. The dorsal fin belongs to the caudal portion of the vertebral column, is opposite and very similar to the anal fin; both approximate to the rounded caudal (with which they are abnormally confluent). Gill-openings wide; pseudobranchiæ none; air-bladder simple or cellular. Stomach without coecal sac; pyloric appendages two.

Large freshwater-fishes of the tropics, whose singular geographical distribution has been noticed above (p. 223).

OSTEOGLOSSUM.—Cleft of the mouth very wide, oblique, with the lower jaw prominent. A pair of barbels at the lower jaw. Abdomen trenchant. Bands of rasp-like teeth on the vomer, palatine and pterygoid bones, on the tongue and hyoid. Pectoral fins elongate.

O. bicirrhosum from Brazil and Guyana, *O. formosum* from Borneo and Sumatra, *O. leichardti* from Queensland.

ARAPAIMA.—Cleft of the mouth wide, with the lower jaw prominent; barbels none. Abdomen rounded. Jaws with an outer series of small conical teeth; broad bands of rasp-like teeth on the vomer, palatines, pterygoids, sphenoid, os linguale, and hyoid. Pectoral fins of moderate length.

Fig. 299.—Arapaima gigas.

The largest freshwater Teleostean known, exceeding a length of 15 feet and a weight of 400 pounds. It is common in the large rivers of Brazil and the Guyanas, and esteemed as an article of food. When salted it is exported in large quantities from the inland fisheries to the seaports.

HETEROTIS.—Cleft of the mouth rather small, with the jaws subequal; barbels none. A single series of small teeth in the jaws; pterygoids and hyoid with a patch of small conical teeth; none on the vomer or palatines.

This fish (*H. niloticus*), which is not uncommon in the Upper Nile and the West African rivers, exhibits several anatomical peculiarities. The fourth branchial arch supports a spiral accessory organ, the function of which is still unexplained. The air-bladder is cellular, and the stomach consists of a membranous and a muscular portion.

TWENTY-SECOND FAMILY—CLUPEIDÆ.

Body covered with scales; head naked; barbels none. Abdomen frequently compressed into a serrated edge. Margin of the upper jaw formed by the intermaxillaries mesially, and by the maxillaries laterally; maxillaries composed of at least three movable pieces. Opercular apparatus complete. Adipose fin none. Dorsal not elongate; anal sometimes very long. Stomach with a blind sac; pyloric appendages numerous. Gill-apparatus much developed, the gill-openings being generally very wide. Pseudobranchiæ generally present. Air-bladder more or less simple.

The family of "Herrings" is probably unsurpassed by any other in the number of individuals, although others comprise a much greater variety of species. The Herrings are principally coast-fishes, or, at least, do not go far from the shore; none belong to the deep-sea fauna; scarcely any have pelagic habits, but many enter or live in fresh waters communicating with the sea. They are spread over all the temperate and tropical zones. Fossil remains of Herrings are numerous, but the pertinence of some of the genera to this family is open to serious doubts, as the remains are too fragmentary to allow of determining whether they belong to Salmonoids or Clupeoids. Therefore, Agassiz comprised both families in one—*Haleeidæ*. Many of the remains belong to recent genera, which are readily recognised, as *Clupea*, *Engraulis* and *Chanos*, principally from the schists of Glaris and Licata, from Monte Bolca and the Lebanon. Others, like *Thrissopater*, from the Gault at Folkestone, *Leptosomus*, *Opisthopteryx*, *Spaniodon*, from the chalk and tertiary formations, can be readily associated with recent genera. But the majority do not show an apparent affinity to the present fauna. Thus, *Halee* from the chalk of Bohemia, *Platinx* and *Coelogaster* from Monte Bolca, *Rhinellus* from Monte Bolca and Mount Lebanon, *Scombroclupea*, with finlets behind the anal, from the Lebanon and Comen, and *Crossognathus* from tertiary Swiss formations, allied to *Megalops*, *Spathodactylus* from the same locality, and *Chiroecentrites* from Mount Lebanon, etc. Finally, a genus recently discovered in tertiary formations of Northern Italy, *Hemitrichas*, has been classed with the Clupeoids, from which, however, it differs by having two short dorsal fins, so that it must be considered, without doubt, to be the representative of a distinct family.

ENGRAULIS (including CETENGRAULIS).—Scales large or of moderate size. Snout more or less conical, projecting beyond the lower jaw. Teeth small or rudimentary. Intermaxillaries very small, hidden; maxillary long, attached to the cheek by a scarcely distensible membrane. Anal fin of moderate or great length. Branchiostegals short, from nine to fourteen in number.

Not less than forty-three different species of "Anchovies" are known from temperate and tropical seas. They exhibit marked differences in the length of their maxillary bone, which sometimes does not reach the gill-opening, whilst in other species it extends far beyond it; and in the number of their anal rays, which varies from 20 to 80. Some have the upper pectoral ray prolonged into a filament, thus leading towards the succeeding genus, *Coilia*. The majority are recognised, besides, by their peculiar structure, by a broad silvery, lateral band, similar to that observed in the Atherines. The most celebrated Anchovy is *E. encrasicholus*, very plentiful in the Mediterranean, but rarely wandering northwards. It is the species which, preserved in salt, is exported to all parts of the world, although similarly lucrative fisheries of Anchovies might be established in Tasmania where the same species occurs, in Chile, China, Japan, California, at Buenos Ayres, each of which countries possesses Anchovies by no means inferior to the Mediterranean species.

COILIA.—Body terminating in a long tapering tail. Scales of moderate size. Snout and jaws as in *Engraulis*. Anal fin exceedingly long, confluent with the caudal. The two or three

Fig. 300.—Coilia clupeoides.

upper pectoral rays are much prolonged, and their branches form four, six, or seven filaments.

Ten species from Indian and Chinese seas.

CHATOËSSUS.—Body compressed; abdomen serrated. Scales of moderate size. Snout obtuse, or obtusely conical, more or less projecting beyond the cleft of the mouth, which is narrow, more or less transverse. Maxillary joined to the ethmoid, its upper portion being behind the intermaxillary. Teeth none. Anal fin rather long; dorsal opposite to the ventrals, or to the space between ventrals and anal. Gill-membranes entirely separate; branchial arches forming two angles, one pointing forward and the other backwards; the fourth branchial arch with an accessory organ; branchiostegals of moderate length, five or six in number.

Ten species from the coasts, brackish and fresh waters of

Central America (one species ranges to New York), Australia, the East Indies, and Japan.

CLUPEA.—Body compressed, with the abdomen serrated, the serrature extending forwards to the thorax. Scales of moderate or large, rarely of small size. Upper jaw not projecting beyond the lower. Cleft of the mouth of moderate width. Teeth, if present, rudimentary and deciduous. Anal fin of moderate extent, with less than thirty rays; dorsal fin opposite to the ventrals. Caudal forked.

This genus comprises more than sixty different species, the geographical distribution of which coincides with that of the family. The majority are of greater or less utility to man, but a few tropical species (*C. thrissa*, *C. venenosa*, and others) acquire, probably from their food, highly poisonous properties, so as to endanger the life of persons eating them. The most noteworthy species are—

1. *C. harengus* (the "Herring").—It is readily recognised by having an ovate patch of very small teeth on the vomer. D. 17-20. A. 16-18. L. lat. 53-59. Vert. 56. Gill-cover smooth, without radiating ridges. It inhabits, in incredible numbers, the German Ocean, the northern parts of the Atlantic, and the seas north of Asia. The Herring of the Atlantic coasts of North America is identical with that of Europe. A second species has been supposed to exist on the British coast (*C. leachii*), but it comprises only individuals of a smaller size, the produce of an early or late spawn. Also the so-called "Whitebait" is not a distinct species, but consists chiefly of the fry or the young of herrings, and is obtained "in perfection" at localities where these small fishes find an abundance of food, as in the estuary of the Thames.

[Separate accounts on the Herring may be found in Cuvier and Valenciennes, "Hist. nat. des Poissons," vol. xx.; J. M. Mitchell, "The Herring, its Natural History and National Importance," Edinb. 1864, 8vo; P. Neucrantz, "De Harengo," Lübeck, 1654; J. S. Dodd, "Essay towards a Natural History of the Herring," Lond. 1768, 8vo; Bock, "Versuch einer vollstændigen Natur-und Handels-Geschichte des Hærings," Königsberg, 1769, 8vo.]

2. *C. mirabilis.*—The Herring of the North Pacific.

3. *C. sprattus.*—The "Sprat." Without vomerine teeth. D. 15-18. A. 17-20. L. lat. 47-48. Vert. 47-49. Gill-cover smooth, without radiating ridges. Abundant on the Atlantic coasts of Europe.

4. *C. thrissa.*—One of the most common West Indian fishes, distinguished by the last dorsal ray being prolonged into a filament. Hyrtl has discovered a small accessory branchial organ in this species.

5. *C. alosa.*—The "Shad" or "Allice Shad," with very fine and long gill-rakers, from 60 to 80 on the horizontal part of the outer branchial arch, and with one or more black lateral blotches. Coasts of Europe, ascending rivers.

6. *C. finta.*—The "Shad" or "Twaite Shad," with stout osseous gill-rakers, from 21 to 27 on the horizontal part of the outer branchial arch, and spotted like the preceding species. Coasts of Europe, ascending rivers, and found in abundance in the Nile.

7. *C. menhaden.*—The "Mossbanker," common on the Atlantic coasts of the United States. The economic value of this fish is surpassed in America only by that of the Gadoids, and derived chiefly from its use as bait for other fishes, and from the oil extracted from it, the annual yield of the latter exceeding that of the whale (from American Fisheries). The refuse of the oil factories supplies a material of much value for artificial manures.

[See G. Brown Goode, "The Natural and Economical History of the American Menhaden," in U.S. Commission of Fish and Fisheries, Part V., Washington, 1879, 8vo.]

8. *C. sapidissima.*—The American Shad, abundant, and an important food-fish on the Atlantic coasts of North America. Spawns in fresh water.

9. *C. mattowocca.*—The "Gaspereau" or "Ale-wife," common on the Atlantic coasts of North America, ascending into

freshwater in early spring, and spawning in ponds and lakes.

10. *C. pilchardus.*—The "Pilchard" or the "Sardine," equally abundant in the British Channel, on the coast of Portugal, and in the Mediterranean, and readily recognised by radiating ridges on the operculum, descending towards the sub-operculum.

11. *C. sagax.*—Representing the Pilchard in the Pacific, and found in equally large shoals on the coasts of California, Chile, New Zealand, and Japan.

12. *C. toli.*—The subject of a very extensive fishery on the coast of Sumatra for the sake of its roes, which are salted and exported to China; the dried fish themselves being sent into the interior of the island. The fish is called "Trubu" by the Malays, about 18 inches long, and it is said that between fourteen and fifteen millions are caught annually.

13. *C. scombrina.*—The "Oil-Sardine" of the eastern coast of the Indian Peninsula.

Other, but less important genera of Clupeoids with serrated abdomen, are *Clupeoides, Pellonula, Clupeichthys, Pellona, Pristigaster*, and *Chirocentrodon* (these three last with very small or without any ventral fins).

ALBULA.—Body oblong, moderately compressed; abdomen flat. Scales of moderate size, adherent; lateral line distinct. Eyes covered with a broad annular adipose membrane. Snout pointed, the upper jaw projecting beyond the lower. Mouth inferior, of moderate width, with villiform teeth; intermaxillary juxtaposed to the upper anterior edge of the maxillary. Dorsal fin opposite to the ventrals; anal fin shorter than dorsal. Gill-membranes entirely separate, with numerous branchiostegals.

One species only (A. *conorhynchus*), ranging over all tropical and sub-tropical seas, and very common in many localities near the coasts. It grows to a length of from two to three feet, and is not valued as food.

ELOPS. — Body rather elongate, moderately compressed; abdomen flat. Scales small, adherent; lateral line distinct. A narrow osseous lamella, attached to the mandibulary symphysis, covers the part between the mandibles. Snout pointed; mouth wide, anterior; intermaxillary short, maxillary forming the lateral part of the mouth. Bands of villiform teeth in the jaws, on the vomer, palatine and pterygoid bones, on the tongue, and on the base of the skull. Dorsal fin opposite to ventrals; anal rather shorter than dorsal. Gill-membranes entirely separate, with very numerous branchiostegals.

Two species, of which one, *E. saurus*, is, like the preced-

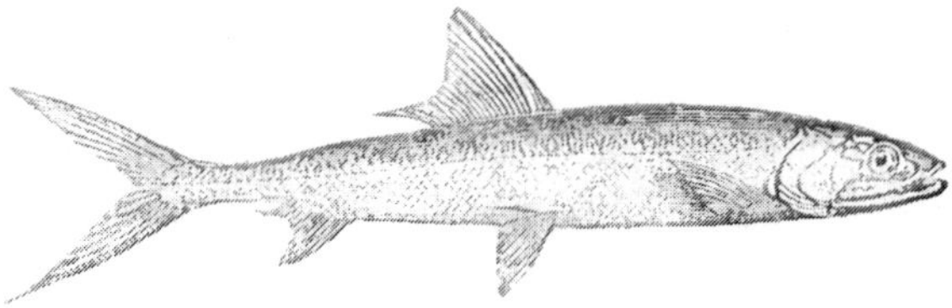

Fig. 301.—Elops saurus.

ing fish, spread over all tropical and sub-tropical seas; it exceeds a length of three feet, and is not esteemed as food.

MEGALOPS.—Body oblong, compressed, abdomen flat. Scales large, adherent; lateral line distinct. A narrow osseous lamella, attached to the mandibulary symphysis, between the mandibles. Snout obtusely conical; mouth anterior, lower jaw prominent; intermaxillary short; maxillary forming the lateral part of the mouth. Bands of villiform teeth in the jaws, on the vomer, palatine and pterygoid bones, on the tongue and on the base of the skull. Dorsal fin opposite to, or immediately behind, the ventrals; anal rather larger than dorsal. Gill-membranes entirely separate, with numerous branchiostegals. Pseudobranchiæ none.

Two species, one belonging to the Indo-Pacific (*M. cyprinoides*), the other to the Atlantic (*M. thrissoides*); they are the largest fishes of this family, exceeding a length of five feet, and excellent eating. Young specimens enter freely fresh waters.

CHANOS.—Body oblong, compressed; abdomen flat. Scales small, striated, adherent; lateral line distinct. Snout depressed; mouth small, anterior, transverse, the lower jaw with a small symphysial tubercle. Intermaxillary in juxtaposition to the upper anterior edge of maxillary. Teeth none. Dorsal fin opposite to the ventrals; anal small, shorter than dorsal; caudal deeply forked. Gill-membranes entirely united below, and free from the isthmus. Branchiostegals four, long. An accessory branchial organ in a cavity behind the gill-cavity proper. Air-bladder divided by a constriction into an anterior and posterior portion. Mucous membrane of the œsophagus raised in a spiral fold. Intestine with many convolutions.

Two species from the Indo-Pacific, of which *Ch. salmoneus* is extremely common; it enters fresh waters, and exceeds a length of four feet; its flesh is highly esteemed. The acces-

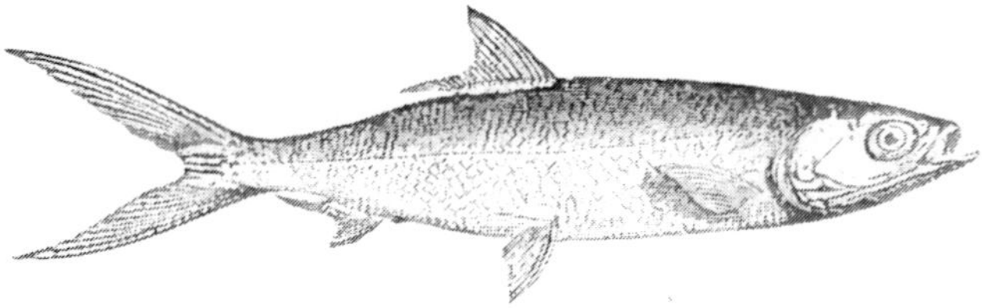

Fig. 302.—Chanos salmoneus.

sory branchial organ and the skeleton have been described by *Müller*, "Bau und Grenzen der Ganoiden," p. 75; and by *Hyrtl*, "Denkschr. Ak. Wiss. Wien." xxi. 1883, p. 1.

The remaining genera belonging to this family are *Spratelloides, Dussumieria,* and *Etrumeus,* which together form a small group, distinguished by an anterior and lateral mouth, by the upper jaw not overlapping the lower, by a rounded abdomen, and by lacking the gular plate of some of the preceding genera.

TWENTY-THIRD FAMILY—BATHYTHRISSIDÆ.

Body oblong, with rounded abdomen, covered with cycloid scales; head naked; barbels none. Margin of the upper jaw formed by the intermaxillaries mesially, and by the maxillaries laterally. Opercular apparatus complete. Adipose fin none; dorsal fin much elongate, many rayed; anal fin short. Stomach with a blind sac; pyloric appendages numerous. Gill-apparatus well developed; pseudobranchiæ; gill-openings wide; an air-bladder. Ova very small; ovaries without duct.

One genus and species only (*Bathythrissa dorsalis*) from deep water (350 fathoms) off the coast of Japan. This remarkable fish has the appearance of a *Coregonus*, and attains to a length of two feet. Nothing is known of its osteology, but possibly a fossil genus from the Gyps of Montmartre; *Notæus*, which has also a long dorsal fin, may prove to belong to the same family.

TWENTY-FOURTH FAMILY—CHIROCENTRIDÆ.

Body covered with thin, deciduous scales; barbels none. Margin of the upper jaw formed by the intermaxillaries mesially, and by the maxillaries laterally, both bones being firmly united, in juxtaposition. Opercular apparatus complete. Adipose fin none; the dorsal fin belongs to the caudal portion of the vertebral column. Stomach with a blind sac; intestine short, the mucous membrane forming a spiral fold; pyloric appendages none. Pseudobranchiæ none; air-bladder incompletely divided into cells; gill-opening wide.

One genus and species only (*Chirocentrus dorab*) is known, which is common in the Indian Ocean, and attains to a length of about three feet; it is not esteemed as food. Remains of fishes similar to *Chirocentrus* are found in the marl slates of Padang, in Sumatra.

TWENTY-FIFTH FAMILY—ALEPOCEPHALIDÆ.

Body with or without scales; head naked; barbels none. Margin of the upper jaw formed by the intermaxillaries and maxillaries, the former being placed along the upper anterior edge of the latter. Opercular apparatus complete. Adipose fin none; the dorsal fin belongs to the caudal portion of the vertebral column. Stomach curved, without blind sac; pyloric appendages in moderate number. Pseudobranchiæ; air-bladder absent. Gill-openings very wide.

Before the voyage of the "Challenger" one species only of this family was known, *Alepocephalus rostratus*, a rare fish from the Mediterranean; now, four genera with seven species are known, and there is no doubt that this family is one of the most characteristic, and will prove to be one of the most generally distributed forms, of the deep-sea. Their vertical range varies between 345 (*Xenodermichthys*) and 2150 (*Bathytroctes*) fathoms. They approach the Salmonoids, but lack invariably the adipose fin. Their dentition is very feeble; their eye large; bones thin. Coloration black.

ALEPOCEPHALUS has thin cycloid scales; a mouth of moderate width, and no teeth on the maxillary.

BATHYTROCTES has cycloid scales, a wide mouth, and teeth on the maxillary as well as intermaxillary.

PLATYTROCTES has small keeled scales and no ventrals.

XENODERMICHTHYS with fine nodules instead of scales.

TWENTY-SIXTH FAMILY—NOTOPTERIDÆ.

Head and body scaly; barbels none. Margin of the upper jaw formed by the intermaxillaries mesially, and by the maxillaries laterally. Opercular apparatus incomplete. Tail prolonged, tapering. Adipose fin none. Dorsal short, belonging to the caudal portion of the vertebral column; anal very long. Stomach without blind sac; two pyloric appendages. Pseudo-

branchiæ none; air-bladder present, divided in the interior. The ova fall into the cavity of the abdomen before exclusion. On each side a parieto-mastoid cavity leading into the interior of the skull.

One genus only (*Notopterus*) with five species which inhabit fresh waters of the East Indies and West Africa. Well-preserved remains of this genus occur in the marl slates of Padang, in Sumatra. Their air-bladder is divided into several compartments, and terminates in two horns anteriorly and posteriorly, the anterior horns being in direct connection with the auditory organ.

TWENTY-SEVENTH FAMILY—HALOSAURIDÆ.

Body covered with cycloid scales; head scaly; barbels none. Margin of the upper jaw formed by the intermaxillaries mesially, and by the maxillaries laterally. Opercular apparatus incomplete. Adipose fin none. The short dorsal belongs to the abdominal part of the vertebral column; anal very long. Stomach with a blind sac; intestine short; pyloric appendages in moderate number. Pseudobranchiæ none. Air-bladder large, simple; gill-openings wide. Ovaries closed.

The only genus belonging to this family was discovered by the Madeiran ichthyologist Johnson, in 1863; but since then the naturalists of the "Challenger" expedition have added four other species, showing that this type is a deep-sea form and widely distributed; the specimens were dredged in depths varying from 560 to 2750 fathoms.

TWENTY-EIGHTH FAMILY—HOPLOPLEURIDÆ.

Body generally with four series of subtriangular scutes, and with intermediate scale-like smaller ones. One (?) dorsal only; head long, with the jaws produced.

Extinct; developed in the chalk and extending into tertiary formations: *Dercetis* (with the upper jaw longest), *Leptotrachelus, Pelargorhynchus, Plinthophorus, Saurorhamphus* (with the lower jaw longest), *Eurypholis; Ischyrocephalus* (?). The latter genus, from cretaceous formations of Westphalia, is said to have two dorsal fins.

Twenty-Ninth Family—Gymnotidæ.

Head scaleless; barbels none. Body elongate, eel-shaped. Margin of the upper jaw formed in the middle by the intermaxillaries, and laterally by the maxillaries. Dorsal fin absent or reduced to an adipose strip; caudal generally absent, the tail terminating in a point. Anal fin exceedingly long. Ventrals none. Extremity of the tapering tail capable of being reproduced. Vent situated at, or at a short distance behind, the throat. Humeral arch attached to the skull. Ribs well developed. Gill-openings rather narrow. Air-bladder present, double. Stomach with a cæcal sac and pyloric appendages. Ovaries with oviducts.

Eel-like freshwater fishes from Tropical America.

Sternarchus.—Tail terminating in a distinct small caudal fin. Teeth small. A rudimentary dorsal fin is indicated by an adipose band fitting into a groove on the back of the tail; it is easily detached, so as to appear as a thong-like appendage fixed in front. Branchiostegals four.

Eight species, some have the snout compressed and of moderate length, like *St. Bonapartii* from the River Amazons; others have it produced into a long tube, as St. oxyrhynchus from the Essequibo.

Rhamphichthys.—Caudal fin none; teeth none; no trace of a dorsal fin. No free orbital margin.

Six species, of which, again, some have a tubiform snout, whilst in the others it is short.

STERNOPYGUS.—Caudal fin none; no trace of a dorsal fin. Both jaws with small villiform teeth; similar teeth on each side of the palate. Body scaly.

Four species, very common, and growing to a length of 30 inches.

CARAPUS.—Caudal fin none; no trace of a dorsal fin. A series of conical teeth in each jaw. Anterior nostrils, wide in the upper lip. Body scaly.

One species (*C. fasciatus*) extremely common, and found all over tropical America, east of the Andes, from 18 to 24 inches long.

GYMNOTUS.—Caudal and dorsal fins absent; anal extending to the end of the tail. Scales none. Teeth conical, in a single series. Eyes exceedingly small.

The "Electric Eel" is the most powerful of electric fishes, growing to a length of six feet, and extremely abundant in certain localities of Brazil and the Guyanas. The electric organ consists of two pairs of longitudinal bodies, situated immediately below the skin, above the muscles; one pair on the back of the tail, and the other pair along the anal fin. Each fasciculus is composed of flat partitions or septa, with transverse divisions between them. The outer edge of the septa appear in nearly parallel lines in the direction of the longitudinal axis of the body, and consist of thin membranes, which are easily torn; they serve the same purpose as the columns in the analogous organ of the Torpedo, making the walls or abutments for the perpendicular and transverse dissepiments, which are exceedingly numerous, and so closely aggregated as to seem almost in contact. The minute prismatic cells, intercepted between these two sorts of plates, contain a gelatinous matter; the septa are about one-thirtieth of an inch from each other, and one inch in length contains a series of 240 cells, giving an enormous surface to the electric organs. The whole apparatus is supplied with more than 200 nerves, which are the continuations of the rami anteriores of the spinal nerves. In their course they give out branches to the muscles of the back, and to the skin of the animal. In the Gymnotus, as in the Torpedo, the nerves supplying the electric organs are much larger than those bestowed on any part for the purposes of sensation or movement.

The graphic description by Humboldt of the capture of Electric Eels by horses driven into the water, which would receive the electric discharges and thus exhaust the fishes, seems to rest either on the imagination of some person who told it to the great traveller or on some isolated incident. Recent travellers have not been able to verify it even in the same parts of the country where the practice was said to exist.

THIRTIETH FAMILY—SYMBRANCHIDÆ.

Body elongate, naked or covered with minute scales; barbels none. Margin of the upper jaw formed by the intermaxillaries only, the well developed maxillaries lying behind and parallel to them. Paired fins none. Vertical fins rudimentary, reduced to more or less distinct cutaneous folds. Vent situated at a great distance behind the head. Ribs present. Gill-openings confluent into one slit situated on the ventral surface. Air-bladder none. Stomach without cæcal sac or pyloric appendages. Ovaries with oviducts.

The fishes of this family consist of freshwater-fishes from tropical America and Asia, which, however, enter also brackish water; and of a truly marine genus from Australia.

AMPHIPNOUS.—Vent in the posterior half of the body, which is covered with minute scales longitudinally arranged.

A common fish (*A. cuchia*) in Bengal, remarkable for its singular respiratory apparatus. It has only three branchial arches, with rudimentary branchial laminæ, and with very narrow slits between the arches. To supplement this insufficient respiratory apparatus, a lung-like sac is developed on each side of the body behind the head, opening between the hyoid and first branchial arch. The interior of the sac is abundantly provided with blood-vessels, the arterial coming from the branchial arteries, whilst those issuing from it unite to form the aorta. *A. cuchia* approaches the Eels in having the humeral arch not attached to the skull.

MONOPTERUS.—Vent in the posterior half of the body, which is naked. Three branchial arches with rudimentary gills, but without breathing sac.

One species (*M. javanicus*), which is extremely common in the East Indian Archipelago and in the eastern parts of the Continent. Upwards of three feet long.

SYMBRANCHUS.—Vent in the posterior half of the body, which is naked. Four branchial arches with well developed gills.

Three species, of which one (*S. marmoratus*) is extremely common in tropical America, and the other (*S. bengalensis*) not less so in the East Indies.

CHILOBRANCHUS.—Vent in the anterior half of the length of the body, which is naked. Vertical fins reduced to a simple cutaneous fold, without rays.

A small fish (*Ch. dorsalis*) from North Western Australia and Tasmania.

THIRTY-FIRST FAMILY—MURÆNIDÆ.

Body elongate, cylindrical or band-shaped, naked or with rudimentary scales. Vent situated at a great distance from the head. Ventral fins none. Vertical fins, if present, confluent, or separated by the projecting tip of the tail. Sides of the upper jaw formed by the tooth-bearing maxillaries, the fore part by the intermaxillary, which is more or less coalescent with the vomer and ethmoid. Humeral arch not attached to the skull. Stomach with a blind sac; no pyloric appendages. Organs of reproduction without efferent ducts.

The "Eels" are spread over almost all fresh waters and seas of the temperate and tropical zones; some descend to the greatest depths of the oceans. The young of some have a limited pelagic existence. (*Leptocephali*, see p. 179.) At Monte Bolca fossil remains are very numerous, belonging to recent genera, *Anguilla*, *Sphagebranchus*, and *Ophichthys*; even larval Leptocephales have been preserved. *Anguilla* has been found also in the chalk of Aix and Oeningen.

In the majority of the species the branchial openings in the pharynx are wide slits (*Murænidæ platyschistæ*); in others, the true Murænæ, (*Murænidæ engyschistæ*) they are narrow.

NEMICHTHYS.—Exceedingly elongate, band-shaped; tail tapering into a point. Vent approximate to the pectorals, but the abdominal cavity extending far behind the vent. Jaws produced into a long slender bill, the upper part being formed by the vomer and intermaxillaries. The inner surface of the bill covered with small tooth-like asperities. Eye large. The nostrils of each side are close together, in a hollow before the eye. Gill-openings wide, nearly confluent. Pectoral and vertical fins well developed.

This very singular type is a deep-sea form, occurring at depths of from 500 to 2500 fathoms. The two species known have hitherto been found in the Atlantic only.

CYEMA.—This genus combines the form of the snout of *Nemichthys*, with the soft and shorter body of a *Leptocephalus*; but the gill-openings are very narrow and close together on the abdominal surface. Vent in about the middle of the length of the body; vertical fins well developed, confined to, and surrounding, the tail. Pectoral fins well developed. Eye very small.

Known from two specimens only, 4½ inches long, dredged in depths of 1500 and 1800 fathoms in the Pacific and Antarctic Oceans.

SACCOPHARYNX.—Deep-sea Congers, with the muscular system very feebly developed, with the bones very thin, soft, and wanting in inorganic matter. Head and gape enormous. Snout very short, pointed, flexible, like an appendage overlapping the gape. Maxillary and mandibulary bones very thin, slender, arched, armed with one or two series of long, slender, curved, widely set teeth, their points being directed inwards; palate toothless. Gill-openings wide, at some distance from the head, at the lower part of the sides; gills very narrow, free, and exposed. Trunk of moderate length. Stomach distensible in an extraordinary degree. Vent at the end of the trunk. Tail band-like, exceedingly long, tapering in a very fine filament. Pectoral small, present. Dorsal and anal fins rudimentary.

This is another extraordinary form of Deep-sea Eels; the muscular system, except on the head, is very feebly developed; the bones are as thin, soft, and wanting in inorganic matter, as in the *Trachypteridæ*. This fish is known from three specimens only, which have been found floating on the surface of the North Atlantic, with their stomachs much distended, having swallowed some other fish, the weight of which many times exceeded that of their destroyer. It attains to the length of several feet.

SYNAPHOBRANCHUS.—Gill-openings ventral, united into a longitudinal slit between the pectoral fins, separate internally. Pectoral and vertical fins well developed. Nostrils lateral, the anterior subtubular, the posterior round, before the lower half of the eye. Cleft of the mouth very wide; teeth small; body scaly. Stomach very distensible.

Deep-sea Congers, with well-developed muscular system, spread over all oceans, and occurring in depths of from 345 to 2000 fathoms. Four species are known. Probably attaining to the same length as the Conger.

ANGUILLA.—Small scales imbedded in the skin. Upper jaw not projecting beyond the lower. Teeth small, forming bands. Gill-openings narrow, at the base of the pectoral fins. The dorsal fin commences at a considerable distance from the occiput.

Some twenty-five species of "Eels" are known from the freshwaters and coasts of the temperate and tropical zones;

none have been found in South America or the west coast of North America and West Africa. The following are the most note-worthy:—The common European species (*A. anguilla*) is spread over Europe to 64° 30′ lat. N., and all round the Mediterranean area, but is not found either in the Danube or in the Black and Caspian Seas; it extends across the Atlantic to North America. The form of the snout varies much, and some naturalists have believed that specimens with a broad and obtuse snout were specifically distinct from those with pointed snout. However, every degree of breadth of the snout may be observed; and a much safer way of recognizing this species, and distinguishing it from other European Eels, is the forward position of the dorsal fin; the distance between the commencement of the dorsal and anal fins being as long as, or somewhat longer than, the head. Eels grow generally to a length of about three feet, but the capture of much larger examples is on record. Their mode of propagation is still unknown. So much only is certain that they do not spawn in fresh water, that many full-grown individuals, but not all, descend rivers during the winter months, and that some of them at least must spawn in brackish water or in deep water in the sea; for in the course of the summer young individuals from three to five inches long ascend rivers in incredible numbers, overcoming all obstacles, ascending vertical walls or floodgates, entering every larger and smaller tributary, and making their way even over terra firma to waters shut off from all communication with rivers. Such immigrations have been long known by the name of "*Eel-fairs.*" The majority of the Eels which migrate to the sea appear to return to fresh water, but not in a body, but irregularly, and throughout the warmer part of the year. No naturalist has ever observed these fishes in the act of spawning, or found mature ova; and the organs of reproduction of individuals caught in fresh water are so little developed and so much alike, that the female organ can be distinguished from the male only with the aid of a microscope.

The second species found in Great Britain, on the coasts of Europe generally, in China, New Zealand, and the West Indies, is (*A. latirostris*) the "Grig" or "Glut," which prefers the neighbourhood of the sea to distant inland-waters, and in which the dorsal fin begins farther backwards, the distance between the commencement of the dorsal and anal fins being shorter than the head; its snout seems to be always broad. On the American side of the Atlantic other species, beside *A. anguilla* are found in abundance: *A. bostoniensis, A. texana.* The largest Eels occur in lakes of the islands of the Indo-Pacific, and they play a conspicuous part in the mythology of the South-Sea Islanders and Maories; individuals of from eight to ten feet in length have been seen, and referred to several species, as *A. mauritiana, fidjiensis, obscura, anciteensis,* etc.

CONGER.—Scaleless. Cleft of the mouth wide, extending at least to below the middle of the eye. Maxillary and mandibulary teeth arranged in series, one of which contains teeth of equal size, and so closely set as to form a cutting edge. No canine teeth. Vomerine band of teeth short. Pectoral and vertical fins well developed, the dorsal commencing behind the root of the pectoral. Gill-openings large, approximate to the abdomen. The posterior nostril opposite to the upper or middle part of the orbit, the anterior in a tube. Eyes well developed.

The "Congers" are marine Eels; the best known species (*C. conger*) seems to be almost cosmopolitan, and is plentiful all round Europe, at St. Helena, in Japan, and Tasmania. It attains to a length of eight feet, and thrives and grows rapidly even in confinement, which is not the case with the fresh-water Eel. Three other species are known, of which *C. marginatus* from the Indian Ocean, is the most common. *Leptocephalus morrisii* is an abnormal larval condition of the Conger.

Genera allied to *Conger* are *Poeciloconger*, *Congromuræna*, *Uroconger*, and *Heteroconger*.

MURÆNESOX.—Scaleless. Snout produced. Jaws with several series of small closely set teeth, anteriorly with canines; vomer with several long series of teeth, the middle of which is formed by large conical or compressed teeth. Gill-openings wide, approximate to the abdomen. Pectoral and vertical fins well developed, the dorsal beginning above the gill-opening. Two pairs of nostrils, the posterior opposite to the upper part or middle of the eye.

Four species from tropical seas, *M. cinereus* being very common in the Indian Ocean, and attaining to a length of six feet.

NETTASTOMA.—Scaleless. Snout much produced, depressed. Jaws and vomer with bands of card-like teeth, those along the median line of the vomer being somewhat the larger. Vertical fins well developed; pectorals none. Gill-openings of moderate width, open. Nostrils on the upper surface of the head, valvular; the anterior near to the end of the snout, the posterior above the anterior angle of the eye.

This genus lives at some depth, the Japanese species (*N. parviceps*) having been obtained at 345 fathoms. *N. melanurum* from the Mediterranean, seems to inhabit a similar depth. *Hyoprorus* is its Leptocephalid form.

Genera allied to *Muraenesox* are *Saurenchelys*, *Oxyconger*, *Hoplunnis*, and *Neoconger*; in all these the nostrils have a superior or lateral position. In other genera the nostrils perforate the upper lip, as in *Myrus*, *Myrophis*, *Paramyrus*, *Chilorhinus*, *Muraenichthys*, and *Ophichthys*, the last genus deserving of particular mention on account of its great range and common occurrence.

OPHICHTHYS.—Nostrils labial; extremity of the tail free, not surrounded by a fin.

More than eighty species are known, many of which are

abundant on the coasts of the tropical and sub-tropical zones. They do not attain to a large size, but many must be extremely voracious and destructive to other fishes, if we draw an inference from the formidable dentition with which their jaws and palate is armed. Other species have much more feeble, and some even obtuse teeth, better adapted for seizing Crustaceans than vigorous and slippery fishes. Some have rudimentary pectoral fins or lack them altogether. Many are highly ornamented with bands or spots, the coloration being apparently very constant in the several species.

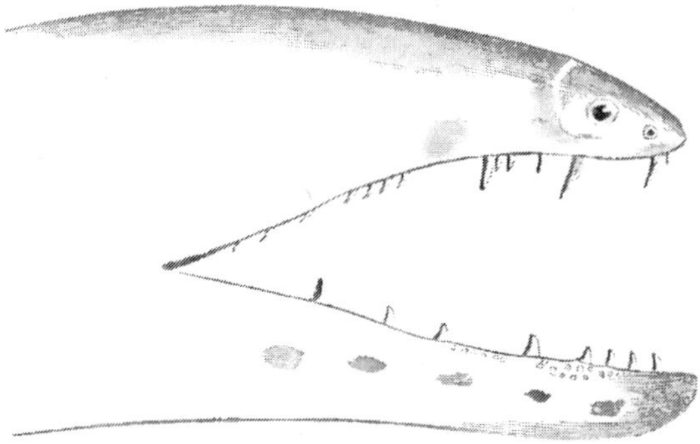

Fig. 303.—Ophichthys crocodilinus, from the Indo-Pacific.

MORINGUA.—Body scaleless, cylindrical, with the trunk much longer than the tail. Pectorals none or small; vertical fins but little developed, limited to the tail. Posterior nostrils in front of the small eye. Cleft of the mouth narrow; teeth uniserial. Heart placed far behind the branchiæ. Gill-openings rather narrow, inferior.

Six species from freshwaters, brackish water, and the coasts of India to the Fiji Islands.

MURÆNA.—Scaleless. Teeth well developed. Gill-openings and clefts between the branchial arches narrow. Pectoral fins none; dorsal and anal fins well developed. Two nostrils on each side of the upper surface of the snout; the posterior a narrow round foramen, with or without tube; the anterior in a tube.

The Muræenas are as abundantly represented in the tropical and sub-tropical zones, and have nearly the same range, as *Ophichthys*. The number of species known exceeds eighty. The majority are armed with formidable pointed teeth, well suited for seizing other fish on which they prey. Large specimens thus armed readily attack persons in and out of the water; and as some species attain a length of some six or eight feet, they are justly feared by fishermen. The minority of species have obtuse and molar-like teeth, their food consisting chiefly of Crustaceans and other hard-shelled

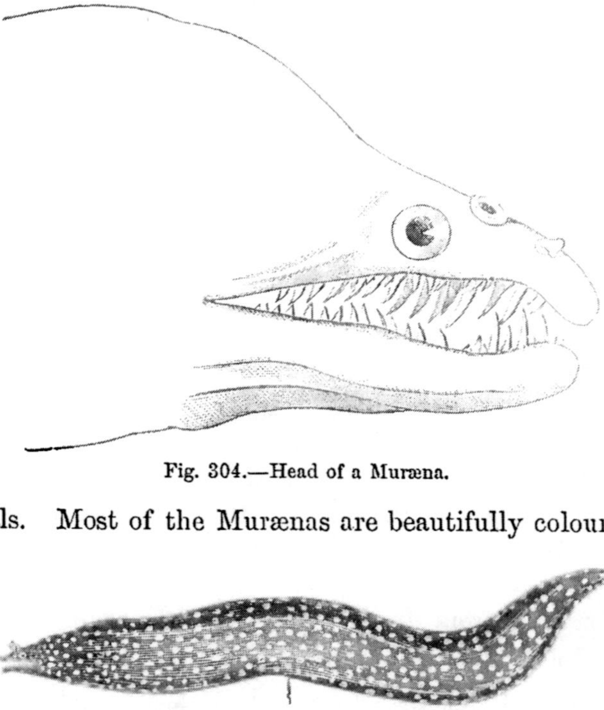

Fig. 304.—Head of a Muræna.

animals. Most of the Muræenas are beautifully coloured and

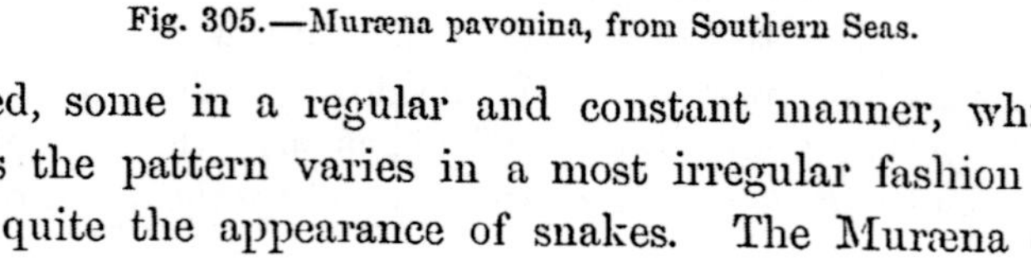

Fig. 305.—Muræna pavonina, from Southern Seas.

spotted, some in a regular and constant manner, whilst in others the pattern varies in a most irregular fashion: they have quite the appearance of snakes. The Muræna of the

Ancient Romans is *Muræna helena*, which is not confined to the Mediterranean, but also found in the Indian Ocean and on the coast of Australia. Its skin is of a rich brown, beautifully marked with large yellowish spots, each of which contains smaller brown spots.

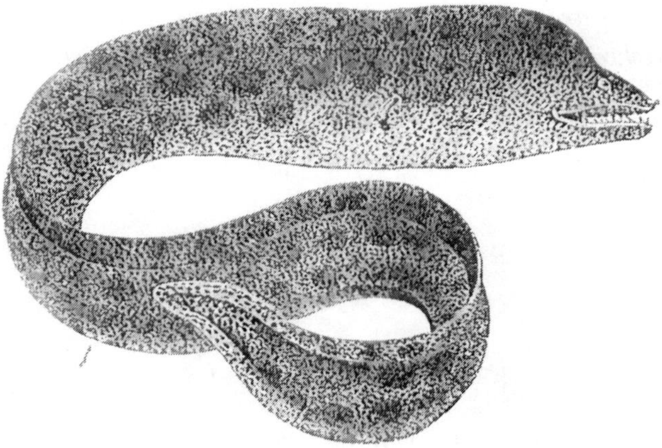

Fig. 306.—Muræna picta, from the Indo-Pacific.

Gymnomuræna differs from *Muræna* in having the fins reduced to a short rudiment near the end of the tail. Six

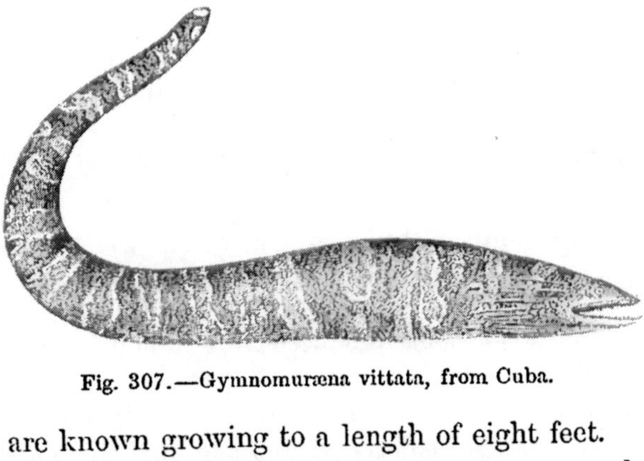

Fig. 307.—Gymnomuræna vittata, from Cuba.

species are known growing to a length of eight feet.

Myroconger and *Enchelycore* belong to the same sub-family as *Muræna*, but the former is provided with pectoral fins, and in the latter the posterior nostril is a long slit, and not round as in the other genera.

FIFTH ORDER—LOPHOBRANCHII.

The gills are not laminated, but composed of small rounded lobes attached to the branchial arches. Gill-cover reduced to a

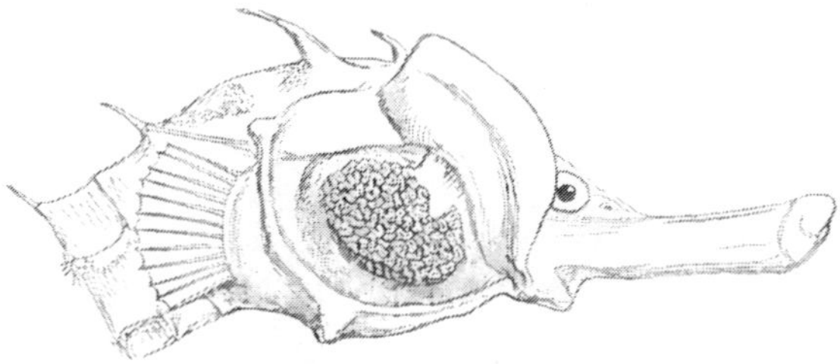

Fig. 308.—Gills of Hippocampus abdominalis.

large simple plate. Air-bladder simple, without pneumatic duct. A dermal skeleton composed of numerous pieces arranged in segments, replaces more or less soft integuments. Muscular system not much developed. Snout prolonged. Mouth terminal, small, toothless, formed as in Acanthopterygians.

FIRST FAMILY—SOLENOSTOMIDÆ.

Gill-openings wide. Two dorsal fins, the rays of the anterior not articulated. All the other fins well developed.

One living genus only is known, which was preceded in the tertiary epoch by *Solenorhynchus* (Monte Postale).

SOLENOSTOMA.—Snout produced into a long tube. Body compressed, with very short tail. All parts covered with thin skin, below which there is a dermal skeleton formed by large star-like ossifications. The soft dorsal and anal fins on elevated bases; caudal fin long. Ventral fins inserted opposite to the anterior dorsal, close together, seven-rayed; they are free in the male, but in the female their inner side coalesces with the integuments of the body, a large pouch for the reception of the eggs being formed thereby. Air-bladder and pseudobranchiæ absent. Branchiostegals four, very thin. Intestinal tract very simple, with a stomachic dilatation, without pyloric appendages. Ova very small.

The dermal skeleton of this singular type is formed by star-like ossifications, four in each horizontal and vertical series on the side of the fore part of the trunk; each consists of four or three radiating branches by which it joins the neighbouring bones; on the hind part of the trunk and tail the series are diminished to two. The dorsal and abdominal profiles in front of the fins are protected by similar bones. The vertebral column is composed of eighteen abdominal and fifteen caudal vertebræ, the vertebræ gradually decreasing in length backwards, so that the shortness of the tail is caused not only by the smaller number of vertebræ, but also by their much lesser length. Neural and hæmal spines are developed. The pelvis consists of two pairs of cartilaginous laminæ, the convex margin of the anterior fitting into an angle of a dermal bone which separates the pelvis from the well-ossified humeral arch.

The singular provision for the retention and protection of the eggs has been described above (p. 162, figs. 73 and 74), and we have only to repeat here that it is the female which takes care of the progeny, and not the male as in the following family. Two or three small species are known from the Indian Ocean; they are beautifully marked, especially the male, which also appears to be of smaller size in this genus than the female.

Second Family—Syngnathidæ.

Gill-openings reduced to a very small opening near the upper posterior angle of the gill-cover. One soft dorsal fin; no ventrals, and, sometimes, one or more of the other fins are also absent.

Small marine fishes, which are abundant on such parts of the coasts of the tropical and temperate zones as offer by their vegetation shelter to these defenceless creatures. They are bad swimmers (the dorsal fin being the principal organ of locomotion), and frequently and resistlessly carried by currents into the open ocean or to distant coasts. All enter brackish water, some fresh water. The strata of Monte Bolca and Licata (Sicily) have yielded evidence of their existence in the tertiary epochs; beside species of *Siphonostoma* and *Syngnathus* (*Pseudosyngnathus*), remains of an extinct genus, *Calamostoma*, allied to *Hippocampus*, but with a distinct caudal fin, have been found. On their propagation see p. 163, Fig. 76.

A. SYNGNATHINA.—The tail is not prehensile, and generally provided with a caudal fin.—*Pipe-Fishes.*

SIPHONOSTOMA.—Body with distinct ridges, the upper caudal ridge continuous with the lateral line, but not with the dorsal ridge of the trunk. Pectoral and caudal fins well developed; dorsal fin of moderate length, opposite to the vent. Humeral bones movable, not united into a "breast-ring." Males with an egg-pouch on the tail, the eggs being covered by cutaneous folds.

Two species, of which *S. typhle* is common on the British, and generally distributed on the European coasts.

SYNGNATHUS.—Body with the ridges more or less distinct, the dorsal ridge of the trunk not being continuous with that of the tail. Pectoral fins well developed; caudal present. Dorsal fin opposite or near to the vent. Humeral bones firmly united into the breast-ring. Egg-pouch as in *Siphonostoma.*

The distribution of this genus nearly coincides with that of the family, some fifty species being known. *S. acus*, the great Pipe-fish (see Fig. 75, p. 163), is one of the most common European fishes, extending across the Atlantic and southwards to the Cape of Good Hope; it attains a length of 18 inches. Another very common species, frequently met at sea, and spread over nearly all the tropical and sub-tropical seas, is *S. pelagicus*, agreeably marked with alternate brown and silvery cross-bars.

DORYICHTHYS.—Body with the ridges well developed. Pectoral and caudal fins present. Dorsal fin long or of moderate length, opposite to the vent. Humeral bones firmly united. Males with the lower ridges of the abdomen dilated, the dilated parts forming a broad groove for the reception of the ova.

In these Pipe-fishes the ova are not received in a completely closed pouch, but glued on to the surface of the abdomen. Twenty species from tropical seas.

NEROPHIS.—Body smooth, rounded, with scarcely any of the ridges distinct. Pectoral fin none, caudal absent or rudimentary, the tail tapering into a point. Dorsal fin of moderate length, opposite to the vent. The ova are attached to the soft integument of the abdomen of the male, and are not covered by lateral folds of the skin.

Seven species from the European seas and the Atlantic. *N. æquoreus* (Ocean Pipe-fish), *N. ophidion* (Straight-nosed Pipe-fish), and *N. lumbriciformis* (Little Pipe-fish), are common on the British coasts.

PROTOCAMPUS.—The whole dermal skeleton is covered with skin. A broad cutaneous fold runs along the back in front and behind the dorsal; a similar fold along the abdomen. Pectoral fin none; caudal very small.

The single species of this remarkable genus, *P. hymenolomus*, occurs in the Falkland Islands. It may be regarded as an embryonal form of *Nerophis*, the median skin-folds being evidently remains of the fringe which surrounds the body of the embryo.

The other genera belonging to this group are, *Ichthyocampus*, *Nannocampus*, *Urocampus*, *Leptoichthys*, *Coelonotus*, and *Stigmatophora*.

HIPPOCAMPINA.—The tail is prehensile, and invariably without caudal fin.—*Sea-horses.*

GASTROTOKEUS.—Body depressed, the lateral line running along the margin of the abdomen. Shields smooth. Tail shorter than the body. Pectoral fins. No pouch is developed for the ova, which are imbedded in the soft integument of the abdomen of the male.

Gastrotokeus biaculeatus, very common in the Indian Ocean to the coasts of Australia.

SOLENOGNATHUS.—Body compressed, deeper than broad. Shields hard, rugose, with round or oval interannular plates; and without elongate processes. Tail shorter than the body. Pectoral fins.

Three species, from the Chinese and Australian Seas; they are the largest of Lophobranchs, *S. hardwickii,* attaining to a length of nearly two feet.

PHYLLOPTERYX.—Body compressed, or as broad as deep. Shields smooth, but some or all of them are provided with pro-

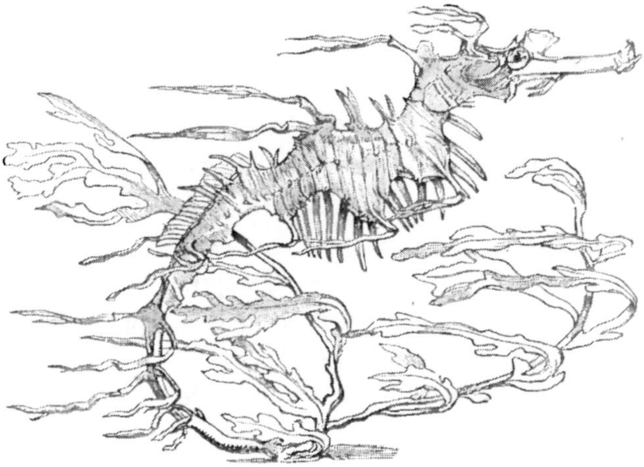

Fig. 309.—Phyllopteryx eques.

minent spines or processes on the edges of the body; some of the processes with cutaneous filaments. A pair of spines on the upper side of the snout and above the orbit. Tail about as long as the body. Pectoral fins. The ova are imbedded in soft membrane on the lower side of the tail, without a pouch being developed.

Three species from the coasts of Australia. The protective resemblances with which many Lophobranchs are furnished, attain to the highest degree of development in the fishes of this genus. Not only their colour closely assimilates that of the particular kind of seaweed which they frequent, but the appendages of their spines seem to be merely part of the fucus to which they are attached. They attain a length of 12 inches.

HIPPOCAMPUS.—Trunk compressed, more or less elevated. Shields with more or less prominent tubercles or spines. Occiput compressed into a crest, terminating at its supero-posterior corner in a prominent knob (coronet). Pectoral fins. The males carry the eggs in a sac at the base of the tail, opening near the vent.

A singular resemblance of the head and fore part of the body to that of a horse, has given to these fishes the name of "Sea-horses." They are abundant between and near the tropics, becoming scarcer in higher latitudes. Some twenty species are known, some of which have a wide geographical range, as they are often carried to great distances with floating objects to which they happen to be attached.—*Acentronura* is a genus closely allied to *Hippocampus*.

SIXTH ORDER—PLECTOGNATHI.

Teleosteous fishes with rough scales, or with ossifications of the cutis in the form of scutes or spines; skin sometimes entirely naked. Skeleton incompletely ossified, with the vertebræ in small number. Gills pectinate; a narrow gill-opening in front of the pectoral fins. Mouth narrow; the bones of the upper jaw generally firmly united. A soft dorsal fin, belonging to the caudal portion of the vertebral column, opposite to the anal; sometimes elements of a spinous dorsal besides. Ventral fin none, or reduced to spines. Air-bladder without pneumatic duct.

FIRST FAMILY—SCLERODERMI.

Snout somewhat produced; jaws armed with distinct teeth in small number. Skin with scutes or rough. The elements of a spinous dorsal and ventral fins generally present.

Marine fishes of moderate or small size, very common in the tropical zone, but scarcer in higher latitudes. They have been found in three localities of tertiary strata, viz., at Monte Bolca, where a species of *Ostracion* occurs, and in the Schists of Glaris, from which two genera have been described, *Acanthoderma* and *Acanthopleurus*, closely allied to *Balistes* and *Triacanthus*. *Glyptocephalus* from the Isle of Sheppey has the skull of a Balistes, but its body is covered with tubercles arranged in regular series. The Sclerodermis may be divided into three very natural groups:—

A. TRIACANTHINA.—The skin is covered with small, rough, scalelike scutes. A spinous dorsal fin with from four to six spines. A pair of strong, movable ventral spines, joined to the pelvic bone.

To this group belong the genera *Triacanthodes*, *Hollardia*, and *Triacanthus*, represented by five species, of which *Triacanthus brevirostris* from the Indian Ocean is the most common.

B. BALISTINA.—Body compressed, covered with movable scutes or rough. Spinous dorsal reduced to one, two, or three spines. Ventral fins reduced to a single pelvic prominence, or entirely absent.

To this group belong the genera *Balistes*, *Monacanthus*, and *Anacanthus*, the last genus being distinguished by a barbel at the lower jaw.

Balistes, or the "File-fishes" proper, inhabit the tropical and sub-tropical seas; shoals of young are not rarely met with in mid-ocean. Some thirty species are known, many attaining a length exceeding two feet; but the majority are much smaller, and frequently beautifully and symmetrically marked. Both jaws are armed with eight strong incisor-like and

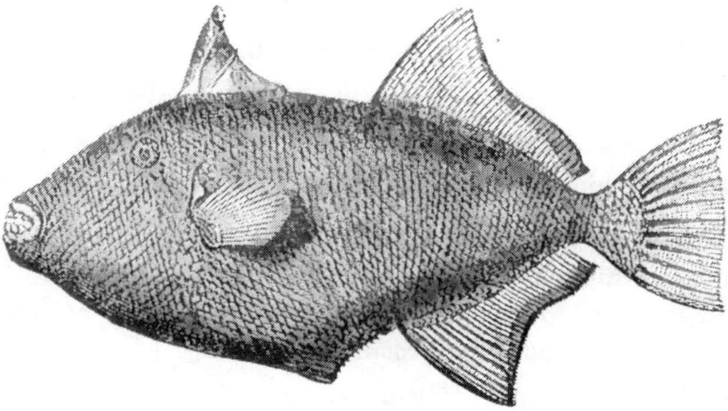

Fig. 310.—Balistes vidua.

obliquely truncated teeth, by which these fishes are enabled to break off pieces of corals on which they feed, or to chisel a hole into the hard shell of Mollusca, in order to extract the soft parts. They destroy an immense number of Mollusks, thus becoming most injurious to the pearl-fisheries. The first of their three dorsal spines is very strong, roughened in front like a file, and hollowed out behind to receive the second much smaller spine, which, besides, has a projection in front, at its base, fitting into a notch of the first. Thus these two spines can only be raised or depressed simultaneously, and the first cannot be forced down, unless the second has been previously depressed. The latter has been compared to a trigger, hence a second name, "Trigger-fish," has been given to these fishes. Some species are armed with a series of short spines or tubercles on each side of the tail. Two species (*B. maculatus* and *B. capriscus*), common in the Atlantic, sometimes wander to the British coasts.

The *Monacanthus* are similarly distributed as the *Balistes*, and still more abundant, some fifty species being known. Their dentition is very similar, but they possess one dorsal spine only, and their rough scales are so small as to give a velvety appearance to the skin (Figs. 17 and 18, p. 48). Adult males of some of the species possess a peculiar armature on each side of the tail, which in females is much less developed or entirely absent. This armature may consist either in simple spines arranged in rows, or in the development of the minute spines of the scales into long stiff bristles, so that the patch on each side of the tail looks like a brush.

C. OSTRACIONTINA.—The integuments of the body form a hard continuous carapace, consisting of hexagonal scutes juxtaposed in mosaic-fashion. A spinous dorsal and ventral fins are absent; but sometimes indicated by protuberances.

The "Coffer-fishes" (*Ostracion*) are too well known to require a lengthened description. Only the snout, the bases of the fins, and the hind part of the tail are covered with soft skin, so as to admit of free action of the muscles moving these parts. The mouth is small, the maxillary and intermaxillary bones coalescent, each jaw being armed with a single series of small slender teeth. The short dorsal fin is opposite to the equally short anal. The vertebral column consists of fourteen vertebræ only, of which the five last are extremely short, the anterior elongate. Ribs none. The carapaces of some species are three-ridged, of others four- and five-ridged, of some provided with long spines. Twenty-two species from tropical and sub-tropical seas are known.

SECOND FAMILY—GYMNODONTES.

Body more or less shortened. The bones of the upper and lower jaw are confluent, forming a beak with a trenchant edge, without teeth, with or without median suture. A soft dorsal, caudal and anal are developed, approximate. No spinous dorsal. Pectoral fins; no ventrals.

Marine fishes of moderate or small size from tropical and sub-tropical seas. A few species live in fresh water. Fossil remains of *Diodon* are not scarce at Monte Bolca and Licata;

a distinct genus, *Enneodon*, has been described from Monte Postale. The Gymnodonts may be divided into three groups:

A. TRIODONTINA.—Tail rather long, with a separate caudal fin. Abdomen dilatable into a very large, compressed, pendent sac, the lower part of which is merely a flap of skin, into which the air does not penetrate, the sac being capable of being expanded by the very long pelvic bone. The upper jaw divided by a median suture, the lower simple.

A single genus and species (*Triodon bursarius*) from the Indian Ocean.

B. TETRODONTINA.—Tail and caudal fin distinct. Part of the œsophagus much distensible, and capable of being filled with air. No pelvic bone.

"Globe-fishes" have a short, thick, cylindrical body, with well developed fins. It is covered with thick scaleless skin, in which, however, spines are imbedded of various sizes. The spines are very small, and but partially distributed over the body in some species, whilst in others they are very large, and occupy equally every part of the body. These fishes have the power of inflating their body by filling their distensible œsophagus with air, and thus assume a more or less globular form. The skin is, then, stretched to its utmost extent, and the spines protrude and form a more or less formidable defensive armour, as in a hedgehog; therefore they are frequently called "Sea-hedgehogs." A fish thus blown out turns over and floats belly upwards, driving before the wind and waves. However, it is probable that the spines are a protection not only when the fish is on the surface and able to take in air, but also when it is under water. Some Diodonts, at any rate, are able to erect the spines about the head by means of cutaneous muscles; and, perhaps, all fill their stomach with water instead of air, for the same purpose and with the same effect. In some Diodonts the spines are fixed, erect, not movable. The Gymnodonts generally, when taken, produce a sound, doubtless by the expulsion of air from the œsophagus. Their vertebral column consists of a small number of vertebræ, from 20 to 29, and their spinal chord is extremely short. All these fishes have a bad reputation, and they are never eaten; indeed, some of them are highly poisonous, and have caused long continued illness and death. Singularly, the poisonous properties of these fishes vary much as regards intensity, only certain individuals of a species, or individuals from a certain locality, or caught at a certain time of the year, being dangerous. Therefore it is probable that they acquire their poisonous quality from their food, which consists in corals and hard-shelled Mollusks and Crustaceans. Their sharp beaks, with broad masticating posterior surface, are admirably adapted for breaking off branchlets of coral-stocks, and for crushing hard substances.

TETRODON (including *Xenopterus*).—Both the upper and lower jaws are divided into two by a mesial suture.

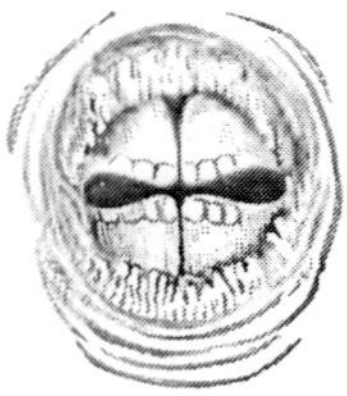

Fig. 311.—Jaws of Tetrodon.

Extremely numerous in tropical and subtropical zones, more than sixty species being known. In some of the species the dermal spines are extremely small, and may be absent altogether. Many are highly ornamented with spots or bands. A few species live in large rivers—thus *T. psittacus* from Brazil; *T. fahaka*, a fish well known to travellers on the Nile, and likewise abundant in West African rivers; *T. fluvi-*

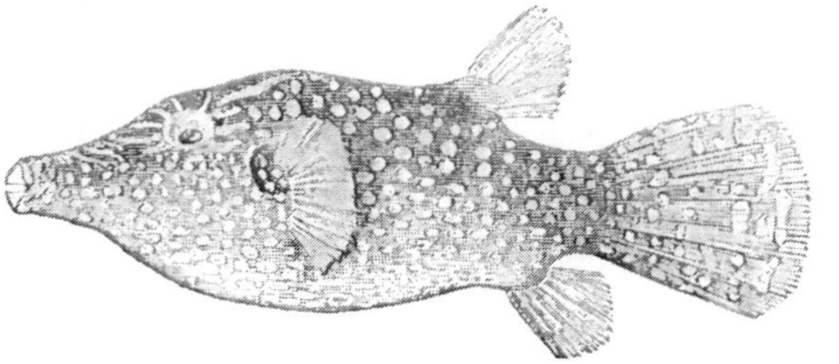

Fig. 312.—Tetrodon margaritatus.

atilis from brackish water and rivers of the East Indies. The species figured is one of the smallest, about six inches long, and common in the Indo-Pacific.

DIODON.—Jaws without mesial suture, so that there is only one undivided dental plate above and one below.

In these fishes, as well as in some closely allied genera, the dermal spines are much more developed than in the Tetrodonts; in some the spines are erectile, as in *Diodon*, *Atopomycterus*, *Trichodiodon*, and *Trichocyclus;* in others they are stiff and immovable, as in *Chilomycterus* and *Dicotylichthys.* Seventeen species are known, of which *Diodon hystrix* is the most common as it is the largest, growing to a length of two feet. It is spread over the Tropical Atlantic as well as Indo-

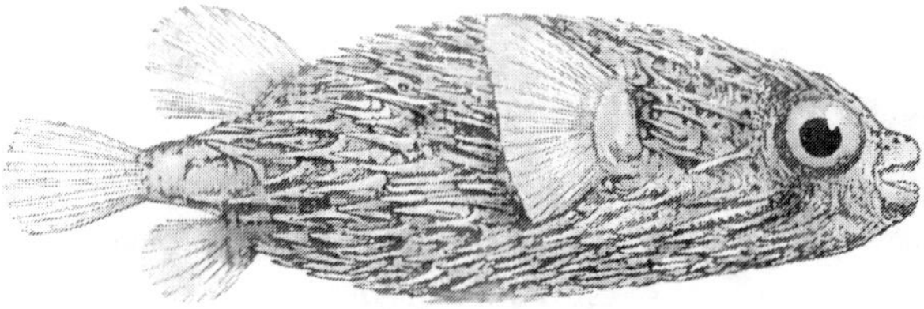

Fig. 313.—Diodon maculatus.

Pacific, as is also a smaller, but almost equally common species, *Diodon maculatus.*

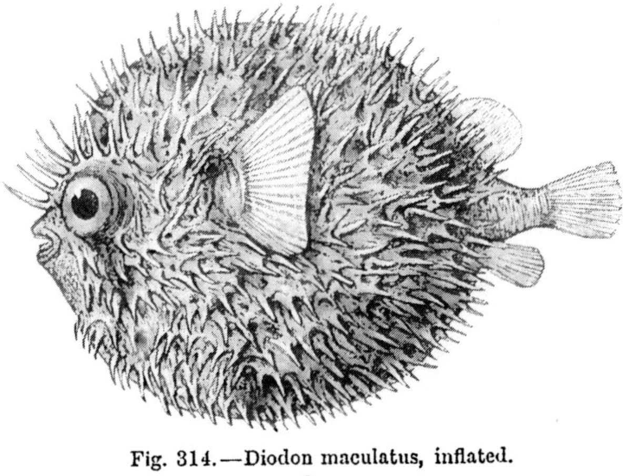

Fig. 314.—Diodon maculatus, inflated.

C. Molina.—Body compressed, very short; tail extremely short, truncate. Vertical fins confluent. No pelvic bone.

The "Sun-fishes" (*Orthagoriscus*) are pelagic fishes, found in every part of the oceans within the tropical and temperate zones. The singular shape of their body and the remarkable changes which they undergo with age, have been noticed above (p. 175, Figs. 93, 94). Their jaws are undivided in the middle, comparatively feeble, but well adapted for masticating their food, which consists of small pelagic Crustaceans. Two species are known. The common Sun-fish, *O. mola*, which attains to a very large size, measuring seven or eight feet, and weighing as many hundredweights. It has a rough, minutely granulated skin. It frequently approaches the southern coasts of England and the coasts of Ireland, and is seen basking in calm weather on the surface. The second species, *O. truncatus*, is distinguished by its smooth, tessellated skin, and one of the scarcest fishes in collections. The shortness of the vertebral column of the Sun-fishes, in which the number of caudal vertebræ is reduced to seven, the total number being seventeen, and the still more reduced length of the spinal chord have been noticed above (p. 96).

THIRD SUB-CLASS—CYCLOSTOMATA.

Skeleton cartilaginous and notochordal, without ribs and without real jaws. Skull not separate from the vertebral column. No limbs. Gills in the form of fixed sacs, without branchial arches, six or seven in number on each side. One nasal aperture only. Heart without bulbus arteriosus. Mouth anterior, surrounded by a circular or sub-circular lip, suctorial. Alimentary canal straight, simple, without coecal appendages, pancreas or spleen. Generative outlet peritoneal. Vertical fins rayed.

The Cyclostomes are most probably a very ancient type. Unfortunately the organs of these creatures are too soft to be preserved, with the exception of the horny denticles with which the mouth of some of them is armed. And, indeed, dental plates, which are very similar to those of *Myxine*, are not uncommon in certain strata of Devonian and Silurian age (see p. 193). The fishes belonging to this sub-class may be divided into two families—

FIRST FAMILY—PETROMYZONTIDÆ.

Body eel-shaped, naked. Subject to a metamorphosis; in the perfect stage with a suctorial mouth armed with teeth, simple or multicuspid, horny, sitting on a soft papilla. Maxillary, mandibulary, lingual, and suctorial teeth may be distinguished. Eyes present (in mature animals). External nasal aperture in the middle of the upper side of the head. The nasal duct terminates without perforating the palate. Seven branchial sacs and apertures on each side behind the head; the inner branchial ducts terminate in a separate common tube. Intestine with a spiral valve. Eggs small. The larvæ without teeth, and with a single continuous vertical fin.

"Lampreys" are found in the rivers and on the coasts of the temperate regions of the northern and southern hemispheres. Their habits are but incompletely known, but so much is certain that at least some of them ascend rivers periodically for the purpose of spawning, and that the young pass several years in rivers, whilst they undergo a metamorphosis (see p. 170). They feed on other fishes, to which they suck themselves fast, scraping off the flesh with their teeth. Whilst thus engaged they are carried about by their victim; Salmon have been captured in the middle course of the Rhine with the Marine Lamprey attached to them.

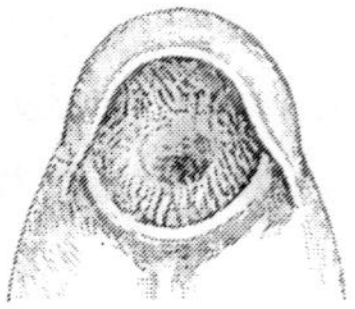

Fig. 315.—Mouth of Larva of Petromyzon branchialis.

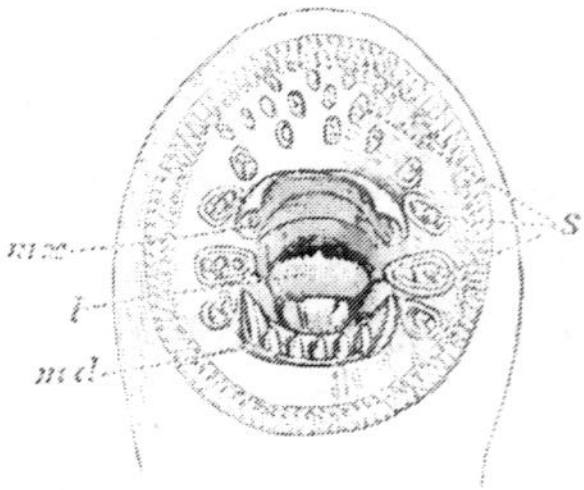

Fig. 316.—Mouth of Petromyzon fluviatilis. *mx*, Maxillary tooth; *md*, Mandibulary tooth; *l*, Lingual tooth; *s*, Suctorial teeth.

PETROMYZON.—Dorsal fins two, the posterior continuous with the caudal. The maxillary dentition consists of two teeth placed close together, or of a transverse bicuspid ridge; lingual teeth serrated.

The Lampreys belonging to this genus are found in the northern hemisphere only; the British species are the Sea-Lamprey (*P. marinus*), exceeding a length of three feet, and not uncommon on the European and North American coasts; the River-Lamprey or Lampern (*P. fluviatilis*), ascending in large numbers the rivers of Europe, North America, and Japan, and scarcely attaining a length of two feet; the "Pride" or "Sand-Piper" or Small Lampern (*P. branchialis*), scarcely twelve inches long, the larva of which has been long known under the name of *Ammocoetes*.

Ichthyomyzon from the western coasts of North America is said to have a tricuspid maxillary tooth.

MORDACIA.—Dorsal fins two, the posterior continuous with the caudal. The maxillary dentition consists of two triangular groups, each with three conical acute cusps; two pairs of serrated lingual teeth.

A Lamprey (*M. mordax*) from the coasts of Chile and

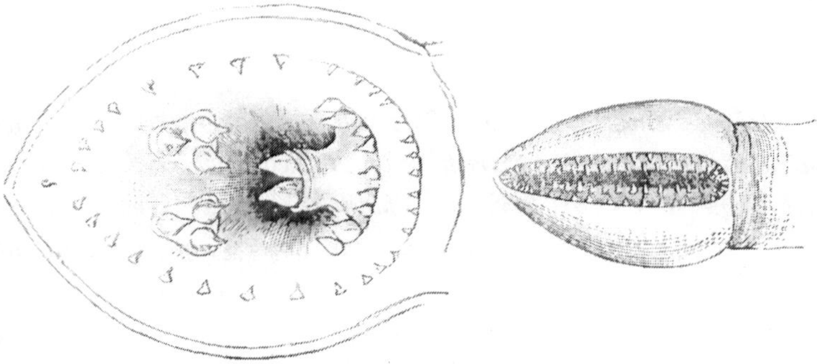

Fig. 317.—Mouth of Mordacia mordax, closed and opened.

Tasmania. This fish seems to be provided sometimes with a gular sac, like the following.[1]

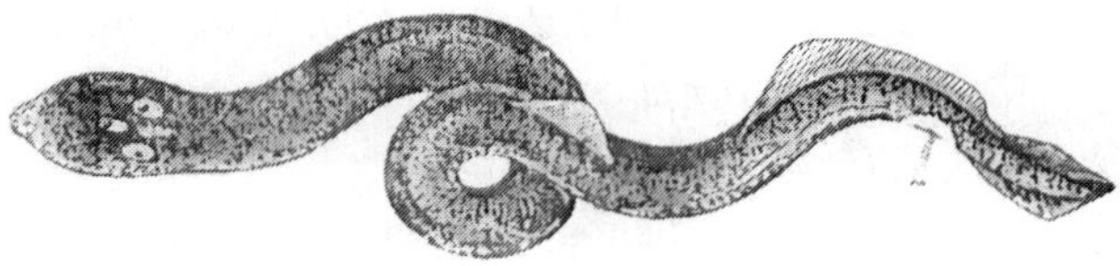

Fig. 318.—Mordacia mordax.

GEOTRIA.—Dorsal fins two, the posterior separate from the

[1] Fig. 317 is taken from a specimen in which the horny covers of the dentition were lost, hence it does not represent accurately the shape of the teeth.

caudal. Maxillary lamina with four sharp flat lobes; a pair of long pointed lingual teeth.

Two species, one from Chile and one from South Australia. They grow to a length of two feet, and in some specimens the skin of the throat is much expanded, forming a large pouch. Its physiological function is not known. The cavity is in the subcutaneous cellular tissue, and does not communicate with the buccal or branchial cavities. Probably it is developed with age, and absent in young individuals. In all the localities in which these Extra-european Lampreys are found, *Ammococtes* forms occur, so that there is little doubt that they undergo a similar metamorphosis as *P. branchialis*.

Second Family—Myxinidæ.

Body eel-shaped, naked. The single nasal aperture is above the mouth, quite at the extremity of the head, which is provided with four pairs of barbels. Mouth without lips. Nasal duct without cartilaginous rings, penetrating the palate. One median tooth on the palate, and two comb-like series of teeth on the tongue (see Fig. 101). *Branchial apertures at a great distance from the head; the inner branchial ducts lead into the œsophagus. A series of mucous sacs along each side of the abdomen. Intestine without spiral valve. Eggs large, with a horny case provided with threads for adhesion.*

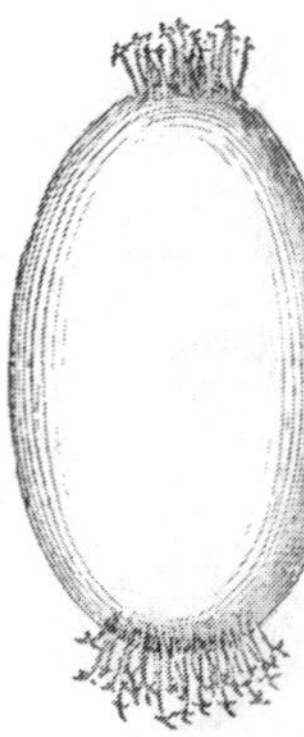

Fig. 319.—Ovum of Myxine glutinosa, enlarged.

The fishes of this family are known by the names of "Hag-Fish," "Glutinous Hag," or "Borer;" they are marine fishes with a similar distribution as the Gadidæ, being most plentiful in the higher latitudes of the temperate zones of the northern and southern hemispheres. They are frequently found buried in the abdominal cavity of other fishes, especially Gadoids, into which they penetrate to feed on their flesh. They secrete a thick glutinous slime in incredible quantities, and are therefore considered by fishermen a great nuisance, seriously damaging the fisheries and interfering with the fishing in localities where they abound. *Myxine* descends to a depth of 345 fathoms, and is generally met with in the Norwegian Fjords at 70 fathoms, sometimes in great abundance.

MYXINE.—One external branchial aperture only on each side of the abdomen, leading by six ducts to six branchial sacs.

Three species from the North Atlantic, Japan, and Magelhæn's Straits.

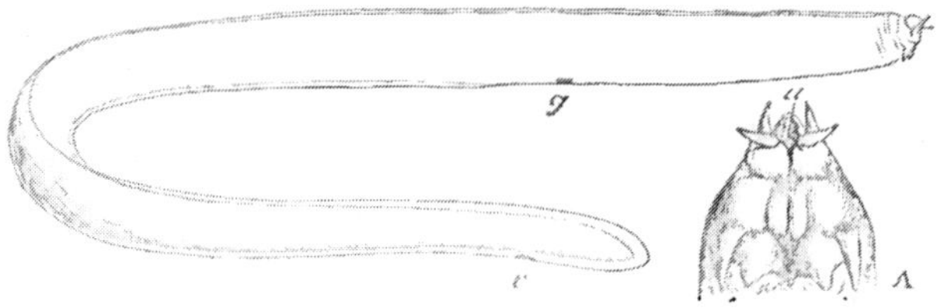

Fig. 320.—Myxine australis. A, Lower aspect of head; *a*, Nasal aperture; *b*, Mouth; *g*, Branchial aperture; *v*, Vent.

BDELLOSTOMA.—Six or more external branchial apertures on each side, each leading by a separate duct to a branchial sac.

Two species from the South Pacific.

FOURTH SUB-CLASS—LEPTOCARDII.

Skeleton membrano-cartilaginous and notochordal, ribless. No brain. Pulsating sinuses in place of a heart. Blood colourless. Respiratory cavity confluent with the abdominal cavity; branchial clefts in great number, the water being expelled by an opening in front of the vent. Jaws none.

This sub-class is represented by a single family (*Cirrostomi*) and by a single genus (*Branchiostoma*);[1] it is the lowest in the scale of fishes, and lacks so many characteristics, not only of this class, but of the vertebrata generally, that Hæckel, with good reason, separates it into a separate class, that of *Acrania*. The various parts of its organisation have been duly noticed in the first part of this work.

The "Lancelet" (*Branchiostoma lanceolatum*, see Fig. 28, p. 63), seems to be almost cosmopolitan within the temperate and tropical zones. Its small size, its transparency, and the rapidity with which it is able to bury itself in the sand, are the causes why it escapes so readily observation, even at localities where it is known to be common. Shallow, sandy parts of the coasts seem to be the places on which it may be looked for. It has been found on many localities of the British, and generally European coasts, in North America, the West Indies, Brazil, Peru, Tasmania, Australia, and Borneo. It rarely exceeds a length of three inches. A smaller species, in which the dorsal fringe is distinctly higher and rayed, and in which the caudal fringe is absent, has been described under the name of *Epigonichthys pulchellus*; it was found in Moreton Bay.

[1] This name is two years older than *Amphioxus*.

APPENDIX.

DIRECTIONS FOR COLLECTING AND PRESERVING FISHES.

Whenever practicable fishes ought to be preserved in spirits.

To insure success in preserving specimens the best and strongest spirits should be procured, which, if necessary, can be reduced to the strength required during the journey with water or weaker spirit. Travellers frequently have great difficulties in procuring spirits during their journey, and therefore it is advisable, especially during sea voyages, that the traveller should take a sufficient quantity with him. Pure spirits of wine is best. Methylated spirits may be recommended on account of their cheapness; however, specimens do not keep equally well in this fluid, and very valuable objects, or such as are destined for minute anatomical examination, should always be kept in pure spirits of wine. If the collector has exhausted his supply of spirits he may use arrack, cognac, or rum, provided that the fluids contain a sufficient quantity of alcohol. Generally speaking, spirits which, without being previously heated, can be ignited by a match or taper, may be used for the purposes of conservation. The best method to test the strength of the spirits is the use of a hydrometer. It is immersed in the fluid to be measured, and the deeper it sinks the stronger is the spirit. On its scale the number 0 signifies what is called proof spirit, the lowest degree of strength which can be used for the conservation of fish for any length of time. Spirits, in which specimens are packed permanently, should be from 40 to 60 above proof. If the hydrometers are made of glass they are easily broken, and therefore the traveller had better provide himself with three or four of them, their cost being very trifling. Further, the collector will find a small distilling apparatus very useful. By its means he is able not only to distil weak and deteriorated spirits or any other fluid containing alcohol, but also, in case of necessity, to prepare a small quantity of drinkable spirits.

Of collecting vessels we mention first those which the collector requires for daily use. Most convenient are four-sided boxes made of zinc, 18 in. high, 12 in. broad, and 5 in. wide. They have a round opening at the top of 4 in. diameter, which can be closed by a strong cover of zinc of 5 in. diameter, the cover being screwed into a raised rim round the opening. In order to render the cover air-tight, an indiarubber ring is fixed below its margin. Each of these zinc boxes fits into a wooden case, the lid of which is provided with hinges and fastenings, and which on each side has a handle of leather or rope, so that the box can be easily shifted from one place to another. These boxes are in fact made from the pattern of the ammunition cases used in the British army, and extremely convenient, because a pair can be easily carried strapped over the shoulders of a man or across the back of a mule. The collector requires at least two, still better four or six, of these boxes. All those specimens which are received during the day are deposited in them, in order to allow them to be thoroughly penetrated by the spirit, which must be renewed from time to time. They remain there for some time under the supervision of the collector, and are left in these boxes until they are hardened and fit for final packing. Of course, other more simple vessels can be used and substituted for the collecting boxes. For instance, common earthenware vessels, closed by a cork or an indiarubber covering, provided they have a wide mouth at the top, which can be closed so that the spirit does not evaporate, and which permits of the specimens being inspected at any moment without trouble. Vessels in which the objects are permanently packed for the home journey are zinc boxes of various sizes, closely fitting into wooden cases. Too large a size should be avoided, because the objects themselves may suffer from the superimposed weight, and the risk of injury to the case increases with its size. It should hold no more than 18 cubic feet at most, and what, in accordance with the size of the specimens, has to be added in length should be deducted in depth or breadth.

The most convenient cases, but not sufficient for all specimens, are boxes 2 feet in length, $1\frac{1}{2}$ foot broad, and 1 foot deep. The traveller may provide himself with such cases ready made, packing in them other articles which he wants during his journey; or he may find it more convenient to take with him only the zinc plates cut to the several sizes, and join them into boxes when they are actually required. The requisite wooden cases can be procured without much difficulty almost everywhere. No collector should be without the apparatus and materials for soldering, and he should be well acquainted with their use. Also a pair of scissors to cut the zinc plates are useful.

Wooden casks are not suitable for the packing of specimens preserved in spirits, at least not in tropical climates. They should be used in cases of necessity only, or for packing of the largest examples, or for objects preserved in salt or brine.

Very small and delicate specimens should never be packed together with larger ones, but separately, in small bottles.

Mode of preserving.—All fishes, with the exception of very large ones (broad kinds exceeding 3–4 feet in length; eel-like kinds more than 6 feet long), should be preserved in spirits. A deep cut should be made in the abdomen between the pectoral fins, another in front of the vent, and one or two more, according to the length of the fish, along the middle line of the abdomen. These cuts are made partly to remove the fluid and easily decomposing contents of the intestinal tract, partly to allow the spirit quickly to penetrate into the interior. In large fleshy fishes several deep incisions should be made with the scalpel into the thickest parts of the dorsal and caudal muscles, to give ready entrance to the spirits. The specimens are then placed in one of the provisional boxes, in order to extract, by means of the spirit, the water of which fishes contain a large quantity. After a few days (in hot climates after 24 or 48 hours) the specimens are transferred into a second box with stronger spirits, and left therein for several days. A similar third and, in hot climates sometimes a fourth, transfer is necessary. This depends entirely on the condition of the specimens. If, after ten or fourteen days of such treatment the specimens are firm and in good condition, they may be left in the spirits last used until they are finally packed. But if they should be soft, very flexible, and discharge a discoloured bloody mucus, they must be put back in spirits at least 20° over proof. Specimens showing distinct signs of decomposition should be thrown away, as they imperil all other specimens in the same vessel. Neither should any specimen in which decomposition has commenced when found, be received for the collecting boxes, unless it be of a very rare species, when the attempt may be made to preserve it separately in the strongest spirits available. The fresher the specimens to be preserved are, the better is the chance of keeping them in a perfect condition. Specimens which have lost their scales, or are otherwise much injured, should not be kept. Herring-like fishes, and others with deciduous scales, are better wrapped in thin paper or linen before being placed in spirits.

The spirits used during this all-important process of preservation loses, of course, gradually in strength. As long as it keeps 10° under proof it may still be used for the first stage of preservation, but weaker spirits should be re-distilled; or, if the collector cannot do this, it should be at least filtered through powdered charcoal before it is mixed with stronger spirits. Many collectors are satisfied with removing the thick sediment collected at the bottom of the vessel, and use their spirits over and over again without removing from it by filtration the decomposing matter with which it has been impregnated, and which entirely neutralises the preserving property of the spirits. The result is generally the loss of the collection on its journey home. The collector can easily detect the vitiated character of his spirits by its bad smell. He must frequently examine his specimens; and attention to the rules given, with a little practice and perseverance, after the possible failure of the first trial, will soon insure to him the safety of his collected treasures. The trouble of collecting specimens in spirits is infinitely less than that of preserving skins or dry specimens of any kind.

When a sufficient number of well-preserved examples have been brought together, they should be sent home by the earliest opportunity. Each specimen should be wrapped separately in a piece of linen, or at least soft paper; the specimens are then packed as close as herrings in the zinc case, so that no free space is left either at the top or on the sides. When the case is full, the lid is soldered on, with a round hole about half an inch in diameter near one of the corners. This hole is left in order to pour the spirit through it into the case. Care is taken to drive out the air which may remain between the specimens, and to surround them completely with spirits, until the case is quite full. Finally, the hole is closed by a small square lid of tin being soldered over it. In order to see whether the case keeps in the spirit perfectly, it is turned upside down and left over night. When all is found to be securely fastened, the zinc case is placed into the wooden box and ready for transport.

Now and then it happens in tropical climates that collectors are unable to keep fishes from decomposition even in the strongest spirits without being able to detect the cause. In such cases a remedy will be found in mixing a small quantity of arsenic or sublimate with the spirits ; but the collector ought to inform his correspondent, or the recipient of the collection, of this admixture having been made.

In former times fishes of every kind, even those of small size, were preserved dry as flat skins or stuffed. Specimens thus prepared admit of a very superficial examination only, and therefore this method of conservation has been abandoned in all larger museums, and should be employed exceptionally only, for instance on long voyages overland, during which, owing to the difficulty of transport, neither spirits nor vessels can be carried. To make up as much as possible for the imperfection of such specimens, the collector ought to sketch the fish before it is skinned, and to colour the sketch if the species is ornamented with colours likely to disappear in the dry example. Collectors who have the requisite time and skill, ought to accompany their collections with drawings coloured from the living fishes ; but at the same time it must be remembered that, valuable as such drawings are if accompanied by the originals from which they were made, they can never replace the latter, and possess a subordinate scientific value only.

Very large fishes can be preserved as skins only ; and collectors are strongly recommended to prepare in this manner the largest examples obtainable, although it will entail some trouble and expense. So very few large examples are exhibited in museums, the majority of the species being known from the young stage only, that the collector will find himself amply recompensed by attending to these desiderata.

Scaly fishes are skinned thus: with a strong pair of scissors an incision is made along the median line of the abdomen from the foremost part of the throat, passing on one side of the base of the ventral and anal fins, to the root of the caudal fin, the cut being continued upwards to the back of the tail close to the base of the caudal. The skin of one side of the fish is then severed with the scalpel from the underlying muscles to the median line of the back; the bones which support the dorsal and caudal are cut through, so that these fins remain attached to the skin. The removal of the skin of the opposite side is easy. More difficult is the preparation of the head and scapulary region; the two halves of the scapular arch which have been severed from each other by the first incision are pressed towards the right and left, and the spine is severed behind the head, so that now only the head and shoulder bones remain attached to the skin. These parts have to be cleaned from the inside, all soft parts, the branchial and hyoid apparatus, and all smaller bones, being cut away with the scissors or scraped off with the scalpel. In many fishes, which are provided with a characteristic dental apparatus in the pharynx (Labroids, Cyprinoids), the pharyngeal bones ought to be preserved, and tied with a thread to the specimen. The skin being now prepared so far, its entire inner surface as well as the inner side of the head are rubbed with arsenical soap; cotton-wool, or some other soft material is inserted into any cavities or hollows, and finally a thin layer of the same material is placed between the two flaps of the skin. The specimen is then dried under a slight weight to keep it from shrinking.

The scales of some fishes, as for instance of many kinds of herrings, are so delicate and deciduous that the mere handling causes them to rub off easily. Such fishes may be covered with thin paper (tissue-paper is the best), which is allowed to dry on them before skinning. There is no need for removing the paper before the specimen has reached its destination.

Scaleless Fishes, as Siluroids and Sturgeons, are skinned in the same manner, but the skin can be rolled up over the head; such skins can also be preserved in spirits, in which case the traveller may save to himself the trouble of cleaning the head.

Some Sharks are known to attain to a length of 30 feet, and some Rays to a width of 20 feet. The preservation of such gigantic specimens is much to be recommended, and although the difficulties of preserving fishes increase with their size, the operation is facilitated, because the skins of all Sharks and Rays can easily be preserved in salt and strong brine. Sharks are skinned much in the same way as ordinary fishes. In Rays an incision is made uot only from the snout to the end of the fleshy part of the tail, but also a second across the widest part of the body. When the skin is removed from the fish, it is placed into a cask with strong brine mixed with alum, the head occupying the upper part of the cask; this is necessary, because this part is most likely to show signs of decomposition, and therefore most requires supervision. When the preserving fluid has become decidedly weaker from the extracted blood and water, it is thrown away and replaced by fresh brine. After a week's or fortnight's soaking the skin is taken out of the cask to allow the fluid to drain off; its inner side is covered with a thin layer of salt, and after being rolled up (the head being inside) it is packed in a cask, the bottom of which is covered with salt; all the interstices and the top are likewise filled with salt. The cask must be perfectly water-tight.

Of all larger examples of which the skin is prepared, the measurements should be taken before skinning so as to guide the taxidermist in stuffing and mounting the specimens.

Skeletons of large osseous fishes are as valuable as their skins. To preserve them it is only necessary to remove the soft parts of the abdominal cavity and the larger masses of muscle, the bones being left in their natural continuity. The remaining flesh is allowed to dry on the bones, and can be removed by proper maceration at home. The fins ought to be as carefully attended to as in a skin, and of scaly fishes so much of the external skin ought to be preserved as is necessary for the determination of the species, as otherwise it is generally impossible to determine more than the genus.

A few remarks may be added as regards those Faunæ, which promise most results to the explorer, with some hints as to desirable information on the life and economic value of fishes.

It is surprising to find how small the number is of the fresh-water faunæ which may be regarded as well explored; the rivers of Central Europe, the Lower Nile, the lower and middle course of the Ganges, and the lower part of the Amazons are almost the only fresh waters in which collections made without discrimination would not reward the naturalist. The oceanic areas are much better known; yet almost everywhere novel forms can be discovered and new observations made. Most promising and partly quite unknown are the following districts:—the Arctic Ocean, all coasts south of 38° lat. S., the Cape of Good Hope, the Persian Gulf, the coasts of Australia (with the exception of Tasmania, New South Wales, and New Zealand), many of the little-visited groups of Pacific islands, the coasts of north-eastern Asia north of 35° lat. N., and the western coasts of North and South America.

No opportunity should be lost to obtain *pelagic* forms, especially the young larva-like stages of development abounding on the surface of the open ocean. They can be obtained without difficulty by means of a small narrow meshed net dragged behind the ship. The sac of the net is about 3 feet deep, and fastened to a strong brass-ring 2 or 2½ feet in diameter. The net is suspended by three lines passing into the strong main line. It can only be used when the vessel moves very slowly, its speed not exceeding three knots an hour, or when a current passes the ship whilst at anchor. To keep the net in a vertical position the ring can be weighted at one point of its circumference; and by using heavier weights two or three drag-nets can be used simultaneously at different depths. This kind of fishing should be tried at night as well as day, as many fishes come to the surface only after sunset. The net must not be left long in the water, from 5 to 20 minutes only, as delicate objects would be sure to be destroyed by the force of the water passing through the meshes.

Objects found floating on the surface, as wood, baskets, seaweed, etc., deserve the attention of the travellers, as they are generally surrounded by small fishes or other marine animals.

It is of the greatest importance to note the longitude and latitude at which the objects were collected in the open ocean.

Fishing in great depths by means of the dredge, can be practised only from vessels specially fitted out for the purpose; and the success which attended the "Challenger," and North American Deep-sea explorations, has developed Deep-sea fishing into such a speciality that the requisite information can be gathered better by consulting the reports of those expeditions than from a general account, such as could be given in the present work.

Fishes offer an extraordinary variety with regard to their habits, growth, etc., so that it is impossible to enumerate in detail the points of interest to which the travellers should pay particular attention. However, the following hints may be useful.

Above all, detailed accounts are desirable of all fishes forming important articles of trade, or capable of becoming more generally useful than they are at present. Therefore, deserving of special attention are the Sturgeons, Gadoids, Thyrsites and Chilodactylus, Salmonoids, Clupeoids. Wherever these fishes are found in sufficient abundance, new sources may be opened to trade.

Exact observations should be made on the fishes the flesh of which is poisonous either constantly or at certain times and certain localities; the cause of the poisonous qualities as well as the nature of the poison should be ascertained. Likewise the poison of fishes provided with special poison-organs requires to be experimentally examined, especially with regard to its effects on other fishes and animals generally.

All observations directed to sex, mode of propagation, and development, will have special interest: thus those relating to secondary sexual characters, hermaphroditism, numeric proportion of the sexes, time of spawning and migration, mode of spawning, construction of nests, care of progeny, change of form during growth, etc.

If the collector is unable to preserve the largest individuals of a species that may come under his observation he should note at least their measurements. There are but few species of fishes of which the limit of growth is known.

The history of Parasitic Fishes is almost unknown, and any

observations with regard to their relation to their host as well as to their early life will prove to be valuable; nothing is known of the propagation of fishes even so common as Echeneis and Fierasfer, much less of the parasitic Freshwater Siluroids.

The temperature of the blood of the larger freshwater and marine species should be exactly measured.

Many pelagic and deep-sea fishes are provided with peculiar small round organs of a mother-of-pearl colour, distributed in series along the side of the body, especially along the abdomen. Some zoologists consider these organs as accessory eyes, others (and it appears to us with better reason) as luminous organs. They deserve an accurate microscopic examination made on fresh specimens; and their function should be ascertained from observation of the living fishes, especially also with regard to the question, whether or not the luminosity (if such be their function) is subject to the will of the fish.

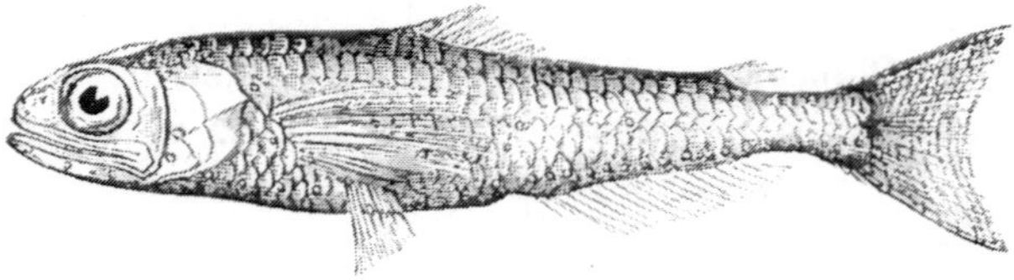

Fig. 321.—Scopelus boops, a pelagic fish, with luminous organs.

INDEX.

THE END.